REACTIVE OXYGEN METABOLITES

METABOLITES

CHEMISTRY AND MEDICAL
CONSEQUENCES

REACTIVE OXYGEN METABOLITES

CHEMISTRY AND MEDICAL CONSEQUENCES

MANFRED K. EBERHARDT, PH.D.

Department of Pathology and Laboratory Medicine
University of Puerto Rico, Medical Sciences Campus
San Juan, Puerto Rico

CRC Press
Boca Raton   London   New York   Washington, D.C.

ABOUT THE SCULPTURES

The front cover of this book shows a sculpture, known as a Mobius strip. The sculpture has two sides but only one surface. Everything in life has two sides: pro or con, good or evil, body or spirit, black or white, day or night, life or death. One side cannot exist without the other. Oxygen clearly shows this duality of pro and con, as expressed by Clerch and Massaro at the beginning of Chapter 1. The field of reactive oxygen metabolites ROM is full of these dualities. Throughout the book the author points out the pros and cons of superoxide radical anion, superoxide dismutase, metal ions, ascorbic acid (antioxidant and prooxidant), α-tocopherol (antioxidant and prooxidant), and many other examples. The sculpture is symbolic for all of ROM chemistry and pathology and for life and death in an oxygen atmosphere. The back cover shows a sculpture named "macrophage". Macrophages are the most prolific producers of reactive oxygen metabolites.

Library of Congress Cataloging-in-Publication Data

Eberhardt, Manfred K.
Reactive oxygen metabolites: chemistry and medical consequences/Manfred K. Eberhardt.
p. cm.
Includes bibliographical references and index.
ISBN 0-8493-0891-7
1. Active oxygen in the body. 2. Active oxygen. 3. Free radicals (Chemistry)--Pathophysiology. I. Title.

QP535.O1 E24 2000
612′.01524—dc21

00-033678
CIP

International Standard Book Number 0-8493-0891-7
Library of Congress Card Number 00-033678
Printed in the United States of America 1 2 3 4 5 6 7 8 9 0
Printed on acid-free paper

Acknowledgements

This book is dedicated to Dr. Angel A. Roman Franco, Professor of Pathology and Dean of the University of Puerto Rico School of Medicine. His unwavering support over many years has given me the time and peace of mind essential for writing this book.

I wish to thank Dr. Eliud Lopez, Chairman of the Department of Pathology and Laboratory Medicine, for his support, and Ms. Myrna Cabán for her expert technical support throughout the preparation of this book and for her dedication in making many last-minute corrections.

I would like to acknowledge the help of numerous librarians whom I kept busy over the years and the help of all the computer experts, who prevented me from smashing my computer on numerous occasions.

I am very grateful to Professor Dr. Waldemar Adam , who despite his busy schedule took the time to carefully read the original manuscript. His comments considerably improved the text.

I am indebted to the staff at CRC Press for a fast production process and especially to Maria Jennings, who proofread the entire manuscript with speed and great expertise.

San Juan, 2000

Manfred K. Eberhardt

Cover design by Myrna Cabán
Sculptures by Manfred K. Eberhardt
Front cover: Beethoven, opus 135, bronce© 1995
Back cover: Macrophage, polyester and marble

TABLE OF CONTENTS

INTRODUCTION

...science is really founded on observations rather than upon
'facts', and so is a continually evolving structure.
John D. Barrow in "The World within the World" [1]

Since the discovery of a free radical by Gomberg in 1900 [2], the field of radical chemistry has remained dormant for almost 50 years. The reason for this state of affairs is clear: research in chemistry was largely influenced by industry, which is interested in synthesizing useful compounds. Most organic chemists were therefore trained to be sophisticated cooks. Due to their high reactivity, radicals were not considered to be suitable for organic synthesis.

During the last several decades, however, the field has undergone an explosive growth. Radicals are highly reactive species and were considered of minor significance in biological systems. However, the oxygen metabolism was found to lead to reactive oxygen species. Evidence for the formation of oxygen-derived radicals in many normal biological as well as pathological processes has been accumulating. At present there is evidence for radical involvement in over 100 diseases [3,4]. This realization can be compared to the discovery of bacteria and viruses as causative agents in human diseases. Yet the subject of radicals and reactive oxygen metabolites (ROMs) is barely touched upon in the curriculum of medical schools. We no longer have to concern ourselves only with outside agents (bacteria, viruses, pollution) that affect our health, but also with damage caused by agents of normal oxygen metabolism. Research in the radical area involves many specialists in different areas of science. The number of journals publishing their results has grown, including journals in chemistry, biochemistry, nutrition, pathology, physiology, pharmacology, toxicology, medicine, environmental sciences, and the *Journal of Free Radicals in Biology and Medicine.*

It is therefore impossible even for the expert in the field to keep abreast of all these new findings. Many reviews have appeared on free radicals in biology and medicine, and new reviews on selected topics appear annually. These reviews are very useful for the expert working in the field of free radical research. A novice to the field with little background in chemistry has a difficult time understanding these complex reactions, and is therefore unaware of the importance of radical chemistry to biology and to the medical profession.

I felt that it was a timely and worthwhile effort to write an introduction to the field of reactive oxygen metabolites with major emphasis on their chemistry.

In order to understand the many possible mechanisms involved in the normal physiology as well as pathology caused by reactive oxygen metabolites, we must have first of all a thorough understanding of their chemistry. We start with the basic chemistry of radicals, their formation, identification and reactions (Chapters 2 and 3). Another group of reactive oxygen species is discussed in Chapter 4. These metabolites are singlet oxygen, and electronically excited carbonyl compounds.

After we have discussed the reactions of radicals, singlet oxygen and excited carbonyl compounds, we have to concern ourselves with the important question: where and how are these reactive species formed *in vivo*? The main source of radicals, reactive oxygen and nitrogen species are the mitochondria (electron transport chain), microsomes, phagocytes, endothelium and neurons (Chapter 5).

In Chapter 6 we apply the knowledge of the chemistry to the reactions of some important biomolecules (fatty acids, proteins and DNA). Damage to membranes (fatty acids) and DNA is most important for the initiation of pathological processes.

I have tried to keep the text as compact as possible without sacrificing clarity. A novice to the field is overwhelmed by the information in the literature. A search of Medline from 1990 to 1996 shows the number of publications dealing with singlet oxygen (roughly 100 per year), superoxide (about 1600 per year) and hydroxyl radical (300-400 per year) at a constant level. However, papers dealing with nitric oxide jumped from 292 in 1990 to 3477 in 1996. This clearly shows where the action is! Nitric oxide was chosen in 1992 by *Science* magazine as "Molecule of the Year" [5]. In 1998 the Nobel Prize in Physiology or Medicine was awarded to 3 scientists (F. Murad, R. Furchgott and L. Ignarro) for their nitric oxide-related work. Due to the ever increasing importance of nitric oxide, I have devoted a separate chapter to the chemistry of nitric oxide (Chapter 7). Since nitric oxide is a metabolite of L-arginine and oxygen, it may be classified as a reactive oxygen metabolite.

During the course of evolution, our bodies have developed defenses against ROMs. These defenses consist of the enzymes superoxide dismutase (SOD), catalase (CAT), glutathione peroxidase (GSH-Px), metal complexing proteins and small molecules which scavenge free radicals or singlet oxygen. These antioxidant defenses are discussed in Chapter 8.

In Chapter 9 I connect the chemistry of these reactive oxygen and nitrogen species to the pathology of various disease processes (aging, cancer, neurodegenerative diseases, lung diseases, atherosclerosis, etc.). The basis of biology is chemistry. Once we understand the chemistry of ROMs, we can better rationalize the pathology. Numerous diseases with diverse clinical symptoms can be understood at the most basic level by the concept of "oxidative stress", which is a disturbance in the prooxidant/ antioxidant balance of a system.

I have tried to compress the huge amount of information about radical chemistry, photochemistry, radiation chemistry, enzymatic oxidations, immunology, aging, cancer, and other diseases into one compact book. It is obvious that in this endeavour I had to be very selective. It is impossible to cite all the original literature and I apologize to anybody who feels that his or her work was not adequately cited. I made extensive use of references to books and reviews, which can serve the reader as a guide for more in depth discussions. This book is intended for people who want to know about the importance of reactive oxygen and nitrogen species in human health and disease, but who do not have the time and inclination, nor background to work their way through the immense original literature. The overwhelming amount of data has not always led to a better understanding, but instead to confusion and erroneous conclusions. The whole field is in a state of rapid evolution. Although we are still in the dark about many details, we can see glimpses of light at the end of the tunnel.

I hope that this introductory text will serve the novice as a guide through the complex world of radical chemistry and pathology and will encourage the reader to learn more about this fascinating subject.

Chapter 1

THE PROS AND CONS OF LIVING IN
AN OXYGEN ATMOSPHERE

Oxygen has long been recognized as a life-sustaining molecule.
Another view, attesting to its now well-recognized lethality, is that
oxygen is the worst environmental pollutant of all time.

Linda Biadasz Clerch
Donald J. Massaro [6]

Oxygen is responsible for providing us with the necessary energy for the daily operation of the cell. The foodstuffs are oxidized (electrons are removed) via the mitochondrial respiratory electron transport chain in a series of steps involving many enzymes. The final electron acceptor in this sequence is oxygen, which is reduced in the last step by 4 electrons to give $2 H_2O$. The energy gained in this series of electron transfers is stored in the form of chemical energy. This is the process of oxidative phosphorylation, which forms adenosine triphosphate (ATP) from adenosine diphosphate (ADP). ATP is used as an energy source for cellular function. Without oxygen life on earth as we know it today would not exist. The atmosphere we breath consists of 21% oxygen. There is a very limited range of oxygen concentration in which we humans can exist. Lowering or increasing the oxygen concentration by just a few percentage points causes serious damage. Why should this be so? Oxygen is reduced to water, which is certainly not toxic, so increased formation of water should be no problem. Electrons are transported along the respiratory chain via 2 e⁻ steps. **Two-electron transfer** does not produce radicals. However nothing in this life is perfect, not even the electron transport chain. Some intermediates along the chain, namely the **flavin semiquinone radical** and the **semiquinone radical of ubiquinone can transfer a single electron to oxygen** to produce the oxygen radical anion (or superoxide radical).

6

The oxygen radical anion is deleterious to the cell. It can initiate lipid peroxidation, or can lead via its dismutation product H_2O_2 to the highly reactive hydroxyl radical. The hydroxyl radical will indiscriminately destroy everything it encounters. High concentrations of hydroxyl radicals will kill us. Horrible examples are Hiroshima and Nagasaki. Most people did not die because of the blast, but because of radiation producing huge concentrations of hydroxyl radicals.

During the course of evolution aerobic cells have developed defense mechanisms against this toxic threat by reactive oxygen metabolites [7-9]. The enzymes dealing with these reactive oxygen metabolites are superoxide dismutase (SOD), catalase (CAT) and glutathione peroxidase (GSH-Px). These enzymes catalyze the sequence: $O_2 \cdot^- \longrightarrow H_2O_2 \longrightarrow H_2O$. These enzymatic defenses are usually sufficient to deal with the toxic threat of $O_2 \cdot^-$ and H_2O_2. However in certain cases when the concentration of these enzymes is low, serious damage by Reactive Oxygen Metabolites (ROMs) can be avoided by the second line of defense against these reactive species. This defense depends on the presence of compounds called antioxidants or radical scavengers. These compounds may be generated naturally in vivo (examples: uric acid, melatonin, carnosine) or may be supplied in the diet. Some of the most important dietary antioxidants are Vitamins A, C, and E.

Life developed in an oxygen-free atmosphere. The blue-green algae did not die. They slowly produced oxygen from water. In order for multicellular organisms to evolve membranes were needed. These membranes consist of polyunsaturated fatty acids, which are easily damaged by oxygen. Therefore, nature has incorporated tocopherol (Vitamin E) into the membrane. Tocopherol is an efficient radical scavenger and singlet oxygen quencher, and therefore affords protection against the damaging effects of oxygen radicals and excited oxygen. The more complex an organism the more may go wrong with it. Remember Murphy's law: If something can go wrong it will go wrong. Single cell organisms are immortal. Death is a consequence of complexity. If something has gone wrong Medical Science tries to "fix" the problem by adding another chemical (drug) to an already complex mix. Sometimes this approach works, but sometimes it ends in a catastrophe (thalidomide!). This is not surprising since we cannot predict the outcome of interfering with a highly complex dynamic system. All prescription drugs contain a pamphlet listing a litany of possible side effects, which may or may not occur. The outcome is very sensitive to initial conditions, which are different for every individual.

If oxygen-derived radicals are the cause for many pathological processes then we may be able to prevent these diseases by the proper dose of antioxidants or radical scavengers. The question is: What is the proper dose? More is not always better. This was already recognized by Paracelsus, who said: Alles ist Gift, es kommt nur auf die Dosis an (everything is poison, it only depends on the dose). Some compounds may even fulfill double roles: Acting as an anticarcinogen at low doses and as a carcinogen at higher doses. One of the best known and promoted radical scavengers is Vitamin C. This compound has been recommended for anything from ingrown toe-nails to cancer [10,11]. It is evident to a chemist that ascorbic acid can act as a radical scavenger, since it has many easily abstractable hydrogen atoms:

$$AH_2 + \cdot OH \longrightarrow AH\cdot + H_2O$$

In this reaction we transform the highly reactive hydroxyl radical into another radical ($AH\cdot$) of much lower reactivity. However $AH\cdot$ can react with oxygen:

$$AH\cdot + O_2 \longrightarrow HO_2\cdot + A \text{ (dehydroascorbic acid)}$$

The $HO_2\cdot$ may initiate lipid peroxidation or give H_2O_2, which in turn produces more hydroxyl radicals.

Since ascorbic acid is a good reducing agent (donates electrons) it causes cellular damage via reduction of Fe^{3+} or Cu^{2+}

$$Fe^{3+} + AH_2 \longrightarrow Fe^{2+} + H^+ + AH\cdot$$

$$Cu^{2+} + AH_2 \longrightarrow Cu^+ + H^+ + AH\cdot$$

Subsequently these reduced metal ions react with H_2O_2 to give OH radicals via a Fenton reaction:

$$Fe^{2+} + H_2O_2 \longrightarrow Fe^{3+} + OH^- + \cdot OH$$

High concentrations of ascorbic acid may be especially damaging in patients with iron overload or in trauma.

Since free radicals have one unpaired electron, they are very reactive and try to find another electron with which they can pair, to saturate their valence.

The thread of pro and con runs through most of free radical biology and medicine. Our cells are carrying out a precarious balancing act like the "Fiddler on the Roof". The concentrations of many chemicals in our bodies, such as Fe(III)/Fe(II), Cu(II)/Cu(I), O_2, H^+, Ca^{2+}, have to be strictly controlled. Ascorbic acid has been shown to be mutagenic in the presence of Cu^{2+} ions [12] , and hydroxyl radicals have been observed in the presence of Fe^{3+} under conditions prevalent in the stomach [13]. I like to call attention to an article by Halliwell with the provocative title: "Vitamin C: the key to health or a slow acting carcinogen?" [14]. Several investigators [13,15] addressed the damaging effects of the iron-ascorbic acid combination, which is often found in multivitamin tablets. We encounter this pro and con situation also with other antioxidant defenses. SOD protects cells from the damaging effect of superoxide, but it produces hydrogen peroxide, which in the absence or low concentrations of catalase or glutathione peroxidase accumulates and subsequently causes cellular damage (most likely via OH). Too much SOD has been shown to increase $O_2^{\cdot-}$ - induced damage [16] (for a discussion on this SOD paradox, see Chapter 8).

Superoxide fulfills a dual purpose. It may be harmful, but it is also an essential intermediate in host defenses against invading bacteria. When leukocytes are stimulated by invading foreign organisms or particles, the leukocytes respond with what has become known as the

respiratory burst. This means increased oxygen consumption and formation of reactive oxygen metabolites ($O_2^{\cdot -}$ and H_2O_2). Subsequently, H_2O_2 reacts with certain metal ions to form the hydroxyl radical, which destroys the invading microorganism. This process is known as **phagocytosis** [17,18]. For example, a macrophage completely envelops the invader and digests it. If, however, digestion is not possible as in inorganic fibers like asbestos, the macrophage produces large quantities of superoxide and hydrogen peroxide, which are released into the microenvironment. The subsequent formation of hydroxyl radicals causes considerable damage to the surrounding tissue. This process is called **failed phagocytosis**. Another mechanism by which $O_2^{\cdot -}$ may be damaging is via combination with nitric oxide ($\cdot NO$) to give peroxynitrite (^-OONO). Peroxynitrite damages many important biomolecules, but on the other hand it represents an additional weapon in the anti-microbial armamentarium of phagocytes.

It has been noted by Watson [19], that "**important biological objects come in pairs" (two strands of DNA, two sexes)**. This tendency towards pairing also extends to the molecular and electronic levels: A pair of electrons of opposite spin form a chemical bond, the electron transport chain in mitochondria proceeds via a series of two-electron transfer steps. It appears that nature has anticipated the destructive power of radicals and is trying to avoid them whenever possible. It has been proposed that the major benefit of sexual reproduction is its promotion of recombinatorial repair in the germ line of organisms during meiosis [20]. Oxidative damage may be the fundamental selective force for the maintenance of sex [20]. The preponderance of sex (**pairs**) may, therefore, be in direct response to the single **unpaired** electron in oxygen radicals.

It took indeed a long time before the importance of radicals in the biological system became accepted by the scientific community mainly due to their destructive potential. On the other hand, radicals are the cause of mutations, which in turn are essential for evolution. Without radicals evolution would come to a dead end. Raold Hoffmann in "The Same and not the Same" points out that "pure" water consists of 18 different kinds of molecules. Hoffmann states, "The little tritiated water in normal water is something we have evolved with over millions of years. It may even be that the chance variation provided by mutations induced in part by this radioactivity was necessary to get us to the present stage of human creative complexity" [21].

A noticeable exception to the concept of pairing is the presence of only one isomer in chiral (asymetric) molecules *in vivo*. Amino acids, the building blocks of proteins, only exist in one enantiomeric form (the L-form) *in vivo*. If, however, these chiral molecules are synthesized in the laboratory under normal conditions, both enantiomeric forms are produced in equal amounts. Pasteur, the discoverer of chirality, placed great importance on the presence of only one enantiomeric form in nature. He believed that this observation distinguished living from non-living matter.

The primary metabolically produced oxygen radicals are the O_2^{-}/ HO_2· and the hydroxyl radical. In addition to these radicals, there are other oxygen derived metabolites, which are not radicals but can have harmful effects on cellular structure and function. These metabolites include hydrogen peroxide, hydroperoxides, epoxides, singlet oxygen and electronically excited states. These reactive species interact with vital cell constituents either directly (epoxides, singlet oxygen and excited states) or indirectly via formation of radicals (hydrogen peroxide and hydroperoxides). I shall first discuss the general properties of radicals and subsequently address the formation and reactions of oxygen radicals both in purely chemical as well as in biological systems.

The latest addition to the pantheon of biologically important radicals is nitric oxide [5]. Since nitric oxide is formed via an oxygen-dependent process from L-arginine, it may be classified as a ROM. This radical is the smallest hormone. Due to its small size and electrical neutrality, it easily penetrates cell walls, which is not the case for the negatively charged superoxide radical anion. Nitric oxide fulfills many important physiological functions: relaxation of vascular smooth muscle, neurotransmission, inhibition of platelet aggregation and phagocytosis. However, nitric oxide also possesses many deleterious effects: it reacts with many biomolecules either by itself or more likely via its oxidation products ·NO_2, N_2O_3, NO_2^{-}, NO_3^{-} or peroxynitrous acid (formed via ·$NO + O_2^{-}$). Nitric oxide and its oxidation products are part of air pollution and are highly damaging to our lungs. On the other hand, the use of nitric oxide may save the lives of newborn babies with pulmonary hypertension, and may also make it possible for an older gentleman to make the baby (the natural way) in the first place. This pro and con situation is also inherent with ozone (O_3), which as part of air pollution is damaging our lungs, but the same molecule high in the atmosphere (the ozone layer) protects us from the damaging effects of ultraviolet radiation.

Accumulated oxidative damage leads to aging (accumulation of lipofuscin) and cancer. Thus it appears that death is inherent with life and programmed already from the very beginning through the oxygen metabolism.

Chapter 2

BASIC CHEMISTRY OF RADICALS [22]

Radicals should not be free, they should be in jail.
                                        Popular saying

A. DEFINITION AND HISTORY

Radicals are chemical fragments which contain an unpaired electron. Radicals are always searching for another electron with which they can form a pair. Since in most chemical environments many electrons are available, radicals are expected to be highly reactive. Some radicals like the hydroxyl radical react with most molecules at close to diffusion controlled rates ($10^{10} M^{-1}s^{-1}$). This means that the ·OH reacts at almost every collision. There are however some not-able exceptions to this rule of high reactivity. The triphenylmethyl radical is highly stable (it can be kept in a test tube). This radical is stabilized by resonance, i.e., the unpaired electron is not localized at a specific atom, but is spread over a large molecular domain. Due to its stability the triphenylmethyl radical was the first radical to be discovered by Gomberg [2] without the use of today's sophisticated analytical instrumentation. The history of the early years of free radicals has been reviewed [23]. Other radicals of relatively low reactivity are oxygen, superoxide radical anion and nitric oxide. We shall discuss these important species later on. Although most radicals are highly reactive they do not (fortunately) react with water at room temperature. This stability of water towards radicals is due to the high H-OH bond dissociation energy (119 Kcal/mole). This bond can only be broken by high energy radiation (γ radiation or X-rays). The radiation chemistry of water and aqueous solutions [24,25] has provided a clean source of hydroxyl radicals and has provided the scientific basis for the whole area of radical biology and medicine.

Radicals are usually described as "free" radicals. The adjective "free" is no longer appropriate. Historically all substituents in a molecule were

14

called "radical". When Gomberg [2] discovered the triphenyl-methyl radical he called it "free", since it was no longer attached to any other atom or group of atoms. The adjective "free" should therefore be eliminated from the literature.* Unfortunately we still have a journal named *Free Radicals in Biology and Medicine*.

B. FORMATION OF RADICALS

Organic chemistry consists of reactions in which bonds are broken and new ones are formed. A simple single bond consists of two electrons with opposite spins:

$$A - B \text{ or } A \uparrow\downarrow B$$

This bond can be broken in 3 different ways:

$$
\begin{array}{llll}
1.) & A - B \longrightarrow A^+ + B^- & \text{heterolytic cleavage} \\
2.) & A - B \longrightarrow A^- + B^+ & \text{heterolytic cleavage} \\
3.) & A - B \longrightarrow A + B & \text{homolytic cleavage}
\end{array}
$$

The charged species indicated in reactions (1) and (2) are usually not free ions, but carry only partial charges and are only present in the transition state. Due to their developing charges these reactions are affected by the polarity of the solvent. Most of organic chemistry deals with these types of reactions. Homolytic cleavage on the other hand produces radicals, which with some exceptions are electrically neutral species. The radicals are therefore affected by the solvent only to a lesser extent [26].

---

* I thank Prof. Dr. Waldemar Adam for calling my attention to this matter.

The homolytic cleavage of bonds requires the input of energy. This energy is known as the bond dissociation energy. Dissociation energies of some common bonds are summarized in Table 1. The energy required for homolytic cleavage can be supplied by the following methods: 1. Thermal energy, 2. Photochemical energy, 3. Ionizing radiation, 4. One-electron transfer reactions, and 5. Molecule assisted homolysis (MAH).

Table 1. Bond dissociation energies [27].

Bond	D^0_{298}/kJ mol^{-1}	Bond	D^0_{298}/kJ mol^{-1}
H-H	435.99	H-Cl	
H-OH	498	H-Br	431.62
CH_3-H	438.5	H-I	366.35
CH_3CH_2-H	422.8		298.407
$(CH_3)_3C$-H	403.5	Cl-Cl	
C_6H_5-H	464.0	Br-Br	242.58
$C_6H_5CH_2$-H	368.2	I-I	192.807
CH_3-CH_3	376.0		151.088
		CH_3-F	
CH_3O-H	436.8	CH_3-Br	472
C_6H_5O-H	361.9	C_6H_5-Br	292.9
HOO-H	369.0	C_6H_5-I	336.8
		Cl_3C-H	273.6
HO-OH	213	Cl_3C-Cl	392.9
CH_3O-OCH_3	157.3		305.9
$(CH_3)_3CO$-$OC(CH_3)_3$	159.0	HS-H	
CH_3COO-$OCOCH_3$	127.2	CH_3-SH	381.6
C_2H_5COO-$OOCC_2H_5$	127.2	CH_3S-H	312.5
		C_6H_5S-H	365.3
		HS-SH	348.5
		CH_3S-SCH_3	276
			272.8

1. Thermal cleavage of bonds

Any bond can be broken by heat energy. The best known process of this type is the technical process of hydrocarbon cracking. In this process high molecular weight hydrocarbons are converted to lower molecular weight fragments which can be used as fuel in internal combustion engines. However most of the thermal methods used by chemists to produce radicals require much lower energies. The amount of energy needed depends on the bond dissociation energy (Table 1).

The highest bond dissociation energy of any bond of biological importance is the H–O bond in H–OH (119 Kcal/mole). Water is very stable and cannot be split homolytically by any chemical reaction. This stability makes water an ideal liquid for the evolution of life as we know it.

Bonds of low dissociation energy are the –O–O– bonds in peroxides (RO–OR) and the C–N bonds in azo-compounds (R–N=N–R)[28]:

$$RO-OR \xrightarrow{\Delta} 2\,RO\cdot$$

$$R-N=N-R \xrightarrow{\Delta} 2\,R\cdot + N_2$$

Some compounds used in these types of reactions are di-tert. butylperoxide:

$$(CH_3)_3C-O-O-C(CH_3)_3 \quad \text{and some azo-compounds like}$$

$$Ph_3C-N=N-Ph \quad \text{or} \quad (CH_3)_2C-N=N-C(CH_3)_2$$
$$\underset{CN}{|} \qquad \underset{CN}{|}$$

In the decomposition of peroxides the first step is the formation of an alkoxy radical:

$$(CH_3)_3CO-OC(CH_3)_3 \longrightarrow 2\ (CH_3)_3CO\cdot$$

The alkoxy radical can either undergo β-cleavage to give acetone and a methyl radical or abstract a hydrogen atom from the solvent:

$$(CH_3)_3CO\cdot \longrightarrow CH_3\cdot + CH_3COCH_3$$

$$(CH_3)_3CO\cdot + R-H \longrightarrow (CH_3)_3COH + R\cdot$$

The ratio of methane/t-butanol depends on the hydrogen-donating ability of the solvent.

Another group of compounds with weak -O-O- bonds are the acylperoxides like benzoylperoxide and acetylperoxide:

$$\phi COOOOC\phi \longrightarrow 2\ \phi COO\cdot \longrightarrow 2\ \phi\cdot + 2\ CO_2$$

$$CH_3COOOCOCH_3 \longrightarrow 2\ CH_3COO\cdot \longrightarrow 2\ CH_3\cdot + 2\ CO_2$$

An important example of the splitting of an -O-O- bond is the decomposition of hydrogen peroxide, which is an intermediate in oxygen metabolism:

$$HO-OH \xrightarrow{\ \Delta\ } 2\ HO\cdot$$

2. Photochemical formation of radicals and excited states [29]

The energy for the photochemical cleavage of bonds is provided by absorption of a photon. Photochemical activation differs from thermal activation in that it may be more specific. Photons can be absorbed by a specific chromophor (like the azo chromophor) within a large molecule:

$$\text{R-N=N-R} \xrightarrow{\text{h}\nu} \text{R-N=N-R*} \longrightarrow 2\,\text{R}\cdot + \text{N}_2$$

The important step is absorption of a photon. If the molecule does not absorb nothing happens. Some molecules (like some azo-compounds) are highly colored and therefore absorb in the visible region of the spectrum, while other molecules are colorless and only absorb photons in the ultraviolet part of the spectrum:

$$\text{RO-OR} \xrightarrow{\text{h}\nu} \text{RO-OR*} \longrightarrow 2\,\text{RO}\cdot$$

Once a molecule has absorbed a photon and is in an electronically excited state, it can undergo a variety of reactions:

$$\text{A*} \longrightarrow \text{A} + \text{h}\nu \text{ (light emission)}$$

$$\text{A*} \longrightarrow \text{A} + \text{heat (radiationless decay)}$$

$$\text{A*} \longrightarrow \text{A*'} \text{ (different excited state)}$$

$$\text{A*} + \text{B} \longrightarrow \text{A} + \text{B*} \text{ (energy transfer)}$$

$$\text{A*} + \text{B} \longrightarrow \text{A}\cdot^{+} + \text{B}\cdot^{-} \text{ (electron transfer)}$$

$$\text{A*} + \text{B} \longrightarrow \text{A}\cdot^{-} + \text{B}\cdot^{+} \text{ (electron transfer)}$$

$$\text{A*} + \text{B} \longrightarrow \text{A-H} + \text{B(-H) (H-abstraction)}$$

$$\text{A*} \longrightarrow \text{unimolecular decomposition to radical fragments}$$

A stable molecule in the ground state has all its molecular orbitals occupied with two electrons of antiparallel spin (molecular oxygen is an exception). We can excite the molecule by moving one electron to a higher unoccupied molecular orbital (MO). In this excited state the molecule has two MOs occupied with one electron each and these electrons have anti-parallel spin. The excitation with simultaneous change of spin is said to be spin forbidden (SF). This first excited state is called a singlet state. The singlet state can undergo intersystem crossing (IC) to a triplet state, in which the two electrons in the half-filled orbitals have parallel spin. This sequence of events can be represented schematically as follows:

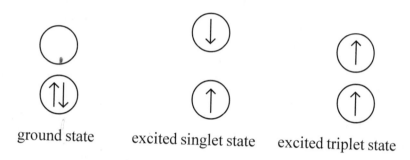

ground state   excited singlet state   excited triplet state

The energy levels of these states are shown in the following simplified diagram:

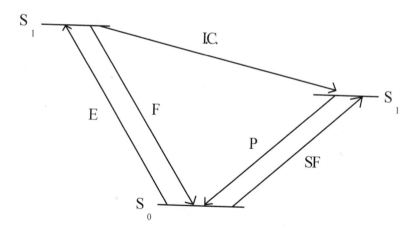

A triplet state is a diradical and can undergo all the typical radical reactions, like addition, H-abstraction or electron transfer. I shall discuss the formation and reactions of triplet excited carbonyl compounds in Chapter 4.

The first formed excited state of a molecule (a singlet state) is meta-stable. One way by which the excited singlet can get rid of its excess energy is via emission of light, which usually is of somewhat longer wave-length (lower energy) than that absorbed. If the emitting excited state is a singlet the emission is called fluorescence (F); if it is a triplet the emission is called phosphorescence (P). Fluorescence involves no change in elec-tron spin (no change in multiplicity), whereas phosphorescence involves simultaneous change in the spin of one electron (change in multiplicity). The singlet state is of higher energy than the triplet and undergoes inter-system crossing to the triplet state. The triplet state has a longer lifetime than the singlet, because its deactivation to the ground state requires a spin inversion (change in multiplicity). Due to their longer lifetimes triplet excited states are the ones we have to concern ourselves with as possible damaging species in biological systems.

Energy transfer. An excited molecule can dispose of its energy via transfer of energy to another molecule B:

$$A^* + B \longrightarrow A + B^*$$

This process actually represents two different types of reactions, depend-ing on your point of view. If the object of the reaction is to protect A^* from decomposition, the process is called quenching and B is called a quencher of the excited state of A. If, however, the object is to produce an excited state of B (which cannot be produced via direct irradiation) then the pro-cess is called photosensitization and A is called a photosensitizer. Both of these processes (quenching and photosensitization) are important pro-cesses in biological systems and have important applications in medicine (photodynamic effect and singlet oxygen quenching).

Electron transfer. Excited carbonyl compounds (may be formed via dioxetanes) interact with DNA via one-electron transfer to give a radical cation, which in turn reacts with water to give 8-hydroxy-guanine among other products [30].

Unimolecular decomposition. Another important process of excited states is the dissociation into radicals. I have already mentioned the dissociation of peroxides and azo-compounds. Another important group of molecules used in the generation of radicals has been the photolysis of ketones [29]:

$$CH_3COC_2H_5{}^* \xrightarrow{1} CH_3\dot{C}{=}O \; + \; \cdot C_2H_5$$
$$\downarrow 2$$
$$CH_3\cdot \; + \; C_2H_5\dot{C}{=}O$$

Towards the long wavelength side of the absorption band reaction 1 predominates (the weakest bond is broken), whereas towards the shorter wavelength the cleavage of the stronger bond (process 2) also occurs.

H-abstraction. In addition to unimolecular decomposition the excited molecule can also react with other molecules via H-abstraction. This type of reaction has been a very important method for the formation of a variety of radicals. The excited species can abstract an H atom from alkanes, alkenes, alcohols and ethers. The following reactions may serve as an example [29]:

$$\phi CO\phi \xrightarrow{h\nu} \phi CO\phi^* \text{ (singlet)} \longrightarrow \phi CO\phi^* \text{ (triplet)}$$

$$\phi CO\phi^* \; + \; (CH_3)_2CH{-}OH \longrightarrow \underset{\overset{|}{OH}}{\phi\dot{C}{-}\phi} \; + (CH_3)_2\dot{C}{-}OH$$
(triplet)

$$(CH_3)_2\dot{C}{-}OH \; + \; \phi CO\phi \longrightarrow CH_3COCH_3 \; + \; (\phi)_2\dot{C}{-}OH$$

$$2\ (\phi)_2\dot{C}\text{-OH} \longrightarrow (\phi)_2\underset{\underset{HO}{|}}{C}\text{-}\underset{\underset{OH}{|}}{C}(\phi)_2$$

Once radicals are formed in a biological system they react rapidly with oxygen to give peroxyl radicals, which in turn can undergo a variety of reactions (see Chapter 3).

The formation of radicals via photochemical activation appears to be of minor importance in biological systems, except in reactions involving epithelial cells. However, as we shall discuss in detail in Chapter 4, electronically excited states (singlet oxygen and carbonyl compounds) can be formed via purely chemical or enzymatic reactions. We can therefore carry out photochemistry without light.

3. Radiation-induced cleavage of bonds [27]

The name ionizing radiation implies the formation of ions via removal of an electron. This process leads to a species with one unpaired electron and a positive charge (a radical cation). The electron can add to some molecules (like O_2) to form a radical anion ($O_2^{\cdot-}$). The cleavage of bonds via ionizing radiation is much less selective than cleavage via photolysis. However, this is not the case in a simple molecule like water where we only have one type of bond . The radiolysis of H_2O and aqueous solutions [24,25] has provided almost all the information concerning the reaction of the hydroxyl radical. The formation of hydroxyl radicals via radiolysis will be discussed in detail in Chapter 3.

4. Bond cleavage via one-electron transfer reactions [31,32]

Some bonds, like the $-O-O-$ bond in $HO-OH$ or $RO-OR$ or $RO-OH$, can be split homolytically via one-electron oxidation-reduction reactions. These reactions involve transition metal ions, like Fe^{2+}, Cu^+, Ti^{3+}, Ce^{3+}, Ce^{4+} or Co^{2+}-EDTA.

Oxidation: $ROOH + Ce^{4+} \longrightarrow Ce^{3+} + H^+ + ROO\cdot$

Reduction: $ROOR + Fe^{2+} \longrightarrow Fe^{3+} + RO^- + RO\cdot$

Since thermal and photochemical cleavage of bonds is unlikely inside cells, the one-electron reduction of H_2O_2 and peroxides (ROOR) or hydroperoxides (ROOH) by Fe^{2+} or Cu^+ represent the most important radical-forming reaction in biological systems. The formation of hydroxyl radicals via one-electron transfer will be discussed in detail in Chapter 3.

5. Molecule-assisted homolysis (MAH)

MAH involves the interaction of two closed-shell molecules to give two radicals. In these types of reactions one molecule is reduced and one is oxidized. The reaction involves peroxides (as electron acceptors) and electron-rich compounds [33]:

$$\begin{array}{c}\diagdown\text{C=C}\diagup + \text{ROOH} \longrightarrow \diagdown\dot{\text{C}}\text{-}\underset{|}{\text{C}}\text{-H} + \text{ROO·}\end{array}$$

$$\begin{array}{c}\diagdown\text{C=C}\diagup + \text{ROOH} \longrightarrow \diagdown\dot{\text{C}}\text{-}\underset{|}{\text{C}}\text{-OH} + \text{RO·}\end{array}$$

$$\phi\text{-N(CH}_3)_2 + \phi\text{COOOCO}\phi \xrightarrow{S_N2} [\phi\text{-}\overset{+}{\underset{|}{\dot{\text{N}}}}\text{-OCO}\phi \quad \phi\text{COO}^-]$$

$$\downarrow \text{electron transfer}$$

$$\phi\text{-}\overset{+\cdot}{\text{N}}\text{(CH}_3)_2 + \phi\text{COO·} + \phi\text{COO}^-$$

In the latter case nucleophilic substitution competes with electron transfer, which amounts to about 18% of total reaction [34-36]. It was suggested by Pryor [33] that these reactions can proceed intramolecularly in lipid hydroperoxides, like linoleate hydroperoxide (has double bonds).

C. REACTIONS OF RADICALS
1. General considerations about the reactivity of radicals

A radical can obtain another electron with which it can form a pair via a variety of reactions, like dimerization, disproportionation, H abstraction, addition to unsaturated bonds, oxidation followed by reaction with a nucleophile, reduction, β-cleavage, and reductive elimi-

24

nation. The reactions of radicals can be classified into 2 types: 1. Reactions which give rise to non-radical or molecular products, and 2. Reactions which produce another radical. Reactions belonging to the first category are: dimerization, disproportionation, oxidation-reduction, and reductive elimination. The second category of reactions include: H-abstraction, addition to multiple bonds, reactions with oxygen, and β-cleavage.

Oxygen radicals, which are of major concern in biology, have the unpaired electron on an oxygen atom, like hydroxyl (·OH), superoxide radical anion ($O_2^{·-}$), alkoxy (RO·), and peroxyl (ROO·) radicals. These radicals are the important intermediates in both normal as well as abnormal (pathological) biological processes.

2. Dimerization and disproportionation

One way for a radical to find another unpaired electron is to react with itself:

$$R· + R· \xrightarrow{2\,k_t} \text{molecular products}$$

One of the products formed is the dimer, for example:

$$CH_3· + CH_3· \xrightarrow{2\,k_t} CH_3-CH_3$$

The rate constants for these types of reactions are in the neighborhood of 10^9 - 10^{10} M^{-1} s^{-1}. Usually the self reaction of radicals involves both dimerization and disproportionation:

$$CH_3CH_2· + CH_3CH_2· \begin{cases} \nearrow CH_3CH_2CH_2CH_3 \\ \searrow CH_2=CH_2 + CH_3CH_3 \end{cases}$$

Disproportionation can be considered an oxidation-reduction reaction. One radical gets oxidized and the other radical is reduced. However, not all radicals undergo combination or disproportionation with rate constants close to the diffusion controlled limit. Some radicals exist in equilibrium with molecular products:

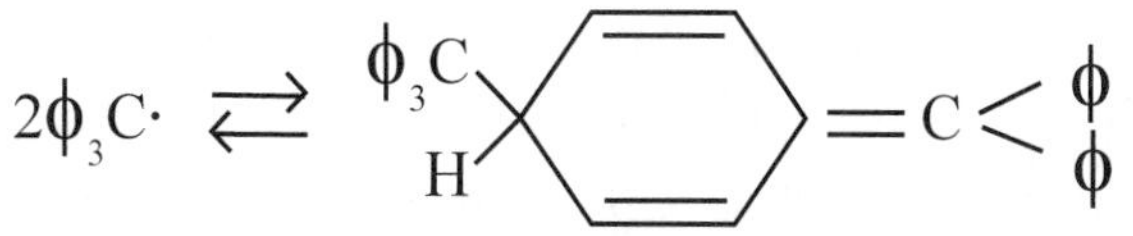

$$2\phi_3C\cdot \rightleftarrows \underset{H}{\overset{\phi_3C}{\diagdown}}\diagup\hspace{-0.2em}\bigcirc\hspace{-0.2em}=C<\begin{matrix}\phi\\\phi\end{matrix}$$

This dimerization was long believed to produce the hexaphenylethane (Ph_3CCPh_3). It took over half a century before the correct structure (above) was determined [37]. Science has a self correcting mechanism. Sometimes it may take some time, but eventually a mistake will be corrected, which is more than can be said for many other human enterprises.

Another group of radicals which undergo slow combination include radicals with unpaired electrons located on oxygen or nitrogen. This group includes sterically hinderd phenoxy radicals and peroxyl radicals.

3. Combination with other radicals

Since oxygen in its ground state is a triplet (i.e., it is a diradical) it can combine with a variety of other radicals:

$$R\cdot \ + \ O_2 \rightleftarrows R-OO\cdot$$

This reaction is one of the most important reactions in radiation biology. The peroxyl radicals can undergo a variety of reactions (elimination of superoxide, H-abstraction, and dimerization), which are biologically important. Due to their importance in both technical (autoxidation) processes as well as in biological systems, the chemistry of peroxyl radicals has been extensively reviewed [38]. Some biologically important reactions of peroxyl radicals will be discussed in detail in Chapter 3, Section C.

4. Hydrogen abstraction

Hydrogen abstraction is an important reaction as a chain propagating step in autoxidation. Very reactive H-abstracting radicals like F·, Cl·, and

·OH react with most organic compounds at close to diffusion controlled rates. The reactivity of a radical depends on its stability, which in turn depends on the presence of substituents attached to the atom carrying the unpaired electron. The radical can be stabilized by resonance with aromatic rings or via hyperconjugation by alkyl groups. In a series of alkyl radicals the order of stability is: tert. >sec.>prim. $((CH_3)_3C\cdot > (CH_3)_2CH\cdot > CH_3CH_2\cdot)$[39]. In the series of phenyl substituted alkyl radicals the stabilty is: $\phi_3C\cdot > \phi_2CH\cdot > \phi CH_2\cdot$. These radicals do not abstract an H-atom from hydrocarbons, they only dimerize or combine with oxygen.

The hydroxyl radical is the most reactive oxygen radical. It reacts with most organic compounds at very high rates $(10^9 - 10^{10} \ M^{-1}s^{-1})$ via H-abstraction [40]:

$$(CH_3)_2CHOH \ + \ \cdot OH \longrightarrow (CH_3)_2\dot{C}OH \ + \ H_2O$$

$$(CH_3)_2\dot{C}OH \ + \ Cu^{2+} \longrightarrow (CH_3)_2C=O \ + H^+ + Cu^+$$
$$84\%$$

$$(CH_3)_3COH \ + \ \cdot OH \longrightarrow (CH_3)_2\underset{\underset{\cdot CH_2}{|}}{C}\!-\!OH \ + \ H_2O$$

$$(CH_3)_2\underset{\underset{\cdot CH_2}{|}}{C}\!-\!OH \ + Cu^{2+} \longrightarrow (CH_3)_2\underset{\underset{CH_2OH}{|}}{C}\!-\!OH \ + Cu^+$$
$$86\%$$

The first radical formed by H-abstraction can be inferred by oxidation to a final stable product.

5. Addition to multiple bonds [41]

The addition of free radicals to olefins often proceeds via chain reactions. They are therefore of great technical importance. The study of free radical mechanisms first started with an addition reaction. Kharasch and Mayo [42] observed that the addition of HBr in presence of benzoylperoxide added to propylene in an anti-Markownikoff fashion. In subsequent studies the reaction was shown to proceed via a chain reaction as follows:

$$HBr + R\cdot \rightarrow R-H + Br\cdot \qquad\qquad (1) \quad \text{initiation}$$

$$Br\cdot + CH_3CH=CH_2 \longrightarrow CH_3\dot{C}H-CH_2Br \qquad (2)$$

propagation

$$CH_3\dot{C}H-CH_2Br + HBr \longrightarrow CH_3CH_2CH_2Br + Br\cdot \quad (3)$$

$$CH_3\dot{C}H-CH_2Br + Br\cdot \longrightarrow CH_3CHBrCH_2Br \qquad (4) \text{ termination}$$

The formation of Br· in reaction (3) (chain transfer step) permits the chain propagation to take place with another molecule of olefin. The repetiton of these two steps (2 and 3) is known as a chain reaction.

An additional reaction of technical importance is the process of polymerization:

$$X\cdot + CH_2=CH_2 \longrightarrow X-CH_2-CH_2\cdot$$

$$X-CH_2-CH_2\cdot + CH_2=CH_2 \longrightarrow X-CH_2CH_2CH_2CH_2\cdot$$

$$\longrightarrow X-(CH_2CH_2)_nCH_2CH_2\cdot \xrightarrow{XY} X-(CH_2CH_2)_nCH_2CH_2-Y$$

The growing polymer chain is terminated if the polymer radical abstracts an atom from molecule XY.

Not all radical additions to unsaturated bonds proceed via a chain reaction. An important example of this type is the addition of hydroxyl radicals to aromatic compounds [43]:

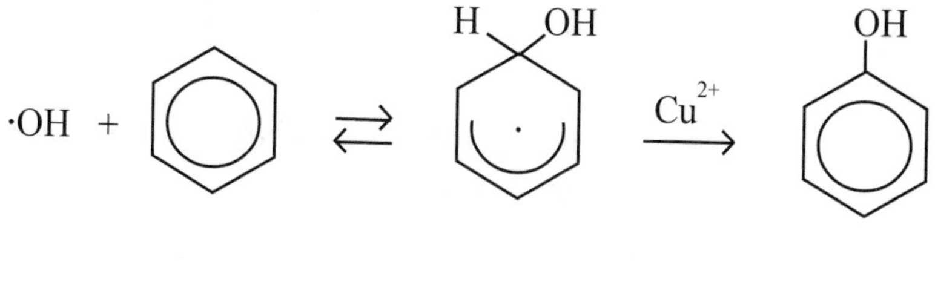

This addition has been shown to be reversible [44]. The addition is very straightforward, but the subsequent reactions are very complex and will be discussed in detail in Chapter 3.

6. Formation of radical cations

The radicals formed by addition of hydroxyl radicals to aromatic rings can (in addition to many other reactions) react with protons (H^+) to produce an aromatic radical cation via elimination of water [45-47]:

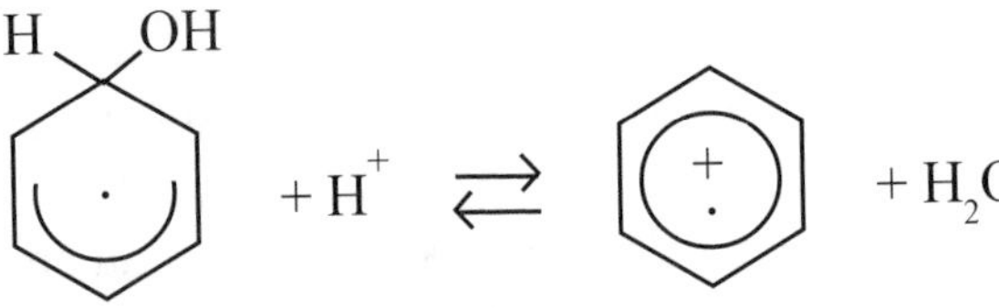

In some cases this elimination is reversible. The reversibility means that hydroxylation of aromatic compounds does not necessarily require hydroxyl radicals, but can also be accomplished via a radical cation. This type of hydroxylation is important in DNA damage by electronically excited states (electron transfer, followed by hydration of DNA·$^+$ [30]). The complex relationship between aromatic radical cations and the isomer distribution obtained in aromatic hydroxylation has been examined in detail [48-50] and has been reviewed [51].

If the radial cation can eliminate a proton the dehydration will lead to a neutral radical [52-55]:

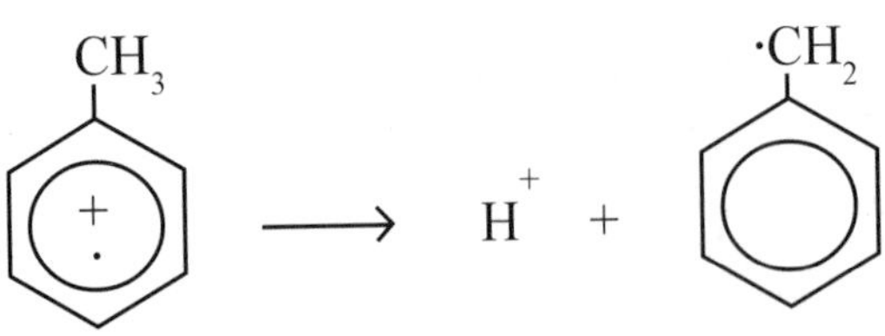

This type of reaction plays an important part in the hydroxylation of thymine, a reaction which I shall discuss in Chapter 6.

7. Fragmentation

The unimolecular decomposition of a radical to yield a molecule and another smaller radical occurs frequently via β-scission. A well known example is the fragmentation of tert.butoxy radical:

$$(CH_3)_3CO\cdot \longrightarrow (CH_3)_2C{=}O + \cdot CH_3$$

The fragmentation can be considered as the reverse of addition to unsaturated bonds:

$$R\cdot + CH_2{=}CH_2 \rightleftarrows R-CH_2-CH_2\cdot$$

The β-scission usually has to compete with H-abstraction:

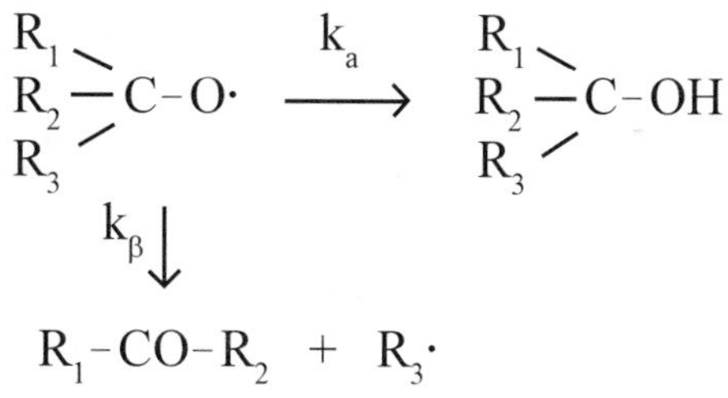

$$\begin{array}{c} R_1 \\ R_2 \\ R_3 \end{array}\!\!\!C-O\cdot \xrightarrow{\;k_a\;} \begin{array}{c} R_1 \\ R_2 \\ R_3 \end{array}\!\!\!C-OH$$

$$k_\beta \downarrow$$

$$R_1-CO-R_2 + R_3\cdot$$

The relative importance of β-scission increases in the order: methyl< ethyl<iso-propyl<t-butyl<benzyl. This sequence corresponds to the increased stability of the radical eliminated.

8. Oxidation reactions [31,32]

Oxidation of a radical, is formally presented by the following equations:

$$R\cdot \longrightarrow R^+ + e^-$$

followed by: $R^+ \longrightarrow R(-H) + H^+$ $\qquad$ oxidative elimination

or: $R^+ + HS \longrightarrow R-S + H^+$ $\qquad$ oxidative substitution

The ionization potential of alkyl radicals decreases in the order: prim.>sec.>tert. The products of oxidation of alkyl radicals are usually alkenes and substitution products; therefore, the standard reactions for oxidation are described more properly as oxidative elimination and oxidative substitution.

Some examples of radical oxidations by Cu^{2+} have already been mentioned above on H-abstraction. The oxidation of radicals leads to stable molecular products, which easily can be identified by conventional methods such as gas chromatography. This method has indeed been used extensively for the identification of radicals prior to the common use of EPR and spin trapping techniques. An example of this type of technique is the reaction of hydroxyl radicals with tert. butyl alcohol in presence of Cu^{2+} [40]:

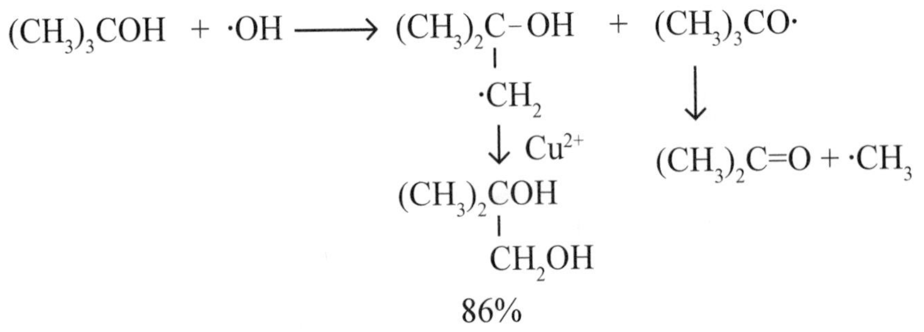

$$(CH_3)_3COH \; + \; \cdot OH \longrightarrow (CH_3)_2\underset{\underset{\cdot CH_2}{|}}{C}\text{-}OH \;\; + \;\; (CH_3)_3CO\cdot$$

$$\downarrow Cu^{2+} \qquad\qquad \downarrow$$

$$(CH_3)_2\underset{\underset{CH_2OH}{|}}{C}OH \qquad (CH_3)_2C{=}O + \cdot CH_3$$

$$86\%$$

9. Reduction reactions [31]

In addition to oxidation a radical can become stabilized by reduction, which formally is represented as follows:

$$R\cdot \; + e^- \longrightarrow R:^-$$

The resulting anion can react with the solvent (H_2O) via protonation:

$$R:^- \; + \; H^+ \longrightarrow R\text{-}H$$

If however the radical R· has a good negative-leaving group (like OH, Hal, $OCOCH_3$) in the β-position, then R· can undergo reductive elimination [56-58]:

$$\overset{\displaystyle CH_3}{\underset{\displaystyle CH_3}{\overset{\displaystyle |}{\underset{\displaystyle |}{\cdot CH_2C\text{-}OH}}}} \;+\; Cu^+ \;\longrightarrow\; CH_2{=}C(CH_3)_2 \;+\; OH^- \;+\; Cu^{2+}$$

In this reductive elimination the transfer of an electron and the elimination proceed in a concerted fashion. Fe^{2+}-EDTA [58] and Cr^{2+} [59] also have been found to be effective in reductive elimination.

Chapter 3

CHEMISTRY OF OXYGEN-DERIVED RADICALS

... the paradox underlying all scientific advances: namely, that scientists love to do experiments that show their colleagues wrong.
Robert Pollack in "Signs of Life" [60]

A. HISTORICAL PERSPECTIVE

Before I discuss the formation and reactions of reactive oxygen metabolites (ROMs), I would like to point out some important milestones in the history of oxygen radical and reactive oxygen metabolite research. These milestones may serve as a guiding light, all the rest is filling in the details.

H. J. H. Fenton discovers the oxidizing properties of Fe^{2+}-H_2O_2 [61].

Haber and Weiss discover the Haber-Weiss reaction and postulate the hydroxyl radical as the reactive species in the Fenton reaction [62].

Radiolysis of water gives the hydroxyl radical [63].

Radiolysis of water gives the solvated electron [64].

Development of pulse radiolysis during the early sixties [65]. The first application of pulse radiolysis to the study of hydroxyl radical reactions was reported in 1962 by Dorfman, Taub and Bühler [43].

Development of EPR spectroscopy and spin trapping [66,67,68]. The first EPR study of hydroxyl radical and reactions with organic compounds was reported by Dixon and Norman in 1962 [69].

Singlet oxygen formation via purely chemical reactions (in absence of light)[70-74].

Chemical formation of electronically excited carbonyl compounds in the absence of light [75-80].

Peroxidase-catalyzed formation of electronically excited states without light [81,82].

Formation of superoxide radical anion by xanthine oxidase [83]

Superoxide dismutase (SOD) dismutates $O_2^{\cdot-}$ to H_2O_2 and O_2 [84].

Formation of superoxide radical and hydrogen peroxide by mitochondria [85] and microsomes [86].

Phagocytosis as a source of reactive oxygen metabolites [17,18].

Discovery of endothelium-derived relaxing factor (EDRF) [87].

Identification of EDRF with nitric oxide ($\cdot$NO) [88-90].

Formation of nitric oxide by platelets, neurons and phagocytes [91].

Formation of peroxynitrite via combination of nitric oxide and superoxide anion [92-95].

Singlet oxygen formation in biological systems *in vivo* [96-98].

The development of GC-MS, HPLC-MS, HPLC-MS-SIM-EC techniques.

It took 40 years before the oxidizing agent described by Fenton was recognized as the hydroxyl radical. Subsequently the radiolysis of water provided an additional clean source for hydroxyl radicals [24,25,63]. The development of pulse radiolysis [65] during the sixties represented another quantum leap forward. Pulse radiolysis provided us with rate constants for thousands of reactions involving hydroxyl radicals [99] and the solvated electron (e_{aq}^-)[100]. These rate constants provide us with a tool for hydroxyl radical identification.

Electron paramagnetic resonance (EPR) and pulse radiolysis were developed around the same time. The first experiments using these tools were both reported in 1962 [43,69]. The spin-trapping technique amounts to a simple trick to transform a highly reactive and short-lived radical like hydroxyl to a more stable radical (the spin adduct). The spin-trapping technique provided a way to detect low concentrations of radicals in biological systems. The pros and cons of spin trapping have been reviewed [67,68] and will be discussed later on.

Formation of singlet oxygen via photosensitization has been known for some time and the reactions of singlet oxygen with many compounds have been studied extensively [101]. These reactions however are unlikely to be important in cells, where no light source is available. Some discoveries during the sixties were responsible for moving the chemistry of reactive oxygen metabolites into the realm of biology. A most significant discovery was the spectroscopic identification of singlet oxygen in a purely chemical reaction (HOCl + H_2O_2) by Khan and Kasha in 1963 [70]. This discovery opened a whole new world for interdisciplinary scientific investigation.

The formation of singlet oxygen via HOCl-H_2O_2 is of no biological significance if no HOCl is formed *in vivo*. Therefore the discovery of a mechanism for HOCl formation via an enzyme-catalyzed reaction was a step forward in establishing singlet oxygen in biological systems [102]. This enzyme, the myeloperoxidase, is present in neutrophils, monocytes and newly activated macrophages, and catalyzes the following reaction:

$$H_2O_2 \; + \; Cl^- \; \xrightarrow{\text{MPO}} \; HOCl \; + \; OH^-$$

Activated neutrophils produce $O_2\cdot^-$ via the NADPH oxidase [17,18] and HOCl via a MPO-catalyzed reaction. They are therefore good candidates for *in vivo* singlet oxygen formation (Chapter 5, phagocytosis).

Subsequently other reactions were found to give singlet oxygen [72-74]. It took over 20 years before these reactions were finally established in biological systems [96-98]. The reason for this long delay is simple: a lack of complete understanding of all the complexities involved in singlet oxygen formation and reactions (Chapter 4).

An analogous development was the formation of electronically excited carbonyl compounds via thermal decomposition of dioxetanes and dioxetanones [75-80]. The decomposition of these highly strained 4-membered ring compounds to yield electronically excited states opened the possibility of carrying out photochemistry without light [76,82]. Soon after the first synthesis of a dioxetane by Kopecky and Mumford in 1969 [75], a possible role for these compounds in biology was suggested [81]. We now know many peroxidase-catalyzed reactions leading to excited states [82]. These oxidations will be discussed in more detail in Chapter 5.

From a biological point of view the most important discovery was made by McCord and Fridovich in 1968 [83] and 1969 [84]. These authors observed 2 things: 1. The enzyme xanthine oxidase produces superoxide radical and 2. Erythrocytes contains an enzyme (now known as the superoxide dismutase) which catalyzes the dismutation of superoxide to hydrogen peroxide and oxygen:

$$2\,O_2\cdot^- \; \xrightarrow[\text{SOD}]{2\,H^+} \; H_2O_2 \; + \; O_2$$

The importance of the $O_2 \cdot^-$/SOD topic is evident from the over 50 reviews written on this topic during the last 25 years (see citations in ref. 9). The different types of SODs have been reviewed [103,104]. The formation of $O_2 \cdot^-$ and the presence of SODs in all aerobic cells gave rise to a hotly debated hypothesis: Superoxide is toxic to cells and SODs are produced by all aerobic cells as a defense against this toxic threat [105].

The main argument against this hypothesis was the fact that $O_2 \cdot^-$ is chemically not very "super" [106]. It was difficult to understand how such a relatively unreactive radical can cause biological damage. The search was on to show that $O_2 \cdot^-$ can indeed damage important biological targets [107]. The $O_2 \cdot^-$ can accomplish this directly (lipid peroxidation) or indirectly via oxidation of metal centers of enzymes. This oxidation leads to H_2O_2, the precursor of the hydroxyl radical. In addition the $O_2 \cdot^-$ may be converted to more reactive species, such as $^1O_2$ or ^-OONO. I shall discuss these reactions later in this chapter (pp. 46-49) and in Chapters 4 and 7. Another approach to demonstrate the toxicity of $O_2 \cdot^-$ is the study of SOD mutants, which do not catalyze the dismutation. Many studies with SOD mutants have been reported and have demonstrated the toxicity of $O_2 \cdot^-$ [108,109]. Although the debates have been rather heated and downright nasty at times, in the end we can say that the controversy has lead to detailed scrutiny of many biochemical phenomena and in the end has increased our appreciation of the complex beauty of biological systems.

After the discovery by McCord and Fridovich [83] the search was on for other enzymatic sources of superoxide radical. The formation of $O_2 \cdot^-$ was observed in mitochondria [85,110-112], in microsomes [86], as well as in activated phagocytes [113-115]. The function of SODs in all aerobic cells provides us with a tool for the identification of superoxide radical. Whenever $O_2 \cdot^-$ is involved in any biological or pathological process, the presence of SOD will affect this process. Extensive use has been made of this technique. The studies of mitochondrial respiration and activated phagocytes have used SOD as a probe for $O_2 \cdot^-$.

Hydrogen peroxide was shown to be a metabolite of mitochondria of different tissues [110-112]. The question was: how is the H_2O_2 formed?

There are two possibilities: one- or two-electron reduction of oxygen.

$$O_2 + 2e^- \xrightarrow{2H^+} H_2O_2$$

$$O_2 + e^- \longrightarrow O_2^{\cdot -} \xrightarrow{\text{dismut.}} H_2O_2$$

Since radicals were not yet as popular as today, the two-electron pathway was suggested by some investigators [110]. However a few years later it was shown (using the SOD probe) that $O_2^{\cdot -}$ is the intermediate in the formation of H_2O_2 [111,112]. Increased oxygen consumption by activated phagocytes was noted many years before [113] the superoxide radical was implicated as an intermediate metabolite [17,18,114,115].

Another quantum jump in reactive oxygen metabolite research came in 1980 with the discovery of endothelium-derived relaxing factor (EDRF) [87]. EDRF was finally recognized as nitric oxide (·NO) in 1988 [88-90]. Nitric oxide is a small neutral radical of relatively low reactivity. Nitric oxide causes relaxation of vascular smooth muscle and was later found to be produced by platelets, phagocytes and neurons [91]. Compared to the hydroxyl radical, nitric oxide as well as superoxide radical are of low reactivity. However it was observed that the combination of ·NO and $O_2^{\cdot -}$ produce the highly oxidative peroxynitrite anion (ONOO⁻) [92-95], which can damage many important biomolecules (see Chapter 7). It has been determined [116] that this radical-radical combination proceeds at a close to diffusion controlled rate (6.7 x $10^9 M^{-1}s^{-1}$). I like to point out the importance of rate data. Since in a biological system many reactions take place simultaneously (competition!) a knowledge of the concentrations and rate constants of the reactants is of utmost importance. Reactions may take place in a test tube, but may be of no significance in a complex biological system.

In order to carry out all the above mentioned investigations we need sensitive and reliable analytical techniques. In addition to pulse radiolysis, EPR spectroscopy and spin trapping already mentioned, other techniques like GC-MS, HPLC-MS, HPLC-MS-SIM, HPLC-EC and the use of microsensors [117] (for NO detection) have contributed to the progress in this field. An interesting non-invasive technique was developed by Boveris et al. [118,119] to detect electronically excited states by low level chemiluminescence measurements in complete organs. The HPLC-EC (electrochemical detection) developed by Floyd [120,121] has been especially useful for the detection of 8-hydroxyguanine as a biological marker for *in vivo* oxidative damage. A relationship between 8-hydroxyguanine formation and carcinogenesis has been established [122-124]. The experimental tools in free radical research have been recently reviewed by Arouma [125].

B. DEFINITIONS

The terms oxygen free radicals, reactive oxygen species (ROS), reactive oxygen metabolites (ROM) and excited oxygen species are often used interchangeably. This is, of course, sloppy scientific terminology. The main metabolites of oxygen can be classified in two groups: 1. Radicals with the unpaired electron on oxygen (like $\cdot OH$, $O_2 \cdot^- \rightleftharpoons HO_2 \cdot$, $RO \cdot$, and $ROO \cdot$) and 2. Molecular oxygen containing products, like H_2O_2, ROOR, ROOH, singlet oxygen, epoxides, endoperoxides and 1, 2-dioxetanes as fleeting intermediates in some enzymic oxidations of aldehydes. With the exception of singlet oxygen and dioxetanes these molecular products are quite stable, lasting for indefinite periods of time (in absence of heat, light, or some metal ions). However, since the molecular products are converted by some metal ions to the highly reactive hydroxyl and alkoxy radicals these molecular products may be called ROMs or ROS. The oxygen radicals are of course also highly reactive and the term ROM or ROS can be considered an all encompassing term, including radicals as well as molecular products. The term excited oxygen species, which has been used to designate superoxide radical anion, hydroxyl radical and hydrogen peroxide, should be reserved for electronically excited species, like singlet oxygen or excited triplet carbonyl compounds.

The latest addition to the realm of biologically important radicals is nitric oxide. Nitric oxide is formed via an oxygen-dependent metabolism from L-arginine [126]. The oxygen atom in ·NO is derived from molecular oxygen and nitric oxide can therefore be designated as a ROM or oxygen-derived radical. Due to the great importance of nitric oxide in both physiological and pathological processes, a separate chapter is devoted to nitric oxide (Chapter 7).

The important oxygen-derived radicals are the following:

1.) $O_2 \cdot^-$, the superoxide radical anion and its conjugate acid $HO_2 \cdot$.

2.) ·OH, the hydroxyl radical.

3.) ROO·, alkylperoxyl radicals formed in autoxidation.

4.) RO·, alkoxy radicals formed from the decomposition of alkylhydroperoxides either thermally or via one-electron transfer from metal ions.

5.) ·NO, nitric oxide.

C. ONE-ELECTRON REDUCTION OF OXYGEN

The addition of an electron to a neutral molecule produces a species which has one unpaired electron and which by definition is a radical:

$$O_2 \; + \; e_{aq}^- \longrightarrow O_2 \cdot^- \xrightarrow{\; H^+ \;} HO_2 \cdot^-$$

The addition of one electron to oxygen yields the superoxide radical. The solvated electron (e_{aq}^-) can be easily generated via radiolysis of water [25], which I shall discuss later in this chapter.

Addition of another electron yields a neutral molecule:

$$O_2 \cdot^- \; + \; e_{aq}^- \xrightarrow{\; 2 \, H^+ \;} H_2O_2 \cdot$$

Further addition of an electron to H_2O_2 gives again a radical:

$$H_2O_2 + e_{aq}^- \longrightarrow OH^- + \cdot OH$$

The hydroxyl radical can be reduced by one-electron to OH^- (or H_2O):

$$\cdot OH + e_{aq}^- \longrightarrow OH^- \xrightarrow{H^+} H_2O$$

We thus have the following sequence of one electron transfer steps from oxygen to water:

$$O_2 \xrightarrow{e^-} O_2 \cdot^- \xrightarrow[2\,H^+]{e^-} H_2O_2 \xrightarrow[H^+ \; (+H_2O) \; H^+]{e^-} \cdot OH \xrightarrow{e^-} H_2O$$

The reduction of oxygen to two moles of H_2O requires the transfer of four electrons. This transfer can take place in one-electron steps (involving intermediate formation of radicals) or in a single four-electron transfer.

A well known four-electron transfer reaction takes place in the mitochondrial membrane from two pairs of cytochrome a,a_3 to oxygen [127]:

$$2 \text{ cyt } (a^{II}a_3^{II}) + O_2 + 4\,H^+ \longrightarrow 2 \text{ cyt } (a^{III}a_3^{III}) + 2\,H_2O$$

The addition of one electron to a neutral molecule yields a radical anion (like $O_2 \cdot^-$). However, some neutral molecules containing some weak bonds can directly break apart without intermediate formation of a radical anion. This process is known as dissociative electron capture [27]:

$$R\text{-Hal} + e_{aq}^- \longrightarrow R\cdot + Hal^- \qquad (1)$$

$$RO\text{-OH} + e_{aq}^- \longrightarrow RO\cdot + OH^- \qquad (2)$$

$$HO\text{-OH} + e_{aq}^- \longrightarrow HO\cdot + OH^- \qquad (3)$$

$$N_2O + e_{aq}^- \longrightarrow N_2 + \cdot O^- \; (+ H^+ \longrightarrow \cdot OH) \quad (4)$$

42

In Reactions (2) and (3) the electron can also be supplied by a one-electron donating metal ion or radical:

$$\text{HO-OH} + Fe^{2+} \longrightarrow Fe^{3+} + OH^- + \cdot OH \qquad (5)$$

$$\text{HO-OH} + O_2^{\cdot -} \longrightarrow O_2 + OH^- + \cdot OH \qquad (6)$$

Reaction (5) is the well known **Fenton reaction**, and Reaction (6) is the **Haber-Weiss reaction**.

In order to understand the toxic effects of these oxygen radicals we first must have a thorough understanding of their formation and their reactions in simple, well-defined chemical systems. Enzyme-catalyzed formation of ROMs will be discussed in Chapter 5.

D. SUPEROXIDE RADICAL ANION [128]

1. Formation of superoxide radical anion *in vitro*
a. Radiolysis of aqueous solutions [24,25]

$$O_2 + e_{aq}^- \longrightarrow O_2^{\cdot -}$$

This reaction is used by radiation oncologists to increase the killing of tumor cells by radiation (increase in oxygen-derived radicals).

b. Autoxidation of metal ions

A great number of different metal ions has been shown to react with oxygen to yield $O_2^{\cdot -}$ [129]. Some of these metal ions are: Ag^+, Cd^+, Co^+, Zn^+ and more important for biological systems Fe^{2+}, Fe^{2+}–EDTA, and Cu^+:

$$M^{+n} + O_2 \longrightarrow M^{+n+1} + O_2^{\cdot -}$$

The formation of $O_2^{\cdot -}$ in these oxidations was studied by pulse radiolysis and the formation of benzoquinone radical anion ($Q + O_2^{\cdot -} \longrightarrow Q^{\cdot -} + O_2$). The superoxide radical can dismutate to give H_2O_2, which in turn can give the hydroxyl radical (discussed later in this chapter). These metal ion-oxygen systems have been known for a long time to hydroxylate aromatic compounds (Udenfriend's reagent) [130,131]. The evidence for intermediate formation of hydroxyl radicals has been reviewed [51].

c. Oxidation of radicals

As already discussed in Chapter 2 radicals have the tendency to get rid of their unpaired electron. This can be accomplished either by accepting another electron or by donating it to another molecule or species. Since oxygen is an oxidizing molecule (it accepts one electron), it is not surprising that some radicals react with oxygen to form superoxide. This can be accomplished via two different pathways: 1. Direct one-electron transfer or 2. Combination to give a peroxyl radical. These peroxyl radicals can undergo a variety of secondary reactions, which are of great biological importance (lipid peroxidations) and will be discussed later in this chapter.

The oxidation of semiquinone-type radicals is the most important source of superoxide in biological systems. It is involved in the electron transport chain in mitochondria [132,133]:

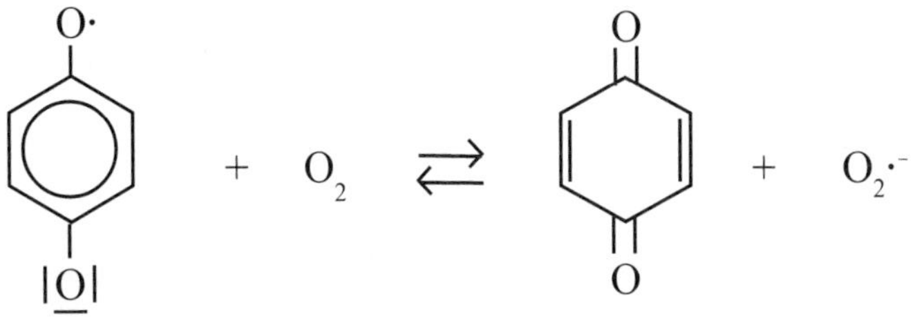

This mode of superoxide formation is not only important in normal metabolism, but also in the metabolism of a number of xenobiotics which contain quinone ring systems, like the drugs adriamycin and mitomycin C. Depending on the structure of the quinone the above reaction is reversible [133]. Many reactions of this type have been studied with the pulse radiolysis technique.

Another radical which has been used as a one-electron donor is the formate radical anion ($\cdot COO^-$). This radical can be produced via radiolysis of aqueous formate solutions [25,134]:

$$HCOO^- + \cdot OH \longrightarrow H_2O + \cdot COO^- \quad k = 2.5 \times 10^9 \ M^{-1}s^{-1}$$

$$\cdot COO^- + O_2 \longrightarrow CO_2 + O_2 \cdot^- \quad k = 2.4 \times 10^9 \ M^{-1}s^{-1}$$

All hydroxyl radicals are converted to $O_2 \cdot^-$.

Another radical of biological importance which can donate one electron to oxygen is the toxic herbicide paraquat radical cation $PQ \cdot^+$ (1,1'-dimethyl-4,4'-bipyridylium)[135]:

$$PQ \cdot^+ + O_2 \longrightarrow PQ^{2+} + O_2 \cdot^-$$

d. Oxidation of some neutral organic compounds

Compounds which have easily abstractable H-atoms can react directly with oxygen. Some of these compounds are polyhydroxy-aromatics, like pyrrogallol [136] (1,2,3-trihydroxybenzene) or 6-hydroxydopamine [137]. In the oxidation of these compounds the formation of a semiquinone radical ($SQ \cdot$) is the first step:

$$6\text{-HODA} + O_2 \longrightarrow SQ \cdot + O_2 \cdot^- + H^+$$

$$SQ \cdot + O_2 \longrightarrow Q + O_2 \cdot^- + H^+$$

6-hydroxydopamine reacted spontaneously with oxygen, but this autoxidation was inhibited by SOD. This result clearly indicates the intermediate formation of the superoxide radical anion in this oxidation. The superoxide radical dismutates to give H_2O_2, which in turn can undergo the Fe(III)-catalyzed Haber-Weiss reaction. The 6-hydroxydopamine causes degeneration of nerve terminals.

Thiols have been shown to undergo autoxidation to give superoxide radical anion [138]:

$$R\text{-}S^- \ + \ O_2 \ \longrightarrow \ RS \cdot \ + \ O_2 \cdot^-$$

e. Formation of superoxide via singlet oxygen

The formation of superoxide radical anion has been observed in the oxidation of electron-rich compounds by singlet oxygen. One such example is the oxidation of electron-rich enamines [139] or tetramethylphenylenediamine [140]:

$$TMPD \ + \ {}^1O_2 \ \longrightarrow \ TMPD \cdot^+ \ + \ O_2 \cdot^-$$

f. Oxidation of hydrogen peroxide

Many transition metal ions can oxidize hydrogen peroxide. A typical EPR spectrum of $HO_2 \cdot$ was observed in the following reaction [141,142]:

$$Ce^{4+} \ + \ H_2O_2 \ \longrightarrow \ Ce^{3+} \ + \ H^+ \ + \ HO_2 \cdot$$

When Ce^{3+} was added a narrow signal was superimposed on the broad $HO_2 \cdot$ band. It was concluded that this signal belonged to the $Ce^{3+}\text{-} HO_2 \cdot$ complex. Many transition metal ions have been found to form this type of complex, i.e., the $HO_2 \cdot$ radical is not 'free'. Many other transition metal ions react in a similiar fashion [141]. Among these Fe^{3+} and Cu^{2+} are of biological interest. This type of oxidation has been shown to be involved in the deactivation of SOD by H_2O_2 and will be discussed later in this chapter.

2. Reactions of superoxide/hydroperoxy radicals
a. Nucleophilic and redox properties

The superoxide radical anion has the following main chemical characteristics:

1. It is a base (i.e. it reacts with protons H^+):

$$H^+ + O_2^{\cdot-} \rightleftarrows HO_2\cdot$$

The conjugate acid $HO_2\cdot$ has a $pK_a = 4.8$. This indicates that at neutral pH the superoxide exists mainly in the unprotonated form. At pH 7.4 only about 1% exists as $HO_2\cdot$.

2. It is a nucleophile (it can undergo nucleophilic substitution):

$$R-X + O_2^{\cdot-} \longrightarrow RO_2\cdot + X^-$$

3. It is a radical. It can undergo all the typical reactions of a radical:

a.) disproportionation: $O_2^{\cdot-} + O_2^{\cdot-} \xrightarrow{2 H^+} H_2O_2 + O_2$

b.) electron transfer: $O_2^{\cdot-} + A \longrightarrow O_2 + A^{\cdot-}$

Superoxide can react as an oxidizing or as a reducing agent. One example in which $O_2^{\cdot-}$ acts as both an oxidizing and a reducing agent is the disproportionation to H_2O_2 and O_2. This reaction demonstrates the dual character of radicals. A radical can get rid of its undesirable unpaired electron either by donating it or by accepting another electron:

$$O_2 \xleftarrow{-e^-} O_2^{\cdot-} \xrightarrow[2 H^+]{+e^-} H_2O_2$$

Some examples are:

$$HO_2\cdot + AH_2 \longrightarrow H_2O_2 + \cdot AH \qquad [143] \text{ oxidation}$$

$$O_2^{\cdot-} + Cu^+ \xrightarrow{2 H^+} H_2O_2 + Cu^{2+} \qquad [144] \text{ oxidation}$$

$$O_2 \cdot^- + Fe^{3+} \longrightarrow Fe^{2+} + O_2 \qquad \text{[145] reduction}$$

The oxidation of Cu^+ was shown by pulse radiolysis to proceed at a fast rate ($k= 10^{10}\,M^{-1}\,s^{-1}$) [144]. The reduction of metal ions is an essential step in the Fe^{3+}-catalyzed Haber-Weiss reaction [145,146]. Another example is the oxidation of oxyhemoglobin and the reduction of methemoglobin [147,148]:

$$O_2 \cdot^- + HbO_2 \xrightarrow{2\,H^+} H_2O_2 + O_2 + metHb$$

$$O_2 \cdot^- + metHb \longrightarrow HbO_2$$

The rate constants (in the order of $10^3 - 10^4\,M^{-1}\,s^{-1}$) have been determined by pulse radiolysis [147] and competition kinetics [148].

The reducing property of $O_2 \cdot^-$ has provided a useful tool for its identification. McCord and Fridovich [83] deduced the formation of $O_2 \cdot^-$ from milk xanthine oxidase by its reaction with cytochrome c:

$$\text{cytochrome c}(Fe^{3+}) + O_2 \cdot^- \longrightarrow \text{cytochrome c }(Fe^{2+}) + O_2$$

The reaction can be followed by a change in absorption and is inhibited by superoxide dismutase.

The reduction of Fe^{3+} by $O_2 \cdot^-$ *in vitro* is involved in the iron-catalyzed Haber-Weiss reaction. However *in vivo* there are many reducing molecules available which can reduce Fe(III) to Fe(II) (GSH, NADPH, ascorbate). It is therefore doubtful whether $O_2 \cdot^-$ exerts its oxidative damage *in vivo* via formation of Fe(II). Another explanation must be sought. It was shown mainly through the work of Fridovich and coworkers that $O_2 \cdot^-$ reacts with some Fe-S containing enzymes via oxidation to H_2O_2 and concomitant release of free iron [149-156]. These enzymes contain [4Fe-4S] clusters and are inactivated by superoxide. This has been demonstrated for dihydroxy acid dehydratase [150], 6-phosphogluconate dehydratase [151], aconitase

[152], and fumarase A and B [153-155]. *Escherichia coli* contains three kinds of fumarases, A, B, and C. Among these, fumarase A and B contain [4Fe-4S] clusters and are oxygen sensitive, wheras fumarase C contains no Fe-S center and is stable [153]. The native enzyme clusters contain two Fe(II) and two Fe(III) atoms, which are oxidized as follows:

$$[2Fe(II)2Fe(III)-4S] + O_2^{\cdot -} + 2H^+ \longrightarrow [Fe(II)3Fe(III)-4S] + H_2O_2$$

The rate constant for these reactions has been determined to be between 10^8 - 10^9 M^{-1} s^{-1}[150-153]. The oxidized cluster then looses Fe(II), which can become complexed to DNA and then form the hydroxyl radical in close proximity to a vital target (site-specific formation of hydroxyl radical). The [4Fe-4S] clusters of quinolinate synthase [156] and fumarase A [154,155] are even oxidized by oxygen with release of Fe(II).

Whenever $O_2^{\cdot -} \rightleftarrows HO_2$ reacts as an oxidizing agent we obtain H_2O_2 which is also a toxic compound. H_2O_2 can react further via one-electron transfer (supplied by some metal ions) to produce the highly destructive hydroxyl radical. During dismutation the O_2 may be formed in the reactive singlet oxygen state. This possibility has been the subject of some controversy. The extracellular production of singlet oxygen by stimulated macrophages as well as the intracellular production by neutrophils via $O_2^{\cdot -}$ dismutation was demonstrated by Khan and coworkers [97,98].

As pointed out by Liochev and Fridovich [157] *in vitro* $O_2^{\cdot -}$ reacts as a reducing agent (the iron-catalyzed Haber-Weiss reaction), but *in vivo* $O_2^{\cdot -}$ reacts as an oxidizing agent. An exception to this rule is the iron release from ferritin by $O_2^{\cdot -}$. It has been suggested by Buettner et al. [158] that this release *in vivo* most likely proceeds via reduction of Fe(III) to Fe(II).

The disproportionation of $O_2^{\cdot -}$ to yield H_2O_2 and oxygen has been investigated in great detail. Due to the basicity of superoxide we have to consider three different dismutation reactions [159,160]:

$$O_2\cdot^- + O_2\cdot^- \xrightarrow{2\ H_2O} H_2O_2 + O_2 + 2\ OH^- \quad k \leq 0.3\ M^{-1}\ s^{-1}$$

$$HO_2\cdot + O_2\cdot^- \xrightarrow{H^+} H_2O_2 + O_2 \quad k = 8.8 \times 10^7\ M^{-1}\ s^{-1}$$

$$HO_2\cdot + HO_2\cdot \longrightarrow H_2O_2 + O_2 \quad k = 7.6 \times 10^5\ M^{-1}\ s^{-1}$$

The spontaneous dismutation is most rapid at pH 4.8. The unprotonated form of superoxide dismutates very slowly, due to the electrostatic repulsion of the two negative charges [159]. If we compare the reaction rate of the uncatalyzed dismutation and the SOD-catalyzed dismutation we have to consider that the uncatalyzed reaction is second order in $O_2\cdot^-$, whereas the SOD-catalyzed reaction is first order in $O_2\cdot^-$. This means that the half-life of the SOD-catalyzed reaction is independent of the $O_2\cdot^-$ concentration. Taking all these aspects into consideration it has been estimated by Fridovich [161] that the superoxide dismutase removes the superoxide faster by a factor of 10^{10}.

The unprotonated form of superoxide (present at neutral pH) is of low reactivity. This is especially true in an aqueous protic environment, where the superoxide is highly solvated. The low reactivity of superoxide [106] gave rise to a serious controversy concerning the importance or unimportance of superoxide in biological oxidative damage. If superoxide is so unreactive why do we need superoxide dismutases to remove this species? However, as pointed out above, $O_2\cdot^-$ can react with Fe-S containing enzymes via formation of H_2O_2 and release of free Fe(II). Another explanation for the need of SODs has recently been proposed by Pryor and Squadrito [162]. SODs prevent the combination of $O_2\cdot^-$ with $\cdot NO$ to produce the highly reactive peroxynitrite (^-OONO). I shall discuss the formation and reactions of peroxynitrous acid in Chapter 7.

There are, however, a number of compounds which react with superoxide at reasonable rates. NADH in aqueous solution reacts very slowly with O_2^- [163], but if it is attached to the enzyme lactate dehydrogenase the reaction proceeds at a reasonable rate [164-166]:

$$LDH\text{-}NADH + O_2^- \rightarrow LDH\text{-}NAD\cdot + HO_2^- \quad k = 1.0 \times 10^5 \ M^{-1} \ s^{-1}$$

$$LDH\text{-}NAD\cdot + O_2 \rightarrow LDH\text{-}NAD^+ + O_2^- \quad k = 10^9 - 10^{10} \ M^{-1} \ s^{-1}$$

The oxidation of NAD· by oxygen leads to a chain reaction, and high concentrations of H_2O_2.

Both catalytic and non-catalytic dismutation lead to the same final products, but proceed via different mechanisms. The non-catalytic dismutation starts with a reaction with a proton donor:

$$O_2^- + HX \rightleftharpoons HO_2\cdot + X^-$$

$$HO_2\cdot + O_2^- \longrightarrow O_2 + HO_2^-$$

$$HO_2\cdot + HO_2\cdot \longrightarrow H_2O_2 + O_2$$

The catalytic dismutation by the Cu-ZnSOD proceeds via two one-electron transfer steps [167,168]:

$$Cu^{2+}ZnSOD + O_2^- \longrightarrow Cu^+ZnSOD + O_2$$

$$Cu^+ZnSOD + O_2^- \overset{2\ H^+}{\rightleftharpoons} Cu^{2+}ZnSOD + H_2O_2$$

It was determined by pulse radiolysis that the rate constant for both reaction steps was $2 \times 10^9 \ M^{-1}s^{-1}$.

SOD transforms one mole of O_2^- to ½ mole of H_2O_2, whereas in absence of SOD in vivo one mole of O_2^- can produce much greater amounts of H_2O_2 by oxidation of the [4Fe-4S] centers of dehydratases or by oxidation of NADH-lactic dehydrogenase. Since H_2O_2 is the precursor of the highly destructive hydroxyl radical it follows that SOD confers protection.

b. Hydrogen abstraction by superoxide and perhydroxyl radicals

The pH has a considerable effect on the reactivity of the superoxide radical. The superoxide radical exists in equilibrium with the perhydroxyl radical:

$$O_2^{\cdot -} + H^+ \rightleftharpoons HO_2 \cdot \qquad pK_a = 4.8$$

At physiological pH of 7.4 the superoxide radical exists mainly in the anionic (non-protonated) form. Only about 1% exist in the protonated form. However at lower pH, as exists for example in phagocytes (neutrophils, macrophages) or in phospholipid bilayers of cell membranes, the concentration of $HO_2 \cdot$ can be quite high. The two species ($O_2^{\cdot -}$ and $HO_2 \cdot$) have different chemical reactivity. The superoxide is a radical and a nucleophile (it can abstract a proton). It acts preferentially as a reducing agent (donate an electron). The perhydroxyl radical on the other hand is a less effective reducing agent and acts mainly as an oxidizing agent. It can abstract a hydrogen atom from many compounds, which have easily abstractable hydrogens, like catechols, hydroquinones [169-171], α-tocopherol [172], and linoleic acid [173,174]:

$$HO_2 \cdot + RH \longrightarrow H_2O_2 + R \cdot$$

The resulting radical $R \cdot$ reacts with oxygen to give $ROO \cdot$ and $ROOH$. The linoleic hydroperoxide has been shown to react with $O_2^{\cdot -}$ in a Haber-Weiss-type reaction [174]:

$$ROOH + O_2^{\cdot -} \longrightarrow RO \cdot + OH^- + O_2$$

The direct abstraction of a hydrogen atom is not commonly possible with superoxide radical. The abstraction is a two-step process [143,175,176]:

$$O_2^{\cdot-} + AH_2 \longrightarrow HO_2^{\cdot} + AH^- \qquad \text{deprotonation}$$

$$HO_2^{\cdot} + AH_2 \longrightarrow H_2O_2 + AH^{\cdot} \qquad \text{abstraction}$$

There is some controversy concerning the possibility of direct H-abstraction by $O_2^{\cdot-}$ [177]. It has been demonstrated that superoxide radical anion in aprotic solvents can react with dihydroxyaromatics via a one-step reaction to yield a semiquinone radical [178,179]:

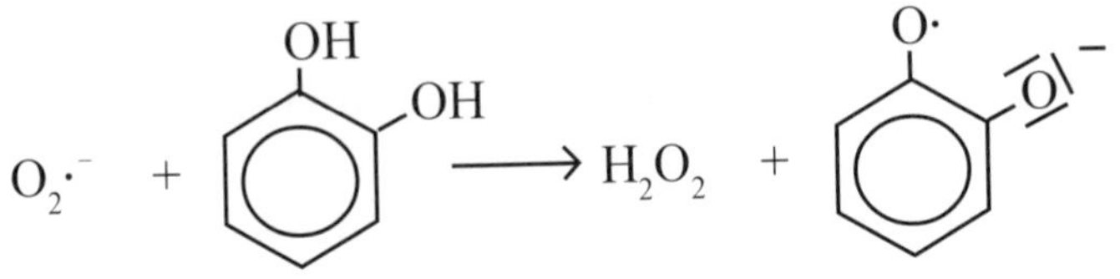

This conclusion was reached because in these reactions no oxygen was observed. If the reaction proceeds via a two-step process (deprotonation-hydrogen abstraction) the $HO_2^{\cdot}$ should be an intermediate and subsequently O_2 should be formed:

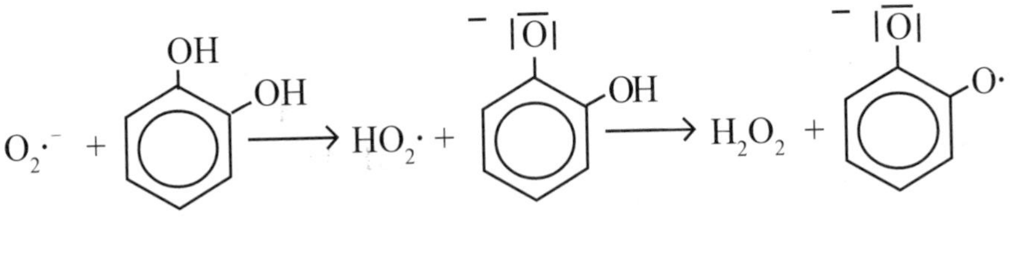

$$HO_2^{\cdot} + O_2^{\cdot-} \longrightarrow HO_2^{-} + O_2$$

However, in some cases (high concentrations of catechol) some oxygen was observed, thus indicating that a two-step process is competing with a one-step process. It appears doubtful whether a one-step process is universal in reactions of $O_2^{\cdot-}$ with all compounds containing two active hydrogen atoms. In the reaction with ascorbic acid a two-step mechanism has been clearly established [143,175,176]. The oxidation of ascorbic acid by O_2 and $O_2^{\cdot-}$ will be discussed in detail in Chapter 8.

c. Superoxide-dependent formation of hydroxyl radicals

Another way by which the superoxide anion can exert its deleterious effect indirectly is via reduction of Fe(III) or Cu(II) followed by a Fenton-type reaction with hydrogen peroxide. However, *in vivo* there are many organic molecules available, which can accomplish the same reduction of metal ions. This competition between $O_2^{\cdot-}$ and other reducing agents has been studied extensively [180-187]. These reducing compounds are thiols (glutathione, cysteamine), NADH, NADPH (reduced forms of nicotinamide adenine dinucleotide and nicotinamide adenine dinucleotide phosphate), ascorbate, catechols and hydroquinones:

$$RS^- + Fe^{3+}(EDTA) \longrightarrow Fe^{2+}(EDTA) + RS\cdot$$

$$NADH + Fe^{3+}(EDTA) \longrightarrow Fe^{2+}(EDTA) + NAD\cdot + H^+$$

These reductions are then followed by the Fenton reaction:

$$Fe^{2+}(EDTA) + H_2O_2 \longrightarrow Fe^{3+}(EDTA) + OH^- + \cdot OH$$

Since superoxide also reduces $Fe^{3+}(EDTA)$ [145,146] the above two examples may proceed via intermediate formation of superoxide radical:

$$RS^- + O_2 \longrightarrow RS\cdot + O_2^{\cdot-}$$

$$O_2^{\cdot-} + Fe^{3+} \rightleftharpoons Fe^{2+} + O_2$$

The formation of superoxide radical in the reaction of oxygen with thiols and the reduction of Cu^{2+} by thiols has been reported by Misra [138].

The superoxide dismutases have been a valuable tool in distinguishing the above two processes (direct reduction of Fe^{3+} or reduction via $O_2^{.-}$). The yield of hydroxyl radicals is decreasing in presence of SOD if $O_2^{.-}$ is an intermediate. Rowley and Halliwell [188] found that NADH-dependent $\cdot OH$ formation is inhibited by SOD, thus indicating the involvement of $O_2^{.-}$. Some reducing compounds like glutathione or cysteamine have been shown to react via both pathways.

Another example is the ascorbate-dependent formation of hydroxyl radicals:

$$AH^- \ + \ Fe^{3+}(EDTA) \longrightarrow Fe^{2+}(EDTA) \ + \ AH\cdot$$

$$\text{or} \quad AH^- \ + \ O_2 \longrightarrow AH\cdot \ + \ O_2^{.-}$$

$$O_2^{.-} \ + \ Fe^{3+}(EDTA) \rightleftharpoons Fe^{2+}(EDTA) \ + \ O_2$$

followed by the Fenton reaction. In this case it was found that the yield of hydroxyl radicals was decreased by SOD to only a minor amount. This observation indicates that both processes are operating with the direct reduction predominant.

E. PEROXYL RADICALS [189]

These radicals were studied long before the importance of radicals in biological systems was recognized. Peroxyl radicals are important intermediates in chain oxidations. Since oxygen in its ground state is a triplet (i.e., it is a diradical) it can combine with a variety of other radicals to form peroxyl radicals:

$$R\cdot \ + \ O_2 \rightleftharpoons R{-}OO\cdot$$

This reaction is reversible when the C-O bond is weak, as for example in the oxidation of the hydroxycyclohexadienyl radical. Peroxyl radicals undergo a variety of unimolecular reactions. One of these reactions, which I have already mentioned, is the elimination of $O_2^{\cdot-}/HO_2\cdot$. The driving force for this elimination is the formation of an aromatic ring system or an unsaturated bond[190-192]:

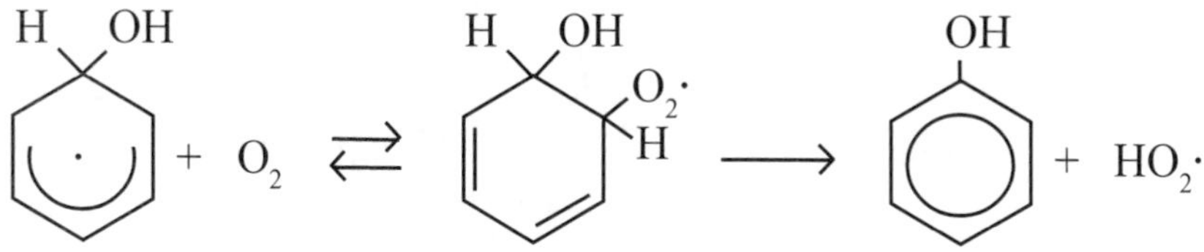

$$R_2CHOH + \cdot OH \longrightarrow H_2O + R_2\dot{C}OH$$

$$R_2\dot{C}OH + O_2 \longrightarrow R_2C(OH)OO\cdot$$

$$R_2C(OH)OO\cdot \longrightarrow O_2^{\cdot-} + H^+ + R_2CO \quad \text{unimolecular}$$
$$\text{decomposition}$$

Another important reaction is the oxidation of the carbon dioxide radical anion [25,134]:

$$\cdot COO^- + O_2 \longrightarrow [^-OOCOO\cdot] \longrightarrow CO_2 + O_2^{\cdot-}$$

Whether this reaction proceeds via an intermediate peroxyl radical or via a direct one-electron transfer is difficult to answer. No intermediate peroxyl radical was observed via pulse radiolysis [134]. This fact, however, does not prove the absence of a peroxyl radical. The peroxyl radical has to have a long enough lifetime to be observed.

Some radicals like phenoxyl-type radicals do not react with oxygen. These radicals only undergo dimerization as has been shown in the case of the tyrosine phenoxy radical[193]:

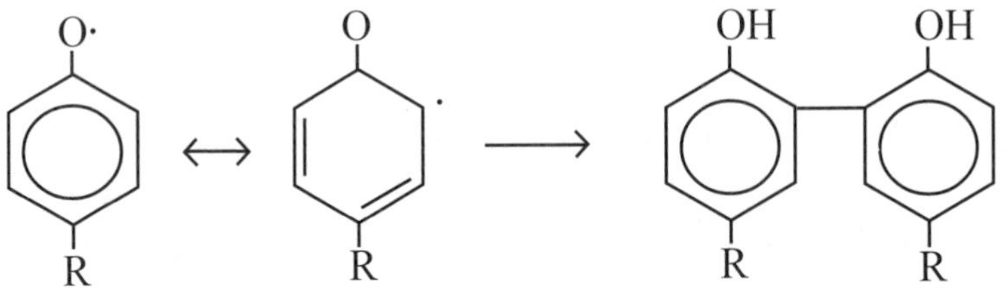

This type of reaction is important in the cross-linking of tyrosine to DNA bases (Chapter 6).

In addition to O_2^- / HO_2· elimination, peroxyl radicals can undergo intramolecular addition to unsaturated bonds. This type of addition has been observed in the oxidation of the hydroxycyclohexadienyl radical and leads to complex secondary reactions [194,195].

Endoperoxides are intermediates in the chain oxidation of polyunsaturated fatty acids [196,197]:

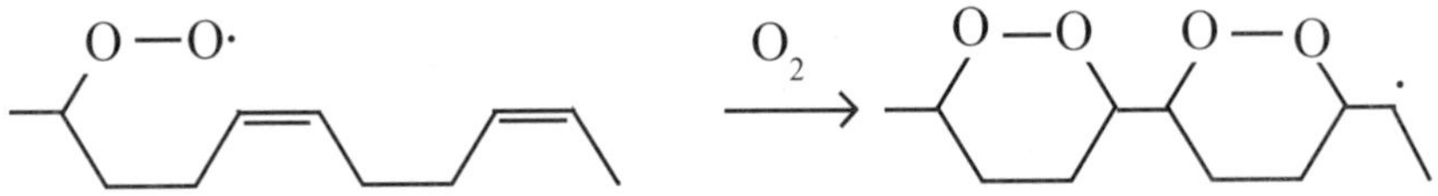

The formation of these endoperoxides can be determined by their degradation product malonaldehyde [198]. Malonaldehyde has been used as a biological marker for membrane damage.

1. Bimolecular reactions of peroxyl radicals

Bimolecular reactions are important as chain-terminating reactions. Peroxyl radicals like any other radical can undergo dimerization:

$$2\ R-OO\cdot \longrightarrow R-OO-OO-R$$

The self reaction of peroxy radicals has been extensively studied due to their importance as a chain-terminating step in hydrocarbon chain oxidation:

$$R-H \longrightarrow R\cdot \qquad \text{initiation}$$

$$R\cdot\ +O_2 \longrightarrow R-O_2\cdot$$

$$\text{propagation}$$

$$R-O_2\cdot + R-H \longrightarrow R-OOH + R\cdot$$

$$R-O_2\cdot + R-O_2\cdot \longrightarrow R-OOOO-R \qquad \text{termination}$$

The combination rate constants for peroxyl radicals are in the range of $10^3 - 10^8\ M^{-1}\ s^{-1}$. The rate constants depend on the structure of the peroxyl radical (primary $RCH_2OO\cdot$ > secondary $(R)_2CHOO\cdot$ > tertiary $(R)_3COO\cdot$). Primary peroxyl radicals are at the high end and tertiary peroxyl radicals are at the low end of this range.

The combination proceeds via the following mechanism:

$$2\ ROO\cdot \longrightarrow ROOOOR \longrightarrow \left[RO\cdot + O_2 + \cdot OR \right]_{cage} \longrightarrow ROOR + O_2$$
$$\downarrow$$
$$2\ RO\cdot\ +\ O_2$$

The ratio of the two competing reactions (combination within the cage versus diffusion out of the cage) depends on the structure of the peroxyl radicals. In the case of tertiary peroxyl radicals diffusion is much faster than combination and thus only a small fraction leads to chain termination. The dimerization of secondary peroxyl radicals is of importance in biological systems.

$$R_2CH-OOOO-CHR_2 \longrightarrow R_2C=O\ +\ R_2CH-OH\ +\ {}^1O_2$$

This reaction was first described by Russell [73] and carries his name. Russell suggested that the reaction proceeded via a concerted process:

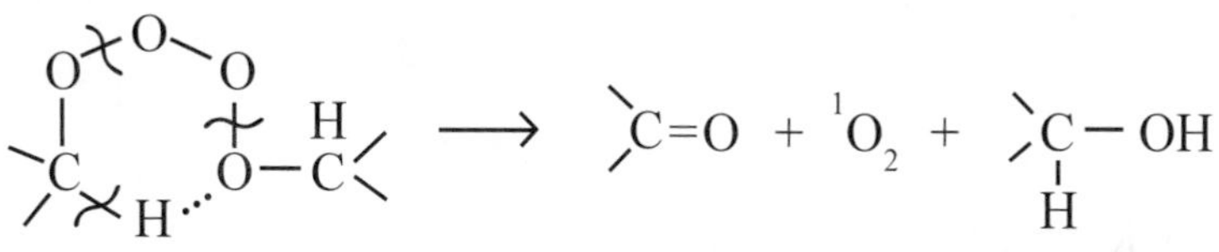

Scientists are great sceptics. Arguments in favor and against the Russell mechanism have been presented [189]. Electronically excited states decay to the ground state via emission of light. This chemiluminescene has been observed and two possibilities for this

58

chemiluminescence have been discussed: 1. Formation of singlet oxygen or 2. Formation of an electronically excited carbonyl compound. We have to consider the following reaction steps:

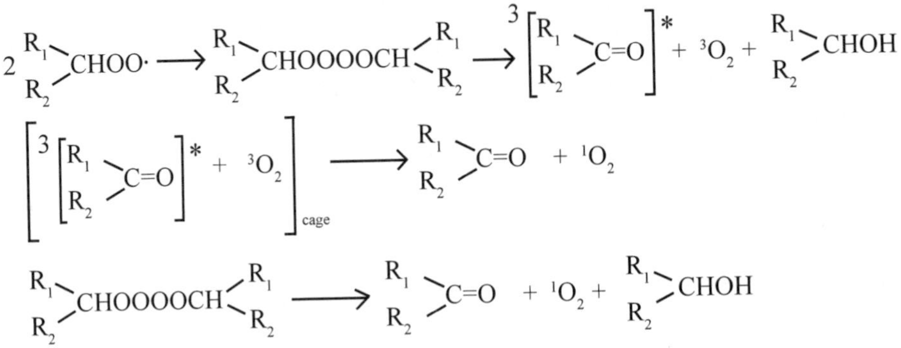

$$2 \; \overset{R_1}{\underset{R_2}{>}}\!CHOO\cdot \longrightarrow \overset{R_1}{\underset{R_2}{>}}\!CHOOOOCH\!\overset{R_1}{\underset{R_2}{<}} \longrightarrow \left[\overset{R_1}{\underset{R_2}{>}}\!C{=}O\right]^{*} + {}^3O_2 + \overset{R_1}{\underset{R_2}{>}}\!CHOH$$

$$\left[\overset{3}{\left[\overset{R_1}{\underset{R_2}{>}}\!C{=}O\right]^{*}} + {}^3O_2\right]_{cage} \longrightarrow \overset{R_1}{\underset{R_2}{>}}\!C{=}O + {}^1O_2$$

$$\overset{R_1}{\underset{R_2}{>}}\!CHOOOOCH\!\overset{R_1}{\underset{R_2}{<}} \longrightarrow \overset{R_1}{\underset{R_2}{>}}\!C{=}O + {}^1O_2 + \overset{R_1}{\underset{R_2}{>}}\!CHOH$$

Present evidence confirms the Russell mechanism and the formation of singlet oxygen and excited carbonyl compounds [199-201]. The yield of singlet oxygen was estimated by Niu and Mendenhall [201] between 3-14%. The yields of excited carbonyl compounds were determined to be considerably lower than that of singlet oxygen [200]. The formation of electronically excited states is of great importance for biological damage to lipids and DNA. Low level chemiluminescence measurements have been used *in situ* in organs as a non-invasive technique for monitoring oxidative stress [118,119].

Summary. The biologically important characteristics of peroxyl radicals are: 1. Production of excited states (singlet oxygen or carbonyls), and 2. H-abstraction to form hydroperoxides (chain oxidation of unsaturated fatty acids).

F. HYDROXYL RADICAL
1. Formation of hydroxyl radicals via radiolysis of water [24,25]

The radiolysis of water is not only the simplest way for the formation of hydroxyl radical, but it has also provided valuable insight into the formation of hydroxyl radicals in complex biological systems. The primary processes are as follows:

$$H_2O \rightsquigarrow H_2O^+ + e^-$$

$$H_2O^+ + H_2O \longrightarrow H_3O^+ + \cdot OH$$

$$e^- + H_2O \longrightarrow e_{aq}^- \text{ (thermalized solvated electron)}$$

In addition to these reactions a small amount of molecular products (H_2 and H_2O_2) are produced. The two reactive intermediates ($\cdot$OH and e_{aq}^-) are formed in approximately equal amounts. We have an oxidizing species (the hydroxyl radical) and a reducing species (the solvated electron). The e_{aq}^- can be considered the simplest radical and it is not "free" in aqueous solutions, but surrounded by a solvation shell. If one wants to study the reactions of hydroxyl radicals one would prefer to have only hydroxyl present in the system. The solvated electron can be converted to the hydroxyl via its reaction with nitrous oxide (N_2O):

$$N_2O + e_{aq}^- \longrightarrow N_2 + O^{\cdot-} (+ H^+ \longrightarrow \cdot OH)$$

The hydroxyl radical reacts with other organic molecules by abstraction or addition, producing a huge number of different radicals, whose reactions (disproportionation, dimerization, oxidation, reduction, β-cleavage, rearrangements, etc.) can be examined. The rate of formation and disappearance of these radicals can be studied by pulse radiolysis. The rate constants determined by pulse radiolysis [99] have provided a valuable tool for the identification of hydroxyl radicals.

The radiolysis of water not only provides us with a reliable source of hydroxyl radicals, but can also be used exclusively to obtain the superoxide radical. This can be accomplished via the addition of formate [134]:

$$\cdot OH + HCOO^- \longrightarrow H_2O + \cdot COO^-$$

$$\cdot COO^- + O_2 \longrightarrow CO_2 + O_2\cdot^-$$

$$e_{aq}^- + O_2 \longrightarrow O_2\cdot^-$$

In this way we can transform the ·OH radical and the solvated electron quantitatively to superoxide radical.

The yields in radiation chemistry are measured in G-values. A G-value is the number of molecules formed per 100 ev of energy absorbed by the system. The G-values of the primary reactive intermediates have been determined by many investigators and have been summarized by Draganic and Draganic [25]. Reliable values for all practical purposes are: $G(OH) = 2.8$ and $G(e_{aq}^-) = 2.8$. The radiolysis of water therefore gives us a clean system producing an exact amount of hydroxyl radicals. In the radiolysis of water in presence of N_2O we have only hydroxyl radicals present and we can study their reactions without interference from any other species. The radiolysis method is therefore the best suited method for the study of hydroxyl radicals. Radiolysis has been used extensively in studies comparing the effects of radiation with the effects of other systems which produce or supposedly produce hydroxyl radicals. Most studies involving chemical reactions discussed later in this book have adopted this approach [202]. Competition kinetics have given some insights into the complexity of hydroxyl radical formation in biological systems (crypto and site-specific formation of hydroxyl radicals).

2. The photo-Fenton reaction

The photochemical generation of hydroxyl radicals has been known for some time. These reactions involve the photolysis of α-azohydroperoxides [203-205] and phthalimide hydroperoxides [206,207]:

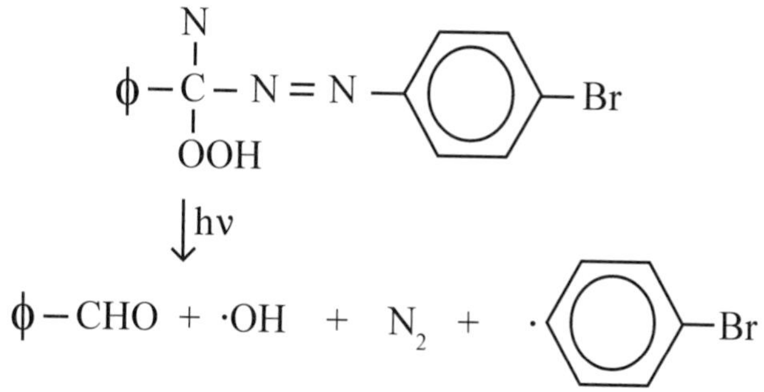

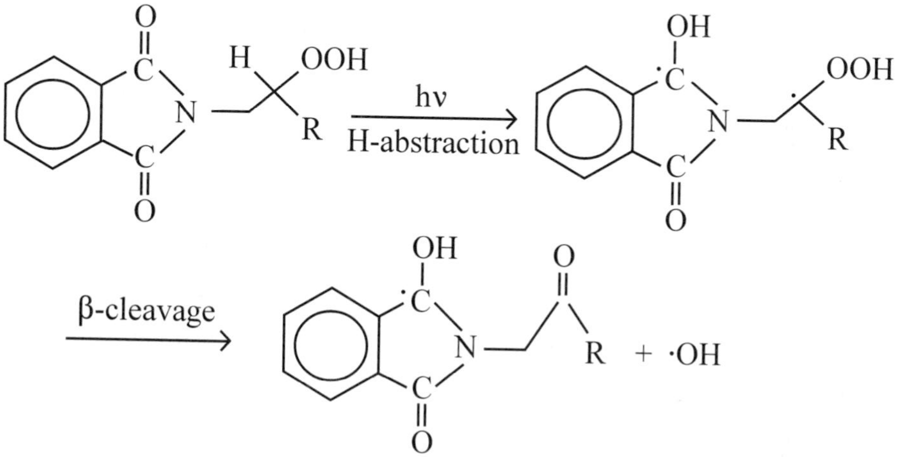

$$\phi-\underset{\underset{OOH}{|}}{\overset{\overset{N}{|}}{C}}-N=N-\!\!\!\bigcirc\!\!\!-Br$$

$$\downarrow h\nu$$

$$\phi-CHO \;+\; \cdot OH \;+\; N_2 \;+\; \cdot\!\!\!\bigcirc\!\!\!-Br$$

(phthalimide hydroperoxide) $\xrightarrow[\text{H-abstraction}]{h\nu}$ (radical)

$\xrightarrow{\beta\text{-cleavage}}$ product $+\ \cdot OH$

The α-azohydroperoxide is not water soluble and has been used for mechanistic studies of aromatic hydroxylation in aprotic solvent [204,205]. The formation of hydroxyl radical from the phthalimide hydroperoxide proceeds via two familiar reaction steps (Chapter 2), namely hydrogen abstraction by an electronically excited carbonyl compound, followed by β-cleavage.

The phthalimide hydroperoxides are also poorly water soluble and are therefore not very well suited for investigations in cellular systems. The discovery of the photochemical decomposition of N-hydroxypyridinethiones to yield hydroxyl radicals [208-211] was therefore an important step forward.

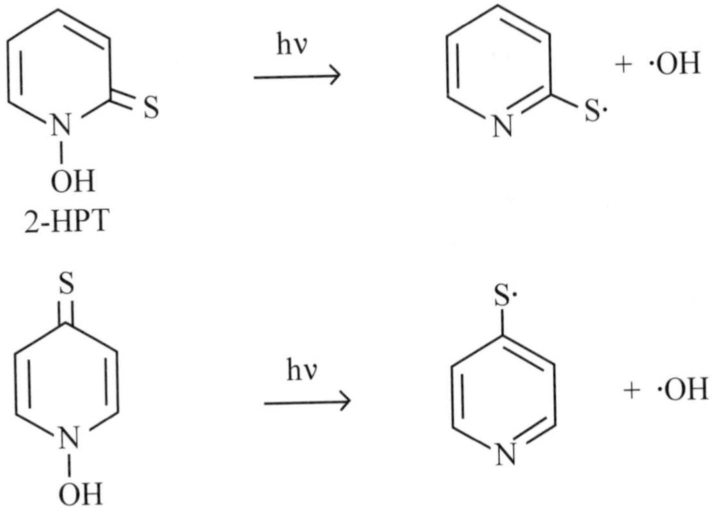

2-HPT

4-HPT

The formation of OH was detected by spin trapping [210,211], as well as by standard chemical assays [209], which I shall discuss later in this chapter. The *N*-hydroxypyridinethiones are water soluble and are therefore suitable for the study of the genotoxic effects of hydroxyl radicals in cellular systems. In particular, they are suitable to correlate DNA damage with its consequences, such as induction of mutations and changes of gene expression [212]. The generation of 8-hydroxyguanine (8-OHG) by photolysis of 2-HPT and 4-HPT in calf thymus DNA was reported [213]. Other photo-Fenton reagents developed by Adam and coworkers [214,215] are the hydroperoxides of furocoumarins. These hydroperoxides of furocoumarins can be intercalated into DNA. These compounds absorb in the UVA region ($\lambda \geq 350$ nm) and provide an effective source of OH radicals for the study of cellular oxidative DNA damage [215].

We may ask: why do we need another source for hydroxyl radical formation since we already have the method of radiolysis of water. I shall discuss the reactions of hydroxyl radicals with biomolecules in Chapter 6.

The method of radiolysis is well suited for studies of cell-free DNA. However, a cell is very different from an aqueous solution. If we irradiate whole cells we not only damage DNA by OH radicals, but also produce many other radicals and reactive oxygen species [216]. This problem is avoided by the use of these photo-Fenton reagents, where the light energy is exclusively absorbed by the photo-Fenton reagent.

2-HPT or 4-HPT on illumination with visible light, induced DNA damage in L1210 mouse leukemia cells and in isolated DNA from bacteriophage PM2 [212]. DNA strand breaks and DNA modifications were quantitatively determined. Specific repair endonucleases have been used to quantify various types of oxidative modifications in cellular and in cell-free DNA [217]. The DNA damage profile serves as a fingerprint of the DNA damaging species. Illumination of cell free DNA in presence of 2-HPT or 4-HPT gave a damage profile characteristic for hydroxyl radicals. Scavenger experiments with t-butanol, CAT, SOD and in D_2O as solvent confirmed that OH radicals are directly responsible for DNA damage. [212]. The DNA damage profile induced by 2-HPT plus light in L1210 mouse leukemia cells was consistent with the assumption that hydroxyl radicals are directly responsible for the cellular DNA damage as well.

3. Formation of OH radicals via chemical reactions

The primary reactive oxygen metabolites (ROMs) are $O_2^{\cdot-}$ and H_2O_2. All other oxygen radicals ($\cdot OH$, $RO\cdot$, $ROO\cdot$) are produced via secondary reactions of these initially formed metabolites. $O_2^{\cdot-}$ and H_2O_2 are formed *in vivo* both via normal metabolism [17,18,83,85,86] and via metabolism of xenobiotics.

The primary oxygen metabolites are chemically quite distinct. The superoxide radical anion ($O_2^{\cdot-}$) is a radical, which also carries a negative charge, while the H_2O_2 is a neutral non-radical species. These two characteristics are important in biology. The superoxide radical cannot pass through membranes, except through special ion channels, whereas H_2O_2 can easily pass through membranes and therefore can cause damage at sites far removed from its site of formation.

There is only one type of reaction, which leads to the formation of hydroxyl radical from hydrogen peroxide. This reaction is one-electron transfer. The electron donating species may be a transition metal ion or an electron donating radical, like $O_2^{\cdot-}$:

$$Fe^{2+} + H_2O_2 \longrightarrow Fe^{3+} + OH^- + \cdot OH \quad \textbf{Fenton reaction}$$

$$O_2^{\cdot-} + H_2O_2 \longrightarrow O_2 + OH^- + \cdot OH \quad \textbf{Haber-Weiss reaction}$$

$$O_2^{\cdot-} + Fe^{3+} \longrightarrow O_2 + Fe^{2+} \quad \textbf{Fe}^{3+}\textbf{-catalyzed}$$
$$Fe^{2+} + H_2O_2 \longrightarrow Fe^{3+} + OH^- + \cdot OH \quad \textbf{Haber-Weiss reaction}$$

The only other transition metal ion of biological significance, which undergoes a Fenton-type reaction, is copper (I) [202]:

$$Cu^+ + H_2O_2 \longrightarrow Cu^{2+} + OH^- + \cdot OH$$

Other reducing radicals have been suggested as sources for hydroxyl radical formation. These radicals include paraquat radical cation $(PQ^{\cdot+})$ [218] and anthracycline semiquinone [219]. However, these suggestions have been repudiated [220,221]. All of these radicals produce hydroxyl radicals via the intermediate formation of $O_2^{\cdot-}$ followed by reduction of Fe^{3+} according to the following sequence of reaction steps:

$$PQ^{\cdot+} + H_2O_2 \xrightarrow{\quad\times\quad} PQ^{2+} + OH^- + \cdot OH$$

$$PQ^{\cdot+} + O_2 \longrightarrow PQ^{2+} + O_2^{\cdot-}$$

$$O_2^{\cdot-} + Fe^{3+} \longrightarrow Fe^{2+} + O_2$$

$$Fe^{2+} + H_2O_2 \longrightarrow Fe^{3+} + OH^- + \cdot OH$$

Another possible source of hydroxyl radicals, which has been discussed, is the reaction of NO with superoxide:

$$\cdot NO \ + \ O_2^{\cdot-} \longrightarrow ONOO^- \xrightarrow{H^+} ONOOH \longrightarrow NO_2 \cdot \ + \ \cdot OH$$

The anion at physiological pH of 7.4 will become protonated to the peroxynitrous acid, which in turn may decompose to yield hydroxyl radicals [222]. This reaction has been quite controversial and will be discussed in detail in the chapter on nitric oxide. According to the latest results the decomposition of peroxynitrous acid is unlikely to yield hydroxyl radicals. Douki and Cadet [223] carried out a comparative study of peroxynitrite and radiolytically-induced DNA damage, and found a different product profile. Another pathway for hydroxyl formation is the reaction of superoxide with hypochlorous acid (produced by stimulated neutrophils) [224,225]:

$$HOCl \ + \ O_2^{\cdot-} \longrightarrow \cdot OH \ + \ {}^1O_2 \ + \ Cl^-$$

This reaction will be discussed in more detail in Chapter 5 under phagocytosis.

The Fenton reaction was discovered over 100 years ago by Fenton (1894). The reaction was given little attention until Haber and Weiss (1934) proposed the hydroxyl radical as an intermediate in the reaction. All one-electron reductions of H_2O_2 are usually referred to as "Fenton-type" reactions. However, the equation for the Fenton reaction as written above has been questioned. Other reaction mechanisms have been proposed and involve formation of a "crypto"-hydroxyl radical or a Fe(IV) (a ferryl) species. The name "crypto" of course only hides our ignorance in explaining certain experimental observations. In order to understand the experimental evidence on which these mechanisms are based, we first must have a detailed understanding of hydroxyl radical reactions and the methods used in the identification of hydroxyl radicals. These complexities involved in the metal ion-H_2O_2 reactions can be avoided by the use of photo-Fenton reagents (pp. 60-63).

4. Detection and reactions of hydroxyl radicals

You shall know them by their deeds.

Swabian saying

In order to discuss the formation and reactions of hydroxyl radicals we first of all need an unequivocal method for hydroxyl radical formation and a reliable and sensitive method for the detection of these fleeting intermediates.

There are three methods, which have been used to identify hydroxyl radicals. These methods are 1. Spin trapping and EPR spectroscopy, 2. Reactions with any molecule, which yields an easily identifiable and hopefully specific product, and 3. Competition kinetics.

The first two of these methods are basically identical. As we have already discussed a radical reacts with a molecule to produce another radical. This radical usually is of lower reactivity than the original radical (i.e., it is more stable). In the method of spin trapping this characteristic of radicals is used to transform a highly reactive radical (like ·OH) to a highly stable radical. Compounds which are used in this type of reaction are N-nitroso compounds, like the 5,5'-dimethylpyrollidine-N-oxide (DMPO):

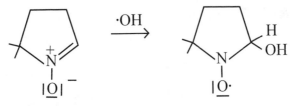

These adducts are stable and can be measured by EPR spectroscopy [66,67,226]. Many spin adducts have been measured and their EPR parameters compiled [227].

In the second method the hydroxyl radical is transformed into another radical, which then undergoes further reactions to yield a stable molecule. The stable molecule can be analyzed by conventional analytical techniques (gas chromatography or high pressure liquid chromatography). Some examples are formation of ethylene from methional [228], formation of methane from dimethylsulfoxide (DMSO) [229] and aromatic hydroxylation [51,229].

In competition kinetics we measure the relative rates of two compounds with the reactive species under investigation.

$$X\cdot + A \xrightarrow{k_a} \text{Product (A)}$$

$$X\cdot + B \xrightarrow{k_b} \text{Product (B)}$$

If k_a/k_b for $\cdot X$ is different from k_a/k_b for $\cdot OH$ we may conclude that $\cdot X$ is not identical to $\cdot OH$. However if the two relative rates are identical it is possible that $\cdot X$ is identical to $\cdot OH$ (or as is often said it is kinetically indistinguishable from $\cdot OH$). In order to apply this technique we must first have a reliable source of hydroxyl radicals. This source of $\cdot OH$ radicals is the radiolysis of water.

The following types of hydroxyl radical reactions will be discussed:
a. H-abstraction combined with EPR spectroscopy.
b. Reactions with spin traps.
c. Reaction with dimethylsulfoxide (DMSO).
d. Reaction with methional.
e. Reaction with deoxyribose.
f. Aromatic hydroxylation.

a. H-abstraction combined with EPR spectroscopy

EPR spectroscopy has long been used for the identification of radicals. It is based on the magnetic moment of the unpaired electron. Depending on the atoms adjacent to the unpaired electron a different spectrum is obtained. EPR spectroscopy therefore can serve as a fingerprint for radicals. The reagents are mixed in a flow system in the cavity of the spectrometer, where the radicals are measured. This method of course requires a certain concentration for the radicals to be detectable.

Since the hydroxyl radical is highly reactive it cannot be observed directly by EPR spectroscopy. The hydroxyl radical is reacted with

68

another molecule to give a more stable secondary radical. We would like to discuss one example, which shows the pitfalls of the EPR technique. The Ti^{3+}-H_2O_2-isopropanol reaction was shown to give the 2-hydroxy-2-propyl radical, whereas the Fe^{2+}-H_2O_2-isopropanol reaction gave the 2-hydroxy-1-propyl radical [230]. This observation led to the conclusion that in the Fe^{2+}-H_2O_2-isopropanol system a species different from the hydroxyl radical was involved (the ferryl ion?). An explanation for these observations was given by Walling [39]:

$$
\begin{array}{c}
CH_3 \\
\diagdown \\
 CH\text{-}OH \\
\diagup \\
CH_3
\end{array}
\xrightarrow[\text{or } Fe^{2+}\text{-}H_2O_2]{Ti^{3+}\text{-}H_2O_2}
\begin{array}{c}
CH_3 \\
\diagdown \\
 \dot{C}\text{-}OH\ + \\
\diagup \\
CH_3
\end{array}
\begin{array}{c}
CH_3 \\
\diagdown \\
 CH\text{-}OH \\
\diagup \\
\cdot CH_2
\end{array}
$$

$$
\text{major radical} \qquad \text{minor radical}
$$

$$
\begin{array}{c}
CH_3 \\
\diagdown \\
 \dot{C}\text{-}OH\ +\ Fe^{3+} \\
\diagup \\
CH_3
\end{array}
\longrightarrow
\begin{array}{c}
CH_3 \\
\diagdown \\
 C{=}O\ \ +\ \ Fe^{2+} \\
\diagup \\
CH_3
\end{array}
$$

In both experiments two radicals are formed with the 2-hydroxyl-2-propyl radical predominant. In the Fe^{2+}-H_2O_2 case the 2-hydroxy-2-propyl radical is rapidly oxidized to acetone, leaving only the 2-hydroxy-1-propyl radical to be detected by EPR.

The moral of the story: do not jump to conclusions on the basis of one experiment!

b. Reactions with spin traps

The spin-trapping technique allows us to concentrate any amount of radicals. This technique is therefore more widely applicable and the literature is very extensive. The spin-trapping technique introduced by Janzen [67,226] has proven to be very useful, although it has its limitations and artifacts. It has been shown that the DMPO/·OH signal can be obtained in a variety of reactions, which do not involve hydroxyl radicals. One such method involves the radiation induced formation of a DMPO radical cation, followed by nucleophilic addition of water [231]:

$$DMPO + X^+ \longrightarrow DMPO^+\cdot + X\cdot$$

$$DMPO^+\cdot + H_2O \longrightarrow DMPO/\cdot OH + H^+$$

where X^+ may be part of a biological redox system.

Another way to obtain the DMPO/·OH adduct is via complexation with Fe^{3+} followed by nucleophilic attack by water [232]:

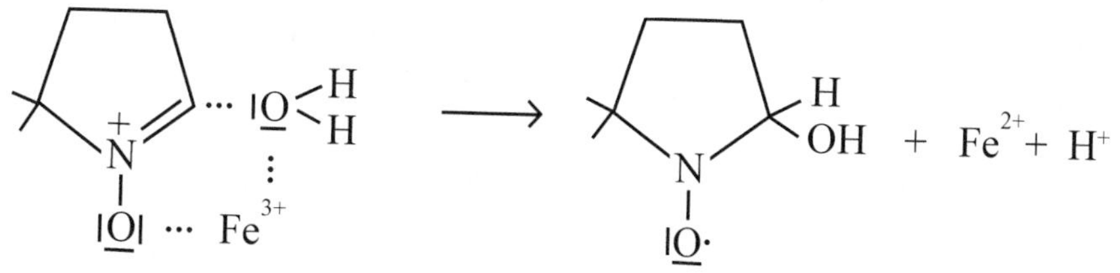

Furthermore the superoxide radical adduct to DMPO (DMPO/·OOH) can produce hydroxyl radicals via thermal cleavage [233]:

$$DMPO/\cdot OOH \longrightarrow DMPO/O\cdot + \cdot OH$$

Another reaction which gives the DMPO/·OH adduct is the reaction of Co^{2+}-ethylenediamine-O_2 [234]. In this case it was shown that the OH group was derived from water and not from oxygen. It therefore most likely also involves a radical cation:

$$DMPO \xrightarrow[O_2]{Co^{2+}\text{-}ED} DMPO^+\cdot \xrightarrow{H_2O} DMPO/\cdot OH$$

If we want to ascertain whether the DMPO/·OH adduct is formed via the hydroxyl radical we can add another hydroxyl radical scavenger to compete with DMPO:

$$DMPO \ + \ \cdot OH \ \longrightarrow \ DMPO/\cdot OH$$

$$RH \ + \ \cdot OH \ \longrightarrow R\cdot$$

$$DMPO \ + \ R\cdot \ \longrightarrow DMPO/\cdot R$$

where RH (detector molecule) reacts with ·OH to give a radical R·. The most commonly used detector molecules in this type of competition is dimethylsulfoxide (DMSO) and ethanol [235,236]:

$$CH_3SOCH_3 \ + \ \cdot OH \ \longrightarrow \ \cdot CH_3 \ + \ CH_3SOOH$$

$$CH_3CH_2OH \ + \ \cdot OH \ \longrightarrow \ CH_3\dot{C}HOH \ + \ H_2O$$

Since the rate constants for all these reactions are known from pulse radiolysis experiments we can calculate the expected decrease in the DMPO/·OH signal. If these calculations agree with the experiments we can say that the radical species is most likely the hydroxyl radical or a species kinetically indistinguishable from hydroxyl. We observe the decrease in the DMPO/·OH signal and the appearance of the DMPO/·CH$_3$ signal.

Although the observation of the DMPO/·CH$_3$ signal is definitive proof for hydroxyl radical its absence does not prove it is truly absent. We encounter this problem in the hydroxyl radical detection by activated phagocytes. It will be discussed in Chapter 5 under phagocytosis.

In conclusion we can state that the observation of the DMPO/·OH adduct is in itself no proof for hydroxyl radicals. The spin-trapping technique has to be used in combination with other hydroxyl radical scavengers. The pros and cons of spin trapping have been periodically reviewed [67,226,235,236].

c. Reaction with dimethylsulfoxide (DMSO)

The formation of methyl radicals in the reaction of $\cdot$OH (produced via Ti^{3+}-H_2O_2) with DMSO was first observed by EPR spectroscopy [237,238]. The reaction yields CH_4 and CH_3CH_3 as final and stable products according to the following reactions:

$$\cdot OH \; + CH_3SOCH_3 \longrightarrow CH_3 \overset{\displaystyle OH}{\underset{\displaystyle O}{\overset{|}{\underset{\downarrow}{-S-}}}} CH_3 \longrightarrow \cdot CH_3 + CH_3SOOH$$

$$\cdot CH_3 \; + \; CH_3SOOH \longrightarrow CH_4 \; + \; CH_3SOO\cdot$$

$$\cdot CH_3 \; + \; CH_3SOCH_3 \longrightarrow CH_4 \; + \; \cdot CH_2SOCH_3$$

$$\cdot CH_3 \; + \; \cdot CH_3 \longrightarrow CH_3CH_3$$

The yield of CH_4 from $\cdot$OH produced via radiolysis of N_2O saturated solutions has been found to be 86% [239]. In order to use product formation as a hydroxyl radical probe the reaction has to be specific, i.e., the product is only formed via hydroxyl radicals. The DMSO probe meets this requirement for specificity. The only other known reaction which interacts with DMSO to yield $\cdot CH_3$ is the reaction with persulfate radical anion $(SO_4^{\cdot -})$ [240]. The DMSO probe therefore appears to be the most reliable method. DMSO has the added advantage of low toxicity and can therefore be used in biological systems as well as in purely chemical systems. The products can easily be detected by gas chromatography. The DMSO probe is especially useful in combination with the DMPO spin trap as discussed above. The decrease in the DMPO/$\cdot$OH signal and the appearance of the DMPO/$CH_3\cdot$ signal is definitive proof for the presence of hydroxyl radicals [236]. In the previously discussed example of the reaction of Co^{2+}-ethylenediamine-oxygen [234] (where no $\cdot$OH is involved) the DMPO/$\cdot$OH signal was not affected by DMSO.

In the presence of oxygen the methyl radicals react to give formaldehyde [241,242]. The DMSO method has been used to detect hydroxyl radicals by enzymes and activated phagocytes [243]. Recently the DMSO method has been used to detect trace amounts of hydroxyl radicals in biological systems [244]. This method traps the $\cdot CH_3$ with a fluorescamine-derivatized nitroxide to produce a stable adduct, which can be analyzed by HPLC and detected fluorometrically.

d. Reaction with methional

The reaction of methional with hydroxyl radicals has been examined via pulse radiolysis [245]. The bimolecular rate constant was determined to be 8.2 x 10^9 $M^{-1}s^{-1}$. The products of the reaction were ethylene, dimethyldisulfide and CO. On the basis of the observed spectra the following mechanism was proposed:

$$CH_3SCH_2CH_2CHO + \cdot OH \longrightarrow CH_3\overset{\overset{OH}{|}}{\underset{\cdot}{S}}CH_2CH_2CHO$$

$$\longrightarrow CH_3\overset{+\cdot}{S}CH_2CH_2CHO + OH^-$$

$$CH_3\overset{+\cdot}{S}CH_2CH_2CHO \longrightarrow CH_3S\cdot + CH_2{=}CH_2 + CO + H^+$$

$$2\ CH_3S\cdot \longrightarrow CH_3S{-}SCH_3$$

This method has been frequently used for the identification of hydroxyl radicals [228,246]. The ethylene is easy to determine by gas chromatography. However, it was recognized early on that this probe may not be reliable due to the complexity of the mechanism and due to the far from quantitative yields of ethylene. It was observed that a variety of alkoxy radicals also react with methional to give low-to-moderate yields of ethylene [247]. The aminoacid methionine also reacts with hydroxyl radicals to give ethylene.

Cytochrome P-450 of rat liver microsomes has been shown to give ethylene from methional, but only very low yields of methane from

DMSO [246]. Since DMSO reacts with radiolytically generated $\cdot$OH to give high yields of CH_4 [239] these results show the dubious value of methional for hydroxyl radical detection.

e. Reaction with deoxyribose

The sugar deoxyribose (2-deoxy-D-ribose) is degraded by hydroxyl radicals which are produced either via radiolysis or via Fenton-type reactions. The resulting complex mixture produces under heating at low pH malonaldehyde, and can be detected by its reaction with thiobarbituric acid. The reaction produces a pink color with an absorption maximum at 532 nm [248-250].

One reaction which was studied using this procedure was the reaction of Fe(II) in oxygen containing aqueous solutions of phosphate buffer at pH 7.4. In this case the deoxyribose degradation was inhibited by hydroxyl radical scavengers like mannitol, thiourea and catechol. It was therefore suggested that the hydroxyl radical is responsible for deoxyribose degradation:

$$Fe^{2+} + O_2 \rightleftharpoons Fe^{2+} O_2 \rightleftharpoons Fe^{3+} + O_2\cdot^-$$

$$2\, O_2\cdot^- + 2\, H^+ \longrightarrow H_2O_2 + O_2$$

$$Fe^{2+} + H_2O_2 \longrightarrow Fe^{3+} + OH^- + \cdot OH$$

A superoxide generating system (the xanthine-xanthine oxidase system) also produced the same chromogen and the formation was inhibited by catalase. This seems to confirm the above reaction sequence. In addition competition kinetics with a number of hydroxyl radical scavengers (ethanol, benzoate, mannitol, etc.) showed that the decrease in deoxyribose degradation was in perfect agreement with the known rate constants for OH reaction [250].

The deoxyribose degradation has therefore been frequently used for the identification of hydroxyl radicals [248-250]. However, it was soon recognized that the deoxyribose degradation was not specific for hydroxyl radicals. In the iron catalyzed Haber-Weiss reaction in phosphate buffer (pH 7.4) and in presence of EDTA deoxyribose degradation was inhibited by a number of hydroxyl scavengers. These results were in good agreement with the known rate constants. However, in absence of EDTA the deoxyribose degradation was not inhibited by DMSO. We have the following observations [251]:

$$Fe^{3+}\text{-EDTA} + H_2O_2 + O_2^{\cdot -} \xrightarrow{\text{pH 7.4}} \text{degradation inhibited by OH scavengers}$$

$$Fe^{3+} + H_2O_2 + O_2^{\cdot -} \xrightarrow{\text{pH 7.4}} \text{degradation not inhibited by DMSO}$$

This last result clearly shows that the deoxyribose is degraded by a species different from hydroxyl radical. This result further shows that DMSO does not react with this species, confirming DMSO as a reliable probe for hydroxyl radicals.

The reaction Fe^{3+}, ascorbic acid, H_2O_2 in phospate buffer at pH 7.4 and in presence of EDTA was also studied using the deoxyribose procedure. The effect of a number of scavengers (ethanol, mannitol, n-butanol, iso-propanol, etc.)was found to be in perfect agreement with their hydroxyl radical rate constants. It was actually proposed that this method could provide a cheap alternative to pulse radiolysis for the determination of hydroxyl radical rate constants [250]. As I have already pointed out, based on competition kinetics we cannot state that hydroxyl radicals are formed, but only that a species kinetically indistinguishable from hydroxyl radical is obtained. Since deoxyribose reacts with species different from OH and alcohols do the same, we may suspect that the good agreement with pulse radiolysis results may have been quite fortuitous.

f. Aromatic hydroxylation [51,229]

Although the hydroxyl radical is very reactive this does not mean it is non-selective in its reactions. The hydroxyl radical (although electrically neutral) is an electrophilic species, i.e., it reacts preferentially at positions of high electron density [51, 604]. The reactions of hydroxyl with monosubstituted benzenes should therefore provide us with a finger-print for the hydroxyl radical. However, the aromatic hydroxylation approach has led to many wrong conclusions concerning the involvement of hydroxyl radicals in a number of reactions. The reason for these wrong conclusions lies in the complexity of the hydroxylation process, leading from the initially formed hydroxycyclo-hexadienyl radicals to the final stable phenols. These complexities are summarized in Fig.1. The first step in the hydroxylation of aromatics by hydroxyl radicals is addition to the aromatic

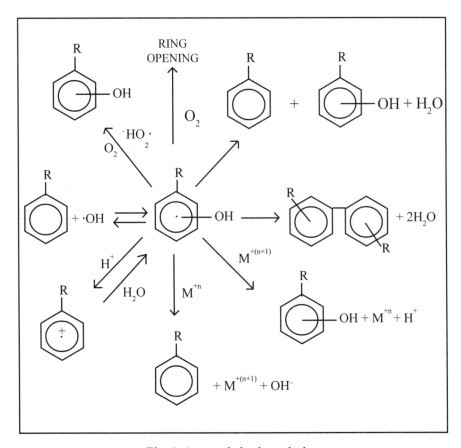

Fig. 1. Aromatic hydroxylation

(from [51] with permission)

ring. The resulting hydroxycyclohexadienyl radical can undergo all the typical radical reactions, discussed in Chapter 2. These reactions include dismutation, dimerization, oxidation, reduction, dehydration, fragmentation and reaction with oxygen.

The reactivity of the different positions (ipso, ortho, meta, para) in monosubstituted benzenes gives a sensitive measure of the reactive species, provided that the final product distribution (isomeric phenols) corresponds to the distribution of the initially formed hydroxy-cyclohexadienyl radicals. This condition is fulfilled whenever we have an oxidizing agent present, which oxidizes the hydroxycyclohexadienyl radicals faster compared to any other reaction (rearrangement, elimination of OH radical, dehydration, reduction, disproportionation) these radicals may undergo. Keeping these complexities in mind the isomer distribution obtained in homolytic aromatic hydroxylation can serve as a valuable tool in identifying the hydroxyl radical [51]. In purely chemical systems the aromatic hydroxylation approach has been very useful. The Cu^+-H_2O_2 reaction will be discussed later in this chapter. Aromatic hydroxylation of benzoic acid has been suggested as a probe for ·OH in humans [252]. The formation of hydroxyl radicals after myocardial ischemia *in vivo* was measured by aromatic hydroxylation of phenylalanine to give the isomeric o-, m-, p-tyrosines [253]. The para-tyrosine is formed via normal metabolism of phenylalanine, but the ortho- and meta-isomers are specific markers for hydroxyl radicals.

Many of the hydroxyl radical probes like methional, methionine, deoxyribose degradation or hydroxylation of phenol have been shown to be non-specific for hydroxyl radicals. These compounds can react with other oxidizing species to give the same product as obtained with hydroxyl radicals. Ethanol, mannitol and formate also have been used in competition experiments to infer the intermediate formation of hydroxyl radicals. However, it has been clearly shown that these compounds not only react with hydroxyl radicals but with other oxidizing species as well.

In summary we can state that the following methods are reliable probes for hydroxyl radicals: 1. DMPO spin trapping (with caution!). 2. Formation of CH_4 and CH_3CH_3 from DMSO or formation of HCHO in presence of oxygen. 3. Aromatic hydroxylation (with careful interpretation of the isomer distribution). 4. Combination of methods 1-3.

5. Formation of hydroxyl radicals involving transition metal ions

Most transition metal ions can exist in more than two oxidation states. The only exception is titanium, which exists in the Ti(III) and the Ti(IV) state. There can be no doubt that the reaction of Ti^{3+} with H_2O_2 proceeds via a one-electron transfer:

$$Ti^{3+} + H_2O_2 \longrightarrow Ti^{4+} + OH^- + \cdot OH$$

This was the first reaction in which the formation of a hydroxyl radical from H_2O_2 was definitely established [69]. The probe which was used in this reaction was aromatic hydroxylation combined with EPR spectroscopy [254,255]. The hydroxyl radical adds to the benzene ring to yield the hydroxycyclohexadienyl radical, which was detected by EPR spectroscopy. The Ti^{3+}- H_2O_2 reaction is of no biological significance. Other transition metal ions which have been shown to react with H_2O_2 to give hydroxyl radicals are: Fe^{2+}, Cu^+, Co^{2+}-EDTA, Cr^{2+}, and Cr^{5+}. The best known of these reactions is the Fenton reaction involving Fe^{2+}. All reactions involving one-electron transfer are referred to as Fenton-type reactions.

a. The Fenton reaction

> All right. But do not plague yourself too anxiously;
> For just where no ideas are
> The proper word is never far.
> Mephisto in Goethe's Faust, Part I. *

The iron ion can exist in a number of different oxidation states. Therefore the oxidation of Fe^{2+} by H_2O_2 can *a priori* proceed via a one-electron or a two-electron transfer:

---

* The quotation from Goethe's Faust is from the translation by Walter Kaufmann, Anchor Books, Doubleday, New York, 1961.

$$Fe^{2+} + H_2O_2 \longrightarrow Fe^{3+} + OH^- + \cdot OH$$

$$Fe^{2+} + H_2O_2 \longrightarrow Fe(IV) + 2\,OH^-$$

The Fenton reaction is an inner-sphere one-electron transfer process. The H_2O_2 forms a complex with Fe^{2+} before electron transfer takes place:

$$Fe^{2+} + H_2O_2 \rightleftarrows [Fe^{2+}H_2O_2] \rightleftarrows Fe\,(OOH)^+ + H^+$$

$$Fe\,(OOH)^+ \longrightarrow FeO^+ + \cdot OH$$

The decomposition of $Fe\,(OOH)^+$ may also produce a ferryl ion:

$$Fe\,(OOH)^+ \xrightarrow{\ H^+\ } FeO^{2+} + H_2O$$

The ferryl ion has been postulated in monooxygenations, dehydrogenations, and dioxygenations in acetonitrile solutions [256].

The formation of a ferryl ion involves a two-electron transfer, since the iron in ferryl is formally in the +4 oxidation state. However, the reaction does not end at the Fe(IV) stage. The final product in the reaction is Fe^{3+}. It has been suggested that the ferryl ion is only stable in aprotic solution, but decomposes in water to yield hydroxyl radicals [257]:

$$Fe^{2+} + H_2O_2 \longrightarrow FeO^{2+} + H_2O \longrightarrow Fe^{3+} + OH^- + \cdot OH$$

I believe that this two-step mechanism can explain all observations. We shall come back to this two-step mechanism later on in the Cu^+ - H_2O_2 reaction.

In the Fenton reaction the hydroxyl radical is formed with Fe(III). It may be argued therefore that the hydroxyl radical is not "free", but complexed to Fe(III). This complex may have a lower reactivity and therefore a longer lifetime than the free hydroxyl formed via radiolysis.

The Fenton reaction was examined in great detail by Walling [39]. In a purely chemical system in dilute aqueous solutions at low pH (0-2) it was observed that the reactivity of a number of organic molecules were in excellent agreement with the rate constants determined by pulse radiolysis. The addition of ClO_4^- to increase the ionic strength had no effect on the rate data, thus indicating that the reactive species is electrically neutral and not a high valency iron species. In conclusion, the work of Walling has shown that the Fenton reaction in acidic solutions produces the hydroxyl radical or a species kinetically indistinguishable from hydroxyl. The formation of hydroxyl radical in a purely chemical system was also demonstrated using DMSO as a probe. Both in absence or presence of EDTA or DETAPAC the reaction gave quantitative yields of methane [239].

There is no doubt that in the Fenton reaction the hydroxyl radical is formed. However the question is: does the reaction occur *in vivo?* The formation of hydroxyl radicals can only be biologically significant if it can occur at the physiological pH of 7.4. To avoid precipitation of iron under these conditions one has to add a complexing agent (EDTA, DETAPAC). It appears that under these conditions the Fenton reaction is more complex than described above. The formation of ·OH in a purely chemical system [39] does not imply that in biological systems the hydroxyl radical is the only damaging free radical. Since the Fenton reaction is an inner-sphere electron transfer there can be many possible intermediates ($Fe(OOH)^+$, FeO^+, FeO^{2+}, $FeOH^{3+}$) which do or do not undergo the typical hydroxyl radical reactions, but nevertheless can react with important biomolecules to initiate cellular damage. Chemists love to argue about mechanistic details until they are blue in the face. The amount of literature on the mechanism of the Fenton reaction is therefore large. The interested reader is referred to some reviews [39,258,259]. The controversy concerning the mechanism of the Fenton reaction continues. Recently Sawyer et al. [260] claimed that there is no such thing as a free hydroxyl radical in Fenton-type reactions. An elegant response to this wild statement has been given by Walling [261] and by Ingold and coworkers [262], which hopefully will put to rest Sawyer's claim. The reaction $Cu^+ - H_2O_2$ with aromatics gave the same isomer distribution as those obtained via radiolysis [202], which is proof for free ·OH as good as it gets.

Very recently Goldstein and Meyerstein [262a] have commented on the complexity of the mechanism of Fenton-like reactions. The authors suggest that there is no general answer to the question of whether hydroxyl radicals are being formed in Fenton-like processes. In addition to hydroxyl radicals there may be other species involved. I shall come back to this problem in the discussion of the $Cu^+-H_2O_2$ reaction (pp. 84-88).

The damage caused by oxygen radicals or reactive oxygen species in a biological system may thus be initiated not only by $O_2^{\cdot -}/HO_2\cdot$ and hydroxyl radicals, but via a number of reactive oxygen species as shown in the reaction sequences above. The concept of "crypto" hydroxyl radical arose from the following observation: in the reaction of paraquat radical cation ($PQ^{+\cdot}$) with H_2O_2, which originally was suggested to produce hydroxyl radicals [218] (but does not), it was observed that in the presence of methionine, ethylene was formed. The formation of ethylene from methionine or methional has been used frequently as a probe for hydroxyl radicals. However in the mentioned reaction the formation of ethylene did not involve hydroxyl radicals, since well known hydroxyl radical scavengers like ethanol, mannitol and benzoate did not decrease the yield of ethylene. The intermediate which produces ethylene must be a species similiar to, but different from hydroxyl radical. It was therefore called "crypto" hydroxyl radical [263]. In summary, the idea of "crypto" hydroxyl radical was based on the following assumptions:

$$PQ^{+\cdot} + H_2O_2 \longrightarrow PQ^{2+} + OH^- + \cdot OH$$

and that the formation of ethylene from methionine is a reliable probe for hydroxyl radicals:

$$CH_3SCH_2CH_2CH(NH_2)COOH + \cdot OH \longrightarrow CH_2{=}CH_2 + products$$

It is interesting to note that both of these assumptions have been shown to be incorrect [220,247]. Therefore the idea of "crypto" hydroxyl radical is part of history [264]. Indeed the concept has been replaced with the idea of site-specific formation of hydroxyl radicals [265-267]. Evidence for a site-specific formation of ·OH was presented by Meneghini and Hoffmann [268] several years before the phrase "site-specific" was coined.

Biological systems, of course, are quite different from dilute aqueous solutions. If we form the hydroxyl radical via radiolysis we have a homogeneous solution of hydroxyl radicals which then can react with a variety of added scavengers. In a biological system the metal ion is associated with a biological macromolecule (a protein or DNA). The reaction with H_2O_2 does not take place in homogeneous solution, but close to the site of complexation. Since hydroxyl radical is highly reactive (diffusion controlled) we do not expect hydroxyl radicals to diffuse far from their site of formation and therefore added hydroxyl radical scavengers do not interfere in this site-specific reaction.

A site-specific metal-catalyzed oxidation of proteins has been suggested by Stadtman and Oliver [269]. These investigators made the following observations: 1.the inactivation of enzymes was insensitive to inhibition by radical scavengers (mannitol, formate, ethanol). 2. only one or only a few amino acid residues in a protein were modified by metal-catalyzed oxidation, whereas almost all amino acids are modified if reacted with hydroxyl radicals produced by radiolysis of water. 3. most of the enzymes which are highly sensitive to metal-catalyzed oxidations, require metal ions for their activity, indicating a metal ion-binding site. 4. deactivation of *E. coli* glutamine synthetase coincided with the loss of a single histidyl and a single arginyl residue per subunit, both of which are situated in close proximity to one of two divalent metal binding sites on the enzyme.

We are fortunate that for the formation of hydroxyl radicals we have available an independent source, namely the radiolysis of water and the photo-Fenton reagents. Ferryl ions (FeO^{2+}) on the other hand are not available from an independent source. The ferryl ion is produced under certain conditions (pH, buffer, complexing agent, solvent) in the Fenton reaction together with the hydroxyl radical. We can only deduce the formation of a species different from hydroxyl by indirect evidence.

If we use the spin-trapping technique we observe the DMPO/·OH signal. If in presence of ethanol this signal decreases and instead we observe the DMPO/$CH_3\dot{C}HOH$ signal we may consider the hydroxyl radical an intermediate. However, it has been observed that in some experiments involving Fe^{2+}-EDTA-H_2O_2 in phosphate buffer at pH 7.4 in presence of ethanol the DMPO/·OH signal decreased, but the DMPO/$CH_3\dot{C}HOH$ signal increased more by a factor of 4 [270]. This observation clearly indicates that the $CH_3\dot{C}HOH$ radical is formed at least partially via a radical different from hydroxyl (possibly the ferryl ion). This problem can be avoided by using DMSO instead of ethanol [236].

b. The Haber-Weiss reaction

As already mentioned there are only two pathways for hydroxyl radical formation from hydrogen peroxide: 1. One-electron transfer from metal ions (Fenton reaction and Fenton-type reactions) and 2. One-electron transfer from superoxide radical (the Haber-Weiss reaction). Both of these processes are important in biological systems. However, it was recognized that the Haber-Weiss reaction is a slow process. Kinetic studies showed that the rate constant for the Haber-Weiss reaction is in the neighborhood of 1-2 $M^{-1}s^{-1}$ [159]. It has been shown that the damaging effect of superoxide producing systems is observed only in presence of trace amounts of iron ions.

Most of the reagents used in biological experiments (like phosphate buffer) contain iron impurities [271,272] which catalyze the Haber-Weiss reaction. The radiolysis method gives us a unique source of superoxide. It was observed that superoxide produced via radiolysis of aqueous solutions of formate produces a reactive intermediate (as measured by deoxyribose degradation) only in presence of EDTA, but not in presence of other complexing agents [251]. The EDTA, by complexing with Fe^{3+}, lowers the redox potential, thus facilitating the one-electron transfer from $O_2^{·-}$ to Fe^{3+}:

$$O_2^{\cdot -} \; + \; Fe^{3+}\text{-EDTA} \longrightarrow O_2 \; + \; Fe^{2+}\text{-EDTA}$$

$$Fe^{2+}\text{-EDTA} \; + \; H_2O_2 \longrightarrow Fe^{3+}\text{-EDTA} \; + \; OH^- \; + \; \cdot OH$$

$$O_2^{\cdot -} \; + \; H_2O_2 \longrightarrow O_2 \; + \; OH^- \; + \; \cdot OH$$

The hydroxyl radical is produced free in solution where it can react with deoxyribose or scavenger. In this case the deoxyribose degradation was inhibited by a number of scavengers in perfect agreement with their hydroxyl radical rate constants.

However, if the superoxide radical is formed via the action of xanthine oxidase on hypoxanthine the results are quite different from the radiolytically generated superoxide radical [251]. In this case deoxyribose degradation was observed in absence of a chelating agent and in presence of EDTA. With other complexing agents, like DETAPAC, ADP, ATP, no degradation was observed. In the absence of EDTA the degradation was not inhibited by benzoate or DMSO, but was inhibited by mannitol, indicating a species different from hydroxyl radical We may call this oxidative species a "crypto" hydroxyl radical or the results can be explained by a site-specific formation of hydroxyl radicals.

In absence of any complexing agent the iron is not necessarily "free". It can complex to deoxyribose and therefore the hydroxyl radical will be formed in close proximity to deoxyribose, causing immediate reaction and degradation. Interference from added hydroxyl radical scavengers will therefore be impossible or will occur only at very high scavenger concentrations. However, if the scavenger can also complex Fe^{3+} then the scavenger will confer some protection to deoxyribose. Such a scavenger is mannitol.

In presence of EDTA, however, the degradation was inhibited by benzoate and DMSO. In the Fe^{3+}-EDTA complex the hydroxyl radical is also formed close to EDTA, but due to the open structure of the EDTA complex [273] the hydroxyl radical can diffuse into the bulk of the solution interacting with deoxyribose or added scavengers like DMSO.

c. *The reaction of Cu⁺ with hydrogen peroxide*

In addition to iron, copper is of biological significance. The reaction of Cu^+ with H_2O_2 has therefore been studied extensively. It is beyond the scope of the present introductory text to discuss all the investigations. Evidence for and against OH formation has been presented. I like to limit the discussion to some studies which show interesting similiarities with the Fenton reaction [202,229]:

$$Cu^+ + H_2O_2 \longrightarrow Cu^{2+} + OH^- + \cdot OH$$

Another mechanism involving a two-electron transfer was suggested [274]:

$$Cu^+ + H_2O_2 \longrightarrow Cu^{III}_{aq} + 2\ OH^-$$

The Cu^{III}_{aq} decomposes in water to give ·OH radical [275]:

$$Cu^{III}_{aq} \longrightarrow Cu^{2+} + H^+ + \cdot OH$$

The Cu^+-H_2O_2 reaction has been examined using two different methods for hydroxyl radical detection: 1. reaction with alcohols to give olefins [276]. 2. aromatic hydroxylation [202]. Both of these studies were carried out in unbuffered dilute aqueous solutions. We will briefly discuss the reactions involved in these two investigations.

Reaction with alcohols. The hydroxyl radicals produced via radiolysis react with ethanol and 2-propanol via H-abstraction [40]:

$$CH_3CH_2OH + \cdot OH \longrightarrow CH_3\dot{C}HOH\ (84.3\%)$$
$$\cdot CH_2CH_2OH\ (13.2\%)$$
$$\text{I}$$

$$(CH_3)_2CHOH + \cdot OH \longrightarrow (CH_3)_2\dot{C}OH\ (85.5\%)$$
$$\cdot CH_2CHOH\ (13.3\%)$$
$$\underset{|}{CH_3}$$
$$\text{II}$$

In presence of Cu^+ radicals I and II undergo reductive elimination (see Chapter 2):

$$\cdot CH_2CH_2OH + Cu^+ \longrightarrow Cu^{2+} + CH_2=CH_2 + OH^-$$

The reaction of II with Cu^+ yields propene [56-58]. It was observed that the ratio ethene/propene formed via radiolytically produced hydroxyl is different from the ratio observed with Cu^+-H_2O_2, thus indicating a species different from hydroxyl [276].

Aromatic hydroxylation [51,229]. The reactions of hydroxyl radicals with aromatics have already been discussed in detail. It was observed that the hydroxylation of nitrobenzene, fluorobenzene, toluene and anisole with Cu^+-H_2O_2 in unbuffered aqueous solutions gave the same isomer distribution as the one obtained via radiolysis under identical conditions [202]. This result clearly establishes the hydroxyl radical as an intermediate.

We have thus two apparently contradictory results. Both results can be reconciled by a two-step process. Like in the Fenton reaction the electron transfer proceeds via an inner-sphere process. Inner-sphere electron transfer proceeds via a bridged activated complex in which contact between the oxidant (electron acceptor) and the reductant (electron donor) is maintained by a ligand which is bonded to both. In our case the ligand is H_2O and the first step in the Cu^+-H_2O_2 reaction is formation of a complex:

$$Cu^+(H_2O)_m + H_2O_2 \longrightarrow Cu^+(H_2O)_m O_2H^- + H^+$$

$$Cu^+(H_2O)_m O_2H^- + H^+ \longrightarrow Cu^{2+}(H_2O)_m + \cdot OH + OH^-$$

Whenever the intermediate complex can react with an added detector molecule no hydroxyl radicals are formed. This is the case with alcohols. I have already discussed the fact that alcohols are not very reliable probes for hydroxyl radicals. On the other hand aromatic compounds are hydroxylated by hydroxyl radicals or via radical cations, which we can exclude in the present case, since they would give a different isomer distri-bution [49-51].

The Cu^+-H_2O_2 reaction leads us to an interesting conclusion: depending on the probe we obtain different answers! Both results can be reconciled by the reaction sequence above (complex formation, followed by decompostion to hydroxyl). The reaction is analogous to the two-step process discussed for the Fenton reaction.

So far we have basically two reactions leading to ·OH formation via transition metal ions: the Fe^{2+}-H_2O_2 and the Cu^+-H_2O_2 reaction. Hydroxyl radicals are expected to lead to DNA damage. The essential role of some metal ions in the formation of radicals and in radical-induced human dis-eases has been reviewed [258,277,278].

Since most metal ions are normally present in the higher oxidation state (Fe^{3+}, Cu^{2+}) these metal ions have to be reduced before they can act as Fenton catalysts. For this reduction there are plenty of compounds available *in vivo*, like NADH, NADPH, ascorbic acid, and thiol com-pounds. Many studies have reported the formation of hydroxyl radicals via metal ions/H_2O_2/reducing agents [180-185,279,280].

Halliwell and his group [281] have studied the DNA damage caused by Fe^{3+}-H_2O_2 and Cu^{2+}-H_2O_2 in absence and in presence of ascorbic acid . The DNA base damage was assessed by measuring typical base hydroxylation products, like cytosine glycol, thymine glycol, 8-hydroxyadenine and especially 8-hydroxyguanine (see Chapter 6). It was proposed that Cu^{2+} binds to DNA and reacts with ascorbic acid and H_2O_2 to give the ·OH radical, which then attacks the DNA base in a site-specific manner:

$$Cu^{2+} + AH_2 \longrightarrow Cu^+ + H^+ + AH·$$

$$Cu^+ + H_2O_2 \longrightarrow Cu^{2+} + OH^- + ·OH$$

Cu^{2+}-H_2O_2 at pH 7.4 causes far greater damage to the bases of DNA than the Fe^{3+}-H_2O_2 system. The addition of ascorbic acid to the Cu^{2+}-H_2O_2 system caused a greater increase in damage than in the Fe^{3+}-

H_2O_2 system. This high activity of Cu^{2+} compared to Fe^{3+} means that the availability of Cu^{2+} *in vivo* must be carefully controlled. Compared to iron-initiated damage, much less attention has been paid to Cu^{2+}.

After the formation of $\cdot OH$ has been clearly established both in dilute aqueous solutions [202] as well as *in vitro* experiments [281] the question arises: does the reaction occur *in vivo*? Experiments carried out by Mason and coworkers [282] have shown that hydroxyl radicals are formed *in vivo*. Rats given ascorbic acid and Cu^{2+} instilled into their stomachs showed OH formation by spin trapping with PBN (phenyl-N-t-butylnitrone). In presence of DMSO the PBN/$\cdot CH_3$ signal was observed, thus indicating that ascorbic acid under conditions prevalent in the stomach and in presence of metal ions produces hydroxyl radicals.

All of the above discussed results demonstrate the dual character of ascorbic acid. Ascorbic acid can act as a radical scavenger, but also as a prooxidant, leading to deleterious long-term effects [14].

Copper(II) is part of the acive site of the CuZnSOD enzyme. This enzyme is damaged by H_2O_2. We shall briefly discuss the mechanism of this deactivation. It was originally proposed by Hodgson and Fridovich [283] that the deactivation does not involve free $\cdot OH$. It was suggested that a Cu^{2+}-hydroxyl radical complex reacted with the histidine of the enzyme. More recent work by Sato et al. [284] showed that the deactivation proceeds in two stages. During the early phase the DMPO/$\cdot OH$ spin adduct was observed. This signal was unaffected by DMSO, and ethanol partly inhibited DMPO/$\cdot OH$ formation. This observation is again a clear indication that no free $\cdot OH$ is formed and that ethanol is not a reliable probe for $\cdot OH$ radical! The absence of any effect by DMSO shows that the DMPO/$\cdot OH$ signal is not formed via free $\cdot OH$. During the latter phase of the deactivation the DMPO/$\cdot OH$ signal was inhibited by DMSO and ethanol. This second phase indicates according to Sato et al. [284] a release of Cu^{2+} from the damaged enzyme, followed by formation of hydroxyl radicals:

$$Cu^{2+} + H_2O_2 \longrightarrow Cu^+ + H^+ + HO_2\cdot$$

$$Cu^+ + H_2O_2 \longrightarrow Cu^{2+} + OH^- + \cdot OH$$

These results show us again to be careful in the interpretation of spin-trapping results. The best probe for free $\cdot OH$ is the formation of the DMPO/$\cdot CH_3$ signal.

It has been shown that mutations to the CuZnSOD enzyme are responsible for the familial dominant form of amyotrophic lateral sclerosis (ALS or Lou Gehrig's disease) [285-287]. This disease is a fatal degenerative disorder of motor neurons in the brain and spinal cord. It has been suggested that the mutant enzyme binds copper ions less tightly and copper ions are therefore released from the enzyme [281,288].

Copper-dependent oxidative DNA damage was also reported for L-DOPA and dopamine. These compounds cause extensive damage in presence of H_2O_2 and traces of copper ions. 8-hydroxyguanine (8-OH-G) was the major product [289]. In these studies it was also observed that Fe^{3+} ions were much less effective in damaging DNA. It was proposed that copper release in presence of L-DOPA may represent an important pathway of neurotoxicity as in Parkinson's disease and ALS.

d. The reaction of Co^{2+} with hydrogen peroxide

Cobalt is an essential nutrient in the human diet, but as with most metals it is toxic at higher concentrations. High concentrations are carcinogenic and cause DNA damage among other effects. The chemical mechanism by which cobalt exerts its toxicity is formation of reactive oxygen species, like the hydroxyl radical.

The Co^{2+}-H_2O_2 reaction was therefore studied under a variety of conditions (pH, phosphate buffer, complexing agents) using a variety of hydroxyl radical detection techniques [290-293].

It is well known that in aqueous solutions containing no complexing agent the oxidation of Co^{2+} to Co^{3+} is very unfavorable [294]:

$$Co(H_2O)_6^{3+} \longrightarrow Co(H_2O)_6^{2+} \qquad E^0 = 1.84 \text{ V}$$

In presence of a complexing agent the redox potential is considerably lowered:

$$Co(NH_3)_6^{3+} \longrightarrow Co(NH_3)_6^{2+} \qquad E^0 = 0.1 \text{ V}$$

$Co(H_2O)_6^{3+}$ in aqueous solutions is such a powerful oxidizing agent that it oxidizes water to oxygen [295]. We therefore do not expect the formation of hydroxyl radicals in the reaction of uncomplexed Co^{2+} with H_2O_2. In presence of EDTA or DETAPAC, however, we observe the oxidation even by visual inspection [293]:

$$Co^{2+}\text{-EDTA} + H_2O_2 \longrightarrow Co^{3+}\text{-EDTA} + \cdot OH + OH^-$$

The Co^{2+}-EDTA has a weak pink color and the Co^{3+}-EDTA has a blue-violet color (absorption maximum at 532 nm). The formation of $\cdot OH$ was detected via its reaction with DMSO to give CH_4 [293] and via its reaction with tert. butyl alcohol to give isobutylene by reductive elimination [58,229]:

$$(CH_3)_3COH + \cdot OH \longrightarrow \cdot CH_2C(CH_3)_2OH + H_2O$$

$$\cdot CH_2C(CH_3)_2OH + Co^{2+}\text{-EDTA} \longrightarrow Co^{3+}\text{-EDTA} + (CH_3)_2C{=}CH_2 \\ + OH^-$$

The two techniques (DMSO and t-butanol) gave the same total yield of hydroxyl radical formation. All of the above reactions were carried out in aqueous solutions without any buffer.

The Co^{2+}-H_2O_2 reaction was also studied in phosphate buffer at pH 7.4 to simulate more closely physiological conditions. The DMPO spin-trapping technique gave the DMPO/$\cdot$OOH adduct in absence of any complexing agent, but the signal was much more intense in presence of nitrilotricarboxylic acid (NTA) [291, 292].

In another study the degradation of deoxyribose together with competition kinetics (mannitol, ethanol ,formate) and hydroxylation of phenol was used as hydroxyl radical probe [290]. Deoxyribose degradation as well as hydroxylation of phenol were observed. The spin-trapping results clearly show that these reactions are not caused by hydroxyl radicals. For purposes of discussion I like to argue for a two-step process as discussed previously for other Fenton-type reactions:

$$Co^{2+} + H_2O_2 \xrightarrow[pH\ 7.4]{PO_4^{3-}} X \longrightarrow \cdot OH$$

In this two-step process the results and conclusions will depend on the detection method employed. We have already discussed the fact that deoxyribose, mannitol and alcohols can react with a species different from hydroxyl radical to give the same products. So if we use deoxyribose degradation as a detector molecule we scavenge X and no ·OH is formed. Addition of competitive scavengers like mannitol or ethanol (which also react with X) will inhibit deoxyribose degradation. From this result we may conclude (wrongly) that ·OH is involved. We may have the following equality:

$k_{XD}/k_{XC} = k_{(OH)D}/k_{(OH)C}$ where k is the rate constant of X or ·OH with deoxyribose (D) or competitor (C)

The above equality shows that we cannot distinguish unequivocally between X and ·OH by competition kinetics.

With deoxyribose the degradation was inhibited by mannitol or ethanol, but higher concentrations of these competitors were needed to afford the same amount of protection in absence of EDTA. It appears that we obtain a different oxidizing species depending on the presence or absence of EDTA. In absence of EDTA we may call this species a 'crypto' hydroxyl radical [290] or the results can be explained by site-specific formation of hydroxyl radicals.

The Co^{2+} can form complexes with PO_4^{3-}, EDTA and deoxyribose. In the Co^{2+}-deoxyribose complex the hydroxyl radical will be formed close to deoxyribose and cause degradation. This site-specific formation of hydroxyl radicals is difficult to protect against. Higher concentrations of scavenger are required and the scavenger has to have the ability to penetrate the inner coordination sphere close to the metal.

An additional complexity may arise from iron contamination of phosphate buffer, and complexing agents like EDTA [271,272]. It was observed [293] that Fe^{3+} in trace amounts catalyzed the formation of Co^{3+} and hydroxyl radicals in the Co^{2+}-EDTA-H_2O_2-DMSO reaction:

$$Fe^{3+} + Co^{2+}\text{-EDTA} \longrightarrow Co^{3+}\text{-EDTA} + Fe^{2+}$$

$$Fe^{2+} + H_2O_2 \longrightarrow Fe^{3+} + OH^- + \cdot OH$$

In summary the formation of reactive oxygen species from Co^{2+} is complex and depends on the conditions, like pH, buffer, complexing agent and iron contamination. The formation of hydroxyl radicals most likely occurs in a site-specific manner, since only a fraction of $\cdot OH$ (ca. 14%) [293] has been found to escape into the bulk of the solution to be captured by DMSO.

e. The reaction of Cr^{2+} and Cr^{5+} with hydrogen peroxide

The Cr^{2+}-H_2O_2 has been examined using the same technique as already described for the Cu^+-H_2O_2 reaction (formation of olefins from ethanol and 2-propanol) [276]. It was determined that at pH 1.0 and 3.0 the Cr^{2+}-H_2O_2 reaction gave propene/ethene ratios identical to the ratio obtained in the radiolysis of aqueous solutions of 2-propanol and ethanol. This observation clearly establishes the hydroxyl radical as intermediate. The initially formed complex does not react with the alcohols, but decomposes to hydroxyl radical:

$$Cr^{2+}(H_2O)_5 \cdot O_2H^- + H_3O^+ \longrightarrow Cr^{3+}(H_2O)_5 + \cdot OH + OH^- + H_2O$$

This example again shows the importance of comparative studies (radiolytically versus chemically produced hydroxyl radicals) in unravelling complex reaction mechanisms.

The reaction of Cr^{5+} with H_2O_2 was studied using the DMPO spin trap [296]. The DMPO/·OH adduct was formed and the signal was decreased in presence of ethanol or formate and instead the DMPO/$CH_3\dot{C}HOH$ or DMPO/$CO_2^{·-}$ was observed. Cr^{VI} in presence of reducing agents, like glutathione or NADPH, has been found to damage DNA. This damage is initiated by reduction to Cr^V followed by a Fenton-type reaction to give hydroxyl radicals [297,298].

f. The reaction of some zero-valent metal powders with hydrogen peroxide

The decomposition of hydrogen peroxide to water and oxygen by a variety of metal powders has been known for a long time. The intermediate formation of the hydroxyl radical has been postulated by Weiss [299]:

$$M^0 + H_2O_2 \longrightarrow M^+ + ·OH + OH^-$$

where M^0 can be Pt, Au, Pd, Ag, and Zn.

More recently we have studied some of these reactions using the DMSO method for hydroxyl radical detection [229]. In the reaction Cu^0 - EDTA - H_2O_2 - DMSO a complete dissolution of the copper powder to form a blue solution of Cu^{2+}-EDTA was observed. The intermediate formation of hydroxyl radical in this reaction was demonstrated by its reaction with DMSO to give CH_4, and via aromatic hydroxylation. The hydroxylation of fluorobenzene gave the same isomer distribution as the reaction of Cu^+-EDTA-H_2O_2 and the radiolysis of fluorobenzene. On the basis of these observations the following mechanism can be postulated:

$$Cu^0 + EDTA + H_2O_2 \longrightarrow Cu^+\text{-EDTA} + ·OH + OH^-$$

$$\text{Cu}^+\text{-EDTA} + \text{H}_2\text{O}_2 \longrightarrow \text{Cu}^{2+}\text{-EDTA} + \cdot\text{OH} + \text{OH}^-$$

This mechanism requires the decomposition of 2 H_2O_2 for each Cu^{2+}-EDTA formed. A ratio of -H_2O_2/Cu^{2+}-EDTA of close to 2 was indeed observed.

In addition to copper the reaction of Cr^0-EDTA- H_2O_2 in the presence of DMSO was found to give high yields of CH_4 and CH_3CH_3. Contrary to the copper experiments the chromium did not go into solution and the mechanism is not as straightforward as in the Cu^0 case.

Whether these reactions involving Cu^0 and Cr^0 are important in promoting the toxic effect of these metal powders inhaled by metal workers has not been investigated. In searching through the literature one finds unfortunately a lot of incorrect terminology. The term 'metal' is ubiquitous in medical literature, when what is really meant is 'metal ions'.

g. The reaction of asbestos fibers with hydrogen peroxide

In addition to the examples of zero valent metal powders, asbestos fibers decompose hydrogen peroxide. The involvement of hydroxyl radicals in this decomposition was shown by spin trapping [300] and by its reaction with DMSO [301]. The reaction involves a one-electron transfer from an active metal containing site on the asbestos surface to hydrogen peroxide:

$$A + \text{H}_2\text{O}_2 \longrightarrow A^{\cdot+} + \cdot\text{OH} + \text{OH}^-$$

Treatment of calf thymus DNA with various kinds of asbestos fibers in the presence of hydrogen peroxide under physiological conditions (pH 7.4, 37°C) resulted in the hydroxylation of the C-8 position of guanine residues. DNA strand scissions were also observed [302].

A relationship between formation of hydroxyl radicals and the ability of these fibers to cause pneumoconiosis has been established [303]. The role of iron and reactive oxygen species in asbestos-induced lung injury has been reviewed [304-308].

94

Summary. The initially formed metabolites ($O_2 \cdot^-$ and H_2O_2) are not very reactive and have to be converted to more reactive species via secondary reactions [268,309-311]. Some of these reactions require catalytic amounts of metal ions (Fe^{2+}, Cu^+). The release of these metal ions from metal storage proteins like ferritin and ceruloplasmin therefore plays an important part in these transformations. Transferrin and ferritin keep iron tightly bound in the unreactive Fe(III) form. Ceruloplasmin binds Cu^{2+} and also oxidizes Fe(II) to Fe(III), which is the unreactive form of iron. The active form of iron is the Fe(II). The reduction can be accomplished by reducing agents, such as ascorbic acid, thiols, glutathione and polyphenolic compounds.
Semiquinone radicals react with oxygen to yield superoxide:

$$SQ \cdot^- + O_2 \longrightarrow Q + O_2 \cdot^-$$

The semiquinone radical of menadione and other redox cycling drugs, like anthracyclin, bipyridyl and nitroaromatic radical anions, as well as $O_2 \cdot^-$, $\cdot NO$ and polyphenols cause release of Fe(II) from ferritin and thus provide the necessary catalyst for $\cdot OH$ formation and DNA damage [312,313].

However, there are some reactions involving $O_2 \cdot^-$ and H_2O_2 which do not require metal ions. These reactions are the formation of $^1O_2$ and the formation of $\cdot OH$ via the Long-Bielski reaction. All of these reactions are summarized in Table 1.

The two oxygen-derived free radicals, the superoxide/hydroperoxyl and the hydroxyl radical have very different chemical characteristics. The hydroxyl radical reacts at very high rates (10^9 -10^{10} $M^{-1}s^{-1}$) with most molecules, whereas $O_2 \cdot^-/HO_2 \cdot$ is of much lower reactivity. The hydroxyl radical basically undergoes three types of reactions: 1. H-abstraction, 2. addition to multiple bonds and 3. one-electron transfer. The first two of these reactions converts the highly reactive hydroxyl into a radical of lower reactivity. The third type of reaction leads in the reaction involving metal ions to non-radical products:

$$\cdot OH + M^{n+} \longrightarrow OH^- + M^{+(n+1)}$$

where M^{n+} can be a number of different metal ions, such as Fe^{2+} or Cu^{2+} [25,99]. Some of these higher valency metal ions convert back to hydroxyl radicals [275]:

$$Cu(III)_{aq} \longrightarrow Cu(II) + H_3O^+ + \cdot OH$$

Table 1. Basic reactions involved in the formation of ROMs.

Reaction	References
$\cdot Fe^{2+} + H_2O_2 \longrightarrow Fe^{3+} + OH^- + \cdot OH$	39, 61
$Cu^+ + H_2O_2 \longrightarrow Cu^{2+} + OH^- + \cdot OH$	202
$O_2^{\cdot -} + H_2O_2 \longrightarrow {}^1O_2 + \cdot OH + OH^-$	62
$HOCl + H_2O_2 \longrightarrow {}^1O_2 + H_2O + HCl$	70
$HOCl + O_2^{\cdot -} \longrightarrow O_2 + \cdot OH + Cl^-$	224,225
$O_2^{\cdot -} + \cdot NO \longrightarrow {}^-OONO$	116
$O_2^{\cdot -} + ferritin \longrightarrow Fe (II) release$	158,314
semiquinone radicals + ferritin $\longrightarrow$ Fe (II) release	312,313
$\cdot NO + ferritin \longrightarrow Fe (II) release$	315
polyphenols + ferritin $\longrightarrow$ Fe (II) release	316

I have already discussed H-abstraction and addition to multiple bonds in Chapter 2. Due to the high reactivity of hydroxyl it is not surprising that hydroxyl radicals cause extensive damage to important biomolecules. H-abstraction from lipids leads to degradation, abstraction from enzymes leads to deactivation, and addition to the bases of DNA or H-abstraction from the sugar moieties leads to DNA damage. There are numerous ways by which $O_2^{\cdot-}$, H_2O_2 and $\cdot OH$ are produced via non-enzymatic reactions, via normal metabolism as well as via metabolism of xenobiotics. Background radiation constantly produces a low level of hydroxyl radicals from the radiolysis of water. Our cells are not only bombarded from radiation from outside our bodies, but also from the water within [21]. The trace amounts of tritium present in normal water provide a low level flux of hydroxyl radicals. It is obvious that radicals like hydroxyl pose a serious threat to the integrity of a cell. In the course of evolution aerobic organisms have developed defenses against this toxic threat. In the metabolic sequence: $O_2 \longrightarrow O_2^{\cdot-} \longrightarrow H_2O_2 \longrightarrow H_2O$ (or $\cdot OH$) we can interfere at various stages. The $O_2^{\cdot-}$ is removed by a series of enzymes known as superoxide dismutases (SODs), which catalyze the disproportionation of superoxide:

$$O_2^{\cdot-} \; + \; O_2^{\cdot-} \; \xrightarrow[\text{SOD}]{2\,H^+} \; H_2O_2 \; + \; O_2$$

The hydrogen peroxide is removed via the enzyme catalase or glutathione peroxidase:

$$H_2O_2 \; + H_2O_2 \; \xrightarrow{\text{CAT}} 2\,H_2O \; + \; O_2$$

$$H_2O_2 \; + \; 2\,GSH \; \xrightarrow{\text{GSH-Px}} \; GSSG \; + \; 2\,H_2O$$

Once the hydroxyl radical is formed via a series of reactions, there is little we can do to prevent its reaction with biomolecules. Since the hydroxyl radical is so highly reactive it will react in close proximity to its site of formation. If this site is close to the active site of an enzyme or close to DNA serious damage will result. This damage, however, can be repaired by a number of enzymatic repair processes. One way to prevent damage is via compounds known as antioxidants (or radical scavengers). These compounds can be present normally *in vivo*, like uric acid, or can be taken up through the diet (Vitamins A, C and E). These antioxidants are effective against ·OH only if they are close to the site of ·OH formation. All these reactions are summarized in Scheme 1.

Scheme 1.

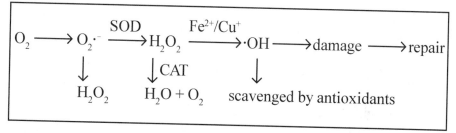

If the concentrations of CAT or GSH-Px are low, H_2O_2 will accumulate and subsequently cause damage via a site-specific formation of hydroxyl radicals. The enzymatic and antioxidant defenses will be discussed in Chapter 8.

Chapter 4

CHEMISTRY OF ELECTRONICALLY EXCITED STATES

The living state is the electronically desaturated state of protein.
When life originated three and a half billion years ago, our globe
was covered by a dense layer of water vapor. There was no light and
no oxygen near its surface.
  Albert Szent Györgyi in "The living state and cancer" [317]

A. SINGLET OXYGEN
1. Reactivity of oxygen
In order to understand the reactivity of molecular oxygen we have
to understand its electronic structure [318]. One of the interesting
characteristics of oxygen is the fact that it has two unpaired electrons
occupying two different molecular orbitals. This type of structure is
called a triplet state and means that oxygen is a diradical. Radicals are
usually very reactive trying to find another electron with which they
can pair. So it is quite surprising that oxygen does not react rapidly
with organic molecules and we do not suffer spontaneous combustion. The reason why this does not happen is the Pauli Exclusion Principle. Most organic molecules exist in the singlet state, which means
that all the electrons occupying the molecular orbitals have electrons
with antiparallel spin. The reaction of a singlet molecule with a triplet
molecule like oxygen to give a stable singlet product is spin forbidden
by the laws of quantum mechanics. This is the Pauli Exclusion Principle,
which states that no more than two electrons can occupy the same orbital
and that these two electrons must have opposite spin.

In order to overcome this spin problem we may move one electron to a higher molecular orbital with inversion of its spin. This
requires input of energy (ca. 23 Kcal/mole). This amount of energy
lies in the near infrared region of the spectrum (1268 nm). However,
triplet oxygen does not absorb in this region, so we cannot supply the
energy directly. We have to use a little trick. We take a dye molecule

(chlorophyll, porphyrins, methylene blue, rose bengal) which absorbs in the visible region and forms an electronically excited dye molecule. The excited dye molecule then transfers its energy to the triplet oxygen. This is called photosensitization [318]. The spin-inverted oxygen is called singlet oxygen or $^1O_2$. The electronic structures of the different oxygen species are oversimplified and schematically shown as follows:

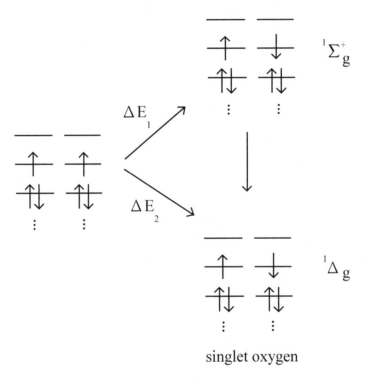

singlet oxygen

In one case the antiparallel spins are located in the same orbital ($^1\Delta_g O_2$), whereas in the second case the two electrons occupy two different orbitals ($^1\Sigma_g^+ O_2$). Both singlet states are formed, but $^1\Sigma_g^+$ decays rapidly to $^1\Delta_g$, which is the singlet oxygen encountered in chemical and biological systems. The electronic structure of triplet oxygen and different excited states of oxygen have been discussed in detail by Kasha and Bradham [318], who criticized the oversimplified diagrams familiar to most chemists. It is beyond the scope of this text to deal with this problem.

The formation of singlet oxygen is responsible for the toxicity of these dyes to cells and tissues. This toxicity is referred to as the photodynamic effect of dyes [319-322]. The dyes most commonly used are chlorophyll, porphyrins, methylene blue, and rose bengal. When living organisms absorb light in the presence of oxygen, oxidation processes may occur and often lead to damaging biological effects. This type of "photodynamic effect" has been observed with many organisms and has been extensively reviewed [320,321,323-325]. In these systems a wide variety of short-lived reactive species are possibly involved in addition to singlet oxygen.

2. Chemical formation, reactions and identification of singlet oxygen [326]

The photosensitized formation of singlet oxygen can hardly occur *in vivo*, except in reactions involving epithelial cells. Thus, the discovery of chemical reactions, which produce singlet oxygen, is of great importance in biology. These reactions give us the possibility of carrying out photochemistry without light. Some of these reactions are of synthetic utility. The methods which might be of biological significance are the hydrogen peroxide-hypochlorite reaction [70], the disproportionation of superoxide radical anion [72] or some peroxyl radicals [73,74,326], and the reaction of hypochlorous acid with superoxide [224,225]:

$$H_2O_2 + OCl^- \longrightarrow H_2O + Cl^- + {}^1O_2$$

$$O_2^{\cdot-} + O_2^{\cdot-} \xrightarrow{2\,H^+} H_2O_2 + {}^1O_2$$

$$ROO\cdot + ROO\cdot \longrightarrow 2RO\cdot + {}^1O_2 \quad [327]$$

$$HOCl + O_2^{\cdot-} \longrightarrow Cl^- + \cdot OH + {}^1O_2$$

Whether these reactions are of significance in biological systems has been the subject of considerable controversy. Progress in answering this question has been slow and did not proceed straightforwardly. The reason for the controversy is the same as we have encountered in the identification of hydroxyl radicals, namely the availability of a sensitive and specific probe for singlet oxygen. The subject has been reviewed periodically [96,328-330].

As with the identification of hydroxyl radical, the most frequently used method for singlet oxygen identification utilizes the formation of an easily identifiable oxidation product. The literature on singlet oxygen reactions is enormous and beyond the scope of the present text [324,331-335].

Singlet oxygen reacts with olefins to give synthetically useful hydroperoxides [336] and endoperoxides [337]:

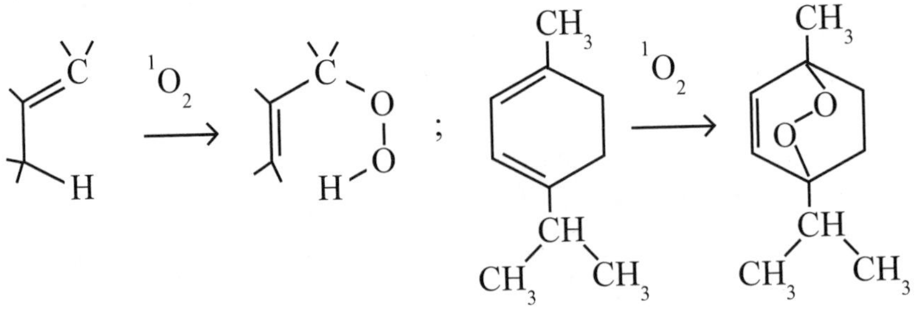

These peroxides react with metal ions to give oxygen radicals, which in turn can initiate further oxidations.

A commonly used reaction for singlet oxygen identification is the reaction with furan derivatives [338,339], which however has fallen in disrepute since ·OH and $SO_4^{·-}$ have been shown to give the same products [340,341].

A very specific detector molecule which has been used is cholesterol [342]. Cholesterol reacts with singlet oxygen to give 3β-hydroxy-5α-cholest-6-ene-5-hydroperoxide:

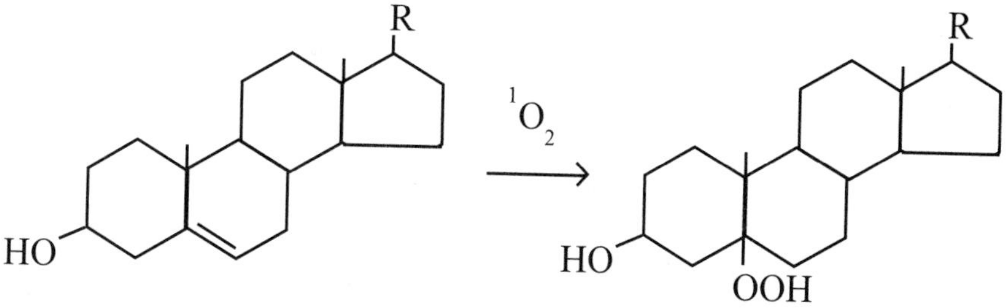

This product has not been formed in any other radical oxidations and has, therefore, been considered as a highly specific monitor of singlet oxygen. However, cholesterol possesses low sensitivity. Activated neutrophils and macrophages in presence of C^{14} labeled cholesterol gave only small yields of the product (ca. 0.2% of oxygen consumed/1.25×10^6 PMNs) [343] leading to the erroneous conclusion that singlet oxygen is at best of minor importance in the respiratory burst of phagocytes.

Singlet oxygen is an electronically excited molecule, which decays to the ground state oxygen with emission of light. The emission (chemiluminescence) in this case is at 1268 nm [318]. There have been many reports in which chemiluminescence has been observed in activated phagocytes. However, in the absence of spectral analysis this only tells us that an excited molecule had been formed, but does not conclusively pinpoint the singlet oxygen. In several instances spectral analysis has been attempted, but no emission at 1268 nm was observed. This absence of the 1268 nm emission may be due to the low sensitivity of the spectroscopic method or competing deactivating effects of quenching.

In the dismutation of superoxide, the absence of a chemiluminescence band at 1268 nm or the low yield of 3β-hydroxy-5α-cholest-6-ene-5-hydroperoxide from cholesterol have been interpreted to mean that singlet oxygen is not involved [343,344]. These conclusions were, however, based on an incomplete understanding of the properties of singlet oxygen and its rate of reaction.

In addition to reactions leading to oxidation products, singlet oxygen may be deactivated by a number of compounds via energy transfer without formation of any product. This process is known as quenching [345]:

$$^1O_2 \xrightarrow{\text{quencher}} \quad ^3O_2 \quad + \quad \text{energy}$$

Some well-known quenchers are β-carotene, 1,4-diazabicyclo[2,2,2] octane (DABCO), superoxide radical anion, and superoxide dismutase.

$$^1O_2 \xrightarrow{O_2^{\cdot-}} \quad ^3O_2 \quad + \quad 22 \text{ kcal/mole}\;;\; k_q = 1.6 \times 10^9 \text{ M}^{-1}\text{ s}^{-1}$$

The quenching rate constant of superoxide is quite high [346], but is even higher for β-carotene (10^{10} M^{-1}s^{-1}) [347]. The formation of 3 β-hydroxy-5α-cholest-6-ene-5-hydroperoxide from cholesterol, although highly sensitive proceeds at a much slower rate ($k = 8.44 \times 10^4$ M^{-1}s^{-1}) [348]. Whenever there is a low rate for product formation, the quenching by superoxide radical anion will dominate and give low yields of oxidation product, thus, leading to the erroneous conclusion of no singlet oxygen formation. In the case of PMNs stimulated by PMA (10 mg/10^7 cells) much larger amounts of $O_2^{\cdot-}$ were obtained, compared to the amount generated by physiological stimuli, such as opsonized bacteria. This example shows the importance of quantitative information, such as concentrations and rate constants of all possible competing reactions.

There are two major pathways for $^1O_2$ quenching, namely energy and electron transfer. Energy transfer is usually very fast (2×10^{10} M^{-1}s^{-1}), whereas electron transfer is slower (10^9 or less) [345]. A well documented example of energy transfer is the quenching by β-carotene, which has a rate constant of 1.3×10^{10}M^{-1}s^{-1} [347]:

$$^1O_2 + Q \longrightarrow Q^* + {}^3O_2$$

Electron transfer involves the interaction between the electron poor $^1O_2$ with an electron rich substrate to form a charge transfer complex, which on electron transfer may lead to the superoxide radical anion, besides physical quenching to afford $^3O_2$, or chemical reaction to the DO_2 oxidation product:

$$D + {}^1O_2 \rightleftharpoons [D^{+\cdots}O_2{}^{\cdot-}] \nearrow D^{\cdot+} + O_2{}^{\cdot-}$$

charge transfer complex $\longrightarrow D + {}^3O_2$

$\searrow DO_2$

This method for the formation of superoxide radical anion has already been discussed (Chapter 3).

Another probe for singlet oxygen is based on the fact that the lifetime of singlet oxygen is very sensitive to the solvent. D_2O compared to H_2O increases the lifetime considerably (by a factor of up to 15) [349], and increases singlet oxygen-induced damage. The replacement of H_2O by D_2O, therefore, provides us with a convenient diagnostic test for the involvement of singlet oxygen in photooxidations, enzymatic oxidations, chemiluminescence and other photophysical processes.

A relatively sensitive method for singlet oxygen detection is its reaction with 9,10-diphenylanthracene:

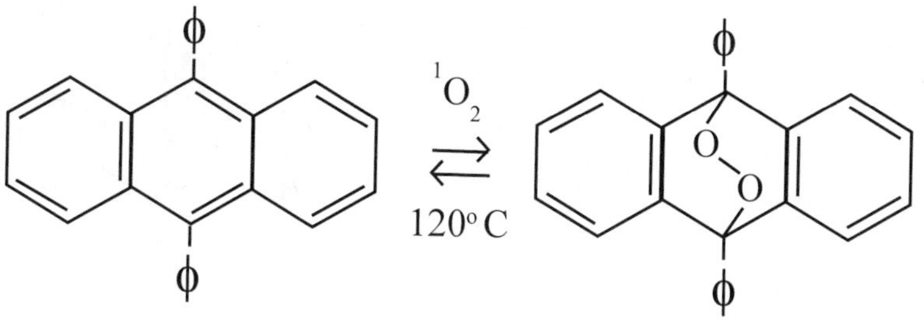

This probe was used by Khan and coworkers [97,98] in their study of activated neutrophils and macrophages (see Chapter 5 under phagocytosis). Singlet oxygen reacts with DPA (k= 1.3 x 10^6 M^{-1}s^{-1}) to give the corresponding endoperoxide (measured as a decrease of the DPA absorption at 355 nm). This reaction rate is about 100 times faster than the rate with cholesterol. The quenching reaction has therefore less chance to compete with the chemical oxidation reaction. Upon heating the endoperoxide to 120°C, the singlet oxygen is regenerated. In the presence of β-carotene, (an efficient quencher of singlet oxygen), no decrease in the DPA absorption was observed. The results showed that activated neutrophils and macrophages produce singlet oxygen in high yields (about 19% of the oxygen consumed) via dismutation of superoxide radical anion. It is interesting to note that the DPA probe was used long ago by Howard and Ingold [74] to demonstrate the formation of singlet oxygen in the dismutation of sec.-butylperoxyl radicals (the Russell mechanism) *in vitro*.

Another sensitive and specific singlet oxygen probe is the formation of the two diasteromers (4R and 4S) of 4,8-dihydro-4-hydroxy-8-

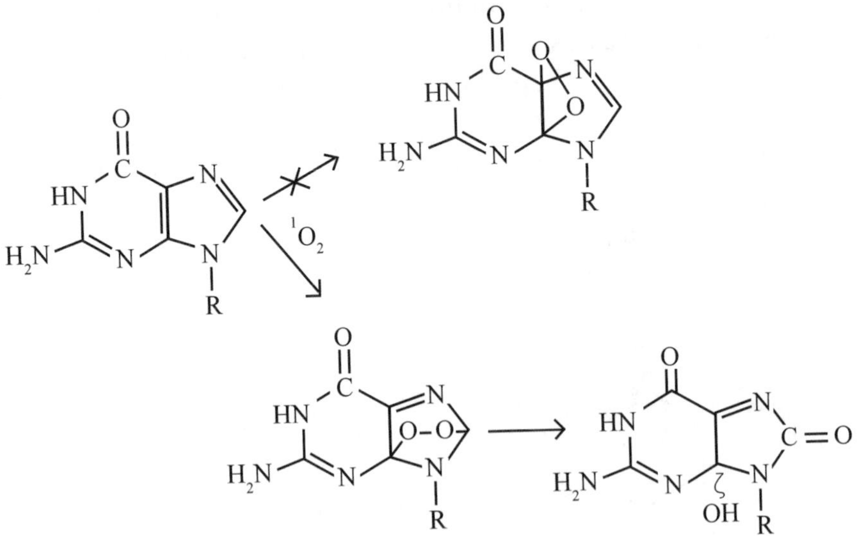

Fig. 1. The reaction of 2'-deoxyguanosine with $^1O_2$.

oxo-2'-deoxyguanosine (4-HO-8-oxo-dGuo) upon reaction of singlet oxygen with 2'-deoxyguanosine [350,351]. Singlet oxygen reacts with DNA almost exclusively at the guanine residue and the 5-membered imidazole ring. Both [4+2] and [2+2] cycloaddition to form endoperoxides and dioxetanes have been suggested [352-354]. However, these intermediates remained speculation until Sheu and Foote identified them [355,356] (see Fig 1).

In the reaction of unsubstituted 2'-deoxyguanosine with singlet oxygen no evidence for [2+2] cycloaddition was observed [355]. However, in the reaction with 8-hydroxy-2'-deoxyguanosine dioxetanes have been observed at -80° C [356]. The latter result is due to the preferred diketo structure of 8-HO-2'-deoxyguanosine [357,358]:

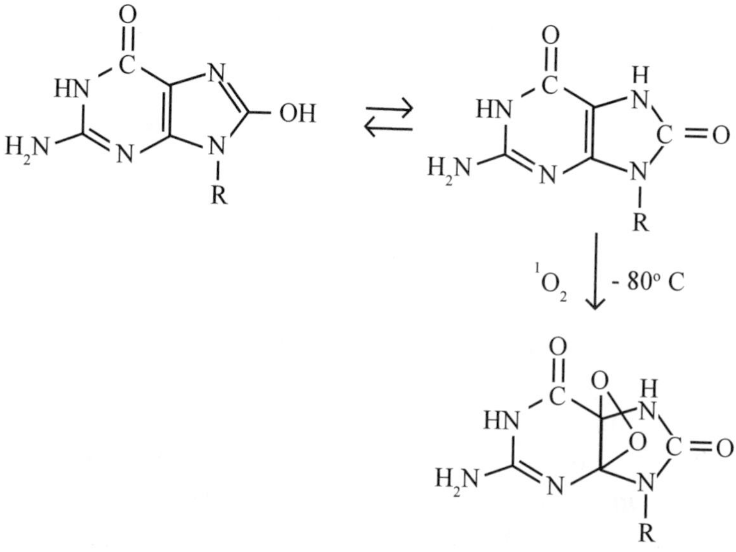

2-diasteromers

The 4-HO-8-oxo-dGuo has been used as a marker for oxidative modification of purine bases, nucleosides and isolated DNA with peroxynitrite [223]. The comparison of the product profile with the one obtained by ionizing radiation (hydroxyl radicals), led to the conclusion that the hydroxyl radical is not the main oxidizing species in the decomposition of peroxynitrite [223]. Using the 4-HO-8-oxo-dGuo as a marker, it has been shown that the reaction of HOONO with H_2O_2 yields singlet oxygen [223].

a. The hydrogen peroxide-hypochlorite reaction

A frequently used chemical method for the formation of singlet oxygen is the hydrogen peroxide-hypochlorite reaction. The chemiluminescence of this reaction has been studied in detail. The two major emissions at 638 nm and 703 nm have been assigned to the so called bimol emission of the singlet oxygen [359]. The reaction has been shown to proceed via the following steps [360]:

$$OCl^- + H_2O_2 \longrightarrow HOCl + HO_2^-$$

$$HOCl + HO_2^- \longrightarrow HOOCl + OH^-$$

$$HOOCl + OH^- \longrightarrow H_2O + {}^-OOCl$$

$$^-OOCl \longrightarrow {}^1O_2 + Cl^-$$

Although chemiluminescence was reported in the hypochlorite-hydrogen peroxide reaction by several investigators (for a review see [326]) it was not until 1963 that Khan and Kasha identified the liberated species as singlet oxygen [70]. The complicated history of rediscoveries and reinterpretations has been carefully traced [334,359,360]. Following the interpretation by Khan and Kasha [70] singlet oxygen became a subject of intense investigations.

This chemical method has been shown to react with a variety of organic compounds to give the same products as those obtained via the photooxygenation method [361,362].

b. Other singlet oxygen forming reactions

A very important *in vivo* reaction for the formation of singlet oxygen is the dismutation of superoxide radical anion. This reaction was first proposed by Khan [72] in a purely chemical reaction (KO_2

in DMSO/H$_2$O). However, the biological importance of this reaction has been questioned. It was concluded by some investigators that the formation of singlet oxygen via dismutation of the superoxide radical anion is only of minor significance in biological systems [343,344]. These conclusions, however, have been shown to be incorrect for reasons already discussed above.

The formation of singlet oxygen via dismutation of sec. butyl peroxyl radicals has been demonstrated using DPA as a singlet oxygen probe [327]. This is the well-known Russell mechanism, which I have already discussed (Chapter 3). Therefore, it appears possible that enzyme-catalyzed formation of singlet oxygen in microsomal lipid peroxidation is formed via dismutation of lipid peroxyl radicals rather than via dismutation of superoxide.

Since phagocytes are the most important source of singlet oxygen I shall discuss the different reactions leading to singlet oxygen *in vivo* in Chapter 5. Although the formation of singlet oxygen *in vivo* has been controversial, it is now an established fact. In enzyme-catalyzed reactions [363-367] as well as in cells, like eosinophils, neutrophils and macrophages [97,98,368], it was shown that singlet oxygen is produced.

B. EXCITED CARBONYL COMPOUNDS
1. Dioxetanes and dioxetanones

Damage to biological systems can arise from the reactions of oxygen radicals like O$_2$·$^-$/HO$_2$· and ·OH, but also via electronically excited triplet states, which also have radical character. There are numerous ways by which these species may be formed in purely chemical as well as in biological systems. I have discussed this in previous chapters. I have also discussed the formation of electronically excited states via photochemical methods. This requires light and it is difficult to see how these excited species may play a role in biological systems except for reactions involving epithelial cells of the skin. The identification of singlet oxygen in a purely chemical reaction by Kasha and Khan [70] was therefore of fundamental importance. It showed that

110

the formation of electronically excited states can take place via a purely chemical reaction without the necessity of light. In other words, photochemistry may be carried out in the dark. Of course, there is considerable evidence available, which indicates that electronically excited species are produced in a biological system. Electronically excited states are responsible for bioluminescence [369]. The concept of a 4-membered ring containing an -O-O- bond was discussed long ago as a possible intermediate in the oxidation of olefins to carbonyl compounds:

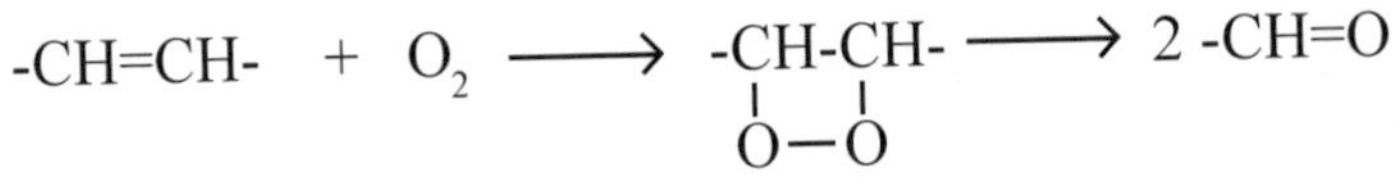

$$-CH=CH- \; + \; O_2 \longrightarrow -\underset{\underset{O-O}{|\quad\;|}}{CH-CH}- \longrightarrow 2\,-CH=O$$

However, all these oxidations have been shown to proceed via radical chain reactions not involving 1,2-dioxetanes, but rather hydroperoxides [370]. Some of these studies involving erroneously identified 1,2-dioxetanes have been reviewed by Schaap and Zaklika [371].

The interaction of $^1O_2$ with alkyl-substituted olefins may result in the formation of allylic hydroperoxides with a shifted double bond:

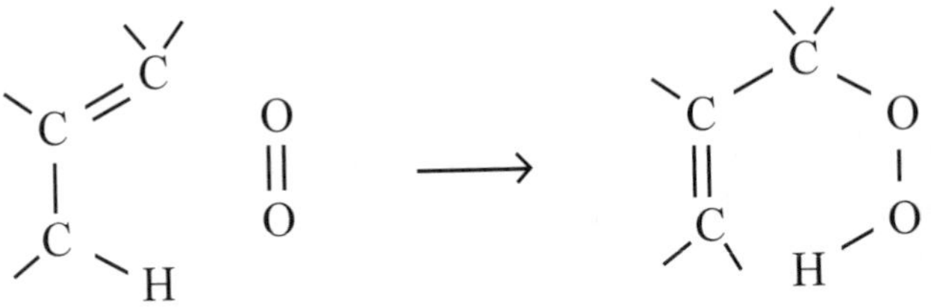

This reaction was discovered by Schenck in 1943 [372] in a patent application and was named by Schönberg [373] as the "Schenck reaction". Nowadays it is mostly referred to as the "-ene-"reaction. *Sic transit gloria mundi.* The inevitably shifted double bond clearly distinguishes the ene-type reaction from the well-known autoxidation reaction (see Chapter 6).

However, other intermediates, the 1,2-dioxetanes, appear to be important in the formation of carbonyl containing products that are frequently obtained in the reaction of conjugated olefins with singlet oxygen:

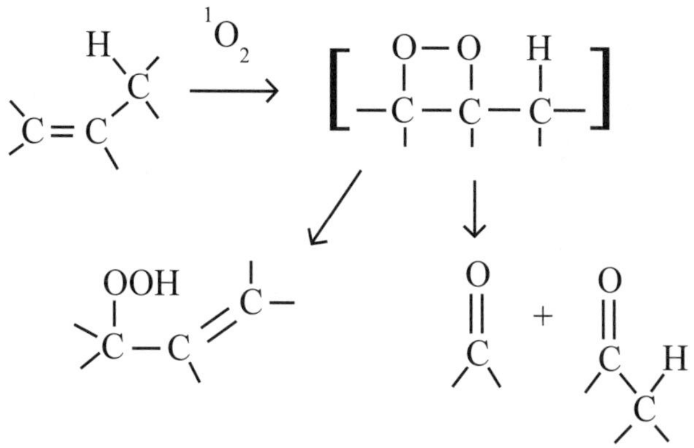

The evidence in favor of 1, 2-dioxetanes has been reviewed by Kearns et al. [374]. Interest in dioxetanes was itensified with the suggestion by McCapra [375] that these strained high energy peroxides might decompose concertedly to form electronically excited products, which would emit light, such as is observed in many chemiluminescence and bioluminescence reactions.

In 1968 and 1969 several groups reported that olefins without allylic atoms were oxidatively cleaved by singlet oxygen [376-378]. One example is the photooxygenation of enamines studied by Foote and Lin [376]:

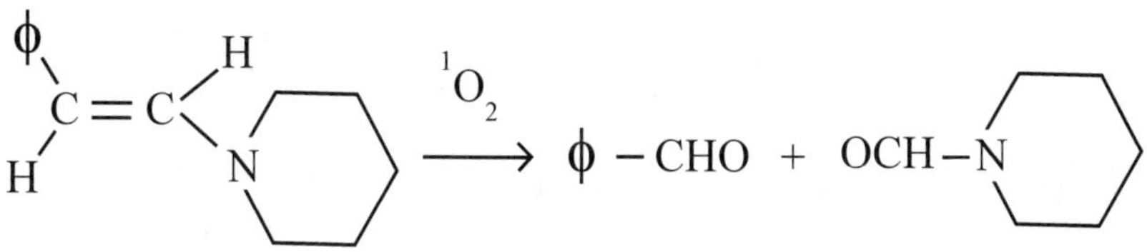

However, the proposals of 1,2-dioxetanes as intermediates in these cleavage reactions remained a hypothesis until the first synthesis of a 1,2-dioxetane by Kopecky and Mumford [75,379].

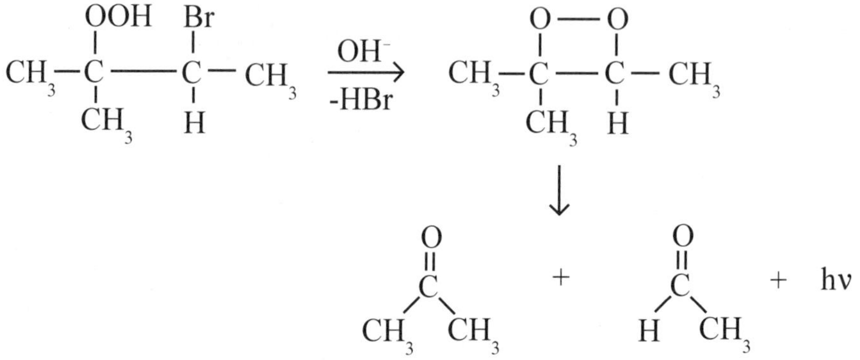

$$
\underset{CH_3-\underset{\underset{CH_3}{|}}{\overset{\overset{OOH}{|}}{C}}-\underset{\underset{H}{|}}{\overset{\overset{Br}{|}}{C}}-CH_3}{} \xrightarrow[-HBr]{OH^-} \underset{CH_3-\underset{\underset{CH_3}{|}}{\overset{\overset{O-O}{|}}{C}}-\underset{\underset{H}{|}}{\overset{\overset{}{|}}{C}}-CH_3}{}
$$

$$\downarrow$$

$$
\underset{CH_3 \quad CH_3}{\overset{O}{\underset{}{\overset{\|}{C}}}} \quad + \quad \underset{H \quad CH_3}{\overset{O}{\underset{}{\overset{\|}{C}}}} \quad + \quad h\nu
$$

The thermal decomposition of this trimethyl-1, 2-dioxetane (TMD) showed fluoroescence ascribed to an excited carbonyl product.

Shortly after Kopecky's report, Bartlett and coworkers [380,381] were successful in isolating a 1,2-dioxetane from the 1,2-cycloaddition of singlet oxygen to an olefin:

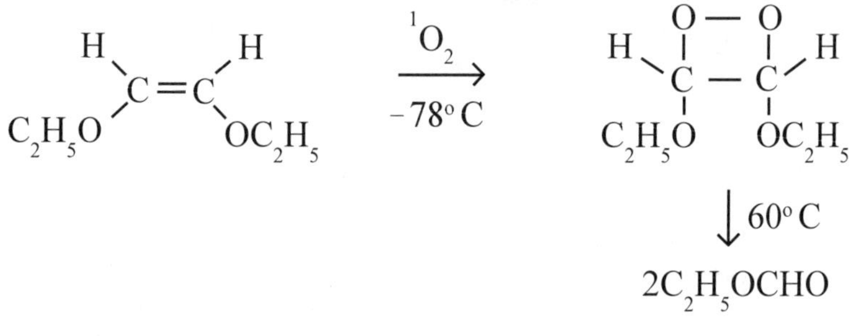

$$
\underset{C_2H_5O \qquad OC_2H_5}{\overset{H \qquad H}{C=C}} \quad \xrightarrow[-78°C]{^1O_2} \quad \underset{C_2H_5O \qquad OC_2H_5}{\overset{H \quad O-O \quad H}{\underset{}{\overset{}{C-C}}}}
$$

$$\downarrow 60°C$$

$$2C_2H_5OCHO$$

Subsequently many 1,2-dioxetanes have been synthesized and their decomposition studied in great detail [331]. Also there are now numerous examples of 1,2-dioxetane formation via the reactions involving olefins and singlet oxygen. The 1,2-dioxetanes cover a wide range of stabilities from only minutes at room temperature to hours at

160°C. This high stability of some dioxetanes is indeed surprising. For the formation of excited triplet states via dioxetanes to be of biological importance one requirement is that the 1,2-dioxetane decomposes at 37°C at a reasonable rate.

The thermal decomposition of trimethyldioxetane produces excited triplet carbonyl compounds. The excited state may undergo a variety of typical excited state reactions (energy transfer, electron transfer, and hydrogen abstraction). In the case of energy transfer the excited acceptor deactivates by emission of light or suffers photochemical change (Scheme 1).

Scheme 1

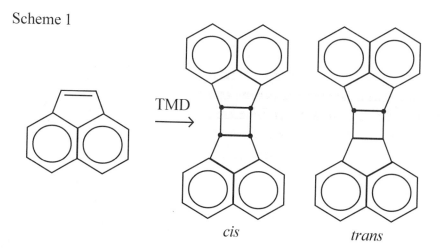

cis   trans

This reaction (dimerization of acenaphthylene) was first used by White in 1969 [76] to carry out "photochemistry without light". The cis/trans ratio was identical to the one obtained via the acetone photosensitized dimerization, thus clearly confirming the involvement of triplet acetone in the reaction.

Recently it has been observed by Adam and coworkers [382] that some excited triplet ketones derived from some hydroxymethyl-substituted 1,2-dioxetanes decompose by α-cleavage to yield radicals, which damage DNA:

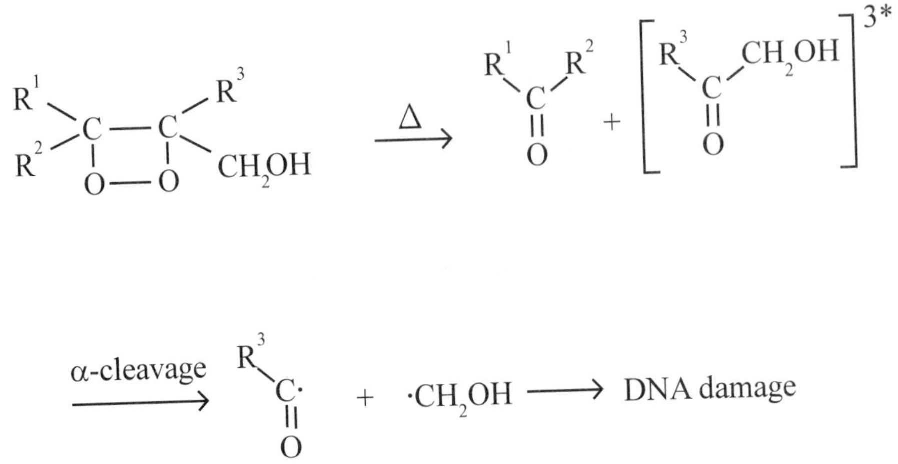

$$R^1\!\!\!\!\diagup^{\!C}\!\!\diagdown\!\!\!\!\diagup^{R^3} \xrightarrow{\;\Delta\;} \underset{O}{\overset{R^1\quad R^2}{C}} + \left[\underset{O}{\overset{R^3\quad CH_2OH}{C}}\right]^{3*}$$

$$\xrightarrow{\alpha\text{-cleavage}} \underset{O}{\overset{R^3}{C\cdot}} + \cdot CH_2OH \longrightarrow DNA\ damage$$

The radicals have been identified by DMPO-spin-trapping. The 1,2-dioxetanes, therefore, not only provide us with electronically excited states, which allow us to carry out photochemistry without light, but also are convenient sources of radicals in biological systems.

The yields of excited singlet (f_s) and triplet (f_t) carbonyl fragments from several dioxetanes have been determined by Turro et al [383] by evaluating the yields of chemiluminescence generated during the thermolysis of the dioxetanes in the presence of added fluorescers, such as 9,10-diphenylanthracene (DPA) and 9,10-dibromo-anthracene (DBA). The yields of excited singlet and triplet acetone from trimethyldioxetane were determined independently by "chemical titration" with *trans*-dicyanoethylene [384,385]. This reagent reacts with ^1A (singlet acetone) to yield only oxetane and with ^3A (triplet acetone) to yield only *cis*-dicyanoethylene and a ground state acetone molecule [385]:

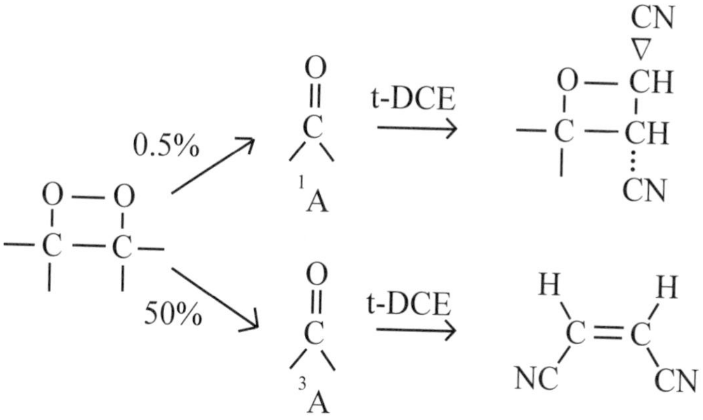

$$
\begin{array}{c}
\text{O}-\text{O} \\
\mid \quad \mid \\
-\text{C}-\text{C}- \\
\mid \quad \mid
\end{array}
\begin{array}{l}
\xrightarrow{0.5\%} \quad {}^{1}\text{A} \quad \xrightarrow{\text{t-DCE}} \\[2em]
\xrightarrow{50\%} \quad {}^{3}\text{A} \quad \xrightarrow{\text{t-DCE}}
\end{array}
$$

The results showed that the decomposition of dioxetanes produces almost exclusively the excited triplet state [384]. More recently the 3-hydroxymethyl-3,4,4-trimethyl-1,2-dioxetane has been shown to yield, in addition to excited carbonyl compounds, low yields of singlet oxygen [386].

Another source of excited carbonyl compounds are the dioxetanones (α-peroxylactones) as characterized by Adam and Liu [78]:

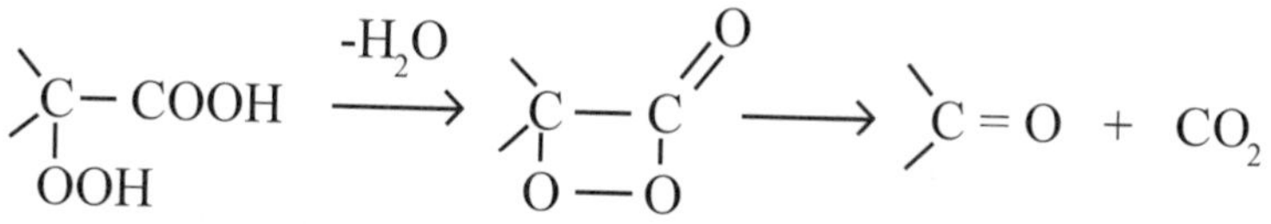

$$
\begin{array}{c}
\diagdown \\
\diagup\text{C}-\text{COOH} \\
\mid \\
\text{OOH}
\end{array}
\xrightarrow{-\text{H}_2\text{O}}
\begin{array}{c}
\diagdown \qquad \diagup\!\!\diagup\text{O} \\
\diagup\text{C}-\text{C} \\
\mid \qquad \mid \\
\text{O}-\text{O}
\end{array}
\longrightarrow
\begin{array}{c}
\diagdown \\
\diagup\text{C}=\text{O} \quad + \quad \text{CO}_2
\end{array}
$$

This type of intermediate has been observed in the enzymatic oxidation of carboxylic acid, which yields the next lower homologous aldehyde [82].

The formation of excited carbonyl compounds via dismutation of sec. alkylperoxyl radicals (the Russell mechanism) has already been discussed (Chapter 3).

Chapter 5

FORMATION OF REACTIVE OXYGEN METABOLITES *IN VIVO*

...the main actors of life are electrons.
...when electrons are passed from donor to acceptor energy will be
liberated and entropy will be increased. It is this that drives life.
Albert Szent-Györgyi in "The living state and cancer"[317]

A. INTRODUCTION

Since triplet oxygen reacts very slowly or not at all with most organic compounds, the reactions of oxygen *in vivo* are catalyzed by a huge variety of oxygen metabolizing enzymes.

Although many enzymes are composed of only a protein, others, including all oxidation-reduction enzymes (oxidoreductases), require the presence of a cofactor or coenzyme [387]. These cofactors or coenzymes are essential for the catalytic activity of the enzyme. Some of these coenzymes are tightly bound (covalently) to the protein. These coenzymes are known as prosthetic groups. Another group of cofactors, which are not covalently linked, is consumed in stochiometric amounts during the reaction. These coenzymes are low molecular weight compounds, which in the case of oxidoreductases are electron or hydrogen donors or acceptors. Many oxidations require the cooxidation of substrate and coenzyme. The most commonly used cofactors in these oxidations are NAD^+ (nicotineamide-adenine-dinucleotide) or $NADP^+$ (nicotineamide-adenine-dinucleotide phosphate).

There are several types of oxidation-reduction enzymes subdivided on the basis of the type of substrate and the type of electron or hydrogen acceptor (or donor) utilized. Enzymes that catalyze dehydrogenation of a substrate and employ NAD^+ or $NADP^+$ as a cosubstrate are called dehydrogenases. When molecular oxygen acts as the hydrogen acceptor (forming H_2O or H_2O_2) the enzymes are known as oxidases. Another group of enzymes that use H_2O_2 or ROOH as the oxidizing agent are called peroxidases or hydroperoxidases. Oxygen can also be incorporated into the substrate. These enzymes are called oxygenases. The oxygenases are subdivided into two groups: Enzymes that catalyze the incorporation of both oxygen atoms of O_2 into the substrate are called dioxygenases, while enzymes that catalyze the incorporation of only one oxygen atom are called monooxygenases:

$$S + O_2 \xrightarrow{\text{dioxygenase}} SO_2$$

$$S + NADH + O_2 + H^+ \xrightarrow{\text{monooxygenase}} SO + NAD^+ + H_2O$$

Monooxygenases require a cosubstrate to accept the second oxygen atom. These enzymes are, therefore, also known as mixed function oxidases.

I now shall discuss the enzymatic formation of superoxide radical anion. The most important source for formation of $O_2 \cdot^-$ and H_2O_2 is the electron transport chain of mitochondria [85,388,389]. Other cellular sources of $O_2 \cdot^-$ and/or H_2O_2 involve the microsomal electron transfer chain that includes NADPH-cytochrome P450 and the NADPH oxidases in phagocytes [390]. One of these oxidases in phagocytes uses NADH as a cofactor and produces $O_2 \cdot^-$ and the other oxidase uses NADPH and produces both $O_2 \cdot^-$ and H_2O_2. The superoxide radical anion formed during the respiratory burst of phagocytes originates mainly from the NADPH-dependent enzyme [390]. In addition to $O_2 \cdot^-$ and H_2O_2 phagocytes produce $\cdot OH$, $^1O_2$, $\cdot NO$ and ^-OONO. Additional sources of $O_2 \cdot^-$ or H_2O_2 or both are xanthine oxidase, aldehyde oxidase, monoamine oxidase, urate oxidase and L- and D-aminoacid oxidases [388].

The total fraction of oxygen reduced to $O_2 \cdot^-$ has been estimated. The present consensus is that a maximum of about 5% is converted to $O_2 \cdot^-$ [7,127]. The compartmentalization of these reactive oxygen metabolites is related to the compartmentalization of the antioxidant defenses (SODs, CAT, GSH-Px). These antioxidant enzymes regulate the steady-state level of $O_2^{-\cdot}$ and H_2O_2. I shall discuss this oxidant-antioxidant balance in Chapter 8. The existence of physiological steady-state levels of oxygen radicals was first proposed by Gerschman on the basis of oxygen poisoning and X-ray irradiation [391]. Imlay and Fridovich [392] suggested the idea that diseases linked to oxidative stress should be confirmed quantitatively by measuring changes in the steady state of these oxidative species. The imbalance between oxidant species and antioxidant defenses is known as oxidative stress [393]. Increased oxidative stress (as in cancer, increased metabolic rate and aging) leads to elimination of modified DNA bases (8-hydroxy-2'-deoxyguanosine, thymine glycol and thymidine glycol) in the urine [394-398].

Oxidative stress modulates many genes. High steady-state levels of $O_2^{-\cdot}$ and H_2O_2 induce the synthesis of proteins in *E. coli* controlled by the oxyR and soxR loci [399,400] and stimulate the activity of repair enzymes [401,402].

B. MITOCHONDRIAL ELECTRON TRANSPORT [85,388,389]

The energy for the operation of cells comes from oxidation-reduction reactions. Oxidation is the transfer of electrons from one molecule to another. The electron-donating molecule is the reducing agent and the electron-accepting molecule is the oxidizing agent.

Some enzymes catalyze the transfer of two, four or even six electrons. In the mitochondrial membrane electrons are transferred from substrate via a series of enzyme catalyzed steps. In biological systems the final electron acceptor is the oxygen molecule. The oxygen molecule is reduced to two moles of H_2O. This reduction requires the transfer of four electrons ($O_2 + 4 e^- + 4 H^+ \longrightarrow 2 H_2O$). The enzyme involved in this final step is the cytochrome oxidase (a hemoprotein), which I shall discuss further on.

There are four major groups of oxidation-reduction enzymes or electron-transferring proteins which participate in the transfer of electrons from substrate to oxygen:

1. The pyridine-linked dehydrogenases, which require either NAD (nicotineamide-adenine-dinucleotide) or NADP (nicotineamide-adenine-dinucleotide phosphate) as coenzyme, 2. The flavin-linked dehydrogenases, which contain flavin adenine dinucleotide (FAD) or flavin mononucleotide (FMN) as prosthetic group, 3. Iron-sulfur proteins, and 4. Cytochromes, which contain an iron-porphyrin prosthetic group. In addition, there is an important coenzyme involved, namely ubiquinone or coenzyme Q.

Electron transport from NADH to O_2 proceeds with a large decrease in free energy, and this energy is conserved via coupled phosphorylation of ADP to ATP. This process is driven by the proton gradient across the membrane and is known as oxidative phosphorylation. The final step is the transfer of four electrons to O_2 to give two moles of H_2O and will be discussed under cytochrome oxidase. All the electron-transfer steps in the mitochondrial respiratory chain (except the last one) involve two-electron transfers. Therefore, we should not have to worry about the formation of oxygen radicals. However nothing in this world is perfect, not even the electron-transport chain. As a minor pathway, the flavoproteins and ubiquinone accept one electron to form a semiquinone radical.

The flavin-linked enzymes, FMN and FAD may undergo reversible one- or two-electron reduction-oxidation reactions:

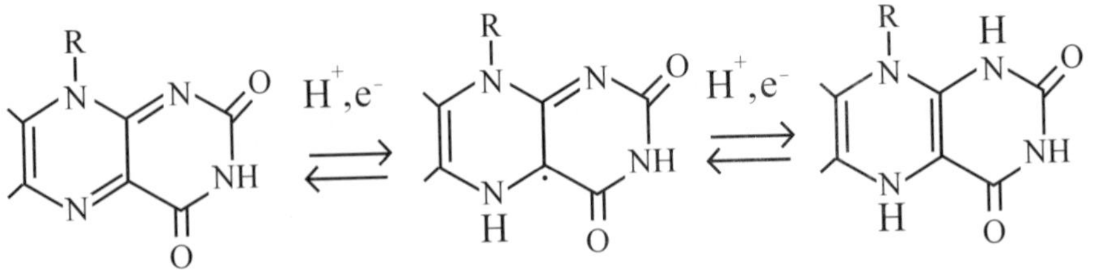

FAD   FADH   FADH$_2$

The flavo-semiquinone radicals react with oxygen to give $O_2\cdot^-$ (one-electron transfer):

$$FADH\cdot + O_2 \longrightarrow FAD + O_2\cdot^- + H^+$$

A similiar reaction is taking place with the coenzyme ubiquinone (coenzyme Q). This coenzyme has a reversibly reducible quinone ring with a long side chain:

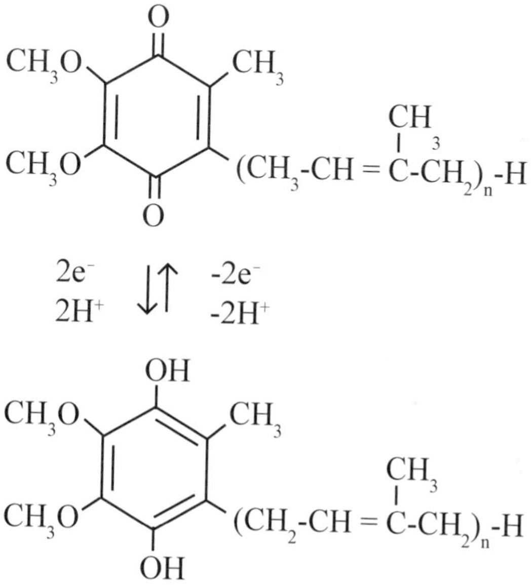

Although the quinone ring may be reduced by two-electron transfer to the hydroquinone, it also accepts only one electron to form the semiquinone radical:

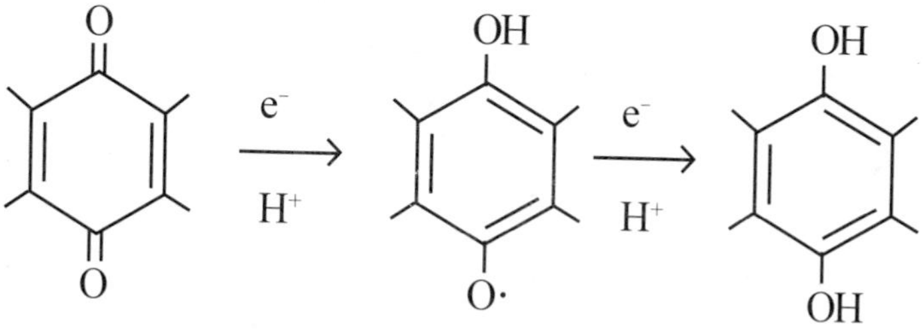

followed by the reoxidation of the semiquinone radical:

$$QH\cdot \;+\; O_2 \longrightarrow Q \;+\; O_2^{\cdot-} \;+\; H^+$$

It has been shown that the one-electron transfer to coenzyme Q is the major source of $O_2^{\cdot-}$ and the one-electron transfer of NADH dehydrogenase is of lesser importance. The fully reduced NADH dehydrogenase produces 0.90 ± 0.07 nmoles $O_2^{\cdot-}$/min per mg of protein, whereas the ubiquinone produces 1.85 ± 0.2 nmoles $O_2^{\cdot-}$/min per mg of protein [403].

C. HEMOPROTEINS [127,389]

These enzymes participate in oxygen transport (hemoglobin, myoglobin), electron transport (cytochromes) and oxygenation reactions (cytochrome P450 and peroxidases). The heme functionality is the prosthetic group in hemoproteins. Hemes are oxidized by oxygen to the corresponding Fe(III) derivative. A cytochrome is a hemoprotein which transfers electrons or hydrogen atoms via reversible Fe(II) $\rightleftharpoons$ Fe(III) transitions. An example of a heme is the iron-protoporphyrin complex shown in Fig. 1.

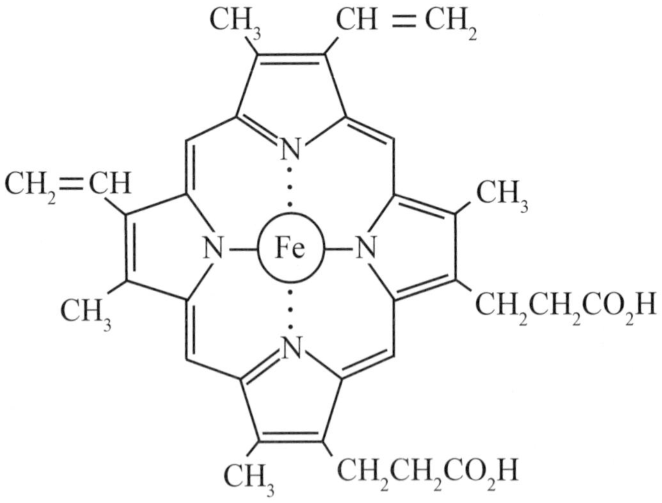

Fig. 1. Structure of iron protoporphyrin IX.

1. Cytochrome oxidase [127]

The cytochrome c oxidase is the final enzyme involved in the mitochondrial respiratory chain. This enzyme reduces molecular oxygen all the way to two moles of H_2O (four-electron reduction). The enzyme is a four-e^- donor and its substrate, the O_2 molecule is a one-e^- acceptor. This represents an interesting problem: The electron transfer can take place *via* one-, two-, or four-electron steps. The cytochrome oxidase consumes about 90% of all the oxygen in living systems.

In addition to act as an oxidase, cytochrome oxidase has been found to have catalase, peroxidase, and superoxide dismutase activities [127]. It has, therefore, been called the "ultimate" or "super" enzyme and has been the most extensively investigated oxidase. The many possible intermediates in the sequence $O_2 \longrightarrow 2\,H_2O$ have been studied by using all the available sophisticated techniques, like magnetic susceptibility, circular dichromism, magnetic circular dichromism, infrared and Raman spectroscopy, Mössbauer absorption, nuclear magnetic resonance, electron nuclear double resonance (ENDOR), and extended X-ray absorption fine structure spectroscopy (EXAFS). Since the reactive oxygen intermediates have been extensively reviewed [127] just a brief summary is given here.

The cytochrome oxidase contains at least four metal atoms per functional unit, i.e., two hemes (cytochrome a and a_3) and two associated copper atoms (Cu_a and Cu_{a3}). The copper atoms are in the bivalent state and the iron atoms in the trivalent one (the resting state). The physiological substrate of cytochrome oxidase is the reduced cytochrome c (Fe^{2+}). The reduced cytochrome c transfers electrons to the Fe_a-Cu_a site, which acts as an electron pool. The Fe_{a3}-Cu_{a3} binuclear center binds O_2 and the Fe_a-Cu_a site transfer four electrons to give two H_2O molecules. In addition to Fe and Cu, cytochrome oxidase has been reported to contain Zn and Mg .

In order for O_2 to become reduced to two H_2O molecules we need four protons (H^+) ($O_2 + 4\,e^- + 4\,H^+ \longrightarrow 2\,H_2O$). Cytochrome oxidase has three substrates, namely cytochrome c, oxygen and protons. Cytochrome oxidase acts as a proton pump. The four-electron transfer produces a proton gradient across the membrane, which in turn drives the production of ATP.

Whether the reduction proceeds via a stepwise electron transfer or a synchronized multiple electron transfer is still unresolved. As I have already pointed out (Chapter 3), the intermediate formation of $O_2\cdot^-$ could *a priori* be detected by SOD. It has been reported that SOD does not inhibit the reaction of cytochrome c oxidation by cytochrome oxidase. However, it should be realized that cytochrome oxidase also has SOD activity. This means that cytochrome oxidase is a superoxide radical scavenger and even if $O_2\cdot^-$ is produced, it will be difficult to detect in solution [127].

2. Peroxidases

The hydroperoxidases and peroxidases consist of a group of enzymes, which are hemoproteins [387]. They catalyze the general reaction:

$$ROOH + SH_2 \xrightarrow{\text{peroxidases}} ROH + H_2O + S$$

Catalase [7,127], which is a very important enzyme in the detoxification of H_2O_2, catalyzes the following reaction:

$$H_2O_2 + H_2O_2 \xrightarrow{\text{catalase}} 2 H_2O + O_2$$

In this case, the substrate is the same as the oxidizing agent. Catalase also catalyzes the oxidation of primary alcohols to aldehydes:

$$RCH_2OH + H_2O_2 \longrightarrow RCHO + 2 H_2O$$

A function similiar to catalase is carried out by glutathione peroxidase (GSH-Px), which contains Se at its active site and uses glutathione (GSH) as the reducing agent [127,404]:

$$H_2O_2 + 2\,GSH \longrightarrow GSSG + 2\,H_2O$$

Catalase requires high concentrations of H_2O_2 to be effective. GSH-Px takes over at low H_2O_2 concentrations. The fact that we have two enzymes to remove H_2O_2 indicates that it is an important oxygen metabolite whose removal is essential for the survival of the cell. Catalase is specific for H_2O_2, while GSH-Px also accepts hydroperoxides (ROOH) as a substrate. Catalase is located within peroxisomes, while glutathione peroxidase is distributed throughout the cytosol. This compartmentalization coupled with the lower K_m of GSH-Px for H_2O_2 [405] suggest that catalase is less important than GSH-Px in the decomposition of endogenously generated H_2O_2 [405]. The GSH may be regenerated from GSSG by the NADPH-dependent glutathione reductase:

$$GSSG + NADPH + H^+ \longrightarrow 2\,GSH + NAD^+$$

NADPH is, therefore, important in maintaining an adequate level of GSH. I shall come back to this function in Chapter 8 (antioxidants).

Cytochrome c peroxidase catalyzes the oxidation of ferrocytochrome c to ferricytochrome c:

$$2\,Fe^{II}(cyt) + H_2O_2 + 2\,H^+ \longrightarrow 2\,Fe^{III}(cyt) + 2\,H_2O$$

Most of the peroxidases contain ferroprotoporphyrin IX as the prosthetic group.

Other important enzymes in this group are the superoxide dismutases (SODs). These enzymes are metal (Cu(II)Zn(II)- or Fe(III)Mn(III))-containing enzymes, which catalyze the dismutation of superoxide [7,9,103,104]:

$$2\,O_2^{\cdot -} + 2\,H^+ \longrightarrow H_2O_2 + O_2$$

These enzymes, together with CAT and GSH-Px, are very important in protecting cells and tissues from the damaging effects of reactive oxygen metabolites.

D. FLAVIN-LINKED OXIDASES [387,389]

As I already discussed, the flavin prosthetic group contains an isoalloxazine ring, which can be reduced via two-electron transfer to the dihydroisoalloxazine ring; however, this type of structure may also form an intermediate radical by one-electron transfer. This type of radical (know as a semiquinone type radical, FADH·) is stabilized by resonance. The semiquinone radical is oxidized by oxygen to give $O_2^{·-}$ /$HO_2·$ and the hydroquinone-type structure ($FADH_2$) is oxidized to give H_2O_2. Many flavoproteins carry out oxidation-reductions, but are not involved in oxidative phosphorylation. Examples of this type of enzymes are xanthine oxidase, the aminoacid oxidases, monoamine oxidase and aldehyde oxidase. These oxidases produce $O_2^{·-}$ or H_2O_2 and both.

Xanthine oxidase oxidizes xanthine to uric acid:

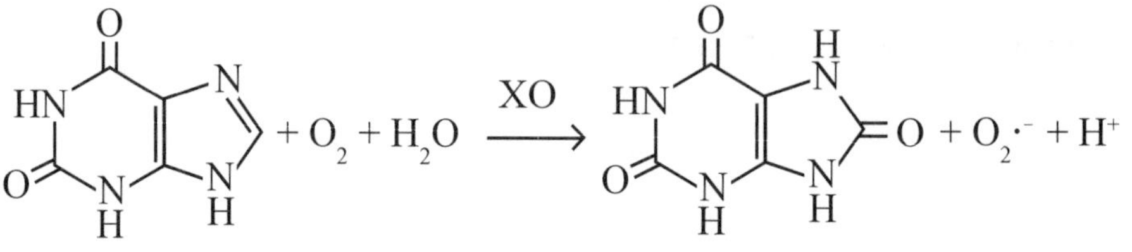

$$\text{xanthine} + O_2 + H_2O \xrightarrow{\text{XO}} \text{uric acid} + O_2^{·-} + H^+$$

xanthine   uric acid

Milk xanthine oxidase (XO) was the first enzyme which was shown to reduce oxygen to give the superoxide radical anion [83]. Ever since this discovery by McCord and Fridovich, the xanthine oxidase has been the most popular enzyme for the generation of superoxide radical anion *in vitro*. McCord and Fridovich used the reduction of cytochrome c as a probe for $O_2^{·-}$. The $O_2^{·-}$ was also identified by spin trapping [406] and later on determined quantitatively by the DMPO spin trap [407].

Since the xanthine oxidase produces both $O_2^{\cdot-}$ and H_2O_2 the above equation is oversimplified. Xanthine oxidase contains four redox centers, consisting of one molybdenum, one flavin group (FAD) and a pair of [2Fe-2S] centers. Its redox chemistry has been studied in great detail [408-410]. The fully reduced xanthine oxidase contains six reducing equivalents and can transfer its electrons by one- and two-electron transfer steps:

$$XO(6) \xrightarrow[H_2O_2]{O_2} XO(4) \xrightarrow[H_2O_2]{O_2} XO(2) \xrightarrow[O_2^{\cdot-}]{O_2} XO(1) \xrightarrow[O_2^{\cdot-}]{O_2} XO(0)$$

Kinetic studies (pulse radiolysis) [411] have shown that the first five electrons are transferred much faster than the last electron. The rate constants for the individual steps have been estimated as 35, 33, 20 and 0.9 s^{-1}. The slow reaction of the one-electron reduced XO is a consequence of both the poor reactivity and small equilibrium concentration of the flavin semiquinone:

$$FADH\cdot + O_2 \xrightarrow{slow} FAD + O_2^{\cdot-} + H^+$$

The complete oxidation of reduced XO yields two moles of H_2O_2 and two moles of $O_2^{\cdot-}$.

E. THE MICROSOMAL ENZYME SYSTEM

Other flavin-containing enzymes are located in the endoplasmic reticulum (microsome fraction) of the liver and kidneys. Some of these enzymes are important in the metabolism of amino acids. Aminoacid oxidases catalyze oxidative deamination of amino acids. The D-aminoacid oxidase contains FAD as the prosthetic group, while the L-aminoacid oxidase contains the flavin mononucleotide (FMN). In their reduced forms, these enzymes are oxidized by oxygen directly to the oxidized form and H_2O_2:

$$\text{D-aminoacid} + H_2O + \text{E- FAD} \longrightarrow \alpha\text{-keto acid} + NH_3 + \text{E-FADH}_2$$

$$\text{E-FADH}_2 + O_2 \longrightarrow \text{E-FAD} + H_2O_2$$

$$\text{L-aminoacid} + H_2O + \text{E-FMN} \longrightarrow \alpha\text{-keto-acid} + NH_3 + \text{E-FMNH}_2$$

$$\text{E-FMNH}_2 + O_2 \longrightarrow \text{E-FMN} + H_2O_2$$

There are many enzymes, which are located in the membrane of the endoplasmic reticulum (microsome fraction) and are non-phosphorylating oxidation-reduction enzymes. Many enzymes involved in detoxification are located in the liver and kidneys. The liver microsomal fraction contains a flavoprotein enzyme, the NADPH-cytochrome P450 reductase and a cytochrome P450 are mostly referred to as P450.

Many organic compounds are oxidized by liver microsomes including steroids, fatty acids, squalene, and some amino acids (aromatic). These liver enzymes also hydroxylate compounds not normally present *in vivo*. All these compounds are sometimes referred to as xenobiotics (foreign to the cell). These include drugs (like phenobarbitol, amphetamines, codeine, morphine) as well as environmental pollutants (like polycyclic aromatic hydrocarbons, PAHs). The hydroxylation of the xenobiotic represents a normal step in their metabolism and detoxification. The hydroxylated products are more water soluble and are excreted. Some of the intermediates in these hydroxylation reactions may, however, be more toxic to the cell than the original xenobiotic, i.e., in the case of PAHs the dihydrodiol-epoxide. In this way, the liver microsomes function as an activator of certain compounds from an inocuous to an active carcinogenic form. The flavoproteins and the P450 cytochrome are inducible enzymes, i.e., they increase in concentration if the animal is treated with certain drugs, like phenobarbitol. Increased production of $O_2^{\cdot-}$ was observed in corticosteroid-treated [412] and phenobarbitol [413] rat liver microsomes.

The NADPH-cytochrome P450 reductase and the cytochrome P450 are prime candidates for the generation of $O_2^{\cdot-}$ in microsomes.

Many studies have reported superoxide formation [86,414,415]. The generation of $O_2 \cdot^-$ is involved in microsomal lipid peroxidation [416]. The oxidation of epinephrine catalyzed by NADPH-cytochrome P450 reductase was shown to be inhibited by SOD [86]. In addition spin trapping by DMPO [417] and PBN [418] showed the superoxide radical adduct. Autoxidation of cytochrome P450 was also shown to produce $O_2 \cdot^-$. In this study, the chemiluminescene decreased in the presence of SOD and singlet-oxygen quenchers [419]. The chemiluminescence was, therefore, ascribed to singlet oxygen produced via dismutation of $O_2 \cdot^-$. Superoxide production was investigated in rat liver microsomes by a highly specific test for $O_2 \cdot^-$, the lucigenin-dependent chemiluminescence [420].

Summary. The major source of oxygen-derived radicals ($O_2 \cdot^-$) is the electron transport chain located in the inner membrane of mitochondria and from flavin-linked enzymes in the endoplasmic reticulum of liver and kidney. This one-electron transfer is due to the unusual stability of the one-electron adduct to quinone and quinone-type structures, which generate a radical stabilized by resonance. Any compound with quinone-type structure is suspect as a source of superoxide radical anion.

F. PHAGOCYTOSIS

Where there is partial knowledge there is controversy.
Roald Hoffmann in 'The Same and not the Same' [21]

1. Introduction

Phagocytosis is the most significant process for the formation of reactive oxygen metabolites. The importance of phagocytosis is evident from the numerous reviews written at perodic intervals [421-437]. These reviews also indicate that we do not have a complete understanding of all the complexities involved in phagocytosis. I shall discuss the evidence (pro and con) for hydroxyl-radical and singlet-oxygen formation.

Phagocytes are cells, which destroy invading microbes or foreign matter. The name phagocyte is therefore an all encompassing term, as is the name leukocytes. Phagocytes come in two varieties, i.e., granulocytes (cells containing granules) and cells without granules (agranulocytes). The most important cells in the first group are neutrophils, eosinophils and basophils. Among the second group, we find the monocytes and the macrophages.

Monocytes are mononuclear phagocytes arising from hematopoetic stem cells in the bone marrow. The monocytes enter the blood and tissue, increase in size, phagocytic activity and lysosomal enzyme content, and become macrophages.

The term polymorphonuclear leukocytes is used to describe fully developed cells of the granulocyte series, which have a nucleus consisting of three or more lobes. The neutrophil is a polymorphonuclear leukocyte.

Phagocytes (neutrophils, eosinophils, basophils, monocytes and macrophages) represent an important first-line defense against invading microorganisms (bacteria and viruses). They also serve as the garbage collectors of the body, removing all dead cells from a point of injury, whether induced mechanically by trauma or chemically by a toxic dosis. Upon stimulation, the phagocyte membrane becomes invaginated and forms the phagosome. The invading microorganisms get trapped in the phagosome and are destroyed. This process is known as phagocytosis.

Increased oxygen consumption in phagocytosis has been recognized long ago [421,438]. Subsequently it was observed that the increased oxygen consumption led to the formation of hydrogen peroxide [439,440]. The discovery of McCord and Fridovich [83] of superoxide dismutase was important as a tool to detect $O_2 \cdot^-$. Stimulated phagocytes were shown to reduce cytochrome c and this reduction was decreased in the presence of SOD [441]. The biochemical basis of the respiratory burst is the activation of a membrane-bound enzyme system [430,442-444], which catalyzes the reduction of oxygen to superoxide radical anion at the expense of NADPH:

$$NADPH \; + \; 2\,O_2 \; \longrightarrow \; NADP^+ \; + \; 2\,O_2{\cdot}^- \; + \; H^+$$

The enzyme which catalyzes this reaction is the NADPH oxidase.

I have already discussed oxygen reduction (Chapter 3) by one- or two-electron transfer to give $O_2{\cdot}^-$ or H_2O_2. The $O_2{\cdot}^-$ dismutates to give H_2O_2 and $^1O_2$. The dismutation catalyzed by SOD is faster by a factor of 10^4 [445], but produces H_2O_2 and oxygen in the ground state. The reduction of O_2 by NADPH oxidase has usually been assumed to proceed via one-electron reduction, followed by dismutation. However, some studies have indicated that in addition to one-electron reduction there is also direct two-electron reduction, depending on the pH and NADPH substrate concentration [446,447]. This is not surprising, since it is known that the completely reduced xanthine oxidase produces both H_2O_2 and $O_2{\cdot}^-$ [408-411]. This direct reduction to H_2O_2 is important in assessing the pathological consequences of these phagocyte-produced ROMs. Superoxide radical anion is far less damaging than H_2O_2. Hydrogen peroxide penetrates membranes and is transformed to $\cdot OH$, HOCl or $^1O_2$ and may therefore act at distant sites.

The respiratory burst occurs whenever phagocytes are exposed to certain stimuli. This stimulus may be a microbe or chemical substance, like phorbol myristate acetate (PMA) or some peptides like FMetLeuPhe or other N-formylated chemotactic peptides. These compounds attach to specific receptor sites at the phagocyte membrane and, thus, activate the NADPH enzyme system. If the phagocytes are stimulated by an inorganic foreign body, like asbestos fibers, the phagocytes release these reactive oxygen metabolites into the external space and the damage occurs in the surrounding tissue. This process is known as failed phagocytosis.

After physical or chemical trauma, phagocytes are attracted to the point of injury by chemotactic substances, formed via lipid oxidation (see Chapter 6). The phagocytes remove the injured cells and stimulate the neighboring healthy cells to proliferate (stimulate mitosis) to promote wound healing. This beneficial function of phagocytes may also have damaging effects. Inflammation and mitosis are essential for carcinogenesis (Chapter 9).

The superoxide initially formed, dismutates rapidly either enzymatically (SOD) or non-enzymatically to yield hydrogen peroxide:

$$2\ O_2^{\cdot-} + 2\ H^+ \longrightarrow H_2O_2 + O_2$$

The rate of the non-catalyzed reaction depends on the pH (Chapter 3). The SOD-catalyzed dismutation gives oxygen in the ground state, whereas non-catalyzed dismutation gives $^1O_2$. This reaction takes place *in vitro* [72,448] and *in vivo* [97,98]. The primary metabolites are $O_2^{\cdot-}$ and H_2O_2 and we have, therefore, the possibility of hydroxyl radical formation via the Haber-Weiss reaction.

The question now arises: How do these initially formed oxygen metabolites exert their microbicidal effect? H_2O_2 has only a weak microbicidal effect and $O_2^{\cdot-}$ none at all. Hydrogen peroxide is a well-known but weak germicidal agent, and its involvement in the microbicidal activity of phagocytes has been proposed long ago. On the other hand, it is also well-known that hydrogen peroxide is a stable compound of low reactivity. In order to exert its damaging effects in cells and tissues, it has to be activated to the more reactive hydroxyl radical, singlet oxygen or some other oxidizing agent. The low reactivity of hydrogen peroxide, however, also permits it to survive long enough to diffuse through membranes and react with lipids or DNA in a site-specific manner. The formation of $O_2^{\cdot-}$ and H_2O_2 presents again the possibility of ·OH formation via the Haber-Weiss reaction. Hydrogen peroxide has been shown [449] to kill *Staphylococcus aureus* by reacting with staphylococcal iron to form hydroxyl radicals.

In addition to the NADPH oxidase system, there are other enzymes present in phagocytes. One of these enzymes is the myeloperoxidase (MPO), which catalyzes the reaction between H_2O_2 and Cl^- to give hypochlorous acid [102]. Another group of enzymes are the NO synthases [126].

There are two ways to increase the reactivity of H_2O_2, i. e., via one-electron transfer to give hydroxyl radicals or via reaction with Cl^- (the myeloperoxidase-H_2O_2-Cl^- system). I have already discussed the first possibility in great detail (Chapter 3). The reactivity of $O_2\cdot^-$ may be increased by combination with nitric oxide ($\cdot$NO) to give the highly reactive peroxynitrite ($^-$OONO), which will be covered in Chapter 7. I now would like to discuss the myeloperoxidase-H_2O_2-Cl^- system.

2. The myeloperoxidase-H_2O_2-Cl^- system

Peroxidases are a group of enzymes (differing in their heme prosthetic groups) with the ability to increase the reactivity of H_2O_2 by many orders of magnitude. Peroxidases alone do not have any antimicrobial effect, but they exert an antimicrobial effect indirectly by catalyzing the conversion of an innocuous substance such as a halide ion to one which is highly toxic. One such peroxidase is the myelo-peroxidase (MPO), which is present in high concentrations in the granules of neutrophils. MPO is green and gives pus its color. The MPO catalyses the following reaction[102]:

$$H_2O_2 \ + \ Cl^- \ \xrightarrow{\text{MPO}} \ HOCl \ + \ OH^-$$

MPO also catalyzes the oxidation of Br^-, I^-, and SCN^-. The rate of oxidation is affected by concentration and pH. The optimum pH is close to the pH prevalent within the phagosome (pH 5.0-5.5) and varies with the concentration and the nature of the halide ion. The enzyme is a hemoprotein with a heme prosthetic group and is inactivated by cyanide (CN^-) and azide (N_3^-). These anions have been used as probes for the involvement of the MPO in the microbicidal activity of phagocytes. Monocytes and macrophages, which are agranular phagocytes, contain considerably lower concentrations of MPO.

The MPO-H_2O_2-Cl^- system was studied extensively because the reaction of HOCl and H_2O_2 is known to produce singlet oxygen [70]. The reaction was scrutinized intensely as a candidate for *in vivo* singlet oxygen production. Arguments pro and contra have been presented. At

134

present the formation of singlet oxygen *in vivo* via a number of reactions is definitely established [97,98]. These reactions will be discussed in the section on singlet oxygen.

3. What is the ultimate bacteriocidal agent?

So far we have $O_2^{\cdot-}$, H_2O_2 and HOCl present in the vacuole. How do these initially formed metabolites cause the microbicidal effect? From a chemical point of view, there are many possibilities:

$$1.\ O_2^{\cdot-} + O_2^{\cdot-} \xrightarrow{\ 2\,H^+\ } H_2O_2 + {}^1O_2 \qquad \text{dismutation [72,97,98]}$$

$$2.\ H_2O_2 + Fe^{2+} \longrightarrow Fe^{3+} + OH^- + {\cdot}OH \qquad \text{Fenton reaction}$$

$$3.\ Fe^{3+} + O_2^{\cdot-} \longrightarrow Fe^{2+} + O_2 \qquad \text{Iron-catalyzed}$$
$$Fe^{2+} + H_2O_2 \longrightarrow Fe^{3+} + {\cdot}OH + OH^! \qquad \text{Haber-Weiss reaction}$$

$$4.\ H_2O_2 + HOCl \longrightarrow {}^1O_2 + HCl + H_2O \qquad [70,338]$$

$$5.\ O_2^{\cdot-} + HOCl \longrightarrow {\cdot}OH + Cl^- + {}^1O_2 \qquad [224,225]$$

In addition to the above reactive oxygen species, phagocytes contain NO synthases. The combination of ·NO with $O_2^{\cdot-}$ forms the highly reactive peroxynitrite (⁻OONO), which is an important part of the antimicrobial defense system of phagocytes. I shall discuss these reactions of peroxynitrite in detail in Chapter 7. I have already discussed Reactions 1 - 3 in Chapter 3. I now would like to consider some of the evidence pro and contra for the involvement of hydroxyl radicals, and singlet oxygen and the reactions of hypochlorous acid in the microbicidal activity of stimulated neutrophils.

a. Hydroxyl radicals

Since the hydroxyl radical is one of the most reactive radicals known, it has been considered as a prime candidate for the microbicidal effect. However, there has been considerable controversy as to whether activated neutrophils exert their microbicidal effect by formation of hydroxyl radicals. Evidence for and against hydroxyl radicals has been presented, sometimes even by the same school of investigators (for a review see [437]). The reasons for this controversy are clear: lack of understanding of the many competing processes and a lack of appreciation of the pros and cons of the analytical techniques used in the identification of these reactive intermediates. I have already addressed the problems of hydroxyl radical (Chapter 3) and singlet oxygen detection (Chapter 4). The literature is full of studies using deoxyribose degradation, formation of ethylene from methional, hydroxylation of phenol or salicylic acid, decarboxylation of benzoic acid and the DMPO spin trap as hydroxyl radical probes. These probes are not specific. Progress in this field went parallel with the development of more sensitive and specific analytical techniques. A simple and specific technique, which so far has not been discredited is the formation of methane from DMSO. The reliability of all the other probes has been questioned.

Another important aspect is the presence of trace-metal contaminants in experiments *in vitro*, which was not truly appreciated until Buettner [271,272] pointed out the importance of trace metals in commercially available buffer solutions. This of course is very important since the Haber-Weiss reaction needs a metal catalyst in order to proceed at a reasonable rate. It has been suggested that lactoferrin, which is secreted into the phagosome upon stimulation, could act as a Haber-Weiss catalyst [450]. Other investigators, however, proposed that lactoferrin actually inhibits formation of hydroxyl radicals [451,452]. On the other hand, it has been demonstrated that $O_2^{\cdot-}$, produced by stimulated PMNs, causes release of free iron from ferritin [158, 314].

There is no doubt that H_2O_2 reacts with Fe^{2+} or some other metal ion (Cu^+) to give hydroxyl radicals. The Haber-Weiss reaction has been well established both in chemical systems as well as by superoxide-generating enzymes (xanthine-xanthine oxidase). However, the question whether these reactions are of any importance in the microbicidal activity of neutrophils has to be evaluated with caution. *In vivo,* the situation is much more complex: many different reactions catalyzed or not can take place simultaneously and the relative importance of any of these reactions depends on the rate constants and the concentrations of the reactants. From a purely chemical point of view all the above-mentioned reactions 1 - 5 can take place in a test tube. Which of these reactions is dominant depends on their relative concentrations and rate constants. The hydrogen peroxide may produce HOCl in much greater quantity than it forms hydroxyl radicals; however, this does not mean that the hydroxyl radicals do not contribute to cell killing. In the case in which the MPO system is inactive (MPO deficiency), the neutrophil retains the microbicidal activity, despite the absence of HOCl formation. The H_2O_2 accumulates to higher levels and the destructive potential is then mediated by the formation of hydroxyl radical via the Fenton or the Haber-Weiss reaction. So the answer to the question what is the ultimate bactericidal agent depends on the specific conditions in each case.

Another possibility for hydroxyl radical formation, which does not require any metal catalyst, has been suggested [224,225]:

$$HOCl + O_2^{\cdot-} \longrightarrow \cdot OH + Cl^- + {}^1O_2$$

This reaction has been observed via pulse radiolysis [224]. The reaction proceeds at appreciable rates, depending on the pH ($k = 10^6 - 10^7 M^{-1}s^{-1}$). The rate constant decreases with increasing pH, thus indicating that the HOCl rather than the ClO⁻ is the reactive species. A weak light emission (chemiluminescence) was observed at pH 8.7, however, no spectral assignment was possible, due to the low intensity of the emission. Nonetheless, in D_2O (instead of H_2O) the light emission was increased fivefold, thus indicating the involvement of singlet oxygen. The reaction of HOCl with superoxide radical was also observed to consume 2,5-dimethylfuran, a commonly used trap for singlet oxygen. However, as I already pointed out, this singlet oxygen probe is not reliable and can also be indicative of hydroxyl radical [340,341]. Direct evidence for hydroxyl radicals in this reaction was missing. More recently the HOCl - $O_2\cdot^-$ reaction was reexamined by Candeias et al. [225]. These investigators have confirmed the formation of hydroxyl radicals by using the hydroxylation of benzoic acid. This reaction is the only presently known reaction giving hydroxyl radical without requiring a metal-ion catalyst.

Although there is little doubt that the reaction occurs *in vitro*, the question is does this hydroxyl radical-forming reaction occur *in vivo*? I would like to discuss some of the results which led to the conclusion that *in vivo* conditions are unfavorable for hydroxyl radical formation via reaction (5). Hydroxyl radicals have been observed in a reaction of H_2O_2 with superoxide (generated by xanthine oxidase) and an Fe(EDTA) complex. They have been detected by deoxyribose degradation, formation of ethylene from methional, and hydroxylation of benzoic acid. Although these methods for hydroxyl radical detection are not reliable and specific, these were the standard methods used in the older literature. Thus neutrophils, stimulated by formyl-Met-Leu-Phe, inhibited this hydroxyl radical formation. Considerably less inhibition was observed with MPO-deficient neutrophils. These results seem to indicate that MPO released by the neutrophils is responsible for the hydroxyl radical decrease. When the neutrophils were stimulated by PMA, little MPO was released and hydroxyl radical production was increased due to the additional superoxide flux generated by the stimulated cells. From these results it may be concluded that under stimulating conditions, when neutrophils release both MPO as well as $O_2\cdot^-$ and H_2O_2, the reaction of H_2O_2 with Cl^-, catalyzed by MPO, provides unfavorable conditions for the generation of hydroxyl radicals (for a review see [437]).

A major difficulty in assessing the role of hydroxyl radicals in phagocytosis is the problem of their detection. As I have pointed out in Chapter 3, the most sensitive and reliable probe is spin trapping in combination with a competitive scavenger, e.g., DMSO or ethanol. These scavengers react with $\cdot$OH to give methyl and α-hydroxyethyl radical, respectively:

$$CH_3SOCH_3 + \cdot OH \longrightarrow \cdot CH_3 + CH_3SOOH$$

$$CH_3CH_2OH + \cdot OH \longrightarrow CH_3\dot{C}HOH + H_2O$$

These carbon-centered radicals react with the spin trap to form a more persistent radical adduct, which may be measured by EPR spectroscopy. The sensitivity of the method depends on two factors: the rate of adduct formation and the persistence of the resulting adduct. The faster the rate of adduct formation, the less chance the radical has to undergo other side reactions. The more stable the adduct (longer lifetime), the more the adduct accumulates during a given time. Both of these factors tend to increase the size of the observed EPR signal. The commonly used DMPO spin trap showed low stability in aqueous solutions [453]. It has been shown that the DMPO/·OH and the DMPO/·CH_3 spin adducts disappeared rapidly in presence of $O_2^{\cdot-}$ [454]. This was a very important discovery, since earlier work concluded from the absence of the DMPO/·OH and DMPO/·CH_3 signal that neutrophils do not produce hydroxyl radicals (see [437]). As was already pointed out (Chapter 3), there are many ways for the formation of the DMPO/·OH adduct including the reaction with $O_2^{\cdot-}$ [233]. These results allow us to proclaim a word of caution: a negative result is no proof of absence and a positive result is no proof of presence!

However, a new spin trap the 4-pyridyl-1-oxide N-tert-butylnitrone (4-POBN) was found to react with the α-hydroxyethyl radical by an order of magnitude faster than the DMPO spin trap [455]. Also the 4-POBN/$CH_3\dot{C}HOH$ adduct was more stable than the corresponding DMPO adduct. The use of 4-POBN was, therefore, a considerable improvement in the sensitivity of the method. This example shows again the importance of a sensitive and specific analytical technique.

The formation of the 4-POBN/$CH_3\dot{C}HOH$ adduct was observed in phorbol 12-myristate-12-acetate (PMA)-stimulated neutrophils and monocytes without addition of supplemental iron [455]. Hydroxyl radical formation was inhibited by SOD, CAT, and azide (N_3^-). Metal chelators did not affect the signal, thus indicating that hydroxyl radicals are formed via a mechanism independent of the transition metal catalyzed Haber-Weiss reaction. Stimulated macrophages (mature forms of monocytes), which lack MPO, required the addition of iron for hydroxyl radical formation. The addition of purified MPO to an enzymatic $O_2^{\cdot-}$ -generating system (xanthine-xanthine oxidase) resulted in the formation of hydroxyl radicals, which depended on the presence of chloride ion, and was inhibited by SOD, CAT and azide (deactivates MPO). A hydroxyl radical-forming reaction, which is consistent with all these observations, is the reaction of HOCl with $O_2^{\cdot-}$ (Reaction 5, page 134).

Stimulation of monocyte-derived macrophages (which upon differentiation from monocytes loose MPO activity) did not lead to detectable amounts of 4-POBN/$CH_3\dot{C}HOH$, except in presence of iron, which then produces ·OH via the Haber-Weiss reaction. These results show that in stimulated neutrophils we have MPO-dependent formation of hydroxyl radicals.

b. Singlet oxygen

Singlet oxygen is of course an *a priori* candidate for the microbicidal effect of phagocytes. As pointed out above, there are several chemical reactions which are candidates for singlet oxygen formation (reaction 1, 4, 5). The question is are these reactions taking place *in vivo*? Since neutrophil granules contain the myeloperoxidase, which catalyzes formation of HOCl from H_2O_2 and Cl^-, the formation of singlet oxygen via reaction 4 appears to be a prime candidate. However, the myeloperoxidase converts H_2O_2 very rapidly to HOCl leaving little H_2O_2 to react according to reaction 4. The myeloperoxidase-H_2O_2-halide system was, therefore, studied in detail.

The formation of singlet oxygen in the respiratory burst of PMNs was suggested by Allen et al. [456,457], based on chemiluminescence studies, and also by Krinsky [458], who showed that bacteria rich in carotenoids resist killing by PMNs. These suggestions were based on the work of Khan and Kasha [70] on the HOCl-H_2O_2 reaction. However, later on Kanofsky et al. [459] by using purified MPO, detected chemiluminescence at 1268 nm which increased 29-fold in D_2O, thus, confirming singlet oxygen formation. The amount of $^1O_2$ was sensitive to the conditions employed: under optimal conditions at pH 5, the

MPO-H_2O_2-Br^- system produced 0.42 mol of $^1O_2$/mol of H_2O_2 consumed, close to the theoretical value of 0.5. The Cl^- was much less efficient, since only 0.09 mol of $^1O_2$/mol of H_2O_2 consumed was found. This optimum yield required pH 4 and 5 mM H_2O_2. At higher pH, the amount of $^1O_2$ dropped to 0.0004 mol/mol of H_2O_2 consumed at pH 7. These authors, therefore, concluded that the MPO-H_2O_2-Cl^- system was not an important source of singlet oxygen formation by stimulated PMNs under normal physiological conditions.

Subsequently it was shown by Kettle and Winterbourne [460] that concentrations of 100 μM H_2O_2 or higher deactivate MPO at pH 7.6. Physiological concentrations of H_2O_2 within the phagosome are probably much less than 100 μM. Therefore, detection of singlet oxygen by the MPO-H_2O_2-Cl^- system led to negative results because high non-physiological concentrations of H_2O_2 (≥ 250 μM) were used.

Activated neutrophils also showed weak chemiluminescence; however, in the absence of a spectral analysis this does not prove the formation of singlet oxygen. Spectral analysis was not possible due to the low intensity of the chemiluminescence. This shows again the importance of reliable, specific and sensitive analytical techniques. We have three candidates for singlet oxygen formation in activated phagocytes:

$$O_2^{\cdot-} + O_2^{\cdot-} \xrightarrow{2\,H^+} H_2O_2 + {}^1O_2 \qquad \text{Khan [448]}$$

$$HOCl + H_2O_2 \longrightarrow HCl + H_2O + {}^1O_2 \qquad \text{Khan and Kasha [70]}$$

$$HOCl + O_2^{\cdot-} \longrightarrow {\cdot}OH + Cl^- + {}^1O_2 \qquad \text{Long and Bielski [224]}$$

Although these three reactions have been shown to occur in purely chemical systems their importance in stimulated PMNs *in vivo* has been questioned. The reasons for this controversy are the same as we encountered with the hydroxyl radical problem. The singlet oxygen traps are not always specific and the same products may be formed by different oxidizing species. Another problem has been competing reactions and the rate of singlet oxygen-trapping reactions. In a trapping experiment the trapping reaction has to compete with quenching:

$$^1O_2 + Q \xrightarrow{k_q} O_2 + Q^*$$

$$^1O_2 + D \xrightarrow{k_d} \text{product}$$

If we want to prove the presence of singlet oxygen via dismutation of superoxide radical anion, we must consider the following. Since superoxide radical anion is an efficient quencher of singlet oxygen [346] ($k = 1.6 \times 10^9$ M^{-1}s^{-1}), high concentrations of $O_2^{\cdot-}$ may obscure the detection of singlet oxygen especially if its reaction with the trap is slow. Such a slow reacting trap is cholesterol. The absence of any cholesterol-derived product (although highly specific) does, therefore, not mean that singlet oxygen is absent.

The superoxide radical anion is produced in large quantity by the xanthine oxidase system and numerous investigations were carried out to obtain evidence for or against singlet oxygen involvement. None of these early studies took into consideration the above mentioned complexities: high concentrations of $O_2^{\cdot-}$ together with a low rate for trapping make the conditions for singlet oxygen detection unfavorable.

In more recent studies by Khan and coworkers [97,98] by using the more effective 9,10-diphenylanthracene (DPA) trap, it was shown that activated neutrophils (stimulated with DPA-coated glass beads and PMA) showed intracellular formation of singlet oxygen. The yield was found to be quite high, about 19% based on the amount of oxygen consumed per 1.25×10^6 neutrophils. This yield is about 100 times higher than the one observed previously for cholesterol as singlet oxygen trap [343,344]. The presence of β-carotene, which is an excellent singlet oxygen quencher ($k = 10^{10}$M^{-1}s^{-1}), no DPA-endoperoxide formation was observed [97,98].

Addition of purified MPO to H_2O_2-Cl^- also showed formation of the DPA-endoperoxide in a somewhat lower yield, although similiar amounts of MPO were present in both systems. This result suggests that in addition to the MPO-dependent pathway for singlet oxygen production there exists an MPO-independent pathway. Such a possible mechanism is, of course, the dismutation of superoxide radical anion [448]. For stimulated macrophages, which produce great quantities of $O_2^{\cdot-}$ via the NADPH complex, but which lack MPO activity, singlet oxygen formation may proceed via such an MPO-independent pathway. In activated macrophages singlet oxygen was observed with the DPA trap [97,98]. In these experiments, SOD decreased the formation of the DPA-endoperoxide, thus, clearly indicating the involvement of superoxide radical anion in the generation of singlet oxygen. SOD converts $O_2^{\cdot-}$ to H_2O_2 and ground state molecular oxygen. Production of high concentrations of $O_2^{\cdot-}$ by stimulated PMNs, however, limit the formation of singlet oxygen, since superoxide is an excellent quencher when present in high concentrations.

A very interesting example of the use of quenchers is β-carotene to treat erythropoetic protoporphyria (EPP), a human photosensitivity disease [462,463]. The photosensitivity of these patients seems to be caused by deposition of photosensitizing porphyrins in the skin. The symptoms of these patients may be largely relieved by oral administration of β-carotene. Therefore, one may conclude that the damaging agent is most likely singlet oxygen.

c. Reactions of HOCl

Although the hydroxyl radical and singlet oxygen are important in cell killing by phagocytes, one should not assume that these species are the only damaging species in phagocytosis. The HOCl is a strong oxidizing agent and can react with many biomolecules. The literature on this subject is enormous and has been reviewed by Klebanoff [435]. The antimicrobial activity of phagocytes was originally thought to occur via the chlorinating ability [464] of the MPO- H_2O_2-Cl^- system.

Hypochlorite has antiseptic properties and has been used for a long time in the purification of drinking water and for the desinfection of swimming pools. Phagocytes, may therefore exert their antimicrobial effect through HOCl.

The hypochlorous acid or its anion can oxidize a variety of biologically important compounds. For example, the OCl⁻ is very reactive towards amino groups and may, therefore, react with amino acids and enzymes. A very potent oxidizing agent is the monochloramine [465-467]:

$$NH_4^+ + OCl^- \longrightarrow NH_2Cl + H_2O$$

In fact monochloramine is a more potent microbicidal agent than OCl⁻. The monochloramine is lipid soluble (has no electric charge) and may therefore, gain easy access to the interior of the microorganism.

Amino acids are deaminated and decarboxylated to yield aldehydes [468]. This reaction proceeds via the following mechanism [469]:

$$\underset{\underset{NH_2}{|}}{R\text{-}CH\text{-}COOH} + HOCl \longrightarrow H_2O + \underset{\underset{NHCl}{|}}{R\text{-}CH\text{-}COOH}$$

$$\underset{\underset{NHCl}{|}}{R\text{-}CH\text{-}COOH} + H_2O \longrightarrow R\text{-}CHO + NH_4Cl + CO_2$$

The formation of aldehydes was suggested as the microbicidal mechanism in the MPO-H_2O_2- Cl⁻ system [464] long before the HOCl was recognized as an intermediate [102]. The identification of HOCl as intermediate opened the possibility of new antibacterial mechanisms, i. e., by the Khan-Kasha [70] and the Long-Bielski [224] reactions.

Monochloramines are further chlorinated to dichloramines. Dichloramines of peptides undergo cleavage at the peptide bond [470,471]:

$$\underset{\underset{R\text{-}CH\text{-}CONHR'}{}}{\overset{NCl_2}{\overset{|}{}}} \xrightarrow{H_2O} RCN + CO_2 + R'NH_2 + 2HCl$$

The HOCl also reacts with Cl$^-$ to yield Cl$_2$:

$$Cl^- + OCl^- + 2 H^+ \longrightarrow Cl_2 + H_2O$$

Similiar reactions take place with Br$^-$ and I$^-$. Tyrosine and other aromatic amino acids are halogenated in this way, e. g., tyrosine yields 2,6-diiodotyrosine [472].

The HOCl reacts with sulfhydryl groups. A number of enzymes requiring a free SH group for their activity are oxidized by the MPO- H$_2$O$_2$-Cl$^-$ system. The SH groups of *E. coli* are oxidized by this system as well as by HOCl [472]. Although only a fraction of the oxidizing equivalents have been accounted for by SH oxidation a relationship between oxidation of SH groups and cell death of *E. coli* has been established. The MPO-H$_2$O$_2$ -Cl$^-$ system oxidizes GSH to GSSG [473].

Although the HOCl/ClO$^-$ reacts with many NH- and SH-containing biomolecules, not all these reactions lead to damage. Human neutrophils contain high concentrations of theβ -amino acid taurine [474]. Taurine modifies the bactericidal effect of the MPO-H$_2$O$_2$ -Cl$^-$ system [464,465] by reacting with HOCl to give a taurine chloride:

$$HOCl + H_2N\text{-}CH_2\text{-}CH_2\text{-}SO_3H \longrightarrow Cl\text{-}NH\text{-}CH_2\text{-}CH_2\text{-}SO_3H + H_2O$$

The taurine chloride is stable and has been used for the analysis of HOCl [475]. Since chloramines slowly regenerate HOCl via hydrolysis [464] in this way neutrophils use the MPO-H$_2$O$_2$-Cl$^-$ system to generate long-lived oxidants (use the chloramines as a storage vessel), which mediate inflammatory effects long after the termination of the respiratory burst [475]. In the formation of acetaldehyde from alanine by the MPO-H$_2$O$_2$-Cl$^-$ system, the yield of CO$_2$ and aldehyde decreased considerably in presence of taurine (taurine competes with alanine for HOCl) [464].

4. Genetic deficiencies in phagocytes

In biological systems, the study of anomalies is of great help in understanding the normal. One of these anomalies in phagocytes is chronic granulomatous disease (CGD) [426,435]. In this condition, phago-cytes cannot manufacture $O_2^{\cdot-}$ either because the NADPH oxidase is not activated or because the oxidase itself is defective. Affected patients suffer frequent, severe and protracted bacterial and fungal infections involving the deep subcutaneous tissues, lymph nodes, lungs, liver and other organs. Children born with this genetic defect usually die in infancy. The history of CGD has been reviewed [426]. Another anomaly of the oxygen-dependent microbicidal killing mechanism is the myeloperoxidase deficiency (MPOD) [435]. This condition is characterized only by a mild impairment of bacterial killing. In MPO deficiency we observe a total absence of MPO in the azurophil granules of neutrophils and monocytes. Individuals with this deficiency are usually healthy and do not suffer recurrent infections as do patients with CGD.

In patients with MPO deficiency, bacterial killing is delayed but is ultimately complete. This delay, however, is not due to slower formation of $O_2^{\cdot-}$ or H_2O_2. On the contrary the respiratory burst is exaggerated. The initial rate of oxidant formation is comparable to normal cells, but in MPO-deficient cells the production of $O_2^{\cdot-}$ and H_2O_2 continues for a longer period of time. This results in a considerable increase in oxygen consumption and $O_2^{\cdot-}$ and H_2O_2 accumulation. The increased formation of $O_2^{\cdot-}$ and H_2O_2 is not due to slower catabolism, since MPO-deficient cells contain normal amounts of catalase and glutathione peroxidase

Another difference to normal cells is that the chemiluminescence is initially decreased in MPO deficiency [476]. However, as time goes by the chemiluminescence becomes greater than in normal cells, most likely due to continued respiratory burst (formation of $O_2^{\cdot-}$ and H_2O_2). This observation supports the idea of singlet oxygen generation (increased chemiluminescence) by dismutation of $O_2^{\cdot-}$. Both MPO-sufficient and MPO-deficient cells show chemiluminescence, whereas CGD cells show no chemiluminescence [476].

In the MPO-deficient phagocyte no HOCl is formed and the mechanism of microbial killing proceeds via a different pathway involving singlet oxygen (produced via dismutation of $O_2^{\cdot-}$) and $\cdot OH$ (produced via the Haber-Weiss reaction), as I have already presented. In a model system generating $O_2 \cdot ^-$ (xanthine oxidase-acetaldehyde and oxygen) [477], it was found that *Staphylococcus aureus* is killed (although slower than in the MPO-H_2O_2-Cl^- system) completely. This result provides strong evidence for the involvement of oxygen radicals or singlet oxygen as a backup system for bacterial killing, whenever the normal pathway (the MPO-dependent) is defective. Nature does not put all her eggs in one basket.

As CGD patients show, $O_2 \cdot ^-$ and H_2O_2 are absolutely essential elements in the mechanism of microbial killing, whereas patients with MPO deficiency show that this enzyme is not essential and that the killing mechanism proceeds via at least two pathways, an MPO-dependent and an MPO-independent pathway. An interesting observation is the fact that the non-essential enzyme MPO is the important one in normal phagocytosis.

Since normal phagocytes produce $\cdot OH$ and $^1O_2$, it is not surprising that they initiate lipid peroxidation (formation of malonaldehyde). Phagocytes from CGD patients showed no malonaldehyde formation [478].

5. Antioxidant defenses of phagocytes

Since phagocytes upon stimulation produce huge amounts of $O_2 \cdot ^-$ and H_2O_2, these reactive oxygen metabolites not only kill the invading microorganism, but impose considerable oxidative stress on the phagocytes themselves. Therefore, phagocytes have a number of antioxidant defense systems, which destroy the ROMs before they do any harm to the phagocytes. These enzymes are superoxide dismutase (SOD), catalase (CAT), and glutathione peroxidase (GSH-Px) as well as some small molecules, which act as antioxidants or radical scavengers. Both catalase and glutathione peroxidase catalyze the decomposition of hydrogen peroxide. We may ask why do we need two enzymes for the same reaction? The rate of hydrogen peroxide destruction by catalase is strictly proportional to the H_2O_2 concentration. Therefore, at the low concentrations normally present in tissue, the catalase is inefficient for the destruction of H_2O_2. The catalase has been considered a backup system whenever the H_2O_2 concentrations reach high levels under situations of high oxidative stress. At this point the glutathione-dependent enzyme system no longer copes with the H_2O_2 overload.

Most of the H_2O_2 to which phagocytes are exposed and the H_2O_2 which leaks out into the cytoplasm during the respiratory burst is eliminated by the glutathione requiring a complex enzyme system, which reduces H_2O_2 to H_2O [404]:

$$2\ GSH\ +\ H_2O_2\ \longrightarrow\ GSSG\ +\ 2\ H_2O$$

This reaction is catalyzed by glutathione peroxidase and the GSH is regenerated by glutathione reductase:

$$GSSG\ +\ NADPH\ +\ H^+\ \longrightarrow\ 2\ GSH\ +\ NADP^+$$

The glutathione peroxidase not only catalyzes the reduction of H_2O_2 but also of alkyl hydroperoxides and the hydroperoxides formed during lipid peroxidation:

$$ROOH\ +\ 2\ GSH\ \xrightarrow{\ GSH\text{-}Px\ }\ GSSG\ +\ ROH\ +\ H_2O$$

From these observations we might expect that during the phagocytic respiratory burst the formation of high H_2O_2 concentrations would lead to the depletion of GSH. However, the GSH levels remain high, indicating a high efficiency of the glutathione reductase. Deficiency in glutathione reductase in humans has been reported [426].

In addition to the enzymatic defense, the neutrophils and monocytes contain a number of small molecules, which are known to act as antioxidants or radical scavengers. These compounds are: taurine, ascorbic acid, and α-tocopherol. As mentioned before taurine is an amine, which reacts with the hypochlorous ion (ClO⁻) to yield an unreactive, stable monochloramine [464,465,473]. Ascorbic acid is an effective radical scavenger, especially in aqueous media:

$$AH_2 \;+\; \cdot OH \longrightarrow AH\cdot \;+\; H_2O$$

$$AH\cdot \;+\; O_2 \longrightarrow A \;+ HO_2\cdot$$

α-Tocopherol is the most important lipid-soluble radical scavenger, which reacts in a similar fashion as ascorbic acid. The tocopheryl radical regenerates α-tocopherol by reacting with ascorbic acid:

$$TO\cdot \;+\; AH_2 \longrightarrow TOH \;+\; AH\cdot$$

The reactions of these and other antioxidants will be covered in Chapter 8.

Summary. The different pathways for the formation of reactive oxygen metabolites by phagocytes are summarized in Fig. 2. The formation of ·OH proceeds via MPO-dependent (Reaction 5) and MPO-independent (Reactions 1 and 2) pathways. MPO deficiency or deactivation of MPO (by N_3^-) leads to a decrease in ·OH or $^1O_2$ via Reactions 4 and 5. It leads at the same time to an increase in H_2O_2 and, therefore, to an increase in ·OH via the Haber-Weiss or Fenton reaction. Activated neutrophils produce singlet oxygen in high yields via an MPO-dependent pathway (Reactions 4 and 5). In addition, there exists a minor MPO-independent pathway for singlet oxygen formation via dismutation of superoxide radicals (Reaction 3). In stimulated macrophages, which are MPO deficient, the formation of singlet oxygen proceeds exclusively via dismutation of superoxide radicals. It is interesting to note that after the discovery of Reaction 3 in a purely chemical system, it took 35 years, before singlet oxygen formation by activated PMNs was definitely solved, hopefully so.

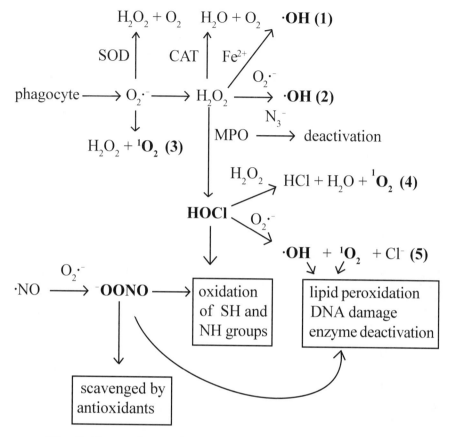

Fig. 2. Formation and reactions of ROMs by phagocytes.

G. ELECTRONICALLY EXCITED CARBONYL COMPOUNDS VIA ENZYMATIC OXIDATIONS

In Chapter 4 I discussed the formation of electronically excited carbonyl compounds via a variety of reactions. These reactions include the thermal decomposition of dioxetanes and dioxetanones as well as the self-reaction of peroxyl radicals (the Russell mechanism). The important question we have to ask is are these highly strained dioxetanes or dioxetanones formed *in vivo*?

Although no dioxetane has ever been directly observed, there is plenty of indirect evidence in support of these intermediates [77,479-481]. Electronically excited states of carbonyl compounds derived from the thermal decomposition of dioxetanes have been shown to cause DNA damage *in vitro* [30] as well as in bacteria and mammalian cells [482,483]. The formation of these excited states allows us to carry out photochemistry without light. The subject encompasses a huge field and a detailed discussion is beyond the scope of this text. I would like to mention just a few examples for the formation of dioxetanes and dioxetanones in some peroxidase-catalyzed reactions.

1. Aldehyde oxidations

The oxidation of isobutyraldehyde in the dark by horseradish peroxidase was found to produce electronically excited acetone. [484,485]. It was subsequently observed that the chemiluminescence depended on the presence of phosphate or arsenate buffer [486] and it was suggested that the substrate for the enzyme is the enolic form of the aldehyde:

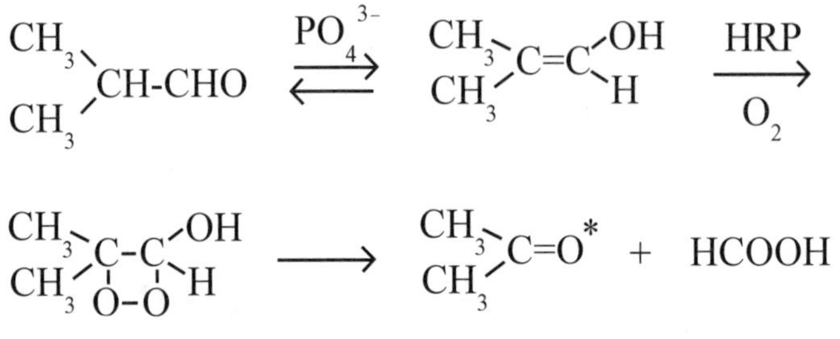

$$\underset{CH_3}{\overset{CH_3}{>}}CH\text{-}CHO \underset{\longleftarrow}{\overset{PO_4^{3-}}{\rightleftharpoons}} \underset{CH_3}{\overset{CH_3}{>}}C\text{=}C\underset{H}{\overset{OH}{<}} \overset{HRP}{\underset{O_2}{\longrightarrow}}$$

$$\underset{CH_3}{\overset{CH_3}{>}}C\text{-}C\underset{O\text{-}O}{\overset{OH}{<}}H \longrightarrow \underset{CH_3}{\overset{CH_3}{>}}C\text{=}O^* + HCOOH$$

The enolic form of isobutanal was independently generated by hydrolysis of the trimethylsilyl enol ether [487]:

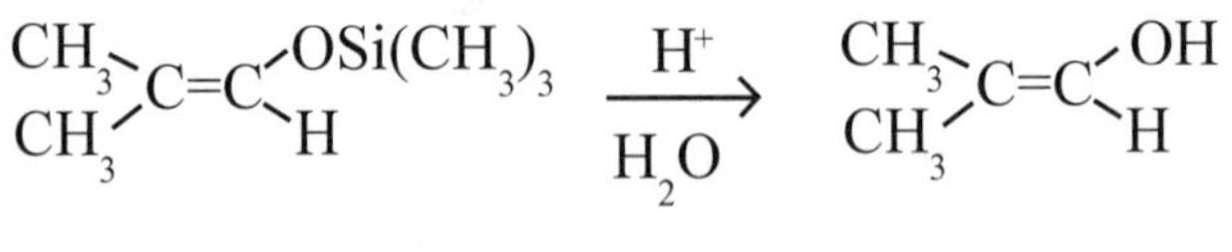

$$\underset{CH_3}{\overset{CH_3}{>}}C\text{=}C\underset{H}{\overset{OSi(CH_3)_3}{<}} \overset{H^+}{\underset{H_2O}{\longrightarrow}} \underset{CH_3}{\overset{CH_3}{>}}C\text{=}C\underset{H}{\overset{OH}{<}}$$

These electronically excited states emit light (phosphorescence) or they are deactivated by quenchers:

$$\begin{matrix} CH_3 \\ CH_3 \end{matrix} C=O^* + Q \longrightarrow \begin{matrix} CH_3 \\ CH_3 \end{matrix} C=O + Q^*$$

Such a quencher is D- and L-tryptophan. These two enantiomeric quenchers were found to quench the triplet acetone at different rates [487]. This observation confirms that the triplet acetone is produced within the enzyme. The D-tryptophan binds more tightly to the enzyme and is, therefore, a more efficient quencher.

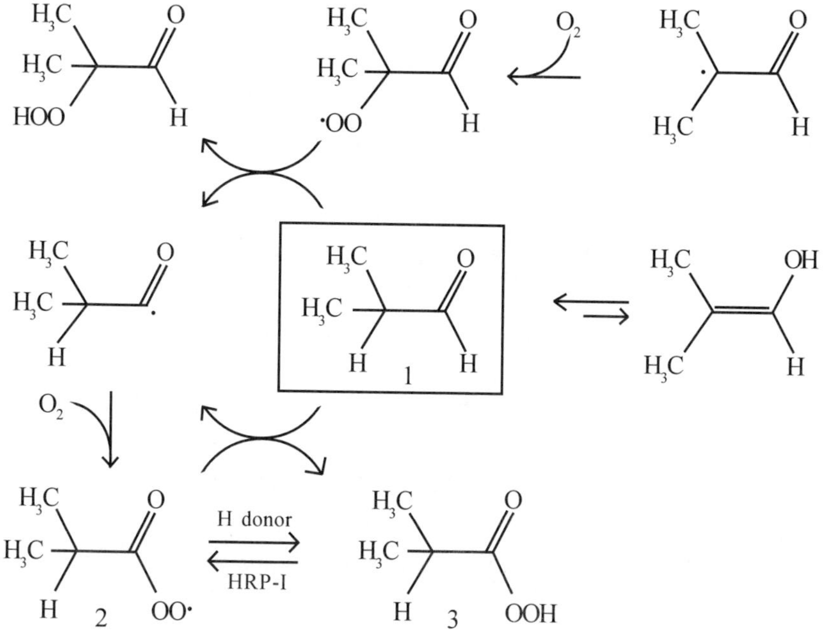

Fig. 3. Generation of acyperoxyl radicals (2) and peroxyisobutyric acid (3) in the HPR-catalyzed autoxidation of isobutanal (1) (reprinted from [488] with permission).

Although electronically excited acetone has been identified in these oxidations *in vitro*, it does not necessarily follow that DNA is damaged exclusively by these excited species. Recently it has been shown by Adam and coworkers [488] that DNA damage in these aldehyde oxidations is mainly mediated by peroxyl radicals derived from the peracid of the aldehyde. The intermediates in this HPR-catalyzed autoxidation are summarized in Fig. 3. These intermediates damage DNA to varying extent (Fig. 4).

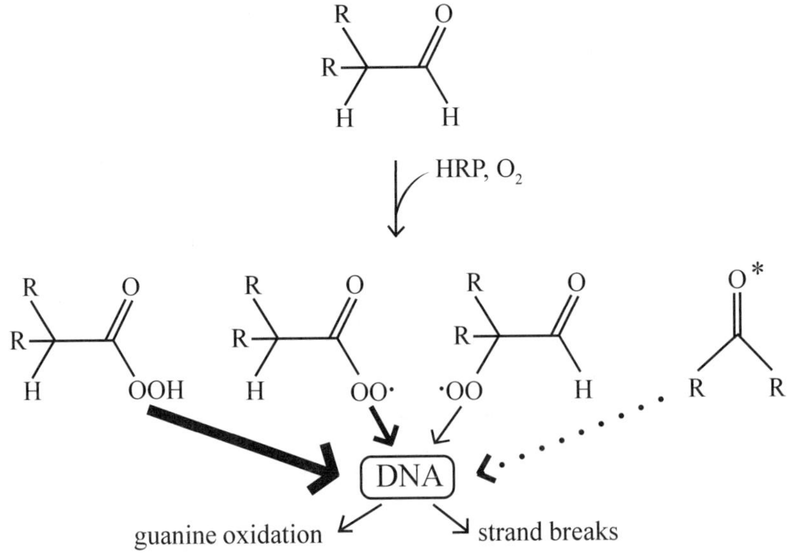

Fig. 4. The major and minor species involved in DNA damage
in the HRP-catalyzed autoxidation of isobutanal
(reprinted from [488] with permission).

Polymorphonuclear leukocytes (PMNs) are rich in peroxidases and the oxidation of isobutanal was studied with PMNs. I have already discussed the formation of electronically excited singlet oxygen by PMNs via the myeloperoxidase halide-H_2O_2 reaction. In addition to the generation of singlet oxygen, PMNs oxidize isobutanal to yield triplet acetone [489]. The triplet acetone causes considerable damage to the PMNs, however, PMNs which lack myeloperoxidase are not affected. In summary we conclude that PMNs not only produce superoxide, hydroxyl radicals and singlet oxygen, but also electronically excited carbonyl compounds, which damage the PMNs.

Another example of enzyme-catalyzed formation of triplet states is the oxidation of phenylacetaldehyde [490], which gives triplet benzaldehyde:

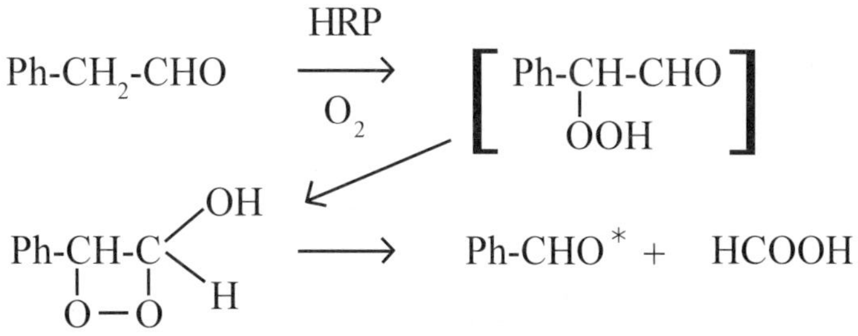

$$Ph\text{-}CH_2\text{-}CHO \xrightarrow[O_2]{HRP} \left[\begin{array}{c} Ph\text{-}CH\text{-}CHO \\ | \\ OOH \end{array} \right]$$

$$Ph\text{-}CH\text{-}C\underset{O-O}{\overset{OH}{\diagdown}}H \longrightarrow Ph\text{-}CHO^* + HCOOH$$

Electronically excited benzaldehyde is generated *in situ* in polymorphonuclear leukocytes through exposure to phenylacetaldehyde through the participation of myeloperoxidase in this process. The reaction initiates lipid peroxidation as shown by malonaldehyde formation [490].

2. Oxidation of fatty acids

Fatty acids are degraded by the α-oxidase system to the next lower aldehyde. This process has been postulated to proceed via the intermediacy of a dioxetanone (α-peroxylactone) [78]:

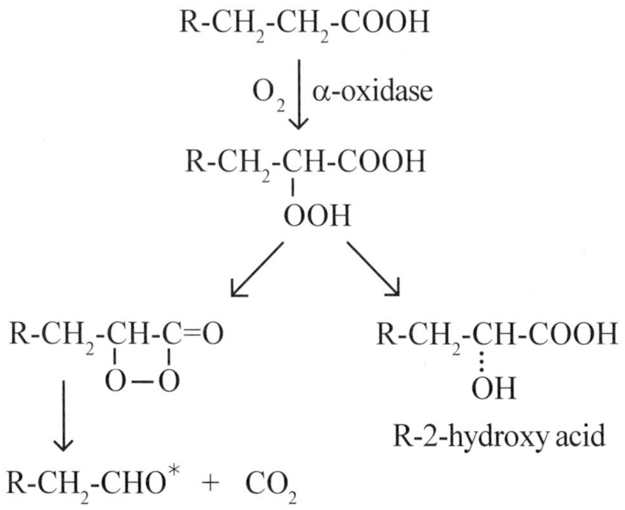

$$R\text{-}CH_2\text{-}CH_2\text{-}COOH$$

$$\downarrow O_2 \quad \alpha\text{-oxidase}$$

$$R\text{-}CH_2\text{-}CH\text{-}COOH$$
$$| \quad OOH$$

$$R\text{-}CH_2\text{-}CH\text{-}C{=}O \qquad R\text{-}CH_2\text{-}CH\text{-}COOH$$
$$O-O \qquad\qquad\qquad OH$$
$$\qquad\qquad\qquad\qquad R\text{-}2\text{-hydroxy acid}$$

$$R\text{-}CH_2\text{-}CHO^* + CO_2$$

In addition to the peroxylactone the initially formed α-hydroper-oxide can also form the α-hydroxy acid. This procedure has recently been used by Adam and coworkers [491] for the synthesis of enantiomerically pure R-2-hydroxy acids. This is the first synthesis of optically pure 2-hydroxy acids by molecular oxygen catalyzed by the α-oxidase of peas (*pisum sativum*).

3. Oxidation of Schiff bases

Aldehydes as well as amines are ubiquitous in biological systems. These two types of substrates react to form a Schiff base:

$$RCHO \; + \; H_2NR' \longrightarrow RCH=NR' \; + \; H_2O$$

The oxidation of Schiff bases by peroxidases has been examined and found to produce triplet carbonyl compounds [492]. Schiff bases have been shown to damage PMNs via the following pathway:

$$RCH_2CH=NR' \rightleftarrows RCH=CH-NHR'$$

$$O_2 \; \downarrow \; \text{peroxidase}$$

$$\underset{\underset{O-O}{|\quad\;|}}{RCH-CH-NHR'} \longrightarrow RCHO^* \; + \; R'NHCHO$$
$$\text{triplet}$$

The enzymatic oxidation of Schiff bases may, therefore, be involved in many pathological processes (liver disease).

4. Oxidation of polycyclic aromatic hydrocarbons

It has been known for a long time that PAHs have to be enzymatically activated in order to exert their carcinogenic potential. Benzo(a)pyrene (B(a)P) is metabolized by the microsomal cytochrome P450 enzyme to give the 7,8-dihydrodiol and subsequently the 7,8-dihydrodiol-9,10-epoxide, which is the ultimate carcinogen.

In the oxidation of B(a)P by cytochrome P450 chemiluminescence was observed [493], clearly indicating the formation of an excited state. It was proposed that this occurs via the intermediate formation of a dioxetane, which decomposes to an excited state of a dialdehyde [494-496]. These enzymatic oxidations of B(a)P are summarized in Fig. 5. The formation of a dioxetane was demonstrated non-enzymatically by reacting the 7,8-diol with singlet oxygen [496], which is a well-known reagent for dioxetane formation [497].

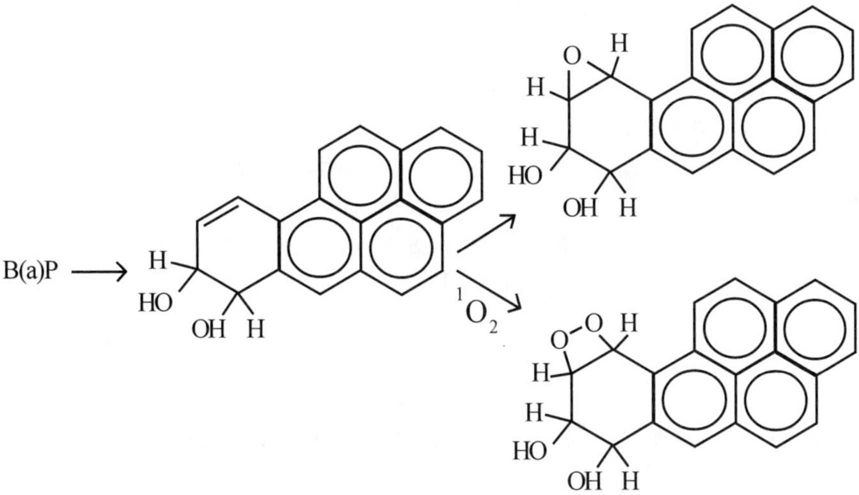

Fig. 5. Oxidation of benzo(a)pyrene

5. Lipid peroxidation

Lipid peroxidation is a very important process in oxidative stress. It has been known for some time that lipid peroxidation is accompanied by low levels of chemiluminescence. The formation of electronically excited carbonyl compounds and singlet oxygen in the reaction of alkyl peroxyl radicals (the Russell mechanism) has been well established.

Several authors have suggested [498-500] that production of electronically excited states in lipid peroxidation proceeds via thermal decomposition of dioxetanes, which are formed in a [2+2] cycloaddition of $^1O_2$ to the unsaturated bonds of polyunsaturated fatty acids (PUFAs). These suggestions are surprising, since it has been well established that dioxetane formation via olefin and $^1O_2$ only occurs with electron-rich or highly substituted olefins [331]. One such example is the oxidation of B(a)P as shown above. Chemiluminescence only tells us that some excited state is produced, but does not pinpoint the structure of the excited intermediate. It does not tell us anything about the mechanism. Recent studies by DiMascio et al. [501] indicate that in lipid peroxidation dioxetanes are not intermediates, in agreement with long- standing chemical knowledge.

Summary. In addition to oxygen-derived radicals and singlet oxygen, some enzymatic oxidations have been shown to produce triplet carbonyl compounds. These oxidations involve common substrates available *in vivo*, like aldehydes, carboxylic acids and Schiff bases, as well as xenobiotics (B(a)P). The triplet excited product initiates many pathological processes. I shall discuss the interaction of dioxetanes with DNA in Chapter 6.

H. ENDOTHELIUM-DERIVED RELAXING FACTOR

1. Introduction

Endothelial cells cover the interior wall of blood vessels. It has been generally assumed that endothelial cells have only a mechanical protective function. In 1980, Furchgott and Zawadzki [87] published their seminal work on isolated rabbit aorta. These authors stimulated the aorta with acetylcholine (a well-known vasodilator) and observed the vasodilation effect. They then removed the endothelial cell layer and observed that stimulation no longer produced relaxation. This finding clearly showed the release of an endothelium-derived relaxing factor (EDRF). It is now well-known that the vascular endothelium is very important in the regulation of blood vessel tone and permeability, the coagulation of blood, and the activity of leukocytes and platelets. The endothelium regulates the vascular tone by releasing powerful vasodilators, such as prostacyclin and

EDRF. EDRF relaxes vascular smooth muscle and inhibits platelet aggregation and adhesion via the elevated production of c-GMP. EDRF is a highly unstable ('half-life' 3-50 s under normal physiological conditions) humoral agent released from the endothelium by a variety of stimulating agents like acetylcholine (ACh), calcium ionophores, ADP, ATP, bradykinin, histamines, serotonin, thrombin, arachidonic acid (AA) and a number of organic nitro compounds (nitroglycerin, nitroprusside, amyl nitrite). Most of the relaxations (except for those produced by AA and bradykinin) are not inhibited by cycloxygenase enzyme inhibitors (i.e., by blocking the prostaglandin biosynthesis).

Like most other fields of scientific investigations the study of EDRF has had its share of unexplainable paradoxes. The reason is quite simply that not all the facts are known and, therefore, there are unresolved puzzles. Acetylcholine, a potent vasodilator *in vivo*, often produced no relaxation or even contraction of isolated preparations of arteries *in vitro*. The results of Furchgott and Zawadzki [87] resolved the paradox. In the isolated preparations the endothelial cells were unintentionally removed during the course of their preparation. Again we have to be cautious and not jump to conclusions until all the relevant facts are known.

The endothelium-dependent relaxation stimulated by ACh was found to increase c-GMP, and cyclic guanylate cyclase was markedly stimulated by hydroperoxides of arachidonic acid and by radicals like NO and ·OH. These observations were made many years before the discovery of EDRF (reviewed by Furchgott [502]) and led some investigators [503] to speculate that EDRF may be a short-lived hydroperoxide or radical resulting from the oxidation of arachidonic acid by a lipoxygenase pathway. It was presumed that this radical stimulates the guanyl cyclase of the arterial smooth muscle and causes an increase in c-GMP, which then somehow activates relaxation.

The proposal that EDRF is a short-lived radical was received with great skepticism. As was the case with $O_2^{\cdot-}/HO_2\cdot$ and ·OH, it was difficult to accept that a short-lived and reactive radical could carry out such important functions as vasodilation and neurotransmission. It took almost a decade before EDRF was identified as nitric oxide [88-90]. In order to decide whether EDRF and NO are identical, a detailed understanding of NO chemistry is essential and a comparison with the chemistry of EDRF is necessary. The chemistry of EDRF/NO will be discussed in Chapter 7.

The relaxant effect of NO on vascular smooth muscle is analogous to that of the nitrovasodilators, which liberate NO, activate guanylate cyclase, and cause an increase in c-GMP [504]. It is now believed that c-GMP-dependent protein phosphorylation and dephosphorylation of myosin light chains in arterial muscle may mediate relaxation.

2. Enzymes for NO synthesis

The enzymes responsible for NO synthesis are known as NO synthases. There are many reviews on this subject [91,505-509]. The NO synthases have been identified and characterized not only in endothelial cells [510-513], but also in macrophages [513-515], neutrophils [516,517], adrenal glands [518], and brain tissue [513,519-522].

NO synthase requires L-arginine as a substrate and yields L-citrulline and nitric oxide as the only products [523-526]. In addition, several coenzymes/cofactors are needed for NO synthase activity. These cofactors are NADPH, FMN, FAD, BH_4, protoporphyrin, calmodulin (a Ca^{2+}-binding protein) [505-509], and oxygen [527].

Two different types of NO synthases have been characterized: one enzyme (the endothelial and neuronal type) is constitutive and depends on Ca^{2+} and calmodulin for its activity, whereas the other (found in macrophages and neutrophils) is inducible and Ca^{2+} and calmodulin independent.

a. Endothelial enzyme

The nitric oxide (NO) synthase has been well characterized in endothelial cells [506-508,512,513]. This enzymatic activity depends on Ca^{2+} and uses L-arginine as the substrate. Several cofactors have been identified: NADPH, BH_4 (tetrahydrobiopterin), FAD and oxygen. Most conditions which lead to an increase in intracellular Ca^{2+} (limited cell injury, ischemia or inactivation of Ca^{2+}-pumps) also lead to the production of NO [502]. L-arginine has been demonstrated to be the only substrate for the endothelial NO synthase, the D-arginine is not effective [524].

NO synthetase may be inhibited in a competitive manner by L-arginine analogs like N^G-monomethyl-L-arginine (L-NMMA) and N^G-nitro-L-arginine (L-NNA) and others [91,505,524]. This inhibition is overcome by addition of more L-arginine. These antagonists are important tools for studying the effects of NO *in vivo*. Whenever a physiological or pathological process is affected by these antagonists, the involvement of NO appears indicated in the process. NO is definitely the simplest hormone. It regulates blood flow and pressure. Whenever endothelial cells are damaged, as in atherosclerosis, by toxic chemicals, or in physical trauma, there is no longer a vasodilator response. Vasodilators such as acetylcholine, arachidonic acid (AA), histamine, vasopressin, serotonin, bradykinin, ATP, thrombin and calcium ionophore A 23187 all mediate vasodilation via the release of EDRF/NO. This in turn activates soluble guanylate cyclase in smooth muscle and produces c-GMP [504,505]. Organic nitrates, many of which are well-known vasodilators, are believed to exert their effect via enzymatic or chemical denitration to yield NO^- and subsequent formation of NO and RSNO. Some of these vasodilators are nitroglycerin and amyl nitrite. Another well known vasodilator is nitroprusside. This complex already has the NO radical incorporated:

$$Na_2[Fe(CN)_5NO] \longrightarrow \cdot NO$$

Drugs which release NO perform a beneficial function during ischemia if given before ischemia. The release of NO from nitroprusside is more complex than the above equation indicates [528,529] and will be discussed in Chapter 7.

Serotonin is released from aggregated platelets acting on endo-thelial cells producing a signal for relaxation of the coronary smooth muscle cells. Aggregated platelets plug arteries, but they induce the endothelial cells to release NO (after reperfusion) thus relaxing the arteries [502].

Arachidonic acid, as well as other unsaturated fatty acids, stimulate endothelial cells to release EDRF [502]. AA is metabolized via the action of lipoxygenase and cycloxygenase to give a variety of vasoactive compounds, e.g., the prostaglandins. However, if this is the effect of AA, then the relaxation should be inhibited by cycloxygenase inhibitors (like aspirin), which was not observed [502]. This result shows that AA does not act via a cycloxygenase pathway, but by activation of the NO synthetase.

b. Macrophage enzyme

Another type of enzyme which converts L-arginine to NO has been found in macrophages [514,530,531]. The properties of this enzyme differ substantially from the endothelial enzyme. Macrophages produce NO at concentrations which are high enough to be cytotoxic [531]. The enzyme is inducible and does not depend on Ca^{2+}. It utilizes NADPH, and L-arginine and as a cofactor, tetrahydrobiopterin (BH_4) and possibly Mg^{2+} [532]. In addition, FAD and GSH have been shown to be cofactors [533]. The most commonly used stimuli for NO production are interferonγ (IFN-γ), $E.$ $coli$ lipopoly-saccharide (LPS) or the tumor necrosis factor (TNF).

Macrophages were stimulated with LPS and it was found that these macrophages produced NO_2^-/NO_3^-. In the presence of morpholine, the formation of N-nitrosomorpholine was observed, thus, indicating the intermediacy of NO [530]. L-arginine was the only amino acid essential for this synthesis. By using ^{15}N-labeled L-arginine (L-[guanido-$^{15}N_2$] arginine) it was established that the NO_2^-/NO_3^- and the NO group in N-nitroso-morpholine was derived exclusively from one or both the terminal guanido nitrogens of L-arginine [515,530] (the importance of reliable and sensitive analytical techniques!).

The formation of NO_2^-/NO_3^- from L-arginine is, of course, an oxidation and we may, therefore, ask whether the well-known respiratory burst, which produces a variety of reactive species ($O_2^{-\cdot}/HO_2^\cdot$, $^1O_2$, H_2O_2, HOCl and $\cdot OH$) may be involved in the synthesis of NO and NO_2^-/NO_3^-. By using macrophage mutations, which could not synthesize superoxide radical and, thus, did not exhibit the respiratory burst, it was shown that the formation of NO_2^-/NO_3^- did not depend on the respiratory burst [530]. Macrophages and neutrophils, therefore, in addition to the respiratory burst, have an additional pathway for the formation of ·NO, which combines with O_2^- to give the highly reactive peroxynitrite. The peroxynitrite is an additional weapon in the antimicrobacterial armamentarium of macrophages and neutrophils.

Macrophages can carry out N-nitrosation of amines [530], thus producing carcinogenic N-nitrosamines, and converting precarcinogens to ultimate carcinogens. Chronic inflammation of the intestine leads to cancer through nitrosation of intestinal amines.

The formation of NO and O_2^- proceeds via two independent pathways, but both pathways are activated by the same stimuli [530]. We have the possibility that these two radicals combine. This very important radical-radical combination is the cause of the deleterious effects of nitric oxide and will be discussed in detail in Chapter 7.

Contrary to the endothelial cell enzyme, the macrophage enzyme does not use exclusively L-arginine as substrate. Substrates of similiar structure, like L-homoarginine, L-arginine-methyl ester, L-arginine-amide and L-arginyl-L-aspartate dipeptide serve as substrates [469]. Antagonists are N^G-nitro-L-arginine (L-NNA), N^G-nitro-L-arginine methylester (L-NAME), and N^G-monomethyl-L-arginine (L-NMMA).

c. Neutrophil enzyme

The macrophage type of NO synthetase has also been found in neutrophils, i.e., it is inducible and Ca^{2+}-independent [516,517]. The NO-synthesizing activity of neutrophils was stimulated by N-formyl-L-methionyl-L-leucyl-L-phenylalanine (FMLP), by platelet-activating factor or by leukotriene B_4.

d. Neuronal NO synthase

Another type of NO synthase is found in neurons. Neurotransmission by acetylcholine (ACh), glutamate and glycine has long been known to be associated with elevated c-GMP levels in the brain and cerebellum (for a review see Garthwaite [534]). It was shown by Garthwaite et al. [519]

that glutamate induces the release of a molecule similiar to EDRF, by acting on the NMDA (N-methylarspartate) receptors on the cerebellar cells. It was found that NO mediates glutamate-linked enhancement of c-GMP levels in the cerebellum [535]. Long ago it was shown that NO stimulates soluble guanylate cyclase in mouse cerebral cortex [536]. These observations, together with the discovery of the L-arginine-NO pathway, led to many studies on the existence of this pathway in the central nervous system.

The isolation and characterization of NO synthase from rat ce-rebellum [537], the development of antibodies to the NO synthase, and the availability of a histochemical method has made possible the localization of this enzyme in a wide variety of cell types within the nervous system [538]. This example shows again the importance of an easy to use, reliable analytical technique for the advance of scientific research. The NO synthase was discovered in the cerebellum, striatum, cerebral cortex, hippocampus, hypothalamus, visual system, and peripheral nervous system [539,540]. The proposal that a simple highly reactive radical like NO serves as a neurotransmitter has been met with great skepticism. We have again a repeat of the oxygen radical story. It was difficult to accept the idea that radicals, which are known to be highly reactive, could play any significant role in a biological system. However, there is no longer any doubt that oxygen radicals, as well as NO, play an important part in biological systems. NO has been established as an intercellular messenger of neurotransmission. The enzyme NO synthase catalyzes the synthesis of NO from L-arginine via a Ca^{2+}/calmodulin-dependent mechanism. Increases in intracellular Ca^{2+} resulting from various causes activates the enzyme. The target for NO action is soluble guanylate cyclase causing increases in c-GMP levels throughout the nervous system. The c-GMP may regulate protein kinases, phosphodiesterases and ion channels. Activation of soluble guanylate cyclase is the mechanism of action of NO in both the vascular and nervous systems.

Studies on rat synoptosomal cytosol have shown that addition of L-arginine in the presence of NADPH increased the formation of NO and L-citrulline and stimulated soluble guanylate cyclase. The nNOS was inactive at the resting free Ca^{2+} concentration of ca. 80 nM [541], but was fully active at Ca^{2+} concentration of ca. 400 nM [520]. From these results we may conclude that similar to the vascular endothelium, an increase in intracellular Ca^{2+} is the physiological mechanism for stimulating NO synthase. However, physiological Ca^{2+} levels, which are essential for NO synthesis, were found to inhibit the soluble guanylate cyclase in the brain [520,542]. This observation was interpreted to mean that guanylate cyclase is not activated in cells which generate NO, but only in effector cells, i.e., NO acts as an intercellular messenger [520]. Neuronal NO synthase has been purified from rat cerebellum and has been shown to be calmodulin-dependent [537].

The NO produced by the different NO synthases has many physiological as well as toxicological effects in the vascular system, in neurotransmission and in the immune system. These reactions have been extensively reviewed [91,502,505,543-553]. A whole issue of the *Annals of Medicine* has been devoted to the role of NO as a modulator, mediator and effector molecule and its role in various diseases [553]. In order to understand these processes a detailed knowledge of the chemistry of NO is necessary, which I shall discuss in Chapter 7.

3. Mechanism of NO formation

There are a number of enzymatic reactions which give rise to NO. These reactions involve azide (N_3^-), ammonia (NH_3), hydroxylamine (NH_2OH), hydrazine (NH_2-NH_2), and L-arginine. The EDRF, however, is exclusively derived from L-arginine and yields NO and L-citrulline as the only two products. The NO reacts with oxygen to give nitrite (NO_2^-) and nitrate (NO_3^-) as final products. Through the use of ^{15}N-labeled L-[guanido ^{15}N] arginine and gas chromatography-mass spectrometry it was shown that all the NO is derived from the terminal guanidino-group of L-arginine both via the endothelial NO synthase [524] and via the macrophage enzyme [515,530]. Several mechanisms have been suggested [508], but the details are still unresolved. One suggested mechanism proceeds as follows [515,532]:

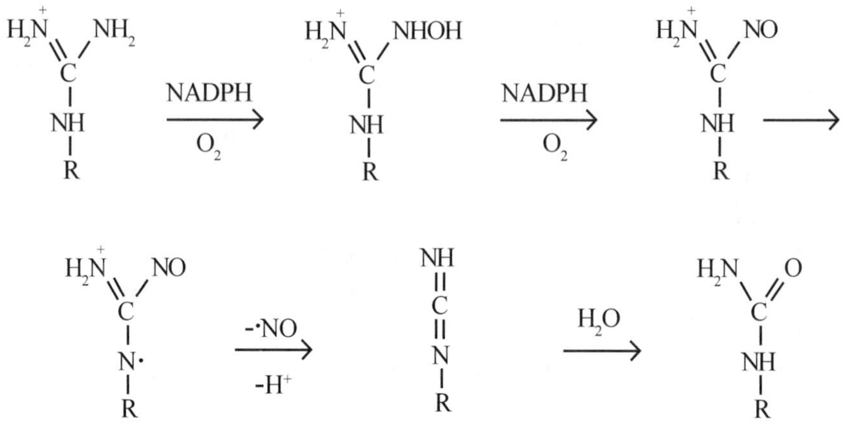

$$\underset{R}{\overset{\overset{\displaystyle H_2\overset{+}{N}\diagdown_{\displaystyle C}\diagup NH_2}{|}}{\underset{NH}{|}}} \quad \xrightarrow[O_2]{NADPH} \quad \underset{R}{\overset{\overset{\displaystyle H_2\overset{+}{N}\diagdown_{\displaystyle C}\diagup NHOH}{|}}{\underset{NH}{|}}} \quad \xrightarrow[O_2]{NADPH} \quad \underset{R}{\overset{\overset{\displaystyle H_2\overset{+}{N}\diagdown_{\displaystyle C}\diagup NO}{|}}{\underset{NH}{|}}} \quad \longrightarrow$$

$$\underset{R}{\overset{\overset{\displaystyle H_2\overset{+}{N}\diagdown_{\displaystyle C}\diagup NO}{|}}{\underset{N\cdot}{|}}} \quad \xrightarrow[-H^+]{-\cdot NO} \quad \underset{R}{\overset{\overset{\displaystyle NH}{\|}}{\underset{N}{\overset{\displaystyle C}{\|}}}} \quad \xrightarrow{H_2O} \quad \underset{R}{\overset{\overset{\displaystyle H_2N\diagdown_{\displaystyle C}\diagup O}{\|}}{\underset{NH}{|}}}$$

The first step in this transformation is hydroxylation. Hydroxy-lations by monooxygenases of aromatic amino acids, like phenylalanine, tyrosine, and tryptophan are well known. In these cases (6R)-tetrahydrobiopterin (BH$_4$) together with NADPH provide the necessary reducing equivalents for the enzyme leading to the activation of oxygen and subsequent hydroxylation of the substrate. The above mechanism appears *a priori* a reasonable assumption, but reasonable does not mean correct. It has been demonstrated [527] by the use of H$_2$O^{18} and gas chromatography-mass spectroscopy that the oxygen in L-citrulline is derived from molecular oxygen and not from water. Although the details of the mechanism leading from L-arginine to NO are yet to be elucidated, we can be sure of one fact: the NO produced by the endo-thelial enzyme is derived exclusively from L-arginine.

Chapter 6

REACTIONS OF REACTIVE OXYGEN METABOLITES WITH IMPORTANT BIOMOLECULES

Who would study and describe the living, starts
By driving the spirit out of the parts:
In the palm of his hand he holds all the sections,
Lacks nothing, except the spirit's connections.
                    Mephisto in Goethe's Faust, Part 1[*]

A. GENERAL REMARKS ABOUT OXYGEN RADICAL INVESTIGATIONS

The investigation of oxygen radical involvement in a biological system proceeds in different stages. First we study a simple chemical reaction under well defined and controlled conditions in dilute aqueous solutions. This is of course a drastic oversimplification. The second step consists of studying the reaction of these oxygen radicals with biologically important molecules, like proteins, enzymes, fatty acids, and DNA. In the third stage we approach more closely an *in vivo* situation. Experiments are carried out with whole cells and tissue cultures. The final stage consists of studies involving whole animals or human subjects.

All of these different stages of investigation require sophisticated analytical equipment. Since most radicals are short lived, highly reactive intermediates we can only rarely observe them directly. Instead we infer their existence from the reaction products they form.

---

[*] This quote from Goethe's Faust may give the impression that the author has joined the growing number of scientists on the anti-reductionist bandwagon. Nothing could be further from the truth. All I want to emphasize is that faced with the tremendous number of observations, we should not lose sight of the whole.

In order to apply this method we need a reaction which gives an easily identifiable and highly specific reaction product and the neces-sary analytical technique. The importance of the availability of an analytical technique for radical chemistry and biology cannot be overemphasized. Progress in the field was slow and had to await the development of sensitive analytical techniques. Once these techniques were available the whole field experienced an explosive growth. Some of these techniques are electron paramagnetic resonance spectroscopy (EPR), spin trapping, pulse radiolysis and more recently HPLC (high performance liquid chromatography), GC-mass spectroscopy, GC-MS-SIM (specific ion monitoring), and HPLC- EC (electrochemical assay). However one should remember that these techniques are only tools. If we use any tool in the wrong way we do not get the desired results and we may be led to the wrong conclusions. Wrong conclusions have, indeed, been drawn from EPR and spin-trapping techniques (see Chapter 3 on identification of hydroxyl radicals).

If we want to prove that radical R· reacts with substrate S to give product P, then we first of all need a clean and unambiguous source of radicals R·. We have, indeed, chemical and physicochemical methods available for the production of oxygen radicals ($O_2^{·-}$/$HO_2^·$, ·OH) and reactive oxygen metabolites (singlet oxygen, excited carbonyl compounds and hydroperoxides). These methods are radiolysis, photolysis, one-electron transfer reactions and photosensitization.

It is fair to say that most of our knowledge about the reactivity of hydroxyl radicals comes from pulse radiolysis studies [65,99]. Once this technique was available the fields of radiation chemistry and biology took off exponentially. Pulse radiolysis made available thousands of rate constants for the reaction of hydroxyl radicals with organic and inorganic compounds, and biological matter [99]. These rate constants gave the scientists working in the field of radical biology another tool for the identification of radicals. This tool is competition kinetics. Let us consider the following reactions:

$$\cdot OH \;+\; S_1 \;\xrightarrow{\;k_1\;}\; P_1$$

$$\cdot OH \;+\; S_2 \;\xrightarrow{\;k_2\;}\; P_2$$

If the reaction of an unknown radical with S_1 and S_2 gives a ratio k_1/k_2 different from $\cdot OH$ we can safely conclude that the unknown radical under examination is not $\cdot OH$. If, however, the ratio is identical we cannot conclude that the unknown radical is actually $\cdot OH$ either, but only that it is kinetically indistinguishable from $\cdot OH$. Like some other tools mentioned above, competition kinetics has led to some wrong conclusions. Since $\cdot OH$ is highly reactive, it never survives long enough to reach the vital biological target, e.g., the DNA. The $\cdot OH$ has to be formed in close proximity to DNA in a "site-specific" reaction (Chapter 3). Competition kinetics only works if the interacting species are "free" to compete.

The three lowest electronic states of molecular oxygen (one triplet and two singlet states) were spectroscopically observed over 60 years ago by the Herzbergs [554-556]. A review of the history of singlet oxygen research has been written by Khan [330] and Kasha [557]. However these discoveries did not draw much attention except among spectroscopists and some astrophysicists, until the second discovery in this field, namely the formation of singlet oxygen ($^1\Delta_g$) via a chemical reaction. Khan and Kasha identified singlet oxygen in the hydrogen peroxide-hypochlorite reaction [70]. This study started an explosive growth in singlet oxygen research. I like to quote Khan [330], "What began as an exercise in the calibration of a spectrograph, became one of the most active interdisciplinary research areas of recent times". A similar story may be told about the hydroxyl radical. The first hint of its formation was obtained in 1898 by Fenton, but it took many decades before the hydroxyl radical was postulated as intermediate in the Fenton reaction. The reactions of hydroxyl radicals were studied seriously during the 1940s when the field of radiation chemistry started during the time of the Manhattan project of World War II. It is amazing that a field of inquiry, which started as a spin-off of the most destructive device ever invented by humankind, has also led to an explosion of research into life processes such as mutation, carcinogenesis, aging, defenses against invading microorganisms, and the cause and development of numerous diseases. It has even led to a possible explanation for the preponderance of sexual versus asexual reproduction [20]. In our budget-conscious and mission-oriented society we often hear the question what is basic science good for? Maybe it is only a "Glasperlenspiel". Well, the story of singlet oxygen, the hydroxyl radical and nitric oxide give us the answer: a field of basic research no matter how esoteric it might appear, can have unforeseen importance in completely unrelated areas of science.

An organism or cell may be compared to a computer. The nucleus of the cell contains the DNA, which is the computer program. Despite a careful screening of DNA from damaging outside influences (by a histone sheath and a nuclear membrane), the DNA sustains damage (computer virus). This damage may be beneficial or harmful and can be partially repaired. However, in order for the computer to carry out all the instructions of the program, it needs energy. This energy is supplied by the mitochondria, the power plant of the body. The mitochondrial respiratory chain transforms oxygen to water in a series of steps, and conserves the released energy in the form of chemical energy. This energy storage device is ATP.

We can damage the computer by damage (virus) to the program or by knocking out the power plant. The theory of aging originally proposed by Harman [558] assumed aging to be due to continuous accumulation of nuclear DNA damage caused by ROMs during the life span of the organism. Recently, evidence has emerged which seems to show that damage to the power plant (mitochondria) is the determining factor in aging [559].

In a chemical system we can study one reaction at a time. If we know the rate constants we can predict how an additive may affect the yield of a product. This method of competition kinetics has been frequently used as a probe for hydroxyl radicals. In a biological system we have a higher level of complexity. Many reactions may and do take place simultaneously. We have to consider many competing reactions.

Competition is a very important concept in biology and chemistry. We have competition between species, among members of the same species, and between superorganisms. Competition is also important at the microscopic level, between cells (cancer cells versus normal cells, phagocytes versus bacteria) all the way down to the molecular level. Antioxidants compete with biomolecules for reactive oxygen species, heme proteins or oxygen compete with superoxide radical for nitric oxide, SOD and NO compete for $O_2^{\cdot-}$, and Fe^{2+} or Cu^+ compete with CAT for H_2O_2 (see Chapters 7 and 8). However, competition does not stop at the molecular level, but is also present at the sub-molecular (electronic) level. Two different atoms connected by a chemical bond do not share the electrons equally. Electrons are attracted towards the more electronegative atom. It is a well established custom among chemists to explain chemical reaction mechanisms by shifting electrons around. Electrons are the main actors on the stage of life, because electrons are the main actors on the stage of chemistry. A world consisting only of hydrocarbons (-C-C-) would indeed be a dull chemical world. Nothing would change, but nothing would be alive. Change is essential for life.

The three principal mechanisms for cell injury and tissue damage induced by radicals and electronically excited states are: 1. the damage is caused directly by the ROMs through the direct attack on sensitive and essential targets, 2. the damage is mediated by the destruction of the lipid bilayer via lipid peroxidation and 3. the damaging effect is due to secondary reactions of the lipid degradation products (aldehydes) with critical targets (enzymes, DNA). These pathways have been extensively investigated and reviewed [560-565]. Since ROMs are formed via normal metabolism, we observe tissue damage even in the absence of oxidative stress. The term "oxidative stress" has been defined by Sies [393] as a shift in the prooxidant/antioxidant status of a system. Another definition has been proposed by Smith [565]: any measurable shift in one or more redox couples to more electron-deficient (oxidized) equilibria or steady states.

Next I shall discuss the reactions of hydroxyl radicals and electronically excited states with some important biomolecules.

B. REACTIONS OF OXYGEN RADICALS WITH FATTY ACIDS

The membrane of cells and subcellular organelles is of utmost importance to the viability of cells and their organelles. The most significant function of membranes is to provide a barrier. The membrane keeps certain compounds in and others out. Without the membrane the many reactions taking place within a cell would lead to chemical chaos and eventual cell death. It is, therefore, not surprising that the reaction of oxygen radicals with membranes and their constituent parts, the fatty acids, have been studied extensively [565-573]. Many of these studies were carried out long before the field of radical biology and medicine became fashionable. Food chemists have been interested in the oxidation of fatty acids, since this process leads to rancification through degradation. As I have already pointed out under conditions of low pH, as exists within the phospholipid bilayer, the perhydroxyl (HOO·) radical contributes significantly to the reactivity of the superoxide radical anion. The perhydroxyl radical can easily abstract a hydrogen from polyunsaturated fatty acids (PUFAs). The important unsaturated fatty acids are linoleic acid (18:2), arachidonic acid (20:4) and docosahexaenoic acid (22:6). These PUFAs all have one or more $-CH_2-$ groups adjacent to two olefinic bonds, i.e., the structural unit $-CH=CH-CH_2-CH=CH-$. Both the hydroxyl and the perhydroxyl radical can abstract a hydrogen from these methylene groups. These reactions are summarized in Scheme 1.

Scheme 1. (from [573] with permission).

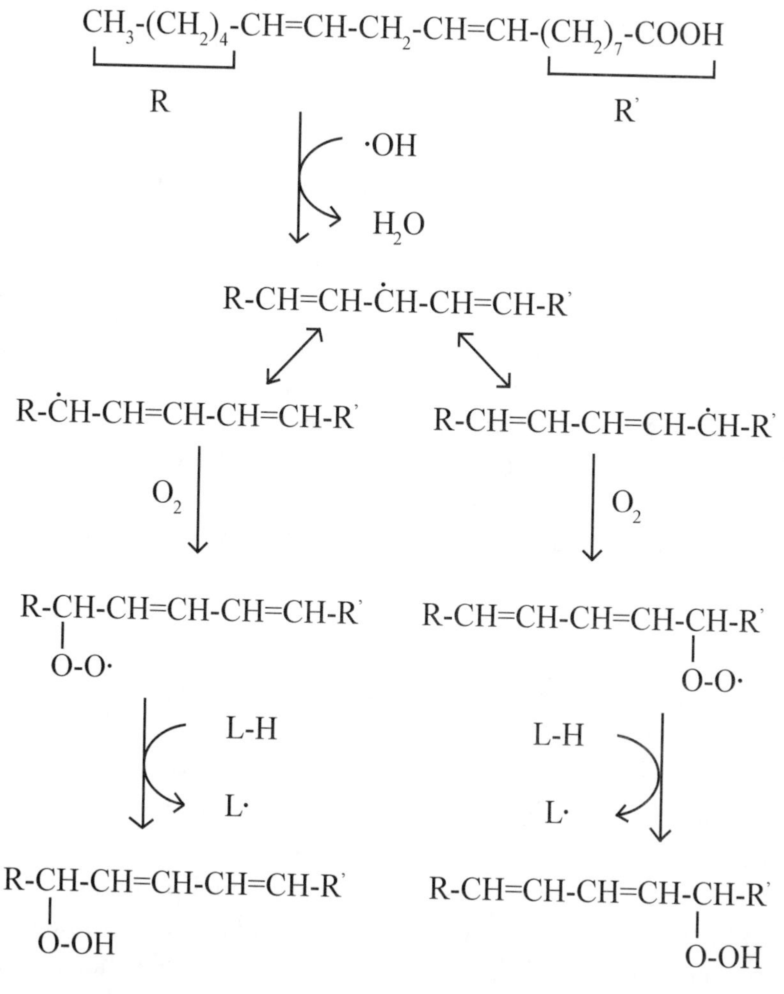

$$CH_3\text{-}(CH_2)_4\text{-}CH=CH\text{-}CH_2\text{-}CH=CH\text{-}(CH_2)_7\text{-}COOH$$

$$\underbrace{\phantom{CH_3\text{-}(CH_2)_4}}_{R} \qquad\qquad \underbrace{\phantom{(CH_2)_7\text{-}COOH}}_{R'}$$

$\cdot OH$

H_2O

$$R\text{-}CH=CH\text{-}\dot{C}H\text{-}CH=CH\text{-}R'$$

$$R\text{-}\dot{C}H\text{-}CH=CH\text{-}CH=CH\text{-}R' \qquad\qquad R\text{-}CH=CH\text{-}CH=CH\text{-}\dot{C}H\text{-}R'$$

$O_2 \qquad\qquad\qquad\qquad\qquad O_2$

$$\begin{array}{l} R\text{-}CH\text{-}CH=CH\text{-}CH=CH\text{-}R' \\ \quad | \\ \quad O\text{-}O\cdot \end{array} \qquad \begin{array}{l} R\text{-}CH=CH\text{-}CH=CH\text{-}CH\text{-}R' \\ \qquad\qquad\qquad\qquad\quad | \\ \qquad\qquad\qquad\qquad\quad O\text{-}O\cdot \end{array}$$

$L\text{-}H \qquad\qquad\qquad L\text{-}H$

$L\cdot \qquad\qquad\qquad L\cdot$

$$\begin{array}{l} R\text{-}CH\text{-}CH=CH\text{-}CH=CH\text{-}R' \\ \quad | \\ \quad O\text{-}OH \end{array} \qquad \begin{array}{l} R\text{-}CH=CH\text{-}CH=CH\text{-}CH\text{-}R' \\ \qquad\qquad\qquad\qquad\quad | \\ \qquad\qquad\qquad\qquad\quad O\text{-}OH \end{array}$$

The initially formed isomeric peroxyl radicals abstract a hydrogen atom from the PUFA (LH). In this case we have a chain reaction, which is known as autoxidation. If the hydrogen is abstracted from another molecule, like glutathione (R-SH) or α-tocopherol (TOH), the chain reaction is terminated. Because of this termination of autoxidation, these compounds are known as inhibitors or antioxidants.

Another fate of the initially formed radical L· is dimerization or dismutation. These reactions only occur if we have a high steady-state concentration of L·:

$$L· \; + \; L· \; \longrightarrow \; L\text{-}L$$

$$L· \; + \; LOO· \; \longrightarrow \; \text{molecular products}$$

$$LOO· \; + \; LOO· \; \longrightarrow \; 2\,LO· \; + \; O_2$$

The dismutation may lead to singlet oxygen or electronically excited carbonyl compounds (Russell mechanism), which in turn may cause further damage to the PUFAs. The lipid peroxyl radicals may also add to a double bond within the same molecule to form cyclic peroxides. (Chapter 3).

Peroxyl radicals also react with some olefins to yield epoxides, which are electrophilic and react with -NH$_2$ groups:

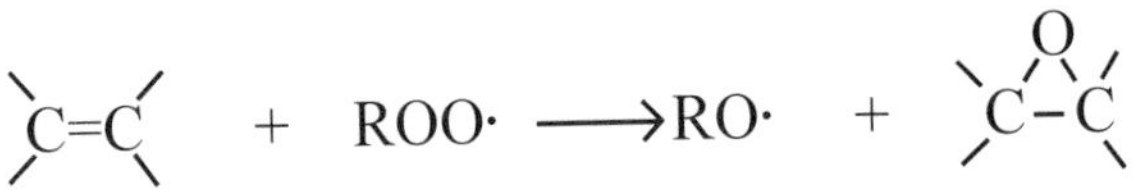

$$\text{C=C} \; + \; ROO· \; \longrightarrow RO· \; + \; \text{C-C (epoxide)}$$

Peroxyl radicals are involved in the activation of some carcinogens (procarcinogen to ultimate carcinogen), which has been extensively reviewed [574,575].

The lipid hydroperoxides are not stable final products, but undergo further reactions to give alkoxyl radicals. The first line of defense against LOOH is the peroxidase enzymes, especially the selenium-dependent glutathione peroxidase, which catalyzes the following reaction:

$$LOOH \; + 2\,GSH \; \longrightarrow \; GSSG \; + \; LOH \; + \; H_2O$$

The GSSG is reduced again to GSH by glutathione reductase which uses NADPH as a hydrogen donor:

$$GSSG + NADPH + H^+ \longrightarrow 2\,GSH + NADP^+$$

The ratio GSH/GSSG has been used as a marker for oxidative stress. The lipid hydroperoxides react in the presence of metal ions (Fe^{2+}, Cu^+) in a Fenton-type reaction:

$$LOOH + Fe^{2+} \longrightarrow LO\cdot + H^+ + Fe^{3+}$$

The complexation of iron with chelating agents often promotes lipid peroxidation, because it keeps the Fe^{2+} in solution. The commonly used complexing agent EDTA promotes lipid peroxidation and degradation, whereas deferoxamine inhibits peroxidation. Some heme proteins like hemoglobin, methemoglobin and cytochrome P450 are effective in promoting lipid hydroperoxide degradation, whereas transferrin and lactoferrin are not.

Iron (III) may also oxidize the lipid hydroperoxide to form again the original peroxyl radical:

$$LOOH + Fe^{3+} \longrightarrow Fe^{2+} + H^+ + LOO\cdot$$

It is, therefore, important for lipid degradation to keep the iron in the Fe(II) state. This type of reduction is accomplished by NADH or ascorbic acid.

The alkoxyl radicals formed in the Fenton-type reaction undergo fragmentation (β-cleavage) to yield a smaller radical fragment and an aldehyde:

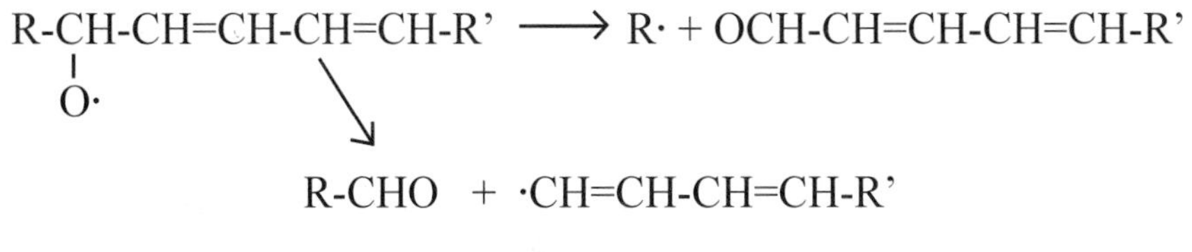

R-CH-CH=CH-CH=CH-R' $\longrightarrow$ R· + OCH-CH=CH-CH=CH-R'
 |
 O·

R-CHO + ·CH=CH-CH=CH-R'

In addition to fragmentation, the alkoxyl radicals may initiate further lipid peroxidation:

$$RO\cdot\ +\ LH\ \longrightarrow\ ROH\ +L\cdot\ \xrightarrow{\ O_2\ }\ LOO\cdot$$

Since each polyunsaturated fatty acid may produce several isomeric hydroperoxides and subsequently several isomeric alkoxyl radicals, which in turn can fragment in two ways, it is obvious that the final products obtained in lipid peroxidation, form indeed a very complex mixture. The use of GC-MS was essential for the analysis of these complex oxidation products [576-579]. The initially formed aldehydes with unsaturated bonds may undergo further oxidations.

Lipid hydroperoxides not only cause damage via formation of RO· and numerous aldehydes, but may also fulfill physiological functions. We again encounter a pro/con situation. Hydroperoxides can activate certain key enzymes, such as phospholipase, cycloxygenase and protein kinase C. The evidence for the physiological role of membrane hydroperoxides has been reviewed by Ursini et al. [580].

C. MECHANISMS OF LIPID PEROXIDATION-INDUCED PATHOGENESIS

A huge amount of evidence has accumulated to support the theory that lipid peroxidation is the major factor in the pathogenesis of many human diseases.

Lipid peroxidation causes damage of the membrane directly or indirectly. The direct effect is caused by a change in the membrane structure and its biophysical properties. The indirect effect is due to the release of a multitude of biologically active metabolites from lipid peroxidation.

The direct effect is due to crosslinking of the lipid chains. This changes the fluidity of the membrane and, thus, the mobility of the membrane-bound enzymes. Due to peroxidation, the membrane becomes leaky and molecules as big as enzymes have been shown to leak out of lysosomes or hepatocytes. In addition to disturbing the ability of membranes to act as a barrier, lipid peroxidation causes loss of ionic homeostasis, thus, destroying the compartmentalization and causing chemical chaos. The most important of these ions is Ca^{2+}. The relationship between oxidative stress and Ca^{2+} homeostasis has been thoroughly investigated and reviewed [581]. The Ca^{2+} levels inside cells are kept low (less than micromolar), whereas the concentration outside cells is in the millimolar range. Calcium is important in the activation of a number of enzymes, e.g., endonucleases, which degrade DNA. Ca^{2+}/calmodulin are important cofactors in the activation of endothelial and neuronal NO synthases (Chapter 5). Loss of Ca^{2+} homeostasis causes cells to lose metabolic control. Lipid peroxidation may also lead to deactivation of membrane ion pumps, which are responsible for ionic homeostasis.

Lipid peroxidation has a direct effect on the activity of membrane-bound enzymes. Enzymes located in the endoplasmic reticulum (cytochrome P-450) and enzymes in mitochondrial membranes or the Golgi apparatus are known to be susceptible to inhibition. Inactivation or activation of membrane-bound enzymes may be caused by alterations in the physical state of the membrane or by the indirect effect of the peroxidation-degradation products.

Indirect effects of lipid peroxidation are those mediated by its products. The products are biologically active and are stable, neutral molecular products (contrary to the peroxyl and alkoxyl radicals) and, therefore, diffuse away from their site of formation to cause damage at remote sites. We have already encountered a similiar state of affairs in considering the superoxide radical anion and H_2O_2. The superoxide radical anion cannot cross the membrane barrier, but H_2O_2 does.

Membranes contain a variety of PUFAs which have varying numbers of olefinic bonds and methylene groups and may, therefore, produce a large number of isomeric hydroperoxides, peroxyl radicals and aldehydes. Not all of these aldehydes will be released from the membrane, but some will remain attached to the phospholipid. The main compounds usually encountered in biological samples are mixture of n-alkanals, 2,4-alkadienals, 4-hydroxyalkenals and malonaldehyde (frequently and incorrectly named malondialdehyde). A simple test for malonaldehyde has been frequently used as a measure for peroxidative damage [582]. We have already encountered this assay as a measure for deoxyribose degradation and as a probe for hydroxyl radical. The most ubiquitous aldehydes found in most tissue samples are malonaldehyde (MA), propanal, hexanal, and 4-hydroxy-2-nonenal (HNE)

The saturated alkanals are of low biological activity, although they may be major products. The 4-hydroxyalkenals, particularly 4-hydroxy-2-nonenal, a product of arachidonic acid peroxidation, are formed in large amounts and are very reactive. These aldehydes and their biological activities have been extensively studied by Esterbauer and coworkers [583-587]. Hydroxyalkenals are highly electrophilic and react with nucleophilic biomolecules that possess SH groups, like glutathione and protein thiols, via the Michael addition:

$$RCH(OH)\text{-}CH=CH\text{-}CHO + H\text{-}X \longrightarrow RCH(OH)\text{-}CH(X)\text{-}CH_2\text{-}CHO$$

HNE also reacts with certain amino groups. These reactions with SH and NH_2 groups are the most important ones in the biochemical effects of HNE, which lead to enzyme deactivation and cytotoxicity. HNE was found to be mutagenic in the Ames test [588]; it reacts with the amino-group of deoxyguanosine *in vitro* [589].

Aldehydes are known to react with amino groups by the formation of a Schiff base:

$$RCHO + R'NH_2 \longrightarrow RCH=NR' + H_2O$$

This reaction of HNE with proteins is responsible for the first step in the pathogenesis of atherosclerosis. Schiff bases undergo enzymatic oxidations to yield electronically excited carbonyl compounds, which in turn may cause photochemical type DNA damage without light (Chapter 5).

At high concentrations, HNE causes degradation of GSH, loss of calcium homeostasis, inhibition of mitochondrial respiration and of nucleic acid and protein synthesis. Hydroxyalkenals are not only cytotoxic, but also show cytostatic properties. It has been known for some time that oxidized fatty acids show carcinostatic activity, which has been shown to be due to the presence of HNE. These observations were made by many investigators (for a summary see [585]). All these properties of hydroxyalkenals may be explained by the high reactivity of these compounds with SH and NH₂ groups. Due to this high reactivity, it is doubtful whether the demonstrated mutagenicity of HNE is due to direct reaction with DNA. The mutagenicity is most likely caused by an indirect pathway. Hydroxyalkenals inhibit DNA repair enzymes. Some of these enzymes have SH groups, which are blocked by HNE.

Another interesting property of hydroxyalkenals is their chemo-tactic property [590,591]. Hydroxyalkenals attract neutrophils. Inflammation causes lipid peroxidation and, thus, formation of HNE. This in turn attracts more neutrophils to the inflammation site, thus amplifying the inflammatory response. We have the following sequence of events:

activated neutrophils ⟶ inflammation ⟶ lipid peroxidation

release of hydroxyalkenals ⟶ attraction of more neutrophils

The study of the deleterious effects of lipid peroxidation products has been mainly limited to 4-hydroxy-2-nonenal, due to its synthetic availability. As mentioned earlier, the peroxidation of PUFAs produces a highly complex mixture of aldehydes, most of which have not been investigated. Considering the toxicity and mutagenicity of HNE and other alkenals it is not surprising that enzymes have evolved to eliminate these toxic metabolites. HNE is oxidized by NADP-dependent aldehyde dehydrogenases [592,593] to the corresponding carboxylic acids and is reduced by NADPH-dependent alcohol dehydrogenase to the alcohol. These secondary products may or may not have biological consequences.

The complex interrelation between ROMs, lipid peroxidation, activated phagocytes, Ca^{2+} homeostasis, GSH depletion, and ischemia reperfusion has been reviewed by Taylor and Shapell [594].

1. Is lipid peroxidation the cause or consequence of tissue injury?

Lipid peroxidation has been implicated in the toxic effects of many chemicals in many tissue injuries and pathological processes. There is no doubt that damage to the lipid bilayer of cell membranes has serious deleterious effects. As I already discussed, this damage may proceed either directly or indirectly (by the products of peroxidation). However, in studying tissue injury, we have to ask if the lipid peroxidation is the cause of tissue injury or a consequence of it. If tissue damage and cell death occur, the membrane is disrupted and, therefore, the structural and functional integrity of the cell is compromised. In that case, we may expect lipid peroxidation. In addition, if the membrane is damaged, metal ions may be released into the microenvironment, and these metal ions catalyze oxidations. It is, therefore, important to determine if in any tissue injury, cell death occurs before lipid peroxidation or vice versa. In order to establish a causal relationship between lipid peroxidation and the pathogenesis of tissue injury, the injury has to meet the following criteria [565]: 1. Lipid peroxidation is detected before tissue necrosis, and 2. Administration of antioxidants (Vitamin E) inhibits lipid peroxidation and cell death.

For a number of halogenated compounds, a causal relationship has been established. The best known of these compounds is the hepatotoxin carbon tetrachloride (CCl_4) [595]. In this case, lipid peroxidation was observed rapidly after carbon tetrachloride injection, whereas necrosis did not occur for several hours. The addition of Vitamin E inhibited both lipid peroxidation and necrosis [596]. The fact that for CCl_4 a causal relationship between lipid peroxidation and tissue damage exists does not imply that other toxins do not exert their toxic effect by other mechanisms. One has to proceed with caution in establishing causal relationships. The role of lipid peroxidation in the hepatotoxicity of a range of compounds was studied by Poli et al. [597]. In the case of menadione (2-methyl-1,4-naphthoquinone) and paraquat, a causal relationship was ruled out [597]. Cell death induced by redox cycling agents like menadione most likely proceeds via GSH depletion and loss of Ca^{2+} homeostasis. In the case of CCl_4, the toxin is metabolized to the reactive radical $\cdot CCl_3$ which initiates lipid peroxidation [595].

In the case of menadione, we have the following sequence of reactions:

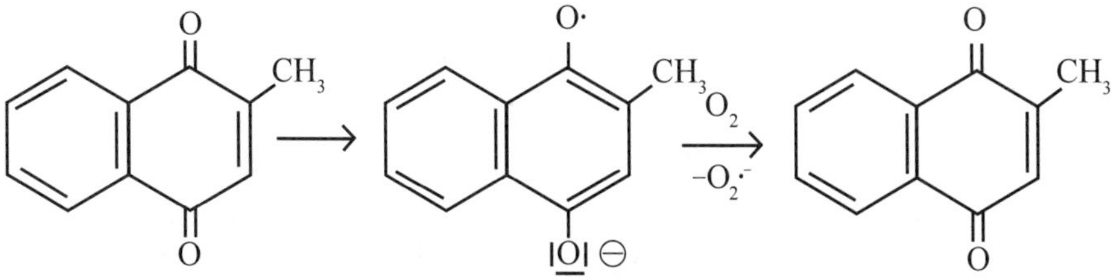

$$2 O_2^{\cdot -} \xrightarrow{2 H^+} H_2O_2 + O_2$$

$$H_2O_2 + 2GSH \longrightarrow GSSG + 2 H_2O$$

We have already discussed the oxidation of semiquinone-type radicals as a source for superoxide radical anion. Superoxide dismutates to hydrogen peroxide, which is removed by glutathione peroxidase. These redox-cycling compounds do not lead to lipid peroxidation unless metal ions are injected simultaneously [597]. Evidence for the participation of lipid peroxidation and iron in diquat-induced hepatic necrosis *in vivo* was presented [598].

It is often difficult to determine if lipid peroxidation is the cause or consequence of the disease. However, I have to point out that even if lipid peroxidation is secondary to injury or disease, it may serve to worsen the condition via the biochemical activity of the degradation products. The chemotactic property of HNE is one such example. HNE attracts neutrophils to the site of injury, thus, increasing the inflammatory response.

D. CHEMISTRY OF RADICAL-INDUCED DAMAGE TO DNA

Reactions of radicals with molecules in the cytoplasm of cells may damage these individual molecules and some metabolic functions or even lead to cell death via disintegration of the cellular membrane. However, in order to cause permanent inheritable damage, such as mutation or disease processes like cancer, the oxygen radicals have to attack the genetic material, the DNA. The intracellular source of oxygen radicals is the respiratory chain of mitochondria, the microsomal metabolism and phagocytosis (Chapter 5). These sources produce the superoxide radical anion and hydrogen peroxide. The H_2O_2 is produced directly (xanthine oxidase, D-aminoacid oxidase) or via dismutation of $O_2^{\cdot-}$. It has been established that neither $O_2^{\cdot-}$ nor H_2O_2 alone can cause damage to DNA [268,309-311]. In addition to internal sources of $O_2^{\cdot-}$ or H_2O_2 there are a number of xenobiotics, like redox-cycling drugs which produce these oxygen metabolites. The species responsible for DNA damage are not $O_2^{\cdot-}$ and H_2O_2. The damage is caused by the hydroxyl radical, singlet oxygen and electronically excited states. The DNA is well protected by the nuclear membrane and by the surrounding proteins (histones). The hydrogen peroxide is the only neutral metabolite which can migrate from its site of formation and react at distant sites. It penetrates the membrane and interacts with DNA. In order to cause damage, it reacts with Fe(II), which is present as a counterion in DNA. The resulting hydroxyl radical reacts close to the site of formation in a site-specific manner (Chapter 3).

The DNA is a double-stranded helix which consists of four different bases and a sugar-phosphate backbone. The sugar moiety is 2-deoxy-D-ribose in DNA and D-ribose in RNA. The sugars are linked by the 3'- and 5'-hydroxyl groups by the phosphate bridges. A schematic presentation of a single strand of DNA is given in Fig. 1.

The hydroxyl radical reacts with the bases and the sugar to produce base eliminations (abasic sites), base modifications, sugar lesions, base-protein crosslinks, single- and double-strand breaks. All these reactions have been extensively reviewed over several decades [599-603]. I am now going to discuss the reactions of the bases (pyrimidines and purines) and the 2-deoxy-D-ribose with hydroxyl radicals. Since the radiolysis of H_2O provides us with the cleanest source of ·OH, it is not surprising that radiation chemists have been at the forefront of these investigations. The technique of pulse radiolysis combined with analysis of final stable products using sensitive analytical techniques (gas chromatography-mass spectroscopy with selective ion monitoring) have given us a detailed understanding of the mechanism, rate constants and characteristics of the radical intermediates involved.

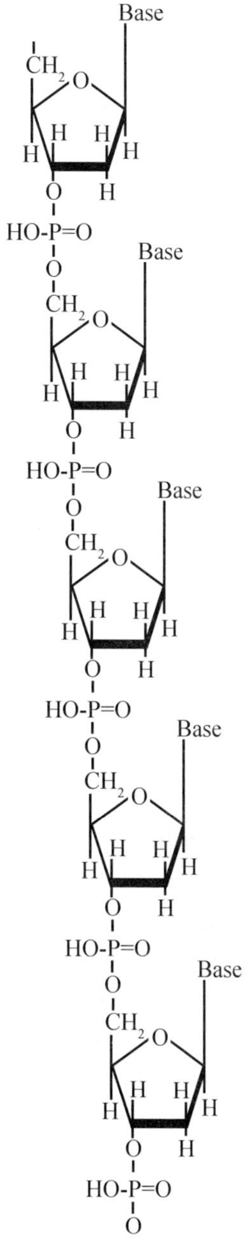

Fig. 1. DNA single strand.

1. The reaction of hydroxyl radical with the bases of DNA

The structures of the four bases of DNA are shown in Fig. 2 .

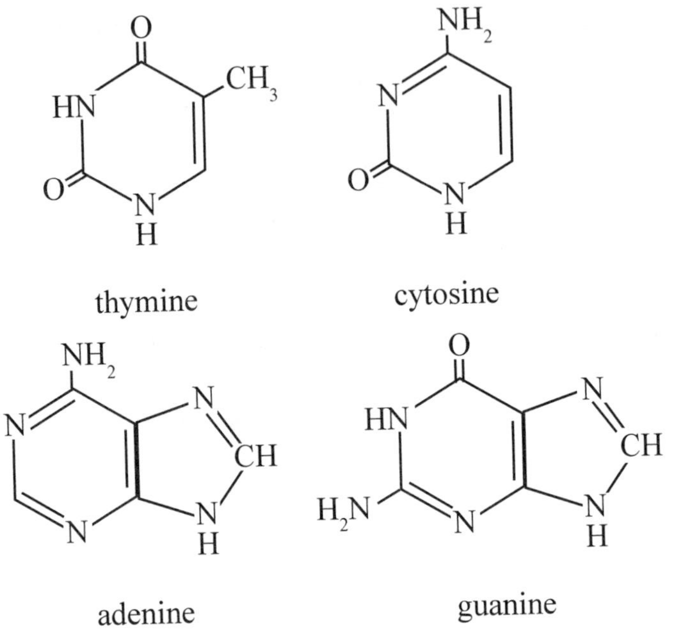

Fig. 2. Pyrimidine and purine bases.

The pyrimidines and purines are electron-rich aromatic compounds. The rate constants of these bases and some nucleosides and nucleotides with hydroxyl radical have been determined by pulse radiolysis and are in the range 4-9 x 10^9 $M^{-1}s^{-1}$. The multitude of reaction products [601-603] obtained in this radiolysis of aqueous solutions of nucleosides and DNA may be scary to the uninitiated. However, all these products are formed via some very basic reaction steps, which I discussed in Chapter 2. We may consider a nucleoside as an aromatic ring with a side chain. Hydroxyl radicals can only undergo two types of reactions: 1. addition to the ring (the purines or pyrimidines) and 2. hydrogen abstraction from the side chain (the sugar). Since ·OH is a highly electrophilic radical [604,605]. It will preferentially attack the electron-rich ring, analogous to the hydroxylation of toluene, which proceeds almost 100% at the ring [55]. In the case of DNA, the relative amount of ring versus side-chain attack depends on whether the single- or double-stranded DNA is involved [606]. The amount of side-chain attack does not exceed 20% [216]. The initial reactions of ·OH radicals with thymine and guanine are shown in Figs. 3 and 4.

The position of hydroxyl radical attack on the pyrimidines is the 5,6 carbon-carbon double bond. These two carbons carry the highest electron density and the electrophilic hydroxyl attacks at these positions. In the purines, the positions of highest electron density are the 4,5, and 8 positions. In the case of thymine, the hydroxyl radical will also react with the side chain by hydrogen abstraction (Fig. 3).

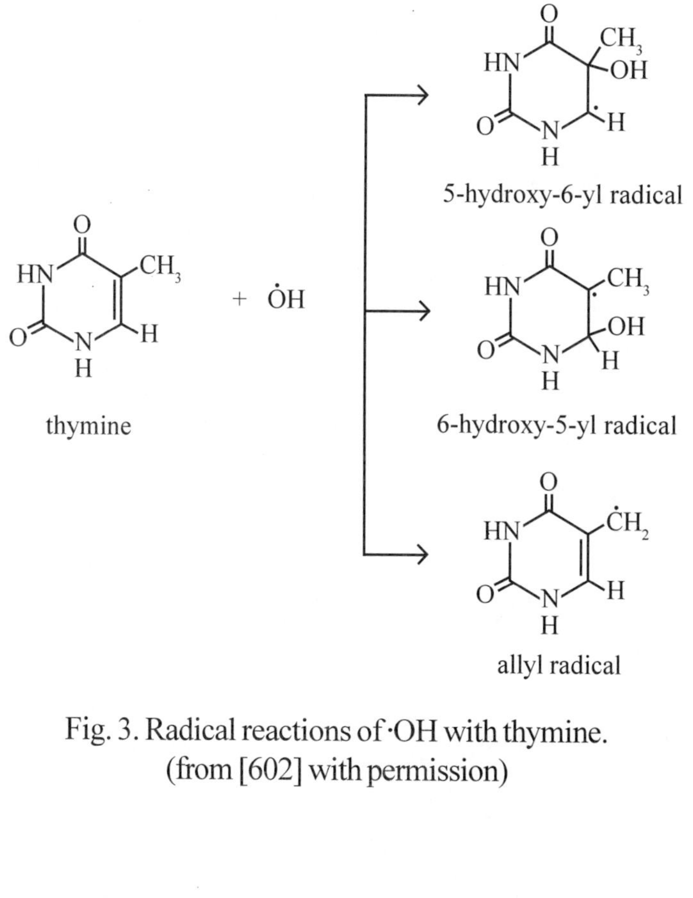

Fig. 3. Radical reactions of ·OH with thymine.
(from [602] with permission)

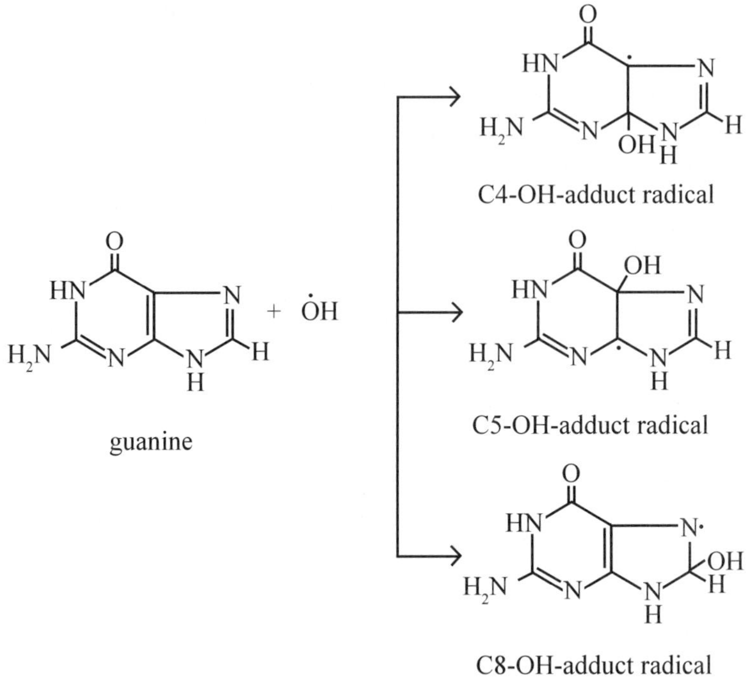

C4-OH-adduct radical

C5-OH-adduct radical

C8-OH-adduct radical

guanine

$+ \ \dot{O}H$

Fig. 4. Reaction of hydroxyl radical with guanine.
(from [602] with permission)

The 5-hydroxyl adduct to thymine may be oxidized by one-electron oxidizing agents, like $Fe(CN)_6^{3-}$ or O_2. The developing positive charge at the 6-position undergoes nucleophilic attack by H_2O to yield the 5,6-dihydroxy-5,6-dihydro-thymine (thymine glycol):

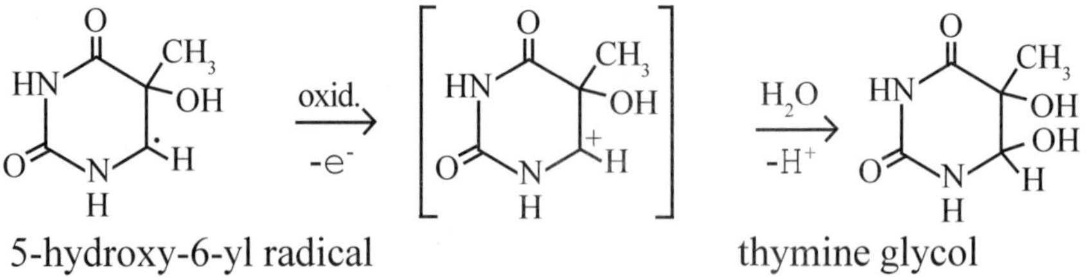

5-hydroxy-6-yl radical

thymine glycol

The same product is obtained from the 6-hydroxy-5-yl radical. If oxygen is the oxidizing agent, we also form the superoxide radical anion ($O_2^{\cdot-}$). I have already discussed this method of superoxide formation from hydroxycyclohexadienyl-type radicals in Chapter 3.

As pointed out in Chapter 2, the hydroxyl radical adduct of some aromatics may give rise to radical cations by acid-catalyzed dehydration:

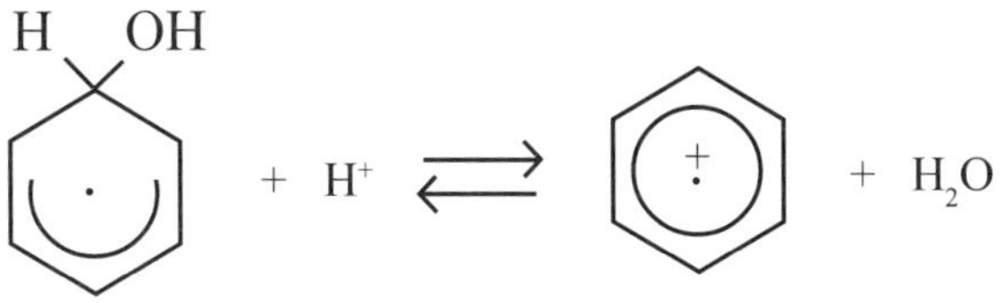

The reversibility of dehydration means that hydroxylation can be accomplished by two distinct pathways: addition of hydroxyl radical or hydration of a radical cation. This radical cation pathway has been responsible for the false positive for hydroxyl radical detection using the DMPO spin trap (Chapter 3). The radical cation reacts with the nucleophilic water to give again the hydroxycyclohexadienyl radical, unless the radical cation undergoes other reactions which compete favorably with nucleophilic addition. This kind of situation is present in the radical cation of toluene, which deprotonates from the CH_3 group (Chapter 2).

The C-8 hydroxyl radical adduct to adenine (8-OHA) or guanine (8-OHG) may undergo the following reactions, depending on the reaction conditions: oxidation, reduction or fragmentation (ring opening), which are shown in Fig. 5.

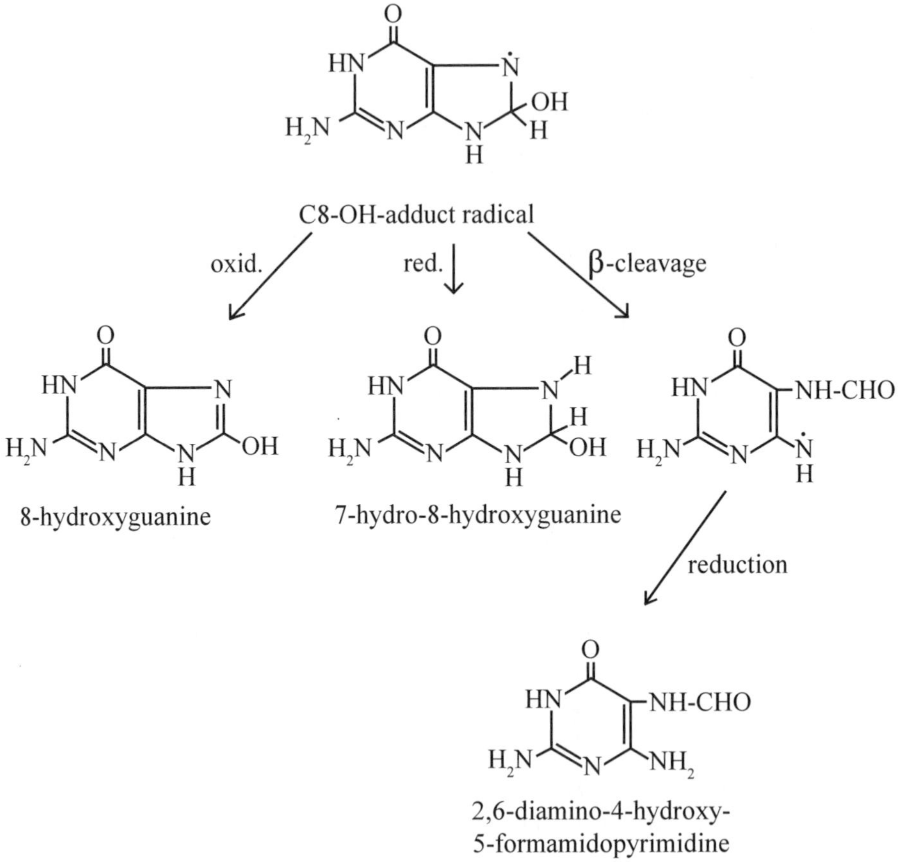

Fig. 5. Reactions of the 8-hydroxyl radical adduct of guanine.
(from [602] with permission)

2. The reaction of hydroxyl radical with 2-deoxy-D-ribose in DNA

Since ·OH is highly reactive and the 2-deoxy-D-ribose has many active C-H bonds, the ·OH abstracts a hydrogen from all the C-atoms [216]. In the nucleosides or nucleotides such reactions occur mainly at the bases (about 80%). Only about 20% of the hydroxyl radicals react with the sugar. While reaction with the bases leads to base modification, the attack on the sugar moiety results in strand breaks [607-609]. In order to illustrate the reactions of the initially formed sugar radical, I use the radical produced by H-abstraction from the 2-position (see Fig. 6).

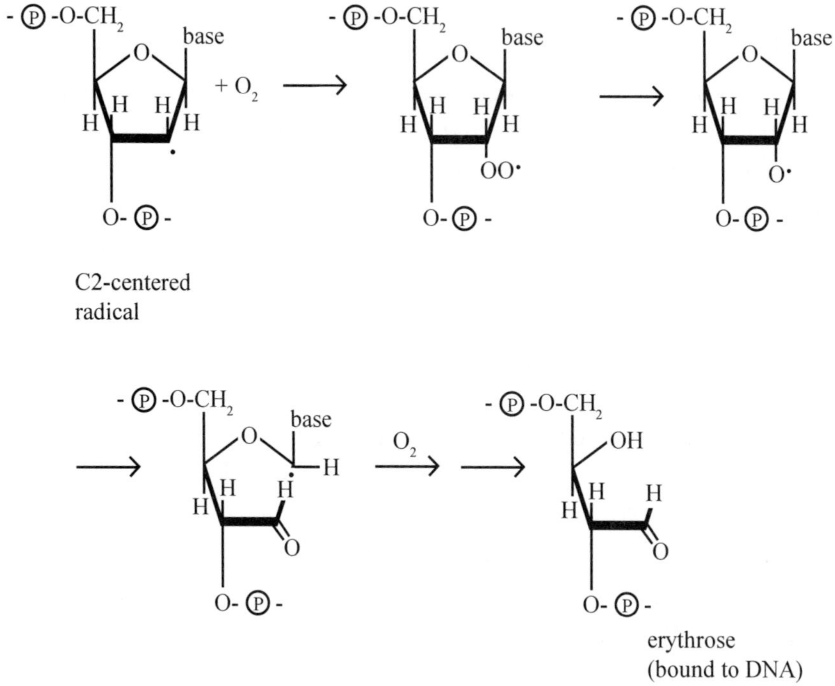

Fig. 6. Hydroxyl radical-induced degradation of DNA
via H-abstraction from 2-deoxy-D-ribose.
(from [602] with permission)

As shown in Fig. 6, the initially formed radicals undergo all the reactions discussed in Chapter 2: oxidation, reduction, β-cleavage, or combination with oxygen to give peroxyl radicals. The main reaction of all the sugar radicals is the combination with molecular oxygen to give a peroxyl radical. This is followed by dismutation to an alkoxyl radical, oxygen and fragmentation. The fragmentation leads to the elimination of bases or to strand breaks. By starting from the C-5' sugar radical in a sequence similar to the one given above, we obtain a strand break. In the absence of oxygen, the C-5' sugar radical may undergo addition to a neighboring guanine base, followed by oxidation to a cyclic product [610].

The numerous products formed in the radiolysis of nucleosides and DNA have been summarized in the reviews by Dizdaroglu [602] and Breen and Murphy [603]. The products and their mechanisms of formation are a great joy for physical organic chemists. It is of course no surprise that we observe 8-OHdG and abstraction products from all the sugar carbons. Although we may be sure that in the radiolysis the 8-OHdG is formed by ·OH addition, it does not necessarily mean that the same mechanism applies *in vivo*. 8-Hydroxydeoxyguanosine has been observed in the reactions with $^1O_2$ [611-613] and with electronically excited carbonyl compounds [30,482,483,614,615] and ^-OONO [223]. The formation of 8-OHdG *in vivo* is, therefore, not a measure of hydroxyl radicals, but of total oxidative damage by various ROMs.

3. Crosslinks between DNA and nucleoproteins

The DNA is surrounded by proteins called histones. These histones are highly basic (contain many basic amino acids) and are, therefore, at neutral pH positively charged. The histones are arranged in the deep groove of the DNA double helix, where they are attached through electrostatic interactions with the negatively charged phosphate groups of DNA.

Another way in which DNA may sustain oxidative damage is by formation of DNA-nucleoprotein crosslinks. The involvement of the hydroxyl radical in these reactions has been demonstrated for radiation-induced crosslinking as well as through hydrogen peroxide and metal ions [616,617]. Crosslinking can be achieved by radical-radical combination, radical addition to an aromatic amino acid or addition of a protein radical to DNA bases. Many DNA-protein crosslinks have been identified in nucleoproteins exposed to ionizing radiation in the absence of molecular oxygen. Some of the amino acids involved are glycine, alanine, valine, leucine, isoleucine, threonine, lysine, and tyrosine.

Radical-radical combinations are of course very sensitive to the presence of molecular oxygen, which reacts with the initially formed radicals to form peroxyl radicals. In dilute aqueous solutions, no radical-

radical combination occurs in the presence of oxygen; however, *in vivo* the conditions are far removed from those of a dilute aqueous solution. The hydroxyl radical reacts close to its site of formation in a site-specific manner. Hydroxyl radical scavengers, like DMSO may, therefore, not always interfere in this site-specific process. The formation of most crosslinks is inhibited by oxygen with one exception [617]: the thymine-tyrosine crosslink is not inhibited by oxygen, which indicates that in this case the thymine radical is formed close to tyrosine, so the radical reacts with tyrosine before oxygen can interfere. The thymine radical, therefore, cannot be called a "free" radical, as I pointed out on a previous occasion.

E. DNA DAMAGE VIA ELECTRONICALLY EXCITED STATES

1. Introduction

Most molecules (with the notable exception of oxygen) exist in the ground state as the singlet state (i.e., all electrons are paired). Absorption of a photon results in the formation of an excited singlet state, which rapidly undergoes intersystem crossing to the more stable triplet state. The triplet state may undergo deactivation to the ground state with emission of light. This emission is known as phosphorescence (see Chapter 2).

The excited molecule in the triplet state is known as the sensitizer. This excited sensitizer (besides phosphorescence) reacts with a substrate by energy transfer, electron transfer or adduct formation.

As I have already discussed (Chapters 4 and 5), excited triplet states may be formed without light by a variety of thermal or enzyme-catalyzed reactions. The resulting triplet carbonyl compounds may damage DNA by two different pathways: 1. direct interaction (type I reaction) and 2. damage mediated by sensitized molecular oxygen (type II reaction). Direct interaction takes place by addition, by energy transfer and by electron transfer:

1. Direct interaction (type I reaction):

$$^3S^* + DNA \longrightarrow \text{adduct formation}$$

$$^3S^* + DNA \longrightarrow DNA^* \longrightarrow \text{UV type damage, e.g., dimer formation}$$

$$^3S^* + DNA \xrightarrow{} DNA^{\cdot +} \xrightarrow{H_2O} \text{radical-type damage (8-OHdG)}$$
$$O_2$$

2. Indirect interaction (type II reaction):
    A triplet excited state may sensitize $^3O_2$ to give singlet oxygen or superoxide radical anion:

$$^3S^* + {}^3O_2 \longrightarrow {}^1S + {}^1O_2 \xrightarrow{DNA} \text{8-OHdG} \qquad (1)$$

$$^3S^* + {}^3O_2 \longrightarrow S^{\cdot +} + O_2^{\cdot -} \xrightarrow{DNA} \text{no direct reaction} \,(2)$$

Singlet oxygen has a long enough lifetime [618] and high enough reactivity that it may modify DNA, proteins and lipids. Superoxide radical on the other hand is of low reactivity and does not modify DNA directly [268,309-311].
    The formation of superoxide radical requires an easily oxidizable sensitizer (Chapter 3). Frequently we obtain back electron transfer between the radical cation and the superoxide radical:

$$S^{\cdot +} + O_2^{\cdot -} \longrightarrow S + O_2 \qquad (3)$$

The combination of Reactions (2) and (3) amounts to the quenching of the excited sensitizer.
    Both type I and type II reactions give rise to 8-OHdG. In order to distinguish the two pathways, we use some simple tests for singlet oxygen. The lifetime of singlet oxygen is very sensitive to the solvent [349]. For example D_2O compared to H_2O increases the lifetime by a factor of 10 or more, and, therefore, may increase singlet-oxygen-induced

damage. The replacement of H_2O by D_2O provides us with an easy diagnostic test for the involvement of singlet oxygen in a reaction, such as photooxidations, enzymatic oxidations, and chemiluminescence.

Other procedures for the identification of singlet oxygen involve the use of quenchers (certain dyes, DABCO, β-carotene, azides, and tocopherols), which are much less reliable. These quenching procedures are subject to numerous complications especially in complex biological systems. The role of these quenchers in the alleviation of diseases will be discussed in Chapter 8.

Photosensitized damage to DNA has been studied extensively. The interested reader is referred to the review by Epe [619]. There are numerous cellular constituents (flavins, porphyrins), as well as certain drugs (tetracycline, thiazides and phenothiazines), which act as photosensitizers and interact with DNA to produce DNA modifications. These compounds present a mutagenic and carcinogenic risk if they are irradiated at the wavelength they absorb. That is why people who take tetracycline should not expose themselves to the sun. These reactions are important under conditions at which light is available and is absorbed. The discovery of purely chemical (Chapter 4) and enzymatic reactions (Chapter 5) for the formation of excited states was of fundamental importance, because now we may carry out "photochemistry in the dark".

2. Singlet oxygen-induced DNA damage

Compared to the previously discussed reactions of DNA and its bases with hydroxyl radicals, singlet oxygen reacts very selectively with the guanosine moiety of DNA [612,613,620]. The products include 8-oxo-7-hydrodeoxyguanosine (8-oxodG), mostly referred to as 8-hydroxyguanosine (8-OHdG); 2,6-diamino-4-hydroxy-5-formamido-pyrimidine (FapyGua) (see Fig. 5); and 4,8-dihydro-4-hydroxy-8-oxo-2'-deoxyguanosine (4-HO-8-oxo-dG) [350,351].

I have already discussed the formation of singlet oxygen in biological systems (Chapters 4 and 5). In order for $^1O_2$ to cause any chemical damage in a biological system, it has to live long enough to compete with physical deactivation. It has been determined that $^1O_2$ has a half-life of 4-50 ms. This means that $^1O_2$ can diffuse within a range of 100 Å from its site of formation [618].

Chemically [611] and photochemically produced $^1O_2$ (methylene blue-O_2) [612] was shown to react with DNA to give 8-hydroxy-deoxyguanosine (8-OHdG). The question is what is the mechanism of 8-OHdG formation? The reaction might involve both type I (direct interaction with excited methylene blue) or type II reactions (via $^1O_2$). Subsequently, the formation of 8-OHdG was observed in a dark reaction by thermal decomposition of an endoperoxide [621,622], thus confirming a type II reaction:

$$dG + {}^1O_2 \longrightarrow 8\text{-OHdG}$$

Although we know the reactants and the products, physical organic chemists are never satisfied with that partial knowledge. They want to know what is the mechanism of the reaction. As I already pointed out (Chapter 3), $^1O_2$ reacts with electron-rich compounds such as guanine by electron transfer to give radical cations, which react with water to give the hydroxyl radical adduct. This adduct (obtained via radiolysis) has been shown to undergo ring-opening to give formamidopyrimidine (Fig. 5) as reviewed by Steenken [601]. Therefore, the intermediate formation of a radical cation is a strong possibility [623], i.e., the transformation involves a one-electron transfer process.

Since the 8-OHdG is also formed by hydroxyl radicals, dioxetanes and peroxynitrite, it is not a specific probe for singlet oxygen in chemical and biological systems. The discovery of the 4-HO-8-oxo-dG as a specific singlet-oxygen probe was therefore of great significance [350,351]. I have already discussed the mechanism of this reaction in Chapter 4.

Evidence for the involvement of $^1O_2$ in DNA damage has been obtained by the use of quenchers, like β-carotene, lycopene, tocopherol and thiols [624-627]. I shall discuss these reactions in Chapter 8. The damaged bases are excised and excreted in the urine [628-630]. The detection of 8-OHdG in the urine has been suggested as an *in vivo* marker for oxidative damage [628].

In addition to hydroxylation, $^1O_2$ leads to single-strand breaks [631-634]. The involvement of $^1O_2$ in these strand breaks was confirmed by the use of D_2O and the use of quenchers. The DNA damage profile induced by $^1O_2$ was shown to be different from the damage induced by OH radicals [614]. Compared to hydroxylation at the 8 position single-strand breaks have been found to be relatively rare events [611,614,635].

A typical difference between $^1O_2$ and ·OH (radiation produced) is their reactivity towards deoxyadenosine. Deoxyadenosine does not react with $^1O_2$ to give the 8-hydroxyadenosine [621], but ·OH radicals give, with deoxyguanosine and deoxyadenosine, the 8-hydroxy derivatives [601,636-638].

3. Reactions of electronically excited states with DNA
a. Introduction

Electronically excited states are usually produced by photosensitization, but also, and more importantly for biological systems, by purely chemical reactions. In these reactions with DNA, the major products are guanine and pyrimidine modification, [2+2] cycloaddition of pyrimidines, adduct formation and to a minor extent strand breaks and base elimination [619].

b. Guanine modification

Hydroxyl radicals are more reactive and less discriminatory in their reactions with DNA compared to electronically excited states. The most frequent kind of photosensitized DNA damage is base modifications of guanine. The 8-hydroxydeoxyguanosine is not only produced by $^1O_2$, but also by direct electron transfer from the guanine moiety to the excited sensitizer. Strand breaks and formation of abasic sites are of lesser importance. Depending on the energy of the excited sensitizer, the base modification may be quite specific. A large number of sensitizers, e.g., methylene blue, proflavin, and hematoporphyrin, selectively modify the guanine base to give 8-hydroxy-deoxyguanosine [612,619,639-641].

8-hydroxydeoxyguanosine has been formed by electronically excited triplet acetone, both in isolated DNA as well as in intact bacterial and mammalian cells [30,482,483]. As I have discussed in Chapters 4 and 5, electronically excited states are produced in absence of light by purely chemical or enzymatic reactions. Many of these reactions involve the formation of dioxetanes, which decompose thermally to give excited carbonyl compounds. The formation of 8-OHdG in the dioxetane-induced oxidation of calf thymus DNA was studied by Adam and coworkers [30,482,483]. Through the use of triplet quenchers, it was clearly established that the DNA damage occurred through triplet excited states by a type I (e⁻-transfer) reaction. The use of D_2O (as a probe for $^1O_2$) did not provide a definite conclusion on the relative importance of type I versus type II reactions.

c. Formation of pyrimidine dimers

If the energy of triplet sensitizers is higher than the triplet energy of the DNA bases, we may observe energy transfer, which leads to a completely different product profile than electron transfer. A well-known example is the reaction of triplet acetone or acetophenone with thymine. This reaction gives by means of a [2+2] cycloaddition a 4-membered-ring compound (cyclobutane derivative) [642,643]:

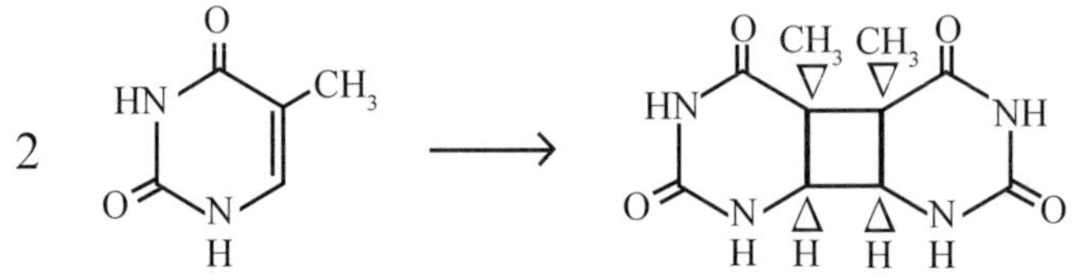

cis-syn dimer

The same product was observed with the photo-produced triplet acetone as well as by thermal decomposition of trimethyldioxetane. This result clearly established triplet acetone as the reactive species in the dioxetane decomposition. Another example is the dimerization of acenaphthalene [76] (see Chapter 4).

The triplet energy of acetophenone is higher than the one of thymine, but lower than the one of cytosine. Therefore, excited acetophenone interacts selectively with thymine.

d. DNA adduct formation

Psoralenes (furocoumarins) are natural products and sensitizers which can bind to DNA. The furocoumarins have been known for a long time for their phototoxic properties. They produce erythema in the skin. All of these compounds have activated double bonds and react by [2+2] cycloaddition to the 5,6 double bond of pyrimidines, mostly thymine. Photoaddition to DNA has also been demonstrated in the dark. Adam and coworkers [615] synthesized a derivative of psoralen with a dioxetane moiety attached:

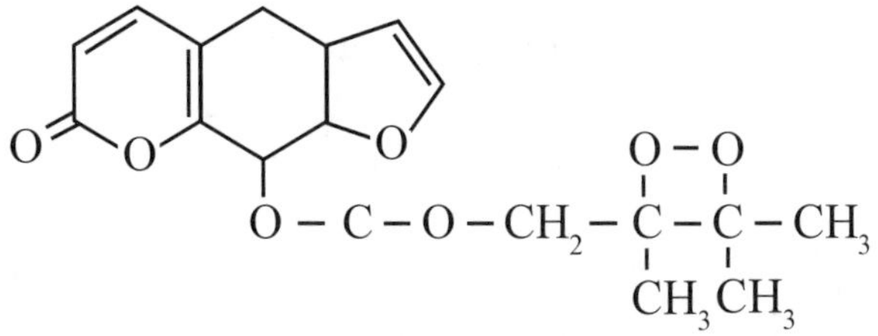

Thermal decomposition of the dioxetane gave an excited triplet state, which in turn was bound to DNA.

e. Strand breaks and base loss

Strand breaks and base loss are much less frequent events than base modifications. In the photosensitized oxidation of DNA by methylene blue, a ratio of 8-OHdG/strand breaks of 17:1 was determined [635]. This low level of strand breaks and base loss contrasts sharply with the high incidence of these reactions with hydroxyl radicals. It has been suggested that these events involve a type II reaction (singlet oxygen). A strong solvent effect (D_2O versus H_2O) has been reported by some but not all investigators (see [619]).

The formation of strand breaks induced by methylene blue has been greatly increased in the presence of Fe(III)-EDTA [614]. Since the ratio strand breaks/8-OHdG is greater in hydroxyl radical attack compared to singlet oxygen, this result may be due to the superoxide-dependent formation of hydroxyl radicals (the Haber-Weiss reaction):

$$2\,O_2^{\cdot-} + 2\,H^+ \longrightarrow H_2O_2 + O_2$$

$$Fe^{3+} + O_2^{\cdot-} \longrightarrow Fe^{2+} + O_2$$

$$Fe^{2+} + H_2O_2 \longrightarrow Fe^{3+} + OH^- + \cdot OH$$

F. BIOLOGICAL MARKERS FOR *IN VIVO* OXIDATIVE DAMAGE

1. 8-Hydroxydeoxyguanosine as a marker for DNA damage

The oxidative damage to DNA bases has been considered to be a significant source of mutations [644] and many degenerative processes like aging and cancer [395,398,558]. The damaged DNA is constantly repaired and the damaged bases are excreted in the urine [396,398]. Studies on the enzyme systems of DNA repair are extensive and have been reviewed [645,646]. I have already discussed the high reactivity of oxygen metabolites in purely chemical systems. There are numerous ways by which DNA or other important biomolecules may be damaged *in vitro*. The question, however, arises whether this damage occurs *in vivo*. In order to answer this question, we need a biological marker of which we are sure it is formed via a reaction involving reactive oxygen species. We also need a specific and sensitive probe to detect such a biological marker [111,112]. One such marker is 8-hydroxydeoxyguanosine (8-OHdG). Biologically the most important property of 8-OHdG is its mutagenicity [647,648]. The 8-OHdG can exist in several tautomeric forms:

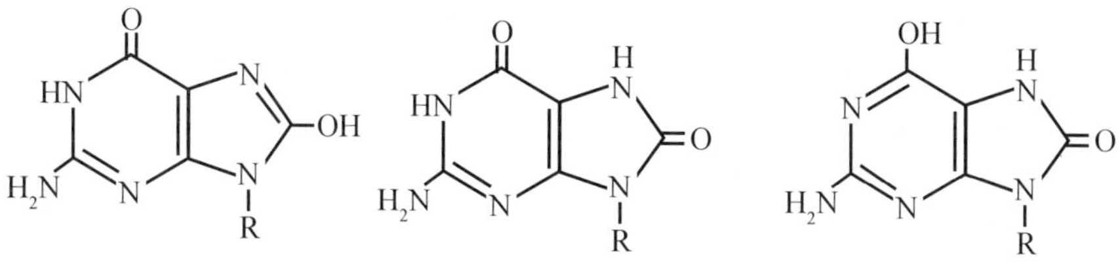

6-keto-8-enol   6,8-diketo   6-enol-8-keto

Quantum mechanical (ab initio) calculation [357] as well as NMR studies [358] have shown that the 6,8-diketo form is the predominant tautomer. The compound should, therefore, be called 8-oxo-7-hydro-guanine; however the common usage is still 8-hydroxyguanine or 8-hydroxydeoxyguanosine (8-OHdG).

The 8-OHdG has been observed in many reactions of DNA *in vitro* as well as *in vivo*. These reactions and their biological significance have been extensively reviewed [114]. Some of these reactions are briefly summarized in Fig. 7.

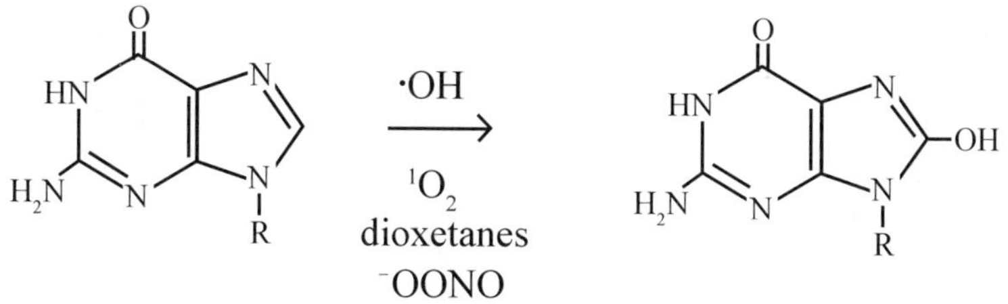

Fig.7. Hydroxylation of guanine (R=H) and deoxyguanosine (R=deoxyribose).

The formation of 8-hydroxydeoxyguanosine was first observed in the reaction of DNA with Fe^{2+}-EDTA-AH_2-O_2 (Udenfriend's reagent) by Kasai and Nishimura [649,650]. Subsequently, a number of reagents, e.g., polyphenols-Fe^{3+}-H_2O_2 [651], X-irradiation [652], γ-irra-diation [637], asbestos-H_2O_2 [302], Fe^{3+}-AH_2-H_2O_2 and Cu^{2+}-AH_2-H_2O_2 [653] were found to react with DNA *in vitro* to give high yields of 8-OHdG. All of these reagents have been shown to produce ·OH radicals (Chapter 3). Although ·OH gives 8-OHdG, the formation of 8-OHdG is no proof of ·OH. As I have already discussed, singlet oxygen [611-613], electronically excited states [30,614,615,654] and peroxynitrite [223] also react with DNA to give high yields of 8-OHdG. The determination of 8-OHdG is, therefore, a marker of total oxidative damage. An easy and sensitive analytical technique (HPLC-EC) developed by Floyd [120] has itensified the use of 8-OHdG as a biological marker.

I already discussed the two possible ways to hydroxylate aromatic compounds: 1. addition of ·OH and 2. electron transfer to give a radical cation, followed by nucleophilic addition of water. These two pathways are summarized in Fig. 8. The reaction with peroxynitrite, $^1O_2$ and excited carbonyl compounds most likely follows such a one-electron transfer pathway:

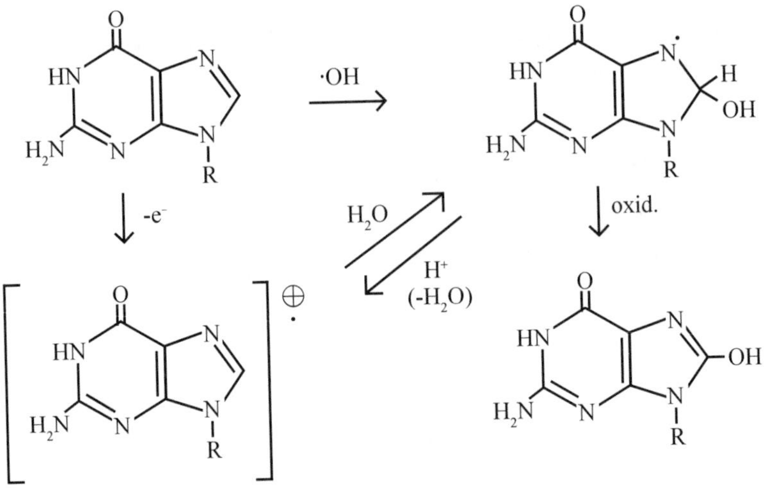

Fig. 8. Mechanism of guanine hydroxylation

The initially formed reactive oxygen metabolites ($O_2^{·-}$ and H_2O_2) are not very reactive and do not react directly with the DNA bases [268,309-311]. These ROMs have to be converted to more reactive species, such as ·OH or $^1O_2$. The formation of ·OH requires metal ions (Chapter 3). Iron plays an important role in the DNA damage produced by ROMs. The DNA has Fe(II) as a counterion and H_2O_2 (which can penetrate membranes and reach the nucleus) may interact with Fe(II) and produce ·OH in a site-specific manner (Chapter 3). Evidence for a site-specific formation of ·OH from H_2O_2 was obtained by Meneghini and Hoffmann [268]. These authors observed that pure isolated DNA does not react with H_2O_2, but DNA from human fibroblasts was degraded (strand scission) by H_2O_2. The factor responsible for ·OH formation could be removed when nuclei were dialyzed against buffer solutions containing EDTA. This result indicates that the H_2O_2 activating factor is an Fe^{2+}-protein or some other Fe^{2+}-macromolecule complex. It is surprising that this paper has been largely ignored. Many years later, the concept of site-specific formation of ·OH was presented as a novel concept. I have already discussed the evidence presented by Stadtmann and Oliver [269] for a site-specific damage to proteins (page 81).

The importance of H_2O_2 in DNA damage has been reviewed by Meneghini and Martins [655]. Although there is strong evidence for hydroxyl radical involvement [655-658], other mechanisms for DNA damage have been discussed. DNA damage could result from an increase in Ca^{2+} in the cytosol [659] or from lipid peroxidation. Lipid peroxides may reach the nucleus, where they are converted to the alkoxyl radicals, which attack DNA [660-662]. The formation of 8-OHdG in the reaction of autoxidized lipids with DNA was shown by Park and Floyd [662]. Although these mechanisms may also contribute to DNA damage, the results of Dizdaroglu et al. [656] demonstrate hydroxyl radical-type damage in H_2O_2-treated mammalian cells.

Neocuproine, a copper-binding chelator, which blocks the copper-H_2O_2 reaction, does not protect cellular DNA in mammalian and human fibroblast cells from H_2O_2-induced damage, but 1,10-phenanthroline (which binds Fe ions) served for protection [663]. This seems to indicate that iron ions and not copper ions are mediating the damage in most *in vivo* cases. Contrary conclusions from *in vitro* experiments were reached by Dizdaroglu et al. [653], Arouma et al. [281], and Park and Floyd [662]. In the reactions with Fe^{3+}-ascorbic acid-H_2O_2 and

Cu^{2+}-ascorbic acid-H_2O_2, the Cu^{2+} system gave higher yields of 8-OHdG [653]. The greater effectiveness of Cu^{2+} was also observed in the DNA damage induced by L-DOPA and dopamine-H_2O_2 [289].

In addition to HPLC-EC, the most common technique is gas chromatography-mass spectroscopy with specific-ion monitoring (GC-MS-SIM). In this procedure, the intact chromatin is hydrolyzed, followed by trimethylsilylation (to make the products more volatile for gas chromatography and mass spectroscopy. The whole spectrum of the base products characteristic of hydroxyl radicals has been determined in chromatin isolated from γ-irradiated human cells in culture [664]. The advantage of the GC-MS-SIM method over the HPLC method is the fact that the GC-MS-SIM method gives structural information, whereas the HPLC method does not. Floyd et al. [120,121] claim a detection limit of one 8-OHdG in 10^6 bases. This is about the same detection limit as in the GC-MS-SIM method [638].

The background concentrations of 8-OHdG have been measured in different tissues and organelles, which were not subjected to oxidative stress. These background levels of 8-OHdG are formed via normal metabolic generation of ROMs (Chapter 5). To assess any effect of any substance on this damage, we have to know the base levels of DNA oxidation. Unfortunately, this is not always known with precision. Different analytical techniques give quite different and varying results. For example, the base levels of 8-OHdG in commercially available calf-thymus DNA vary between 8 and 320 8-OHdG per 10^6 DNA bases. A summary of the base-line variations of different DNAs is found in a review by Halliwell [665]. Not surprisingly, mitochondrial DNA has higher base levels of 8-OHdG than nuclear DNA [666].

The GC-MS-SIM technique requires hydrolysis and derivatization, and these procedures are prone to artifacts (remember spin trapping!). The pros and cons of the different techniques of DNA-damage detection have been critically reviewed by Halliwell [665]. The relationship between aging, carcinogenesis and 8-OHdG formation has been thoroughly investigated. I shall discuss these studies in Chapter 9.

Due to the importance of some transition-metal ions in hydroxyl radical formation it is not surprising to find a relationship between iron body stores and colorectal and lung cancer [667,668]. On the basis of these results one may be tempted to assume that high body iron stores contribute to other diseases as well. Studies by Salonen et al. [669] on Finnish men indeed showed that high levels of iron stores represent a risk factor for myocardial infarction. However, studies by Sempos et al. [670] on a large U.S. population showed no increased risk for CAD. On the contrary, the results showed an inverse relationship between iron body stores and overall mortality and with mortality from cardiovascular causes.

2. 3-Nitrotyrosine

The peroxynitrite anion plays a central role in oxidative damage. This compound is formed by the rapid combination of $\cdot NO$ and $O_2^{\cdot -}$ [116]. Peroxynitrite reacts with tyrosine to produce 3-nitrotyrosine. I shall discuss these reactions in detail in Chapter 7. The 3-nitrotyrosine is a stable product, which is excreted in the urine. It has, therefore, been suggested as a marker for total oxidative damage caused by reactive nitrogen species (RNSs) [671]. 3-Nitrotyrosine was found in blood serum and in synovial fluid of patients with inflammatory joint disease (rheumatoid arthritis). Blood and synovial fluid from normal subjects contained no detectable levels of 3-nitrotyrosine [671].

The formation of nitro-aromatic compounds by radiolysis of aqueous nitrate solutions has been known for some time [672]. The formation of 3-nitrotyrosine, therefore, could be the result of an attack by $\cdot OH$ to form the phenoxy radical, followed by combination with $\cdot NO_2$:

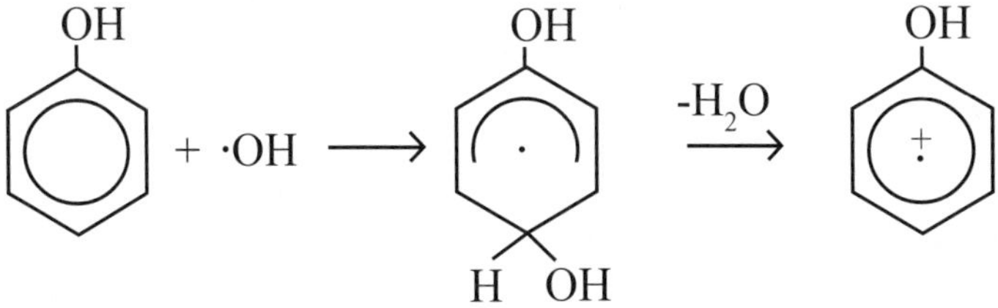

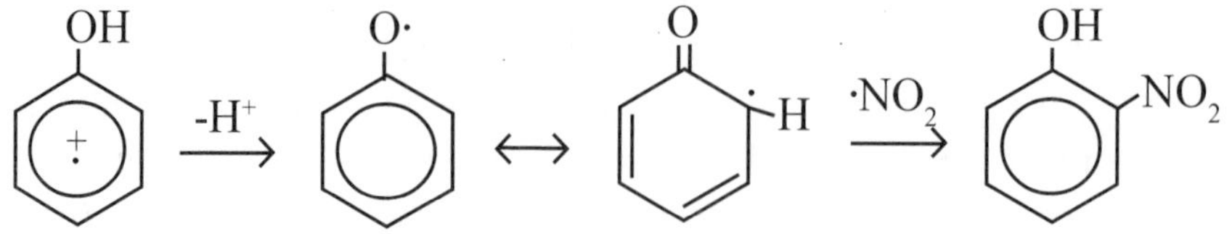

Originally it was suggested that the peroxynitrous acid decomposes to yield hydroxyl radicals:

$$HOONO \longrightarrow \cdot OH + \cdot NO_2$$

This proposal has come under intense dispute. I shall discuss this controversy later in Chapter 7. At present it appears unlikely that peroxynitrous acid produces free hydroxyl radicals [223].

The formation of 3-nitrotyrosine does not necessarily require hydroxyl radicals. The mechanism most likely proceeds via a one-electron transfer:

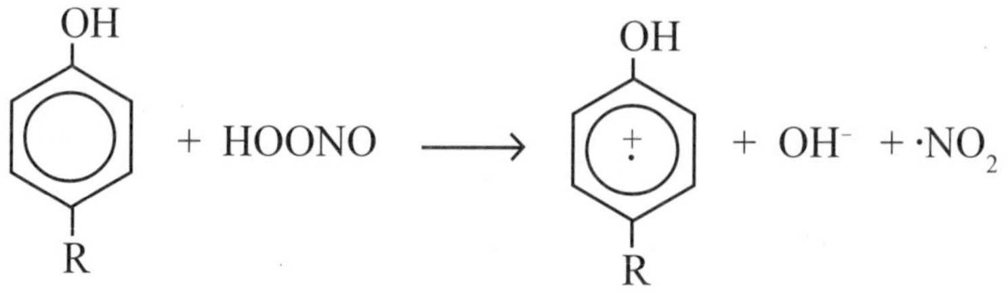

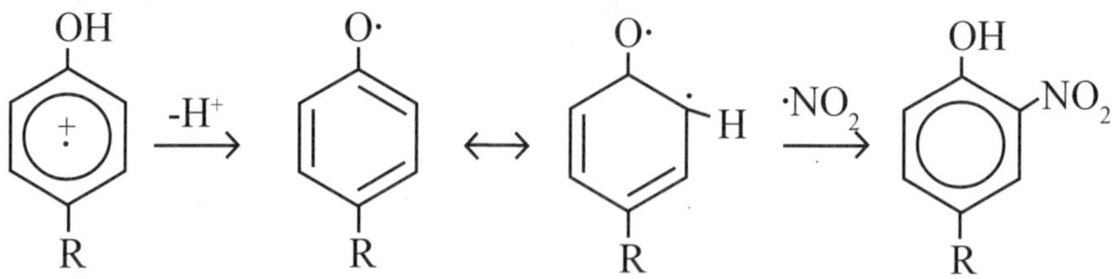

3. Allantoin

The reaction of a number of reactive oxygen species with uric acid was found by Ames et al. [673] to produce allantoin:

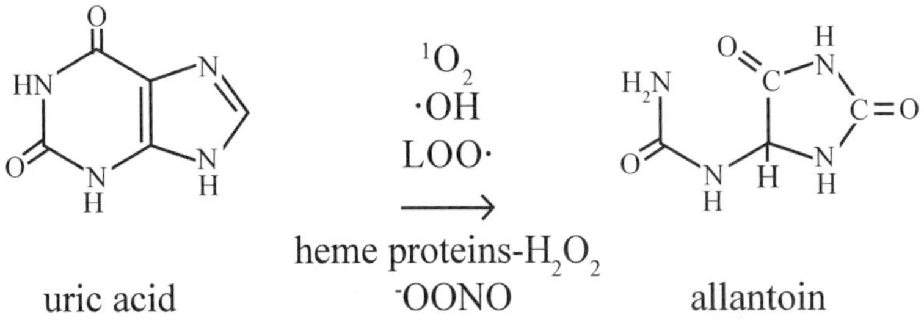

uric acid $^1O_2$ / ·OH / LOO· $\longrightarrow$ heme proteins-H_2O_2 / ^-OONO allantoin

It was, therefore, suggested by these authors that uric acid is an *in vivo* antioxidant and the determination of allantoin and uric acid may serve as a marker for oxidative stress *in vivo*. Grootveld and Halliwell [674] measured allantoin and uric acid in human body fluids (serum and synovial fluid).

If uric acid acts *in vivo* as an antioxidant, it should be possible to detect its reaction product allantoin in human body fluids after exposure to oxidative stress. Allantoin and uric acid were measured in healthy human subjects and in patients with rheumatoid disease [674]. Uric acid decreased and allantoin increased in serum and synovial fluid of patients with rheumatoid arthritis. It is well known that free radical reactions play an important role in the pathology of inflammatory joint disease [267,277]. The importance of uric acid as an antioxidant will be discussed in Chapter 8.

4. Aldehydes and alkanes

Aldehydes are the major products of lipid peroxidation. The most biologically important aldehydes are malonaldehyde and 4-hydroxy-2-nonenal. Aldehydes are involved in mutagenesis and tissue damage, and their chemical and biological properties have been reviewed [583-587,675]. In addition to aldehydes, lipid peroxidation has been shown to yield ethane as well as other alkanes and alkenes. The determination of ethane (CH_3CH_3) and pentane (C_5H_{12}) in exhaled breath has been suggested as a measure of lipid peroxidation *in vivo* [676,677].

5. The GSH/GSSG couple

Glutathione peroxidase eliminates H_2O_2 at the expense of GSH and formation of GSSG. The GSSG is reduced back to GSH by the glutathione reductase, which uses NADPH as the hydrogen donor. Therefore, the formation of GSSG has been used as a marker for increased production of reactive oxygen metabolites. The GSH/GSSG ratio is normally high to protect cells from oxidative stress. In the liver, the ratio has been found to be ca. 300/1 [678]. The decrease of GSH/GSSG is one of the definitions of oxidative stress given by Smith [565]. However, it must be noted, that even massive shifts in GSH/GSSG are not necessarily accompanied by visible tissue damage *in vivo* [565]. A decrease in GSH can also be due to mechanisms not involving radicals or ROMs. The cell injury caused by a decrease in GSH/GSSG is possibly mediated by reactions of GSSG with critical protein-SH groups [565]:

$$PSH + GSSG \longrightarrow PSSG + GSH$$

Evidence for this type of reaction has been obtained *in vitro* [679], but not *in vivo* [680].

6. Aromatic hydroxylation

In addition to the hydroxylation of DNA to yield 8-OHdG, the hydroxylation of benzoic acid and phenylalanine has been used as a test for *in vivo* formation of hydroxyl radicals [252,253]. These reactions have already been discussed in Chapter 3.

Chapter 7

NITRIC OXIDE

There are many ways to skin a cat.

A. CHEMICAL FORMATION OF NITRIC OXIDE

Nitric oxide is formed by a number of purely chemical reactions [294]:

$$Fe^{2+} + NO_2^- + 2H^+ \longrightarrow Fe^{3+} + NO + H_2O \quad (1)$$

$$2NO_2^- + 2I^- + 2H^+ \longrightarrow 2NO + I_2 + H_2O \quad (2)$$

Reaction (2) may be used for the quantitative iodometric determination of nitrite. Both of these reactions are reductions of NO_2^-. However, it is well known that nitrite (NO_2^-) can react under usually acidic conditions to give NO via the following pathway:

$$NO_2^- + H^+ \longrightarrow HONO$$

$$3HONO \longrightarrow HNO_3 + 2NO + H_2O$$

These reactions occur readily in the acid environment of the stomach, where nitrites produce nitrosamines, which are highly carcinogenic. The formation of NO occurs at only slightly acidic or even neutral pH to a limited extent. The pH in smooth muscle cells is between 5.9 and 7.0, within the range for these reactions. HONO can also give rise to N_2O_3 ($2HONO \longrightarrow N_2O_3 + H_2O$), which is the reactive intermediate in the nitrosation of secondary amines:

$$R_2NH + N_2O_3 \longrightarrow R_2N\text{-}NO + HONO$$

Another reaction for NO formation is the reduction of NO_2^- by ascorbic acid (AH_2):

$$AH_2 + HNO_2 \longrightarrow AH\cdot + \cdot NO + H_2O$$

This reaction is of physiological significance in the acid environment of the stomach. Ascorbic acid inhibits N-nitrosamine formation in the stomach [681] by converting HNO_2 to $\cdot NO$. Nitric oxide does not nitrosate amines directly, but only after oxidation to N_2O_3. This reduction by ascorbic acid has been suggested as a means by which neurons regulate their own oxygen supply [682].

Other common reducing agents are thiols. It has been reported by Pryor et al. [683] that thiols reduce $\cdot NO_2$ via the following reactions:

$$2\,RSH + \cdot NO_2 \longrightarrow RS\cdot + RS\text{-}N{\overset{\displaystyle OH}{\underset{\displaystyle OH}{\diagup}}} \longrightarrow RSNO + H_2O$$

$$RSNO \longrightarrow RS\cdot + \cdot NO$$

S-nitroso compounds are unstable and decompose to yield nitric oxide. Nitrosothiols are storage vessels for $\cdot NO$.

From a chemical point of view NO may be synthesized from ammonia (NH_3) by several two- and one-electron oxidations:

$$^-NH_3 \xrightarrow{2\,e^-} NH_2OH \xrightarrow{2\,e^-} HNO \xrightarrow{e^-} \cdot NO$$

It has indeed been known for a long time that hydroxylamine (NH_2OH) is transformed enzymatically to nitric oxide [684]. This reaction was first observed in the catalase-induced decomposition of hydrogen peroxide. In this reaction the formation of a catalase-Fe^{2+}-NO complex was observed [685]:

$$\text{catalase-Fe}^{III}\text{- H}_2\text{O}_2 \xrightarrow{} \text{cat [Fe}^V\text{=O]} \xrightarrow{\text{NH}_2\text{OH}} \text{cat [Fe}^{III}\text{HNO]} \xrightarrow{-\text{H}^+}$$

$$\text{cat [Fe}^{II}\cdot\text{NO]} \longrightarrow \text{cat [Fe}^{II}\text{]} + \cdot\text{NO} \quad \text{followed by:}$$

$$\text{cat [Fe}^{II}\text{]} + \text{O}_2 \longrightarrow \text{cat [Fe}^{III}\text{]} + \text{O}_2\cdot^-$$

However, this process is not involved in the *in vivo* formation of EDRF. The nitric oxide originates exclusively from the two terminal N-containing groups of L-arginine. The mechanism of this reaction has already been discussed in Chapter 5.

B. CHEMISTRY OF NITRIC OXIDE

1. Introduction

Nitric oxide is a ubiquitous chemical in our environment. It is produced by combustion engines and fossil fuel burning power plants. Nitric oxide is an electrically neutral radical with the following electronic structure:

$$\cdot\underline{\text{N}} = \underline{\text{O}}| \quad \longleftrightarrow \quad \underline{\text{N}} \overset{\cdots}{=} \underline{\text{O}}|$$

The unpaired electron is delocalized between nitrogen and oxygen. Nitric oxide can therefore be designated as a nitrogen or an oxygen radical. If the unpaired electron would be located exclusively at nitrogen, we should expect that NO dimerizes to N_2O_2, which is not the case. Also measurements of the N-O bond distance has shown the distance to be 1.14 Å, which is between a double bond (1.18 Å) and a triple bond (1.06 Å) [294].

The chemical reactivity of NO depends on its reaction partner. Some reactions occur rapidly (high rate constants), while others are slow. Nitric oxide by itself in deoxygenated aqueous solution is stable indefinitely [686]. The reactions of biochemical importance are fast (thus limiting its period of action) and include reactions with oxygen, superoxide radical anion, oxyhemoglobin, oxymyoglobin, guanylate cyclase, SOD and iron and iron-heme proteins. These reactions are of great physiological as well as pathological importance. On the other hand, NO is of relatively low reactivity towards many biological targets (amines and thiols) and reacts with these compounds not directly, but via its oxidation products, like NO_2, N_2O_3, N_2O_4, HONO, and ^-OONO. This low reactivity of $\cdot NO$ is in stark contrast to the high reactivity of the hydroxyl radical.

The hydroxyl radical reacts close to its site of formation (site-specific). It cannot be transported by any means over large distances. Therefore its precursor, hydrogen peroxide, is transported instead and the hydroxyl radical is then formed in a site-specific manner. The nitric oxide, on the other hand, may be transported over larger distances by hitching a ride on a thiol compound (like GSH, cysteine and SH-containing proteins). The NO is then released from the S-nitroso-compound wherever needed.

Hydroxyl radicals are highly reactive and destructive in biological systems. Nitric oxide in itself is far less reactive. Nitric oxide causes its damaging effect mainly by its oxidation products. On the other hand, these highly toxic species are also fulfilling very important physiological functions. The hydroxyl radical is part of the phagocytic defense system and NO is the most simple hormone, regulating blood flow and pressure and serves as a messenger in neurotransmission. At the same time NO (via its oxidation products) reacts with amines to form the carcinogenic N-nitrosamines. We may speculate whether this duality of function is part of the economy of design of living forms.

In order to determine whether EDRF is identical to NO we need a reliable and specific analytical technique for nitric oxide detection. Since NO is a free radical it cannot be observed directly, but only indirectly by the reactions it undergoes and the products it forms. We again encounter the same problem as we already discussed in the detection of hydroxyl radicals. We need a probe which is sensitive and specific. Nitric oxide and nitrogen dioxide are an important part of air pollution and a number of methods for their detection are available. Next I will discuss some of these reactions. NO can be determined by means of spectroscopy

[505,687,688], chemiluminescence [90,689], bioassay techniques [690] and its breakdown products NO_2^- and NO_3^-. Nitric oxide and nitrite can be determined by diazotation of sulfanilic acid. Nitrate does not react with sulfanilic acid; first it has to be reduced to nitrite.

The chemistry of nitric oxide was studied long before its importance in biology was recognized. Numerous reviews on nitric oxide chemistry have been written. For references see the review by Fukuto [691].

I am now going to discuss some reactions of nitric oxide, which are important in understanding its physiological as well as pathological character.

2. Reaction with oxygen

The reaction with oxygen is the normal pathway of NO deactivation in aqueous solutions. Since NO is a radical it reacts with other radicals and with oxygen (which is a diradical):

$$\cdot N{=}O \ + \ O_2 \longrightarrow \cdot OO\text{-}NO \tag{1}$$

Under normal physiological conditions NO disappears via its reaction with oxygen leading to the formation of nitrite (NO_2^-) and nitrate (NO_3^-):

$$\cdot OO\text{-}NO \ + \ \cdot N{=}O \longrightarrow O{=}N\text{-}O\text{-}O\text{-}N{=}O \longrightarrow 2\,\cdot NO_2 \tag{2}$$

$$2\,\cdot NO_2 \longrightarrow N_2O_4 \tag{3}$$

$$N_2O_4 \ + \ H_2O \longrightarrow NO_2^- \ + \ NO_3^- \ + 2\,H^+ \tag{4}$$

The Reactions (1) to (4) show that the disappearance of NO is second order in NO and first order in oxygen. The rate expression is:

$$-d\,[NO]/dt = 4\,k\,[NO]^2\,[O_2] \text{ with } k = 2 \times 10^{6.}\,M^{-2}s^{-1} \text{ [692-695]}$$

From this rate expression it follows that the half life of NO depends on its concentration. This fact has been often overlooked in the literature, where usually a half life of 3-5 sec is quoted. This half life is only correct for a specific NO concentration. Reaction (2) is an interesting example of a radical-radical combination, followed by dissociation. The dissociation is favored by the weak -O-O- bond.

From Equations (1)-(4) it follows that NO forms equal amounts of NO_2^- and NO_3^-, which is hardly ever the case. The formation of NO_2^- and NO_3^- requires the combination of 2 molecules of $\cdot NO_2$. This bimolecular reaction has to compete with the combination of $\cdot NO$ with $\cdot NO_2$:

$$\cdot NO_2 \ + \ \cdot NO \longrightarrow N_2O_3 \tag{5}$$

$$N_2O_3 \ + \ H_2O \longrightarrow 2\,NO_2^- \ + \ 2\,H^+ \tag{6}$$

The decomposition of N_2O_3 in water gives nitrite exclusively.

From the Equations (1)-(6) we see that in aqueous oxygenated solutions of NO we have a great variety of nitrogen-oxygen species present ($\cdot NO$, $\cdot OONO$, $\cdot NO_2$, N_2O_3, N_2O_4, NO_2^-, NO_3^-) which all may undergo a variety of reactions with many biomolecules. The chemistry of NO under physiological conditions is, therefore, quite complex. Nitric oxide in deoxygenated aqueous solutions is stable indefinitely [686].

The intermediates in the oxidation of NO to nitrite and nitrate (NO_2, N_2O_3, N_2O_4) are all powerful nitrosating agents. They nitrosylate primary and secondary amines to highly carcinogenic N-nitrosamines:

$$R_2NH \ + \ N_2O_3 \longrightarrow R_2N\text{-}NO \ + \ HONO$$

Intestinal flora produces large amounts of dimethylamine from choline. There are many different amines formed by colonic bacteria. A compilation of these amines can be found in the book by Grisham [696]. In patients with chronic inflammation of the colon the formation of N-nitrosamines may contribute to the high incidence of colon cancer in these patients.

The formation of nitrite and nitrate by mammalian metabolism [697] has been known for a long time, although the mechanism was not recognized as involving NO.

3. Reaction with ozone

Another more direct measurement of NO is to measure the chemiluminescence produced in the reaction of NO with ozone [689]:

$$NO + O_3 \longrightarrow NO_2^* + O_2$$

$$NO_2^* \longrightarrow NO_2 + hv$$

This reaction is extremely sensitive. It can detect NO at levels as low as $< 10^{-13}M$. This method has been used by Palmer, Ferrige and Moncada [90] in their identification of EDRF, and in the formation of NO by activated neutrophils [516,517].

4. Reaction with oxyhemoglobin

Another method for NO detection is its reaction with oxyhemo-globin [505,687,688]:

$$HbO_2 + NO \xrightarrow{k} MetHb + NO_3^- \quad k = 8 \times 10^7 \ M^{-1}s^{-1}$$

The high rate constant means that the reaction is complete within 100 ms in presence of mM HbO_2. The difference spectrum of oxy-hemoglobin and methemoglobin is recorded with a double beam spectrophotometer (maximum at 401 nm). The reaction is very sensitive and specific for NO. Oxygen, nitrate or nitrite do not interfere. The accuracy of the method has been tested with the chemilumunescence technique involving ozone. The only interference which was noted was in presence of high concentrations of H_2O_2 [698]. This occurs in the measurements of NO from macrophages, which generate high concentrations of H_2O_2:

$$H_2O_2 + MetHb \longrightarrow HbO_2 + H_2O$$

5. Diazotation of sulfanilic acid

Nitric oxide and nitrite react with sulfanilic acid to give a diazonium salt, which subsequently may be coupled to N-(1-naphthyl)-ethylenediamine to produce a colored azo-compound [89]:

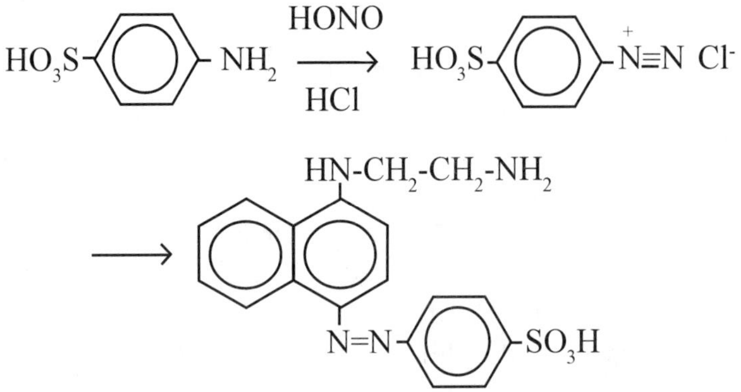

The azo-compound has a maximum absorbance at 540 nm. This procedure does not respond to NO_3^- and was used by Ignarro et al. [89] in the identification of EDRF as nitric oxide. Nitrate is determined after its reduction to nitrite.

The diazotation of aromatic amines has been used extensively in organic synthesis to synthesize a variety of substitution products, like halogenated aromatics or phenols. The diazonium salt decomposes to give phenols [699]:

$$Ar-\overset{+}{N}\equiv N \quad Cl^- \overset{H_2O}{\longrightarrow} Ar-OH + N_2 + HCl$$

Nitric oxide in oxygenated aqueous solutions (formation of HONO) deaminates DNA and its constituents [700].

6. Nitric oxide detection by spin trapping and by microsensors

As pointed out before, the development of any research area depends on the availability of a reliable and sensitive analytical technique. This is especially true for the investigation of radicals, which have a short lifetime.

The method of spin trapping has been discussed in Chapter 3 and I have pointed out the problems associated with this technique in the detection of hydroxyl and superoxide anion radicals. Similar problems arise in the detection of nitric oxide. A number of different spin traps have been found to produce artifacts. These problems have been discussed in detail by Arroyo and Kohno [701].

A highly sophisticated and sensitive method for NO detection was developed by Malinski and Taha [117]. These authors used an electrochemical method. They used a carbon fiber coated with a layer of polymeric porphyrin. Electrochemical oxidation of NO on the sensor involves one-electron transfer:

$$\cdot NO \longrightarrow NO^+ + e^-$$

Using this method NO production in a single cell could be determined. The method could detect as little as 10^{-20} moles of NO. This detection limit is 2 - 4 orders of magnitude smaller than the estimated amount of NO released by a single cell. The characteristics of this method (size, detection limit, range, response time, selectivity and applicability to *in situ* measurements) are indeed unique among NO detection methods.

Summary. In the NO detection by chemical reactions we encounter the same problem as we already discussed in the detection of hydroxyl radicals. Although many procedures are easy to use and sensitive they are not always specific. The further development of microsensors appears to hold great promise.

C. IMPORTANT REACTIONS IN THE DEACTIVATION OF NITRIC OXIDE

1. Introduction

The ability of EDRF/NO to act as an inter- and intracellular messenger depends on its limited period of action. EDRF/NO has a very short half-life (3-5 sec) in aqueous solutions under physiological conditions of temperature, pH and oxygen concentration. Nitric oxide, although a radical, is of comparatively low reactivity and exerts its damaging effect mainly through its reaction products like $\cdot NO_2$, N_2O_3, NO_2^-, NO_3^-, HONO and ^-OONO. The most important reactions of $\cdot NO$ *in vivo* are the reaction with oxygen, oxyhemoglobin, oxymyoglobin, superoxide radical anion, cytochrome c, guanylate cyclase, and SH- and NH-containing compounds. Nitric oxide exerts its muscle relaxing effect by activation of soluble guanylate cyclase to yield c-GMP, which in turn phosphorylates myosin light chains and causes muscle relaxation.

The chemical properties of EDRF and authentic NO have played an important part in establishing the identity of EDRF as $\cdot NO$. There are a number of compounds which inactivate both EDRF and $\cdot NO$ and it has been demonstrated that SOD is enhancing the relaxing effect.

The short half life is an advantage in terms of its role as a local vasodilator, but on the other hand, the numerous reactions $\cdot NO$ may undergo makes it difficult to determine exactly which reactions are important in its deactivation *in vivo*. This also makes the task of unravelling the mechanism of deactivation a complex task.

I will now discuss some of these deactivating reactions.

2. Reaction of NO with superoxide radical anion

The most important combination reaction of NO is its reaction with superoxide radical anion. Superoxide is the most ubiquitous radical in biological systems and combines with NO to form the peroxynitrite:

$$\cdot NO \ + \ O_2\cdot^- \ \xrightarrow{\ k\ } \ ^-OONO \ \overset{H^+}{\rightleftarrows} \ HOONO \quad pK_a = 6.8 \quad (7)$$

Like most radical-radical combinations this reaction has a high rate constant of $6.7 \times 10^9 \ M^{-1}s^{-1}$ [116]. Originally [702] this rate constant was claimed to be much lower ($k = 2.7 \times 10^7 \ M^{-1}s^{-1}$). These quite different rate constants can teach us something. I, therefore, like to discuss how these results were obtained. The standard method for rate constant determination is pulse radiolysis or flash photolysis. We have to consider the following reactions:

Pulse radiolysis:

$$HCOO^- \ + \ \cdot OH \longrightarrow H_2O \ + \ COO\cdot^- \quad (8)$$

$$COO\cdot^- \ + \ O_2 \longrightarrow CO_2 \ + \ O_2\cdot^- \quad (9)$$

$$\cdot NO \ + \ O_2\cdot^- \longrightarrow \ ^-OONO \quad (10)$$

However in this case the peroxynitrite was also produced via competing reactions:

$$\cdot NO \ + \ e_{aq}^- \longrightarrow NO^- \quad (11)$$

$$NO^- \ + \ O_2 \longrightarrow \ ^-OONO \quad (12)$$

The pulse radiolysis method measures the build-up of the peroxy-nitrite anion, which in this case is produced via the slower combination in Reaction (12).

Flash photolysis:

$$NO_2^- \ \xrightarrow{\ h\nu\ } \ \cdot NO \ + \ O\cdot^- \quad (13)$$

216

$$O^{\cdot-} + H^+ \longrightarrow {}^\cdot OH \qquad (14)$$

$${}^\cdot OH + HCOO^- \longrightarrow H_2O + COO^{\cdot-} \qquad (15)$$

$$COO^{\cdot-} + O_2 \longrightarrow CO_2 + O_2^{\cdot-} \qquad (16)$$

$${}^\cdot NO + O_2^{\cdot-} \longrightarrow {}^-OONO \quad k = 6.8 \times 10^9 \text{ M}^{-1}\text{s}^{-1} \qquad (17)$$

Pulse radiolysis is the most reliable and most commonly used method for the determination of radical rate constants. Most of our knowledge of hydroxyl radical reactions comes from pulse radiolysis. Thousands of rate constants have been compiled. As I have pointed out before (Chapter 3, EPR spectroscopy) a reliable and sensitive analytical technique is no guarantee for the correct answer. Unless we have a complete understanding of all the reactions involved, one may easily be led to the wrong conclusions.

3. Reaction with Fe^{2+}

In aqueous buffer Fe^{2+} deactivates EDRF/NO. We have already discussed the formation of $O_2^{\cdot-}$ by autoxidation of Fe^{2+} (Chapter 3):

$$Fe^{2+} + O_2 \rightleftharpoons Fe^{3+} + O_2^{\cdot-}$$

This reaction is followed by:

$${}^\cdot NO + O_2^{\cdot-} \longrightarrow {}^-OONO$$

Another way to deactivate NO is by formation of an iron-NO complex. In presence of anions (inorganic or organic) Fe^{2+} can react with NO to form complexes of the general structure $Fe(NO)_2(A^-)_2$, where A^- can be Cl^-, OH^-, HPO_4^{2-} or RS^- [703]:

$$Fe^{2+} + 2 {}^\cdot NO + 2 Cl^- \longrightarrow Fe(NO)_2(Cl^-)_2$$

One well known iron-NO complex is the vasodilator nitroprusside $Na_2[Fe(CN)_5NO]$, which slowly releases NO.

Since autoxidation of Fe^{2+} is reversible, high concentrations of $O_2^{\cdot-}$ reduce Fe^{3+} to Fe^{2+}. Superoxide radical anion may deactivate NO directly by combination or by reduction of Fe^{3+} to Fe^{2+} and subsequent formation of an iron-NO complex.

4. Reactions of NO with iron-heme proteins

The ability of NO to form complexes with metal ions has been known for a long time. This ability of NO to form complexes is important as an analytical tool for NO detection as well as in NO deactivation (oxyhemoglobin, oxymyoglobin, cytochrome c).

Reduced iron (Fe^{2+}) complexed with protoporphyrin IX has a high affinity for NO. Hemoproteins, such as hemoglobin, myoglobin, cytochrome c, and soluble guanylate cyclase react rapidly with NO to yield nitrosyl-heme (NO-heme) adducts [704-706]. Solutions of reduced heme are red, whereas oxidized hemes are brown and NO-heme adducts are bright pink-red. This formation of a colored nitrosyl compound has been used as the basis for an analytical technique. However this method is not very specific. The nitrosylation reaction has been used in the food industry (addition of $NaNO_2$ and ascorbic acid) to preserve the pink-red color of fresh meat. In presence of ascorbic acid, nitrite generates HONO and NO, which reacts with hemoglobin to give nitroso-hemoglobin. This treatment makes the meat look fresh, but at the same time (especially under the strongly acid conditions in the stomach) can produce N-nitrosamines, which are highly carcinogenic.

Hemoglobin is the important oxygen carrying protein in the blood. It contains a heme prosthetic group which binds iron in the +2 oxidation state. This protein can bind oxygen reversibly to form oxyhemoglobin without any change in the Fe oxidation state. The Fe^{II} however can be oxidized by a variety of oxidizing agents. One such agent is $\cdot$NO [707]:

$$Hb(Fe^{II}) + O_2 \longrightarrow HbO_2(Fe^{II})$$

$$HbO_2(Fe^{II}) + \cdot NO \longrightarrow Hb(Fe^{III}) + NO_3^-$$

This reaction is frequently used as a spectroscopic assay for NO. From spectroscopic evidence it has been determined that the oxygen in oxyhemoglobin is attached to the Fe center in the form of $O_2^{\cdot-}$ [708]:

$$[Hb(Fe^{II} O_2) \leftrightarrow Hb(Fe^{III} O_2^{\cdot-})] + \cdot NO \rightarrow Hb(Fe^{III}) + {}^-OONO \rightarrow NO_3^-$$

The $Fe^{III} O_2^{\cdot-}$ form contributes significantly to the resonance hybrid. Since NO may act as an electron donor or an acceptor, it can react with both Fe^{II} and Fe^{III}. If NO forms a complex with Fe^{III} - heme it acts as an electron donor:

$$Fe^{III} + \cdot NO \longrightarrow [Fe^{III} - NO \leftrightarrow Fe^{II} - {}^+NO] \xrightarrow{Nu:^-} Fe^{II} + NuNO$$

The partial positive charge on NO makes the complex prone to attack by nucleophiles, like thiols, amines or water [709]. These reactions yield nitrosothiols, N-nitrosamines and nitrous acid.

Oxyhemoglobin (HbO_2) is commonly used to deactivate NO. With normal concentrations of HbO_2 of 10mM the NO has a half life of 0.8 ms. This means that *in vivo* (in presence of HbO_2) NO disappears about three to four orders of magnitude faster than in ordinary aqueous oxygenated buffer solutions ($NO \longrightarrow NO_2^- + NO_3^-$). This fast reaction *in vivo* clearly demonstrates the effectiveness of HbO_2 in inhibiting the relaxing effect of NO. It also explains the absence of any specific deactivating enzyme for the control of NO concentration.

The ability of NO to form complexes with heme-iron centers is very important in the physiological role of nitric oxide. Nitric oxide activates guanylate cyclase by formation of a heme-Fe^{II}-NO complex. This complex then releases a histidine ligand, which is a general acid-base catalyst [704-706,710]. These reactions are schematically shown as follows:

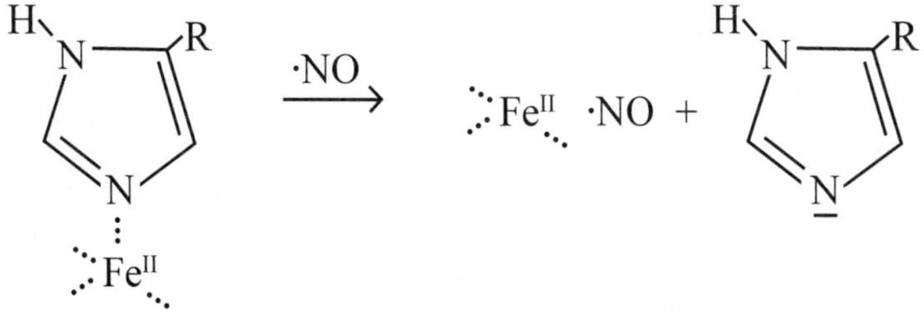

5. Reactions with easily oxidizable compounds (hydroquinone, pyrrogallol, SH and NH-containing compounds)

Any substance which produces $O_2^{\cdot-}$ will deactivate NO via combination to give peroxynitrite. Hydroquinone and pyrogallol are easily autoxidizable substances, which produce superoxide radical anion (see Chapter 3). The autoxidations are catalyzed by trace amounts of metal ions. Although the detailed mechanism of these autoxidations is not completely understood, we can envisage a mechanism as shown in Fig. 1.

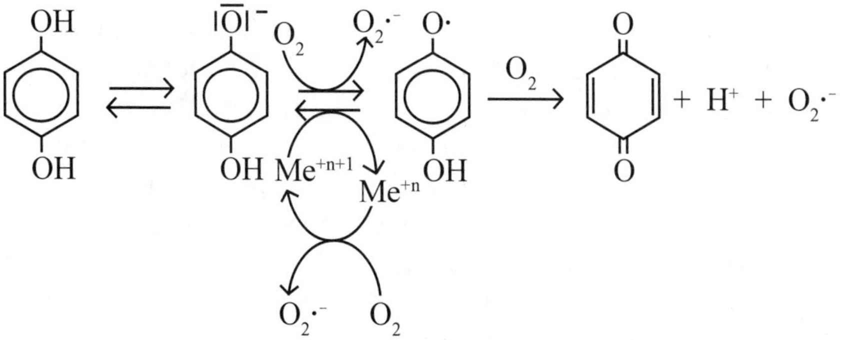

Fig. 1. Autoxidation of hydroquinone.

Another group of oxidizable compounds contain SH and NH groups. S-nitrosothiols play an important role in vasodilation. The question is are the S-nitrosothiols formed *in vivo* and if the answer is yes, how are they formed? It has been suggested that NO reacts with thiols directly to give nitrosothiol [544], however, this reaction does not occur. Nitrosylation only occurs in presence of oxygen [694]. I have already discussed the reaction of NO with oxygen. In the aqueous phase N_2O_3 is the major product. The N_2O_3 nitrosylates RSH:

$$RSH \quad + \quad N_2O_3 \longrightarrow RSNO \ + H^+ \ + \ NO_2^- \qquad (1)$$

In order for RSNO formation to take place Reaction (1) has to compete with Reaction (2):

$$N_2O_3 \quad + \quad H_2O \longrightarrow 2 \, NO_2^- \ + \ 2 \, H^+ \qquad (2)$$

A detailed kinetic analysis of these reactions has been carried out by Wink et al. [694]. These authors observed that under physiological conditions Reaction (1) involving cysteine or glutathione proceed faster than Reaction (2) by several orders of magnitude.

·NO reacts with oxygen to form NO_x (a number of different nitrogen oxides) and with $O_2 \cdot^-$ to form ^-OONO. Both of these species are quite damaging in many ways. Glutathione and other thiol-containing proteins may serve as scavengers of NO_x via Reaction (1). Since glutathione levels in cells are as high as 10 mM [711] glutathione is a good candidate for Reaction (1). Cysteine residues have an affinity for NO_x three orders of magnitude greater than non-SH-containing aminoacids, and $>10^6$ times greater than the amino groups of DNA [694]. Chinese hamster V79 fibroblasts depleted of glutathione showed enhanced cytotoxicity towards ·NO [694]. Glutathione is therefore a universal antioxidant scavenging not only $O_2 \cdot^-$, H_2O_2, and ·OH, but also NO_x. On the other hand reactive NO_x species can also react with other thiol-containing proteins, which may lead to the inactivation or inhibition of these enzymes.

In vivo the formation of S-nitrosoglutathione depends on the concentrations of ·NO and glutathione. We have to consider the following competing reactions:

$$NO_3^- \xleftarrow{HbO_2} \cdot NO \xrightarrow[\text{second order in } \cdot NO]{O_2} N_2O_3$$

$$\downarrow RSH$$

$$RSNO + HONO$$

Since the autoxidation of NO is second order in NO, at low NO concentrations the half life of NO may be quite long and no S-nitroso-glutathione is formed. At low NO concentrations NO may become deactivated by other pathways, such as the direct reaction with HbO_2. This deactivation pathway may regulate the NO concentration and thus prevent formation of toxic NO_x species within cells and tissues. However, with increasing NO concentration the NO_x formation predominates and glutathione serves as a scavenger to form S-nitrosoglutathione.

In summary S-nitrosylation of proteins is a reaction which takes place under physiological conditions and stabilizes NO in a bioactive form. This pathway may represent a mechanism for modification of proteins with important pharmacological and biological consequences.[712].

Another pathway for S-nitrosothiol formation is via autoxidation of thiols. The autoxidation of cysteine was investigated by Saez et al. [713]. These authors used the DMPO spin trap to determine the radical intermediates and they observed both thiyl and hydroxyl radicals. These radicals are produced according to the following mechanism:

$$RSH + Me^{n+} \longrightarrow RS\cdot + H^+ + Me^{+(n-1)}$$

$$RS\cdot + RSH \longrightarrow RSSR^{\cdot-} + H^+$$

$$RSSR^{\cdot-} + O_2 \longrightarrow RSSR + O_2^{\cdot-}$$

$$RSH + O_2^{\cdot-} + H^+ \longrightarrow RS\cdot + H_2O_2$$

$$RSSR^{\cdot-} + H_2O_2 \longrightarrow RSSR + OH^- + {}^-OH$$

In this autoxidation we obtain both RS· and $O_2^{·-}$. Nitric oxide can react with both of these species and become deactivated:

$$RS· \; + \; ·NO \longrightarrow RSNO$$

$$O_2^{·-} \; + \; ·NO \longrightarrow {}^-OONO$$

Vascular relaxation induced by ·NO and EDRF is inhibited by sulf-hydryl compounds [714]. According to Jia and Furchgott, these observations mean that free ·NO not RSNO is responsible for the activation of guanylate cyclase (GC).

Nitric oxide does not react directly with amines but only via its oxidation product N_2O_3 or HONO to give N-nitrosamines:

$$R_2NH \; + \; N_2O_3 \longrightarrow R_2NNO \; + \; HONO$$

$$R_2NH \; + \; HONO \longrightarrow R_2NNO \; + \; H_2O$$

Amines are formed by bacteria in the gut [696], where NO can then form the highly carcinogenic N-nitrosamines. In chronic inflammation, which has been implicated in increased cancer incidence, activated macrophages or neutrophils produce high concentrations of NO and $O_2^{·-}$ (H_2O_2). In addition to forming the N-nitrosamines the NO can react with $O_2^{·-}$ to form the highly reactive ^-OONO, which may then initiate the carcinogenic process.

Although the ^-OONO is a strong oxidizing agent, it does not nitrosate amines. It was observed that $O_2^{·-}$ decreased the ·NO-induced N-nitrosylation of secondary amines [715]. We have the following competing reactions:

$$
\begin{array}{ccc}
& O_2^{·-} & O_2 \\
{}^-OONO \longleftarrow & ·NO & \longrightarrow N_2O_3 + \text{other oxides} \\
\downarrow R_2NH & & \downarrow R_2NH \\
\text{no } R_2NNO & & R_2NNO
\end{array}
$$

6. Formation of alkoxy and hydroxyl radicals

It has been reported that t-butyl-hydroperoxide and n-butyl-hydroperoxide react with NO in benzene solutions at room temperature to yield alkoxy radicals [716] :

$$R_3C\text{-}O\text{-}OH \;+\; NO \longrightarrow R_3C\text{-}O\cdot \;+\; HONO$$

Whether this type of radical forming reaction is of any biological significance has not been demonstrated.

A similiar reaction with hydrogen peroxide has been studied in detail in the gas phase, where the following reaction was postulated [717]:

$$\cdot NO \;+\; H_2O_2 \longrightarrow HONO \;+\; \cdot OH$$

The reaction was also studied in aqueous solutions using DMPO as a spin trap [718]. The NO was produced via photolysis of N-methyl-N'-nitro-N-nitrosoguanidine (MNNG):

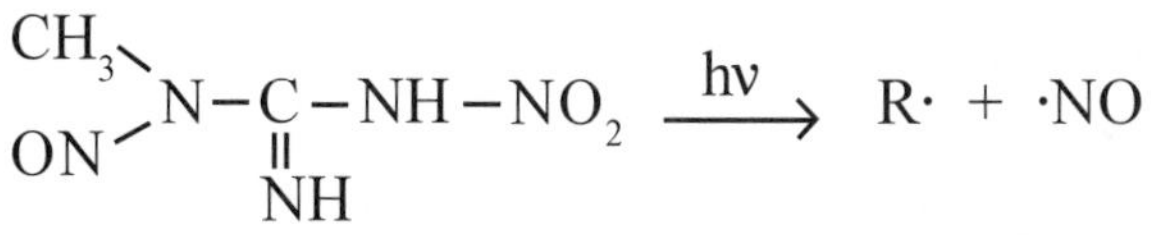

$$\begin{array}{c} CH_3 \\ \diagdown \\ \diagup \\ ON \end{array} N-\underset{\underset{NH}{\parallel}}{C}-NH-NO_2 \xrightarrow{\;h\nu\;} R\cdot \;+\; \cdot NO$$

The DMPO/·OH signal, and in presence of ethanol, the DMPO/CH$_3$CHOH signal were observed. This result seems to indicate that the NO-H$_2$O$_2$ reaction established in the gas phase also takes place in the aqueous phase. So far so good! However, almost a decade later the same group of investigators observed the DMPO/·OH signal in absence of H$_2$O$_2$ [719]. The OH production did not depend on the intermediate formation of O$_2$·$^-$ or H$_2$O$_2$, since SOD or CAT did have no effect on the DMPO/·OH signal. This is strange indeed, but things are getting stranger still. The reaction of NO (released from some N-nitroso-compounds) with DMSO in aqueous solutions gave CH$_3$SOOH and the reaction with DNA gave 8-OHdG, 8-OHdA and the deamination products xanthine, hypoxanthine and uracil [720]. These observations, according to the authors, seem to indicate the intermediate formation of hydroxyl radicals. However, as I have already discussed (Chapters 3 and 6) these products are also formed by one-electron transfer, followed by reaction with H_2O:

$$CH_3SOCH_3 \longrightarrow CH_3\overset{+\cdot}{S}OCH_3 \xrightarrow{H_2O} CH_3SOOH + CH_3 + H^+$$

The mechanism of these (·OH)-forming reactions has not been elucidated.

It is doubtful whether these hydroxyl radical forming reactions are of any biological significance. The H_2O_2 has to compete with the normal deactivation of NO in oxygenated aqueous solutions. However, in presence of high concentrations of H_2O_2 (as present in activated macrophages) these reactions may take place and contribute to the cytotoxic activity of macrophages.

7. Reactions of nitric oxide with nucleosides, nucleotides and DNA *in vitro* and *in vivo*

I have already discussed the reaction of aromatic amines with nitrous acid (HONO) to give diazonium salts. These diazonium salts have been known to organic chemists for a long time for the synthesis of a number of substituted aromatics via the following steps [699]:

$$Ar\text{-}NH_2 + HONO \longrightarrow Ar\text{-}\overset{+}{N}\equiv N \quad OH^- + H_2O$$

$$Ar\text{-}\overset{+}{N}\equiv N \quad OH^- \longrightarrow Ar\text{-}OH + N_2$$

The diazotation reaction is carried out at acidic pH and it may therefore appear to be of no importance in a biological system, except in the acid environment of the stomach or within macrophages. However, NO in presence of O_2 (formation of nitrogen oxides) was found to deaminate DNA bases at pH 7.4 [700,720,721].

This type of reaction converts deoxycytosine to deoxyuracil:.

$$dC + \cdot NO \xrightarrow[O_2]{-e^-} dU + N_2 + H^+$$

Deamination required both NO and O_2 [700] (deamination was negligible under anaerobic conditions). Deamination of guanine and adenine lead to xanthine and hypoxanthine, respectively. The relative yields of xanthine/hypoxanthine was variable, but with the xanthine usually predominant. Ratios of xanthine/hypoxanthine of 80 have been reported [721]. We have already encountered the special role of guanine in ROM-induced damage (Chapter 6). The yields of xanthine and hypoxanthine were much higher in intact nucleic acids than from free guanine or adenine. The yield of xanthine from guanine was 0.3 nmol/mg, but 550 nmol/mg from intact calf thymus RNA [721]. It has been suggested that these NO-O_2-induced deaminations are responsible for the mutagenicity of NO towards bacteria and mammalian cells.

The deaminating ability of NO-O_2 was also studied *in vivo* [700,722,723]. Treatment of *Salmonella typhimurium* strain TA1535 with NO-releasing agents (nitroglycerin) caused mutations. DNA sequence damage was > 99% C $\longrightarrow$ T transition in the *his* G46 (CCC) target codon, consistent with a cytosine deamination reaction [700]. Arroyo et al. [722] using the *Salmonella typhimurium* TA1535 strain showed a dramatic protection against NO-induced mutations by lipid-soluble antioxidants (β-carotene and α-tocopherol). Interestingly enough, ascorbate afforded no protection. Mutations and chromosome aberrations were observed in rats exposed to nitrogen oxides [723].

In presence of $O_2^{\cdot-}$ the NO can combine to form peroxynitrite. It was observed that peroxynitrite does not nitrosate amines [715] and it seems that in this case NO does not cause deamination of DNA bases. However peroxynitrite can react with DNA via one-electron transfer to give DNA$^{\cdot+}$ [223,724], which in turn reacts with H_2O to give hydroxylated bases (8-HO-G). Thus, this also leads to mutations, but via a different mechanism. One of the most abundant base modifications in mammalian DNA is the 8-hydroxyguanine, which I already discussed in Chapter 6.

We again encounter competition between different reactions. These reactions are summarized in Fig. 2. SOD removes $O_2^{\cdot-}$ and thus favors the oxidation of NO to nitrogen oxides and mutations via base deaminations. SOD however also can catalyze the reversible one-electron reduction of NO to NO⁻. SOD protects NO from deactivation by removing $O_2^{.-}$ or by reducing NO to NO⁻. Since both NO and NO⁻ may activate guanylate cyclase this reduction amounts to a protection.

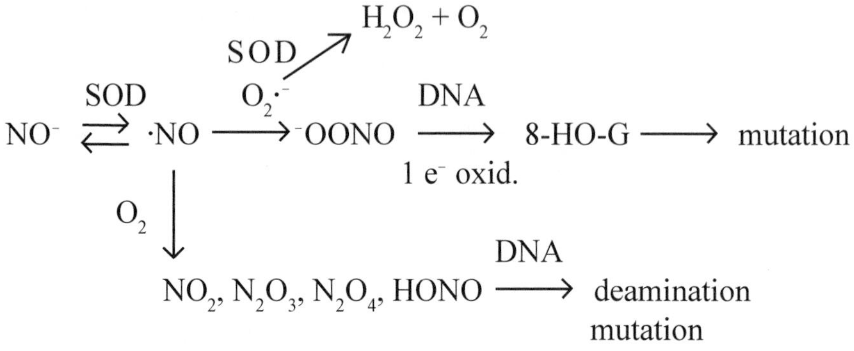

$$NO^- \underset{SOD}{\overset{SOD}{\rightleftarrows}} \cdot NO \xrightarrow{O_2^{.-}} {}^-OONO \xrightarrow[\text{1 e}^-\text{ oxid.}]{\text{DNA}} \text{8-HO-G} \longrightarrow \text{mutation}$$

with SOD branch → $H_2O_2 + O_2$

and O_2 branch ↓ → $NO_2, N_2O_3, N_2O_4, \text{HONO} \xrightarrow{\text{DNA}} \begin{array}{l}\text{deamination}\\\text{mutation}\end{array}$

Fig. 2. Oxidation-reduction reactions of nitric oxide.

D. REACTIONS OF NITROGEN DIOXIDE

Nitrogen dioxide is also a radical with the following structure:

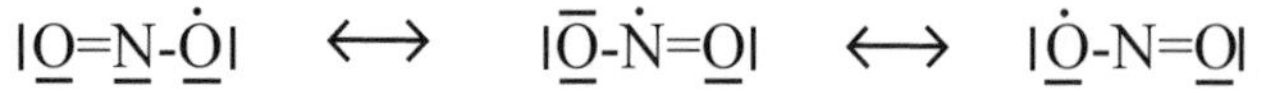

$$|\underline{O}=\underline{N}-\dot{\underline{O}}| \quad \longleftrightarrow \quad |\overline{\underline{O}}-\dot{N}=\underline{O}| \quad \longleftrightarrow \quad |\dot{\underline{O}}-N=\underline{O}|$$

The N-O bond length (1.197 Å) is intermediate between the length of a N-O single bond (1.43 Å) and a N=O double bond (1.18 Å). EPR results have indicated that the odd electron is localized to the extent of 48% on nitrogen, leaving 52% to be shared by the 2 oxygen atoms. Nitrogen dioxide therefore reacts with another radical both at nitrogen and at oxygen [725]:

$$R\cdot \quad + \quad \cdot NO_2 \longrightarrow R\text{-}NO_2$$

$$R\cdot \quad + \quad \cdot ONO \longrightarrow R\text{-}ONO \xrightarrow[H_2O]{H^+} R\text{-}OH \quad + \quad HONO$$

In addition to combination with another radical nitrogen dioxide can undergo H-abstraction or addition to multiple bonds [726]:

$$RH + \cdot NO_2 \longrightarrow R\cdot + HONO$$

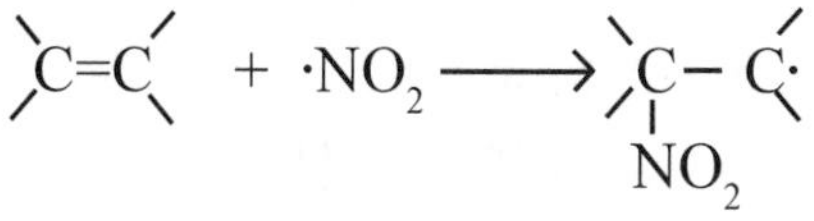

$$\text{\textbackslash}C=C\diagup + \cdot NO_2 \longrightarrow \diagup C - C\diagdown$$
$$NO_2$$

Nitrogen dioxide has been shown to initiate lipid peroxidation and formation of nitrosothiols [726]

Like most radicals nitrogen dioxide can dimerize:

$$2 \cdot NO_2 \longrightarrow N_2O_4$$

The N_2O_4 can react with hydroperoxides to give alkoxy radicals [727]:

$$R_3COOH + N_2O_4 \longrightarrow R_3COONO + HNO_2$$

$$R_3COONO \longrightarrow R_3CO\cdot + \cdot NO_2$$

In this reaction $\cdot NO_2$ functions like a metal ion:

$$ROOH + Me^{n+} \longrightarrow RO\cdot + OH^- + Me^{n+1}$$

In the gas phase $\cdot NO_2$ reacts with hydrogen peroxide to give hydroxyl radical [717]:

$$\cdot NO_2 + H_2O_2 \longrightarrow HONO_2 + \cdot OH$$

E. NITRIC OXIDE AND SUPEROXIDE DISMUTASE

Since $\cdot NO$ reacts rapidly ($6.7 \times 10^9 M^{-1}s^{-1}$) with $O_2^{\cdot-}$ and thus can become deactivated, SOD will confer protection against deactivation by any compound which produces $O_2^{\cdot-}$ (hydroquinone, pyrrogallol, thiols). This effect has been consistently observed and has been interpreted by dismutation of $O_2\cdot^-$ [93,728,729].

However, many objections have been raised to this interpretation on the basis of the wrong rate constant for the $\cdot NO$ - $O_2\cdot^-$ reaction (for a review see Noack and Murphy [505]). The SOD effect on nitric oxide deactivation shows again the importance of quantitative rate data.

In order for NO to react with $O_2\cdot^-$ it has to compete with many other reactions:

$$\cdot NO\ +\ O_2\cdot^- \xrightarrow{\quad} {^-}OONO \xrightarrow{\text{isomerization}} NO_3^- \qquad (1)$$

$$NO\ +\ O_2 \longrightarrow \cdots \longrightarrow \cdot NO_2, N_2O_3, N_2O_4, HONO \qquad (2)$$

$$\cdot NO\ +\ HbO_2\ (Fe(II)) \longrightarrow Hb(Fe(III))\ +\ NO_3^- \qquad (3)$$

Since Reactions (2) and (3) are fast, Reaction (1), although it has a high rate constant, can only compete at very high concentrations of $O_2\cdot^-$. These high concentrations are only present in activated phagocytes, where $^-$OONO is part of the antimicrobial armamentarium of phagocytes. Nitric oxide perfused through the heart gives exclusively NO_2^- ($N_2O_3 + H_2O$ 2 $NO_2^- + 2\ H^+$) and not the expected NO_3^- (Equation 1). Kelm and Schrader [730] speculated that NO_2^- is formed as follows:

$$\cdot NO\ +\ \cdot OH \longrightarrow HONO \longrightarrow H^+\ +\ NO_2^-$$

This is, of course, mission impossible. Hydroxyl radicals react on almost every collision and do not wait around until a $\cdot NO$ comes along. Work by Garlick et al. [731] on the reperfused rat heart (after ischemia) has shown the formation of radicals using the PBN spin trap. However, no $\cdot OH$ was detected, but instead carbon-centered radicals and alkoxy radicals which are formed via secondary reactions of $\cdot OH$ with membrane lipids were detected. The $\cdot OH$ does not survive long enough to react with PBN.

Nevertheless SOD did confer protection on NO. The protection of NO by SOD was even observed in buffer solutions in absence of any cells [732], which is strange indeed. It was suggested that the buffer contained $O_2^{\cdot-}$. Since $O_2^{\cdot-}$ can be formed via autoxidation of some metal ions (Fe^{2+}, Cu^+) and since these metal ions may be present in buffer as trace impurities [271,272] the presence of $O_2^{\cdot-}$ is a distinct possibility. Another possibility for the SOD effect has been suggested [733]. The SOD produces hydrogen peroxide, which can stimulate NO synthase and thus increase NO concentration

One of the characteristics of radicals is one-electron oxidation and reduction. For a radical to get rid of its unpaired electron, it can either obtain another one or donate its unpaired electron to another molecule. In the case of nitric oxide we have the following redox reactions:

$$NO^- \underset{-e^-}{\overset{+e^-}{\rightleftharpoons}} \cdot NO \underset{+e^-}{\overset{-e^-}{\rightleftharpoons}} NO^+$$

I have already mentioned the fact that in addition to NO a variety of other organic nitrites and nitrates can activate guanylate cyclase. Whether NO^- can activate guanylate cyclase may depend on the oxidation state of its Fe-heme prosthetic group [734]. When the iron is in the Fe(III) state NO^- might bind directly to yield the Fe(II) heme-NO complex, which is the activated form of the enzyme:

$$Fe(III)\text{-heme} + NO^- \longrightarrow Fe(II)\text{-heme-NO}$$

$$Fe(II)\text{-heme} + \cdot NO \longrightarrow Fe(II)\text{-heme-NO}$$

Evidence in support of NO^- has come from ion exchange experiments, which have indicated that EDRF has a negative charge and is hydrophilic [735,736]. The answer to the question is EDRF identical to NO or NO^-, may therefore be both NO and NO^- depending on the specific conditions. Under physiological conditions, $\cdot NO$ can be interconverted among different redox forms with distinctive chemistries [737]. The appropriate packaging of nitric oxide might facilitate its transport, prolong its life in the blood and tissue, target its delivery to specific effectors, and mitigate its adverse cytotoxic effects. It has been shown that NO^- also activates guanylate cyclase [738].

Any one-electron reduction of $\cdot NO$ to NO^- would therefore represent not a deactivation, but a protection. Murphy and Sies [734] have shown that SOD can catalyze the reversible reduction of $\cdot NO$ to NO^-:

$$\cdot NO + SOD\,[Cu(I)] \rightleftarrows NO^- + SOD\,[Cu(II)]$$

This redox reaction competes with the reaction with $O_2^{\cdot -}$:

$$O_2^{\cdot -} + SOD[Cu(I)] \xrightarrow{2\,H^+} H_2O_2 + SOD[Cu(II)]$$

$$O_2^{\cdot -} + SOD[Cu(II)] \longrightarrow O_2 + SOD[Cu(I)]$$

SOD donates one electron and one proton to reduce $O_2^{\cdot -}$ to H_2O_2, and accepts one electron from $O_2^{\cdot -}$ to give O_2.

The oxidation of NO^- was even observed with aqueous Cu^{2+} as well as with oxygen [738]:

$$NO^- + Cu^{2+} \longrightarrow \cdot NO + Cu^+$$

$$NO^- + O_2 \longrightarrow \cdot NO + O_2^{\cdot -}$$

NO^- acts as a reducing agent similiar to $O_2^{\cdot -}$ (the metal ion catalyzed Haber-Weiss reaction). Nitric oxide is synthesized from L-arginine via NO synthase (NOS). It was observed by Hobbs et al. [739] that SOD caused a marked increase in the production of free NO by mechanisms unrelated to the dismutation of superoxide or activation of NOS. It was suggested that SOD accelerates the conversion of an intermediate in the L-arginine $\longrightarrow$ $\cdot NO$ pathway, such as oxidation of NO^- to $\cdot NO$. These observations are in agreement with the oxidation-reduction reaction postulated by Murphy and Sies [734].

Since superoxide radical anion is chemically not very "super", the purpose of SOD to remove $O_2^{\cdot-}$ has been extensively and heatedly debated [105-109]. Nitric oxide reacts rapidly with $O_2^{\cdot-}$ to form the highly destructive peroxynitrite ($^-$OONO). It has been recently suggested [740] that the purpose of SOD might be the prevention of peroxynitrite formation. We have the following competing reactions:

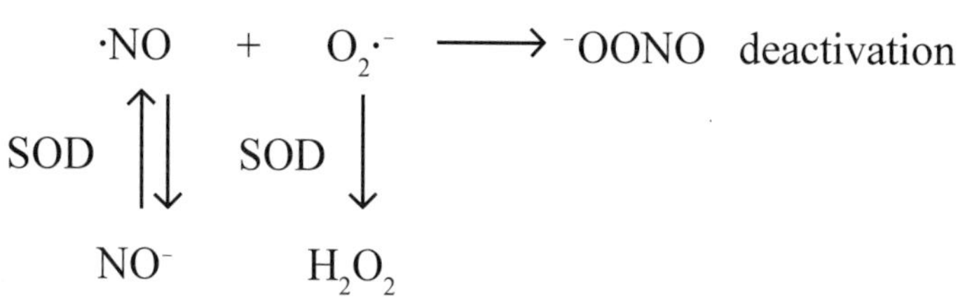

$$\cdot NO \; + \; O_2^{\cdot-} \; \longrightarrow \; ^-OONO \quad \text{deactivation}$$

$$SOD \updownarrow \qquad SOD \downarrow$$

$$NO^- \qquad H_2O_2$$

To prevent formation of $^-$OONO and deactivation, we can either remove $O_2^{\cdot-}$ or $\cdot$NO. SOD can accomplish both of these tasks. Which of the two competing reactions predominates depends on their relative rates. Both reactions lead to protection since both $\cdot$NO and NO$^-$ are vasorelaxants and act on guanylate cyclase [738]. The observation that SOD protects the vasorelaxing effect of nitric oxide does not tell us which of these two pathways is predominant (it of course depends on the steady state concentrations of $\cdot$NO and $O_2^{\cdot-}$ (among other factors).

The one-electron oxidation product of $\cdot$NO, the nitrosonium ion (NO$^+$) has been discussed as an intermediate in nitrosylation of thiols and amines [737]. I have already discussed the reaction of NO with transition- state metal ions and metaloproteins. NO reacts with Fe(III) porphyrin to give the Fe(III)$\cdot$NO complex in which the NO carries a slightly positive charge. This complex may be called a "crypto NO$^+$" ion and can nitrosylate thiols and amines.

The formation of NO$^+$ can take place at low pH:

$$HONO \; + \; H^+ \; \longrightarrow \; H_2O \; + \; NO^+$$

This reaction is analogous to the well known formation of NO_2^+ which can be found in any organic chemistry textbook. However, the formation of

NO^+ in a biological system at pH 7.4 is highly unlikely. *In vivo* the reaction of ·NO with transition metal ions has to compete with the fast reaction with oxygen to form a number of nitrogen oxides. N_2O_3 can nitrosylate thiols and amines [693,694].

F. THE CHEMICAL NATURE OF EDRF

There is overwhelming evidence that EDRF is identical to NO. Several years before the discovery of EDRF by Furchgott and Zawadzki [87] it was shown by Murad and coworkers [741,742] that some nitroso-compounds, nitroprusside, organic nitrates and nitrites (nitroglycerin, amylnitrite), nitric oxide and inorganic nitrite cause cyclic GMP accumulation in various tissue preparations. These organic and inorganic nitro-compounds have been used extensively in the treatment of coronary heart disease. Their pharmacological effect is attributed to the release of NO. Subsequently Ignarro and coworkers [743] established a relationship between activation of guanylate cyclase and vasorelaxation.

In order to demonstrate that EDRF and NO are identical one has to determine whether these two compounds undergo the same chemical reactions and whether their biological effects (relaxation, c-GMP levels) are affected by some agents in the same way.

EDRF as well as NO are very labile, having a half life (in dilute oxygenated aqueous solutions) of a few seconds. Both EDRF and NO are inactivated by pyrogallol, hydroquinone, superoxide radical, oxyhemoglobin, and methylene blue and are stabilized by SOD [89]. Both EDRF and NO cause considerable increases in c-GMP and this increase was inhibited in both cases by pyrogallol, methylene blue and oxyhemoglobin. Two chemical reactions which gave identical products are the reaction with hemoglobin to give nitrosyl-hemoglobin and the diazotation reaction of sulfanilic acid and coupling with N-(1-naphthyl)-ethylenediamine to yield a highly colored azo-compound [89].

Since scientific minds work along the same lines it is not surprising that the above results and conclusions were reached simultaneously by several groups of investigators [88-90]. Quantitative measurements of NO release (using the HbO_2 technique) from a variety of nitrovasodilators were carried out by Feelisch and Noack [687]. These authors found that the rate of NO release was linearly related to the activation of guanylate cyclase (GC). Further evidence for NO as the active species in the activation of GC came from results by Jia and Furchgott [714], who found that thiols inhibited the vasorelaxant effect of EDRF and NO in a bioassay. This observation was explained by the formation of superoxide radical anion from thiol autoxidation. The results indicate that NO not RSNO activates guanylate cyclase. However, as pointed out by the authors it does not settle the question is EDRF, released from endothelial cells, free NO or some sort of nitrosothiol, which then decomposes to give NO.

However, evidence has emerged indicating that EDRF may be different from NO. Some observations showed that EDRF specifically relaxes vascular smooth muscle whereas NO also relaxes nonvascular muscles (gastrointestinal, bronchial etc.) [744,745]. However, these observations have been questioned and could not be confirmed [746]. A more serious discrepancy between EDRF and NO was noted in anion exchange experiments. It was concluded that EDRF has a negative charge and is hydrophilic [735,736,747]. These results seem to indicate that EDRF is NO^-.

Another important difference between EDRF and NO is their different half lifes. NO is known to stimulate guanylate cyclase by forming a nitrosyl-heme complex at the active site of the enzyme [704-706,710]. However, we have to ask the following question how is NO transported from its site of formation (the endothelium) to the guanylate cyclase. NO reacts very rapidly with oxygen [692-694], superoxide [116], and heme [707] as well as non-heme iron [703]. These reactants are all abundant in plasma and cells. It is therefore unlikely that NO reaches guanylate cyclase just by simple diffusion. EDRF has a relatively long half life of several seconds in contrast to the *in vivo* half life of authentic NO of only 0.1 seconds [730]. These observations suggest that NO is stabilized *in vivo* by a reaction with a molecule, which increases its halflife, but at the same time preserves its biological activity. Prime candidates for this type of molecule are thiols, which have been shown to form S-nitrosothiols under physiological conditions [694,712]. S-nitrosothiols are known to be potent vasodilators and have been shown to activate guanylate cyclase [748]. However, the results of Feelisch and Noack [687] and Jia and Furchgott [714] have clearly demonstrated that S-nitrosothiols do not activate GC directly, but only after the release of free NO. Nitric oxide is a small neutral species, which can easily penetrate cellular membranes [749]. Most scientists agree so far. The question arises what is the detailed mechanism for the formation of free NO from S-nitrosothiols?

One may be tempted to answer who cares. The important fact is that NO is released and activates guanylate cyclase. However, if we know the detailed mechanism of NO release, we may be able to design better vasorelaxants.

It is generally assumed that S-nitroso-compounds decompose spontaneously in aqueous solutions to give free ·NO:

$$2\ RSNO \longrightarrow RSSR\ +\ 2\ \cdot NO$$

This reaction sequence shows that the half life of S-nitrosothiols is dependent on its concentration. The half life can be considerable at low concentrations. The stability of S-nitrosothiols varies over a wide range from seconds to hours [544]. S-nitrosocysteine in dilute (≤ 1 mM) oxygenated solutions has a half life of 30 s. The simple spontaneous decomposition has been questioned by several investigators [750-752]. RSNOs have a considerably longer half life than NO, and possess vasorelaxant properties *in vivo* resembling those of EDRF [712,751]. It was reported by Myers et al. [751] that the biological half life of EDRF resembles more closely that of S-nitrosocysteine than of free NO. In a bioassay S-nitrosocysteine was found to be 80 times more active than ·NO.

A careful study of a number of S-nitrosothiols, their rate of decomposition and activation of guanylate cyclase showed no relationship between stability and activation [753]. Experiments in which the formed NO was purged from the solutions did not affect GC stimulation [753].

Bates et al. [754] measured NO release from S-nitrosocysteine by bioassay and found that the S-nitrosocysteine had greater potency than authentic ·NO in agreement with previous results by Myers et al. [751]. This observation can only be explained if NO is less stable and does not survive to the end of the bioassay. Bates et al. therefore suggest that the ·NO is not free in solution, but the S-nitrosothiol decomposes either enzymatically or non-enzymatically on the surface of the smooth muscle cell to give ·NO, which then penetrates the cell and activates guanylate cyclase. The same conclusions were reached by Matthews and Kerr [753]. RSNOs relax vascular and non-vascular smooth muscles, inhibit platelet aggregation and stimulate guanylate cyclase. All of these effects may be mediated by ·NO, but the RSNOs cannot simply be considered as NO-releasing agents. The structure of the R-group plays a significant part in determining the biological activity in a manner, which cannot be predicted on the basis of its stability in aqueous solutions [751]. However, as pointed out by the authors the release of ·NO on the cell surface may not be correct in all cases. In some cases the target cell could have a transport system allowing entry of RSNOs into the cell without decomposition. The detailed mechanism of the decomposition on the cell surface is still not completely understood. The major S-nitroso compound present in human plasma is the S-nitroso-albumin [755,756]. The half life of this nitroso-protein was determined to be ca. 40 min. under normal physiological conditions. S-nitroso-albumin is present at a concentration of about 1 mM, whereas the concentration of free NO is 3 nM [755]. The S-nitroso-albumin is a huge molecule, which can not penetrate membranes and therefore cannot activate guanylate cyclase.

It was suggested by Stamler et al. [755] that S-nitroso-albumin is a storage vessel for the NO radical. The S-nitroso-albumin can slowly release NO or transfer it to other -SH compounds (cysteine or glutathione) and thus preserve its vasorelaxant activity. At the same time albumin protects cells from the many deleterious effects of nitric oxide (via reaction with oxygen or superoxide). NO can be liberated whenever needed and the NO does not have to be synthesized from scratch (via activation of NO synthases).

The S-nitroso-albumin can transfer its NO moiety to another lower molecular weight thiol like cysteine or glutathione in a process called transnitrosylation [757]:

$$\text{RS-NO} + \text{R'S-H} \longrightarrow \text{RS-H} + \text{R'S-NO}$$

The S-nitroso-cysteine has been identified as a product of endothelial cells [751]. RSNOs, mainly as S-nitroso-glutathione were detected in high concentrations (up to mM) in alveolar fluid [758]. S-nitroso-glutathione is a comparatively long lived relaxant of bronchial smooth muscle and provides a possible mechanism by which NO is protected from deactivation by reactive oxygen species present in the lung [759].

Many pathological processes are associated with abnormal vascular relaxation. Some of these are hypertension [760], diabetes [761], ischemia-reperfusion [762], and atherosclerosis [763]. These disease processes are also associated with formation of oxygen radicals, which can react with SH compounds (cysteine, glutathione etc.) to form the disulfides and thus deplete the antioxidant defenses. The formation of S-nitrosothiols is no longer possible and instead the endothelial cells release ·NO, which is a far less powerful vaso-relaxant than the RSNOs [752].

A complicating factor in vascular relaxation and constriction is the observation that endothelial cells, stimulated by the same stimuli which stimulate NO synthases, produce contracting factors (EDCF) and relaxing factors, which are different from NO [511]. One of these contracting factors has been characterized as a 21-amino acid peptide, named endothelin. Endothelin is a very potent vasoconstrictor which becomes active under acute hypoxia and requires extracellular Ca^{2+} for its activity. On the other hand EDRF/NO requires oxygen for its formation.

Prostacyclins, which are products of the lipoxygenase and cycloxygenase pathways, are also released from stimulated endothelial cells. Some of these products constrict veins, while others dilate arteries [511]. The story of vasorelaxation and constriction is not over yet and will keep scientists from the cardinal sin of thinking that they know it all.

Summary. NO synthase using L-arginine as a substrate produces ·NO or NO⁻ (NO⁻ ⟶ ·NO). Nitric oxide is released from endothelial cells and converted to S-nitrosothiols (S-nitrosoalbumin, S-nitroso-glutathione, S-nitrosocysteine and possibly others). The S-nitrosothiol diffuses through the plasma until it reaches the vascular smooth muscle. On the cell surface by some unknown mechanism the S-nitrosothiol releases ·NO, which penetrates the cell wall and activates guanylate cyclase. The important biological reactions of nitric oxide discussed so far are summarized in Fig. 3.

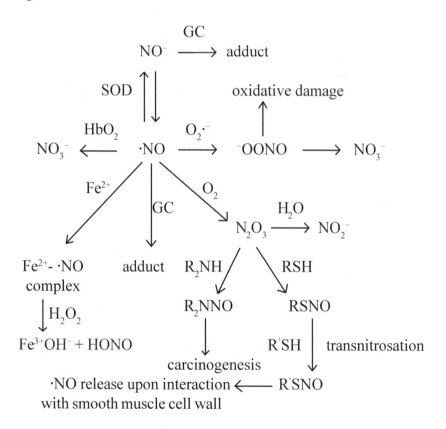

Fig. 3. Some important biological reactions of nitric oxide.

G. CHEMISTRY OF PEROXYNITRITE/PEROXYNITROUS ACID

1. Does the decomposition of peroxynitrous acid give hydroxyl radicals?

The decomposition of peroxynitrous acid was first studied by Beckman and coworkers [764]. They suggested that peroxynitrous acid decomposes to give hydroxyl radicals:

$$HOONO \longrightarrow HO\cdot + \cdot NO_2$$

We have already discussed the different probes for the detection of hydroxyl radicals. The most reliable probes are spin trapping, reaction with DMSO to give methyl radical, combination of spin trapping and reaction with DMSO or formate, and aromatic hydroxylation.

All of these methods have been applied in the study of peroxynitrous acid decomposition. Different investigators however reached different conclusions. It has been generally assumed that the formation of peroxynitrous acid is the major pathway for nitric oxide toxicity. Because of the importance of peroxynitrous acid decomposition, I would like to discuss the evidence for and against hydroxyl radical formation in some detail. The original paper by Beckman et al. [764] used the reaction with deoxyribose (to give malonaldehyde) and the reaction with DMSO to give formaldehyde as hydroxyl radical probes. Deoxyribose degradation is definitely not a reliable hydroxyl radical probe and it is indeed surprising that this method is still routinely used (Chapter 3). Although the formation of methyl radical from DMSO in presence of the DMPO spin trap has been shown to be highly reliable, I have already pointed out that methyl radicals are also formed in the reaction of DMSO with peroxydisulfate where a radical cation is the intermediate on the way to methyl radicals:

$$CH_3SOCH_3 + {}^-OS(O_2)O\text{-}O(O_2)SO^- \longrightarrow$$

$$CH_3\overset{+\cdot}{S}OCH_3 + SO_4^{2-} + SO_4^{-\cdot}, \text{ followed by}$$

$$CH_3\overset{+\cdot}{S}OCH_3 + H_2O \longrightarrow CH_3\overset{\overset{OH}{|}}{S}OCH_3 \longrightarrow CH_3SOOH + \cdot CH_3$$

We can envisage a similiar reaction with peroxynitrous acid:

$$CH_3SOCH_3 + HO\text{-}ONO \xrightarrow{\text{1 e}^- \text{ transfer}} CH_3\overset{\overset{OH}{|}}{S}OCH_3 + \cdot NO_2$$

$$\downarrow \text{2 e}^- \text{ transfer}$$

$$CH_3SO_2CH_3 + HONO \qquad\qquad CH_3SOOH + \cdot CH_3$$

The observed yield in the deoxyribose experiments was only 5.1 %, and in the DMSO case was 24.3 %. These low yields obviously are trying to tell us something (that we really do not understand all that is going on). Another red warning flag was the result of competition kinetics. The rate constant for the reaction with DMSO was $1.3 \times 10^9 M^{-1}s^{-1}$ instead of $7 \times 10^9 M^{-1}s^{-1}$ as determined by pulse radiolysis. On the other hand competition kinetics with deoxyribose gave a rate constant of $1.8 \times 10^9 M^{-1}s^{-1}$ in perfect agreement with the hydroxyl radical rate constant (from pulse radiolysis). Since in the meantime the formation of free hydroxyl has been ruled out, we can conclude that deoxyribose is not a reliable probe for hydroxyl radicals (see Chapter 3).

In 1994 a series of papers appeared giving support to the hydroxyl radical hypothesis. One of these studies used aromatic hydroxylation [765]. In the reaction of peroxynitrous acid with phenylalanine ortho, meta and para-tyrosine were obtained in very low yield (ca. 1%). The isomer distribution was 33 % o, 35 % m, 32 % p, independent of peroxynitrite concentration. The isomer distribution obtained is identical to the distribution obtained via radiolysis of aqueous solutions in presence of N_2O [766]. However, the formation of a phenol from an aromatic compound is in itself no proof of hydroxyl radicals. Hydroxylation may also be achieved via radical cations, which however lead to different isomer distributions [48-51,229]. Isomer distributions are only reliable probes for $\cdot$OH if we obtain a close to quantitative conversion (see Chapter 3).

In addition to phenylalanine, tyrosine was used as a probe for hydroxyl and $\cdot NO_2$. In this case the product was 3-nitrotyrosine. This product has been detected in the urine and has been recommended as a probe for nitric oxide and peroxynitrous acid formation *in vivo*. The formation of 3-nitrotyrosine, however, does not imply the involvement of hydroxyl radicals. The nitration of tyrosine was studied by pulse radiolysis involving the following reactions [767]

$$H_2O \rightsquigarrow \cdot OH, \ e_{aq}^-, \ \text{and molecular products}$$

$$\cdot OH + NO_2^- \longrightarrow OH^- + \cdot NO_2$$

$$e_{aq}^- + NO_3^- \longrightarrow NO_3^{2-} \xrightarrow{H_2O} 2\,OH^- + \cdot NO_2$$

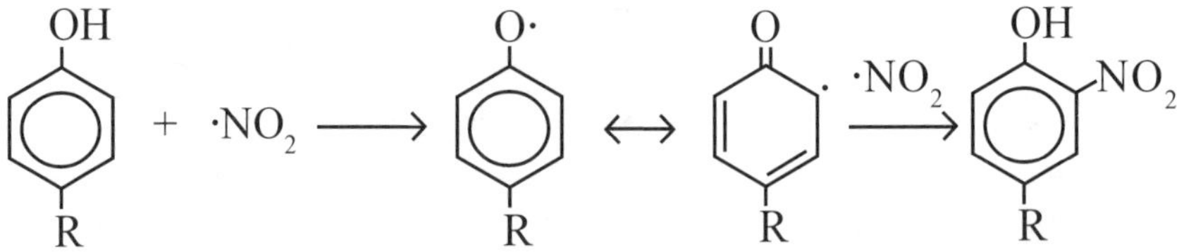

$$\underset{R}{\overset{OH}{\bigcirc}} + \cdot NO_2 \longrightarrow \underset{R}{\overset{O\cdot}{\bigcirc}} \longleftrightarrow \underset{R}{\overset{O}{\bigcirc}} \cdot NO_2 \longrightarrow \underset{R}{\overset{OH}{\bigcirc}} NO_2$$

At pH 7.5 the rate constant for H-abstraction was 3.2×10^5 M^{-1}s^{-1} and the rate constant for combination was 3×10^9 M^{-1}s^{-1} [767]. Since from pulse radiolysis data all the above rate constants are known the concentrations of NO_2^- and NO_3^- were chosen so that all $\cdot OH$ and all solvated electrons were scavenged. Under these conditions the phenoxy radical cannot be formed via hydroxyl radical and the formation of the final product, the 3-nitrotyrosine does not involve hydroxyl radical. Nevertheless, the detection of 3-nitrotyrosine in the urine may be a good *in vivo* indicator for peroxynitrite-induced damage.

It is well known that the $\cdot NO_2$ radical does not nitrate a benzene ring by the normal homolytic aromatic substitution process [768]. The radiolysis of aqueous solutions of nitrate and benzene however yielded high yields of phenol and nitrobenzene in equal amounts. Since the nitrobenzene cannot be formed via the normal homolytic substitution pathway the following mechanism has been suggested [672]:

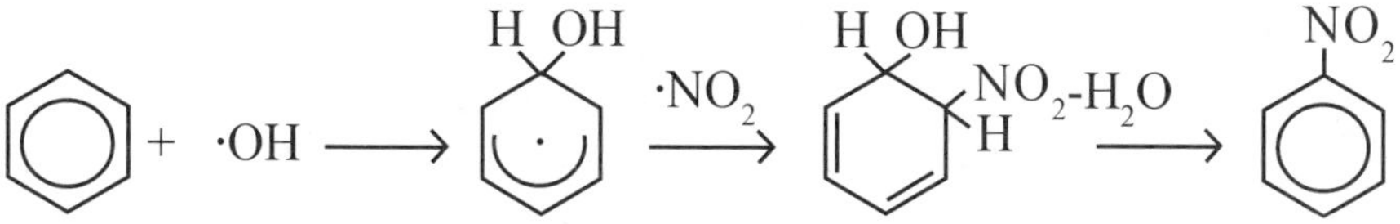

According to this mechanism the reaction of peroxynitrous acid (if it decomposes to $\cdot OH$ and $\cdot NO_2$) with phenylalanine should give 3-nitrophenylalanine, a compound which has not been reported. The formation of 4-nitrophenylalanine, which was observed, requires the formation of the nitronium ion (NO_2^+). The nitronium ion is the well known intermediate in the nitration of aromatics, which is achieved with a mixture of conc. sulfuric and nitric acid. It is not obvious how the nitronium ion is formed at neutral pH. Beckman and his group [769] have pointed out the following possibility:

$$^-OONO + SOD(Cu^{2+}) \rightarrow SOD(Cu^+\text{-}OONO) \xrightarrow{H^+} SOD(HO\text{-}Cu^+) + NO_2^+$$

In this reaction the nitration of tyrosine residue was observed. This type of nitration may be involved in the pathology of peroxynitrite by nitrating critical tyrosine containing targets.

Another possibilility for the formation of NO_2^+ was recently suggested by Pryor and coworkers [770]. These authors found that ^-OONO reacts rapidly with CO_2 according to the following reaction sequence:

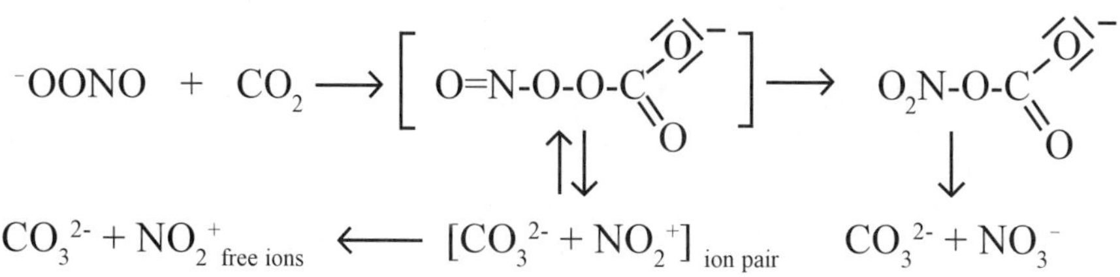

242

In addition to free ions the formation of free radicals ($CO_3^{\cdot -}$ + $\cdot NO_2$) was suggested. Since the CO_2 is ever present in biological systems, this reaction may be important *in vitro* and *in vivo*.

So far the evidence in favor of hydroxyl radical formation has been less than overwhelming. In 1994 two papers appeared, which used the DMPO spin-trapping technique as a probe for hydroxyl radicals [771,772]. Unfortunately, one group claimed formation of DMPO/·OH [771], whereas the other group reported the formation of the 5,5-dimethyl-pyrrolidone-2-oxy-(1) radical [772], and con-cluded that in the decomposition of peroxynitrous acid no hydroxyl radicals are formed. I have already discussed the possible pitfalls of the spin-trapping technique in Chapter 3 (pp.68-70).

Thermodynamic and kinetic arguments against the formation of hydroxyl radicals have been presented [773]. The peroxynitrous acid rearranges to nitric acid:

$$ONOOH \longrightarrow NO_3^- H^+$$

The proposed decomposition to ·OH and ·NO_2 has to compete with this rearrangement to nitric acid. Based on thermodynamic calculations it was concluded that no such decomposition is possible, but that instead there exists along the reaction coordinate from peroxynitrous acid to nitric acid an excited state of ONOOH, which is the reactive oxidant. This process is schematically depicted in the following diagram:

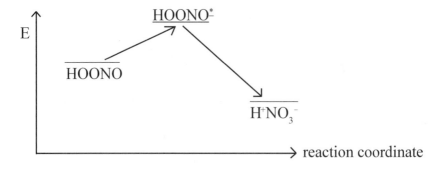

However, thermodynamic calculations by Merenyi and Lind [774] suggest that HOONO can undergo homolysis to form hydroxyl radical and nitrogen dioxide, contrary to previous conclusions by Koppenol et al. [773]. Calculations are based on the assumption that the parameters used in the calculations are the correct ones. These parameters are not always known with precision. Koppenol and Kissner [775] have recalculated the dissociation and have estimated the following rate constants for the two competing reactions:

$$HOONO \xrightarrow{k_d} \cdot OH + \cdot NO_2 \qquad k_d = 1 \times 10^{-2}\ s^{-1}$$

$$HOONO \rightleftharpoons H^+ + {}^-OONO \xrightarrow{k_{iso}} NO_3^- \qquad k_{iso} = 1.2\ s^{-1}$$

Based on these results the authors concluded that homolysis of HOONO to $\cdot OH$ and $\cdot NO_2$ is unlikely.

Peroxy-compounds have weak -O-O- bonds, which undergo substitution reactions (S_N2) or a one-electron transfer reaction with many nucleophiles:

$$Nu: + RO\text{-}OR' \begin{cases} \xrightarrow{S_N2} Nu^+\text{-}OR + R'O^- \\ \xrightarrow{ET} Nu^+\cdot + RO\cdot + R'O^- \end{cases}$$

cage escape to give 'free' radicals

These two types of competing reactions are difficult to distinguish, since the kinetics and products are similar in both pathways. Many examples of these types of reactions have been documented [33-36, 776,777]. Since we start with two neutral molecules (Nu: and ROOR') and obtain two radicals, this type of reaction is referred to as molecule-assisted homolysis

(MAH) (Chapter 2). It has been suggested that HOONO* reacts with a variety of substances by a one-electron transfer process:

$$HOONO + S \longrightarrow S^{+\cdot} + OH^- + \cdot NO_2$$

example:

$$tyrosine + HOONO \longrightarrow tyr^{+\cdot} + OH^- + \cdot NO_2$$

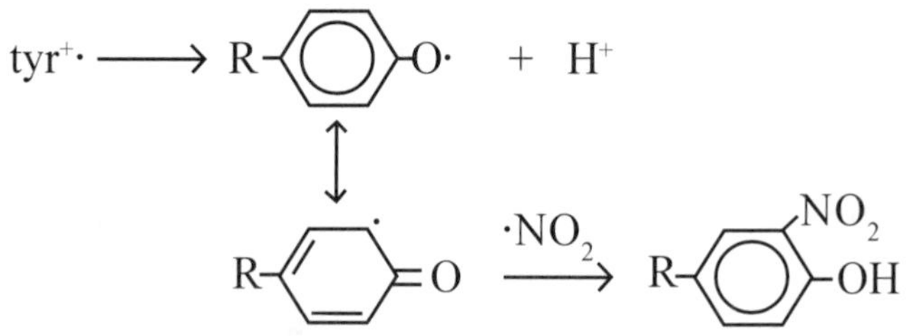

This reaction does not require the intermediate formation of hydroxyl radical. However, in all of the above reactions $\cdot NO_2$ is produced. $\cdot NO_2$ is a reactive radical which can initiate lipid peroxidation [683,726] and causes depletion of antioxidant defenses (uric acid, ascorbic acid, tocopherol and RSH compounds) in human plasma [778].

The formation of 8-hydroxyguanidine has been accomplished via the radical cation of guanidine (discussed in Chapter 6). 8-HO-G has been frequently used as a biological marker for oxidative damage by hydroxyl radicals. It still is a marker for oxidative damage, but it no longer indicates hydroxyl radical-induced damage. Guanidine reacts with peroxynitrite to yield 8-HO-G:

$$G + {}^-OONO \xrightarrow{2\,H^+} G^{+\cdot} + H_2O + \cdot NO_2$$

Another study by Douki and Cadet [223], indicating no hydroxyl radical formation, has already been discussed in Chapter 3 (page 65). Considering all the evidence we must conclude that HOONO does not decompose to give free hydroxyl radical.

2. Reactions of peroxynitrite/peroxynitrous acid with some biomolecules
a. Reactions with thiols

Sulfhydryl groups are frequent targets for radical attack and so are amines, aromatic aminoacids, antioxidants and unsaturated lipids. Thiols are critical to the active site of many enzymes.

If we want to examine the reaction of peroxynitrite with thiols we have to consider the following equilibria:

$$RSH \rightleftarrows RS^- + H^+ \qquad pK_a \text{ (cysteine)} = 8.3 \text{ at } 37° C$$

$$HOONO \rightleftarrows {}^-OONO + H^+ \qquad pK_a = 6.8 \text{ at } 37° C$$

Which species reacts depends on the pH. At pH 7.4 the two reactants exist as RSH and ^-OONO:

$$RSH + {}^-OONO \xrightarrow{k_2} RS \cdot + HO^- + \cdot NO_2 \qquad (1)$$

The following rate expression has been derived [779]:

$$k_2' = k_2 \frac{K_{a1} \qquad [H^+]}{(K_{a1} + [H^+]) \quad (K_{a2} + [H^+])}$$

The apparent rate constant k_2' depends on the pH, the dissociation constants of HOONO (K_{a1}) and RSH (K_{a2}) and the second order rate constant k_2 of cysteine reacting with peroxynitrite. It was determined that k_2 for cysteine is approx. $5.9 \times 10^3 \text{ M}^{-1}\text{s}^{-1}$ and for BSA (bovine serum albumin) about

2.6×10^3 $M^{-1}s^{-1}$ at pH 7.4 and 37° C. These rate constants are 3 orders of magnitude times greater than the rate constants for oxidation by H_2O_2. We have the following competing reactions:

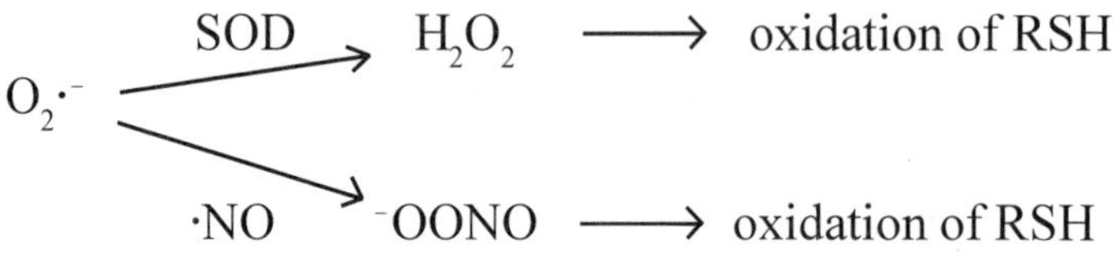

$$O_2^{\cdot-} \begin{array}{c} \xrightarrow{SOD} H_2O_2 \longrightarrow \text{oxidation of RSH} \\[2em] \xrightarrow[\cdot NO]{} {}^-OONO \longrightarrow \text{oxidation of RSH} \end{array}$$

SOD and NO compete for the superoxide radical anion. Additional complexity arises from the fact that nitric oxide can also react with SOD, a reaction which I have already discussed.

Reaction (1) is followed by addition of thiyl radical to another molecule of thiol:

$$RS\cdot \ + \ RSH \longrightarrow [RSSR]^{\cdot-} \ + \ H^+$$

$$[RSSR]^{\cdot-} \ + \ O_2 \longrightarrow RSSR \ + \ O_2^{\cdot-}$$

The final product of the reaction is cystine, but at the same time we produce more superoxide radical anion.

b. Reactions with antioxidants (ascorbate and α-tocopherol)

In the reaction of ascorbic acic with peroxynitrite we have to consider again the same type of acid-base equilibria as discussed above:

$$AH_2 \ \rightleftharpoons \ AH^- \ + \ H^+ \quad pK_a = 4.04$$

$$HOONO \ \rightleftharpoons \ {}^-OONO \ + \ H^+ \quad pK_a = 6.8$$

The rate of the reaction depends on the pH. The maximum rate was observed at pH 5.8 and at this pH the most abundant species are AH^- and HOONO [780]:

$$AH^- + HOONO \xrightarrow{\text{k}} A\cdot^- + H_2O + \cdot NO_2 \qquad k = 2.35 \times 10^2 \, M^{-1}s^{-1}$$

This oxidation represents a one e^- transfer to split the -O-O- bond. The oxidation of SH compounds (cysteine) is faster by a factor of 10 [779]. It was, therefore, concluded that ascorbate is unlikely to play a role in the defense against peroxynitrous acid.

Ascorbic acid is the most important water soluble antioxidant and α-tocopherol is the most important lipid soluble antioxidant. The α-tocopherol, located within the lipid bilayer, inhibits lipid peroxi-dation:

$$\alpha\text{-TOH} + \text{LOO}\cdot \longrightarrow \alpha\text{-TO}\cdot + \text{LOOH}$$

α-Tocopherol is oxidized by peroxynitrite by two different mechanisms: 1. one-electron transfer and 2. two-electron transfer. These reactions are shown in the following scheme [781]:

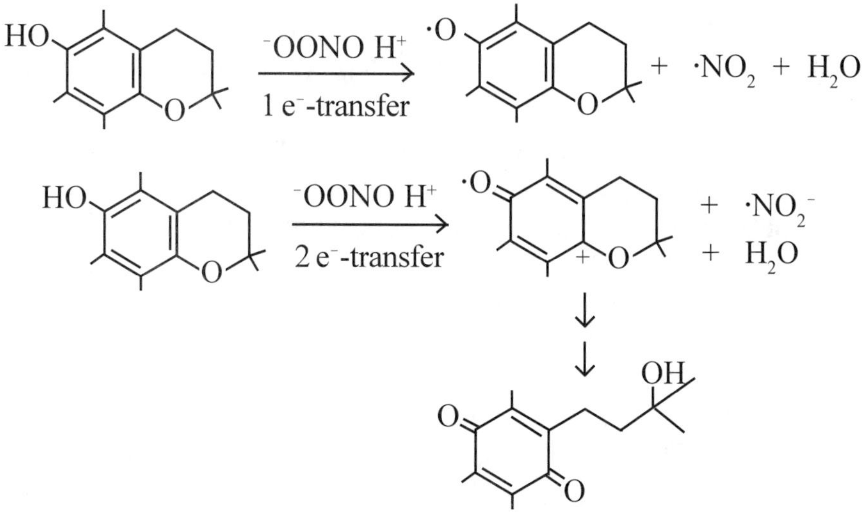

The initially formed α-tocopheroxyl radical and the α-tocopherone cation undergo a series of straightforward secondary reactions. One-electron transfer of course leads to a radical and two-electron transfer yields a cation. Other examples of two-electron oxidations by peroxynitrous acid have been reported. Pryor et al. [782] have shown that peroxynitrite oxidizes methionine in protein to methionine sulfoxide:

$$CH_3\text{-}\underline{S}\text{-}CH_2\text{-}CH_2\text{-}\underset{\underset{\oplus}{NH_3}}{CH}\text{-}COO^- \xrightarrow{^-OONO} CH_3\text{-}S(O)\text{-}CH_2\text{-}CH_2\text{-}\underset{\underset{\oplus}{NH_3}}{CH}\text{-}COO^-$$

In the oxidation of α-tocopherol it was determined that the major pathway is by a two-electron oxidation to yield the α-tocopherone cation, which leads to α-tocopherylquinone as the final product. It has been shown that in aqueous solutions the tocopheroxyl radical is regenerated by reacting with ascorbic acid [783]:

$$\alpha\text{-TO·} \;+\; AH^- \longrightarrow \alpha\text{-TOH} \;+\; A^{·-} \qquad k = 1.55 \times 10^6 \; M^{-1}s^{-1}$$

The two-electron oxidation product of tocopherol however cannot regenerate α-TOH in this way [784]. The two-electron oxidation of tocopherol by peroxynitrite effectively reduces the antioxidant effect of ascorbate. It was speculated by Hogg et al. [781] that the tocopherol content of low-density lipoprotein (LDL), trapped within the arterial wall, may be irreversibly oxidized by peroxynitrite with pathological consequences.

c. Reactions with lipids

Like the thiols, unsaturated fatty acids are critical targets for oxygen-derived radicals ($O_2^{·-}$, ·OH) and reactive oxygen species (singlet oxygen). Since superoxide is not very reactive, the superoxide-mediated damage of polyunsaturated fatty acids has been explained by its more reactive conjugate acid, the hydroperoxyl radical, or through OH radical formation via the Fe-catalyzed Haber-Weiss reaction.

Another possible mechanism of lipid peroxidation is via oxidation by peroxynitrite. It has been postulated that $O_2\cdot^-$ reacts in or near the membrane aqueous-lipid interface to yield peroxynitrite, which in turn initiates lipid peroxidation (as determined by malonaldehyde formation). Contrary to lipid peroxidation by oxygen radicals the oxidation via peroxynitrite does not require the presence of metal ions [785]. The first step in the oxidation is hydrogen abstraction with formation of nitrogen dioxide:

$$L\text{-}H \;+\; {}^-OONO \longrightarrow L\cdot \;+\; {}^-OH \;+\; \cdot NO_2$$

The resulting $\cdot NO_2$ is a small easily diffusible, lipid soluble radical, which can react with another lipid either via H-abstraction or addition to a double bond [726,767]. The rate constant for H-abstraction from PUFA by $\cdot NO_2$ is between 10^5-10^6 $M^{-1}s^{-1}$ [767], which is of course five orders of magnitude less than the abstraction by hydroxyl radical.

d. Reaction of peroxynitrite with DNA

The oxidation profile obtained in the reaction of peroxynitrite with isolated DNA was quite different from the profile obtained by γ-radiolysis [223]. These results by Douki and Cadet clearly show that the damaging species is not the hydroxyl radical. The major product is 8-OHdG, which I have already discussed as a biological marker for oxidative damage (Chapter 6).

H. THE ROLE OF PEROXYNITRITE/PEROXYNITROUS ACID *IN VIVO*

Peroxynitrite and its conjugate acid are powerful oxidizing agents, and react with a variety of important biomolecules, such as thiols (cysteine, glutathione), antioxidants (ascorbate, α-tocopherol), aro-matic aminoacids (phenylalanine, tyrosine, tryptophan), and unsa-turated lipids in purely chemical systems. These compounds all play a role in metabolism and damage to these molecules will result in serious consequences. It has, therefore, been assumed that ${}^-OONO$ and HOONO play an important role in the cytotoxicity of nitric oxide (which by itself is of low reactivity) *in vivo*. Although this hypothesis is quite attractive it is not necessarily correct.

In order to determine whether $^-$OONO/HOONO plays a role *in vivo*, we first must determine if it is formed under *in vivo* conditions. We have discussed the formation in a purely chemical system:

$$\cdot NO \ + \ O_2\cdot^- \ \longrightarrow \ ^-OONO \ \longrightarrow \ HOONO$$

In order for this reaction to occur it has to compete with all the other possible reactions of NO (reaction with oxygen, oxyhemoglobin, and Fe-heme proteins). The formation of $^-$OONO *in vivo* was deduced from experiments with SOD. SOD was found to prevent nitric oxide deactivation and this effect was ascribed to the removal of $O_2\cdot^-$ by SOD ($2\,O_2\cdot^- + 2\,H^+ \longrightarrow H_2O_2 + O_2$). For a discussion on other explanations of this SOD effect see Section E, this chapter.

The rate constant for the $\cdot NO$-$O_2\cdot^-$ reaction was originally determined as $5 \times 10^7\,M^{-1}s^{-1}$ [702]. Based on this rate constant and the rate of NO oxidation by oxygen it was concluded that the concentration of $O_2\cdot^-$ has to be very high in order to compete. The upward revision of the rate constant to $7 \times 10^9\,M^{-1}s^{-1}$, however, makes the formation of $^-$OONO possible in many tissues. On the basis of competition kinetics, Squadrito and Pryor [786] concluded that concentrations of nitric oxide as low as 3 nM are expected to compete efficiently for superoxide near the surface of endothelial cells.

Quantitative measurement of peroxynitrite formation by alveolar macrophages (stimulated by phorbol 12-myristate 13-acetate) has been carried out by Beckman and colleagues [787]. They determined a rate of 0.11×10^{-9} moles $\times 10^6$ cells^{-1} min^{-1} and concluded that the major part of NO produced by these macrophages was converted to peroxynitrite.

Malinski et al. [788,789] reported concentrations of nitric oxide of 450 nM at the endothelial cell surface after stimulation with bradykinin and concentration of up to 4 mM during middle cerebral artery occlusion in the rat. These results imply that nitric oxide reacts with superoxide radical anion 150 and 1300 times more rapidly than the superoxide decays by all other pathways. Therefore under these conditions peroxynitrite formation does occur *in vivo*.

These examples show the importance of quantitative data. In order to answer the question is peroxynitrite formed *in vivo*, we have to know the rate of formation of the reactants, their concentrations and their rate constants with many other reactants. A difference of 100 in the rate constants has a dramatic effect on the answer.

After the formation of $^-$OONO *in vivo* has been established it does by no means follow that peroxynitrite is responsible for cytotoxicity of nitric oxide. The $^-$OONO/HOONO pair can react with many biomolecules, but these reactions have to compete with the unimolecular isomerization of $^-$OONO to nitrate:

$$\text{reaction with thiols, aromatic aminoacids, lipids, DNA, antioxidants} \longleftarrow \text{$^-$OONO} \xrightarrow{\text{inactivation}} \text{NO}_3^-$$
$$t_{1/2} = 1\text{ s at pH 7.4}$$

The reactions of $^-$OONO with biomolecules are bimolecular and depend on the concentration of both reactants, while the isomerization to nitrate (an inactive metabolite) is concentration independent. At low concentrations of peroxynitrite the deactivation may gain the upper hand.

Another reaction affecting the life-time of peroxynitrite is its reaction with excess NO or $O_2{}^{\cdot-}$. In order for a radical-radical combination to occur (NO + $O_2{}^{\cdot-}$) both species have to be produced in high concentrations and at the same time and location. Excess NO or $O_2{}^{\cdot-}$ react with $^-$OONO to give a yet unknown species, significantly decreasing certain oxidative processes [790].

The optimum yield of $^-$OONO is produced if both NO and $O_2{}^{\cdot-}$ are formed at the same site and at equal concentrations. The time-concentration profile of NO and $O_2{}^{\cdot-}$ is not always equal. These two species are formed by two independent pathways. For example, phorbol ester-treated neutrophils produce 10 times higher $O_2{}^{\cdot-}$ concentrations than $\cdot$NO during the first few minutes. However, the $O_2{}^{\cdot-}$ flux rapidly subsides, whereas

NO production continues for several hours [715]. The time during which the ratio $NO/O_2\cdot^-$ is one to one is brief and therefore the amount of ^-OONO is relatively small.

Cytokine-stimulated alveolar macrophages, however, produce both NO and $O_2\cdot^-$ at the same rate and for extended periods of time [787], thus increasing the chances for ^-OONO formation in this particular cell line. In the macrophage the ^-OONO is an important part in the antimicrobial defense system.

It has been suggested that *in vivo,* excess NO or $O_2\cdot^-$ act as endogenous modulators of peroxynitrite-induced tissue damage [790]. Although the detailed chemistry has not been elucidated the following reactions have been suggested as being thermodynamically feasible [773,791]:

$$NO_3^- \longleftarrow {}^-OONO + \cdot NO \longrightarrow NO_2 + NO_2^- \qquad (1)$$

$$HOONO + O_2\cdot^- + H^+ \longrightarrow NO_2 + O_2 + H_2O \qquad (2)$$

Reaction (1) may offer some explanation for the exclusive formation of NO_2^- if $\cdot NO$ is perfused through the heart [730]. At high $\cdot NO$ concentrations we produce exclusively NO_2^-, but at low $\cdot NO$ concentrations ^-OONO can isomerize to NO_3^-.

I. PROTECTIVE EFFECT OF NITRIC OXIDE IN OXYGEN RADICAL-INDUCED DAMAGE

Since nitric oxide is a radical it can combine with other radicals. It is, therefore, a radical scavenger or antioxidant. I have already discussed many deleterious effects of nitric oxide metabolites. I would now like to discuss some examples where NO exerts a protective effect on damage caused by oxygen-derived radicals.

Kanner et al. [792] studied the hydroxylation of benzoic acid by a complexed Fe^{2+} and hydrogen peroxide. In presence of nitric oxide the formation of salicylic acid was decreased. This was explained via the following hydrogen peroxide decomposition:

$$X\text{-}Fe^{2+}\text{-}NO + H_2O_2 \longrightarrow X\text{-}Fe^{3+}\text{-}NO \longrightarrow X\text{-}Fe^{3+}\text{-}OH^- + HONO$$
$$\quad\quad\quad\quad\quad\quad\quad\quad\quad\quad \underset{OH^-\ OH}{\mid\ \mid}$$

where X is some chelator, like phosphate, EDTA or protoporphyrin. The Fe^{2+}-NO complex decomposes hydrogen peroxide via a pathway not involving free hydroxyl radicals, leading to an Fe^{3+} complex and nitrous acid. In the same reaction system nitric oxide also protected methionine oxidation (to give ethylene) and lipid peroxidation (as measured by thiobarbituric acid reactive product).

Normally Fe^{2+} reacts with H_2O_2 by one-electron transfer (the Fenton reaction), but in the above Fe^{2+}-NO complex the nitric oxide reacts with the incipient hydroxyl radical. Formally, we can consider that nitric oxide acts as a scavenger of hydroxyl radical. In this reaction however neither the ·NO nor the ·OH are "free".

The reactivity of nitric oxide (toxic or protective) depends on the time concentration profile. If we have high concentrations of nitric oxide it will react rapidly with oxygen to produce a variety of reactive products or it will react with $O_2\cdot^-$ to give peroxynitrite. However, if NO is produced continuously at low concentrations over extended periods of time, the NO can have a protective effect.

Hydrogen peroxide and superoxide have been known for some time to cause cell death in both eukaryotic and prokaryotic organisms [655,656]. The reaction of H_2O_2 with a number of transition metal ions leads to the formation of the highly destructive hydroxyl radical which can damage DNA, causing eventual cell death [657]. Since NO is oxidized to a variety of reactive species, which can damage biomolecules we should expect that NO increases the H_2O_2 and $O_2\cdot^-$ -induced damage (especially since NO and $O_2\cdot^-$ can form the highly reactive ^-OONO). However, as pointed out before the recombination of NO and $O_2\cdot^-$ has to compete with many other reactions of NO and $O_2\cdot^-$. Since these two species are produced by two independent pathways their concentration profiles are not identical. A radical-radical combination is only effective at high radical concentrations. At low concentrations of NO it was found that it is not cytotoxic, but rather inhibits the toxic effects produced by H_2O_2 and $O_2\cdot^-$ [793].

It is rather difficult to control the concentration of NO in a chemical system. However, a series of compounds known as NONOates spontaneously release NO in a controlled manner [794]. These NONOates have the following structure:

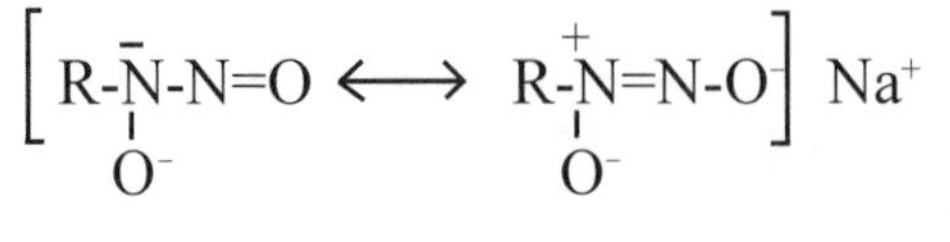

$$\left[R\text{-}\overset{-}{N}\text{-}N{=}O \longleftrightarrow R\text{-}\overset{+}{N}{=}N\text{-}O^{-} \right] Na^{+}$$
$$\overset{|}{O^{-}}\overset{|}{O^{-}}$$

where R is some nucleophile, usually some amine. Depending on the nature of the substituents these NONOates release NO over periods as short as a minute and as long as 3 days. With the use of these NONOates it was determined [795] that NO can be protective in various biological systems. Exposure of Chinese hamster V 79 cells to varying concentrations of hydrogen peroxide resulted in cell death. Exposure to $Et_2N[N(O)NO]^-Na^+$ (DEA/NO with a half life of 2.1 min.) resulted in no toxicity by itself. However exposure of DEA/NO and H_2O_2 resulted in abatement of the H_2O_2 toxicity. Nitric oxide also proved to be protective against damage induced by alkyl peroxides [796].

Nitric oxide affords protection against $O_2^{\cdot -}$ and $^-OONO\text{-}$ induced lipid peroxidation [797]. Nitric oxide acts as a prooxidant via its recombination with $O_2^{\cdot -}$ to produce ^-OONO, and as a antioxidant as a consequence of its direct reaction with peroxyl radicals formed during lipid peroxidation. The $\cdot NO$ acts as a terminator of lipid radical chain propagation:

$$L\text{-}H \longrightarrow L\cdot \qquad\qquad \text{initiation}$$

$$\left.\begin{array}{l} L\cdot \ + \ O_2 \longrightarrow L\text{-}OO\cdot \\[1em] L\text{-}OO\cdot \ + \ L\text{-}H \longrightarrow LOOH \ + \ L\cdot \end{array}\right\} \quad \text{propagation}$$

$$L\text{-}OO\cdot \ + \ \cdot NO \longrightarrow LOONO \qquad \text{termination}$$

The recombination of a number of peroxyl radicals with ·NO has been determined by Padmaja and Huie [798] to proceed with a rate constant of ca. $1\text{-}3 \times 10^9 \, M^{-1}s^{-1}$. However, as pointed out by the authors the product, peroxynitrites, are not final stable products. A number of peroxynitrites were found to decay unimolecularly with a rate constant $k = 0.1\text{-}0.3 \, s^{-1}$. The products of this decay can be the less reactive alkylnitrate or the more reactive alkoxy radical:

$$ROONO \nearrow \begin{array}{l} R^+ \, NO_3^- \\ \\ RO\cdot \;\; + \;\; \cdot NO_2 \end{array}$$

Alkoxy radicals of course can initiate further lipid peroxidation or undergo β-cleavage to give smaller alkyl radicals.

Nitric oxide has antiatherogenic properties. NO inhibits platelet aggregation, leukocyte adhesion and vascular smooth muscle proliferation [91]. During hypercholesterolemia one of the early events is a reduction in the ability of vessels to respond to EDRF. It has been demonstrated that this reduction in EDRF responsiveness is not due to decreased NO synthase activity, but due to increased $O_2^{\cdot-}$ production [799-801]. It has been shown by Hogg et al. [802] that low density lipoprotein (LDL) oxidation is inhibited by ·NO. Therefore, the reaction of ·NO with $O_2^{\cdot-}$ is removing the antioxidant ·NO, generating the prooxidant ^-OONO and inhibiting vascular relaxation. These combined effects may increase the oxidative stress level and lead to foam cell formation [802].

Activated neutrophils are present at sites of inflammation and migrate across vascular endothelium. Since stimulated endothelium releases NO, which is a radical scavenger, it appears possible that NO protects the endothelium from neutrophil-mediated injury. Results by Clancy et al. [803] suggest that NO inhibits superoxide generation by neutrophils via a direct action on a component of the NADPH oxidase complex. This inhibition may involve the nitrosylation of a SH containing protein of the NADPH oxidase and may represent a mechanism of endothelial defense against neutrophil-dependent tissue injury during inflammation [803].

Conclusion. NO at low concentrations is protective against oxygen-derived free radical damage, while at high concentrations NO is cytotoxic. So NO can be good or bad depending on its site of formation and concentration. We encounter a similar situation with $O_2^{\cdot-}$, which will be discussed in Chapter 8 (under SOD paradox) and Chapter 9 (under ischemia-reperfusion).

J. TOXICITY OF NITRIC OXIDE VIA RELEASE OF IRON(II) FROM FERRITIN

Fe^{2+} is the essential catalyst for the formation of hydroxyl radicals, either via the Fenton or the Haber-Weiss reaction. The concentrations of free Fe^{2+} *in vivo* are therefore tightly regulated. The iron is stored in ferritin and is mobilized whenever needed for some enzymatic reaction. However, it has been observed that nitric oxide can release Fe^{2+} from ferritin [315]. Superoxide radical anion, as well as hydrogen peroxide, are chemically not very reactive [268,309,310] and have to be converted to more reactive species (the hydroxyl radical). The release of Fe^{2+} from ferritin therefore is expected to have a profound effect on the oxidative damage caused by these oxygen metabolites. In addition to ·NO many other ROMs cause release of Fe^{2+} from ferritin. Ascorbic acid and superoxide radical anion, as well as many superoxide radical anion-generating compounds (xenobiotics) like paraquat, bleomycin, adriamycin, and alloxan, cause reduction of the Fe(III) to the Fe(II) state and release of free Fe^{2+} [312-314]. The released iron increases oxidative damage via a Fenton-type reaction.

K. PROTEIN-DERIVED RADICALS AND NITRIC OXIDE

In Chapter 6 I placed emphasis on damage caused by ROMs on DNA and lipids and damage caused by their degradation products. However, damage to enzymes plays an important part in the overall picture. The relationship between protein oxidation and aging has been explored [804].

For a long time it has been assumed that radicals are highly destructive and do not fulfill any physiological role. The discovery of nitric oxide of course has taught us otherwise and has opened new horizons for radicals in biological systems. It has been recognized that protein radicals in some enzymes are important for their catalytic activity. The aminoacids most likely involved in these oxidations are those which have easily oxidizable groups, like tyrosine, cysteine, glutathione, methionine, tryptophan and cytochrome c. The studies on the catalytic function of radicals within enzymes are extensive and have been reviewed by Stubbe [805,806]. Many enzymes have been reported to contain radicals at their active sites. The tyrosyl radical has been found at the active site of ribonucleotide reductase [807-809], and the tryptophan radical at the active site of DNA photolyase [810] (repairs 4-membered ring dimers) and of cytochrome c peroxidase [811]. I will briefly discuss the ribonucleotide reductase. This enzyme is involved in DNA synthesis by reducing the ribonucleotide to deoxyribonucleotide:

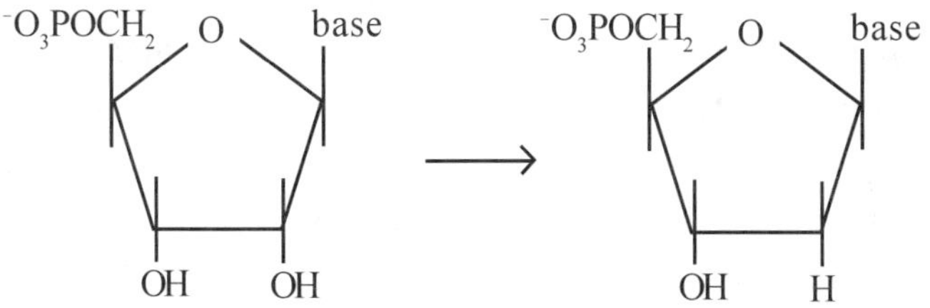

The enzyme contains two subunits. The B1 subunit contains two SH groups, which act as two-electron donors and the B2 subunit contains the tyrosyl radical, which acts as a one-electron acceptor. The following steps have been proposed [809]:

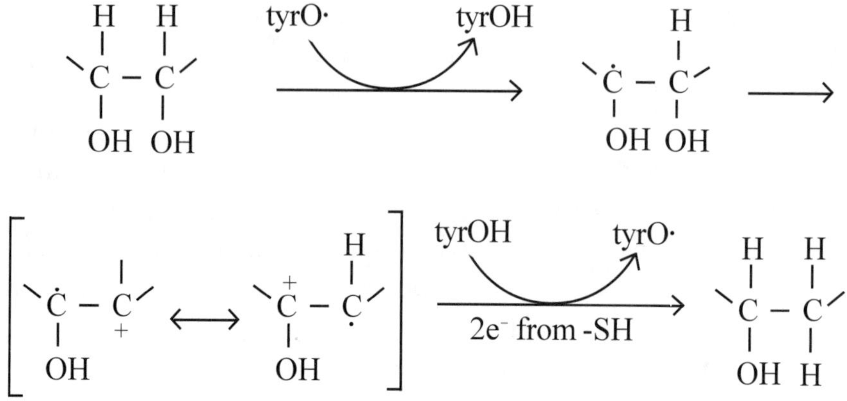

Nitric oxide provides us with an ideal tool to explore these radicals at active sites of enzymes. The ·NO as a radical combines very rapidly with other radicals, like $O_2^{·-}$ [116], ROO· [798], tyrosyl and trypto-phan radicals [812]. The tyrosyl radical is buried deep within the ribonucleotide reductase [813]. Due to its small size ·NO can penetrate the active site and combine. The rate constants for these combination reactions have been determined by pulse radiolysis [812] in dilute N_2O saturated solutions. These reactions proceed as follows:

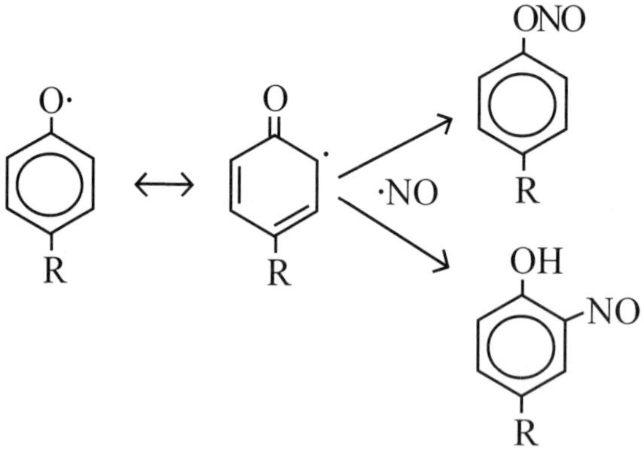

$$k = 1 \times 10^9 \, M^{-1}s^{-1}$$

$$k = 1.4 \times 10^9 \, M^{-1}s^{-1}$$

Rate constants are important in order to assess the importance of competing reactions. A radical like tyrosyl can be reduced by other reducing agents, like ascorbate, urate or thiols, but these compounds are too bulky to easily penetrate the active site of the enzyme. The ·NO has no competition. It has been shown that ribonucleotide reductase is deactivated by ·NO via combination with the tyrosyl radical [814,815].

The inhibition of enzymes involved in DNA synthesis or repair may be put to some beneficial purpose. Stimulated macrophages (produce·NO) have been found to inhibit the growth of tumor cells *in vitro* [816,817]. It has been suggested that the inhibition of ribonucleotide reductase may contribute to the ability of activated macrophages to inhibit the proliferation of bacteria, fungi, protozoa and helminths [818].

Summary. From a chemical point of view there are numerous pathways by which ·NO may exert its toxic effects. It reacts with Fe centers of Fe-heme proteins (may be toxic or beneficial), with Fe-S centers or radical centers of enzymes or via its oxidation products (NO_2, N_2O_3, N_2O_4, HONO, and $^-OONO/HOONO$) with lipids, NH and SH groups, aromatic aminoacids, antioxidants, and DNA. It may also release Fe^{2+} from ferritin and thus lead to formation of hydroxyl radicals (via the Fenton reaction or the Fe^{3+}-catalyzed Haber-Weiss reaction). Whether any of these reactions occur *in vivo* depends on the site and rate of formation and the concentrations and rate constants of many competing reactions. There are indeed many ways by which ·NO may exert its damaging effect, which eventually lead to pathological conditions.

Chapter 8

ANTIOXIDANTS

An ounce of prevention is worth a pound of cure.

A. INTRODUCTION

If oxygen metabolites are indeed involved in many pathological processes, we should be able to prevent these processes by addition of antioxidants or radical scavengers. Extensive use has been made of this technique. As I have already pointed out, our bodies are constantly bombarded by a variety of toxic substances, both from our environment as well as compounds produced by our own metabolism. During the course of evolution our bodies have developed defenses against this toxic threat [8]. It has been determined that the steady-state concentration of $O_2^{\cdot-}$ in cells containing normal levels of SOD is about 10^{-12} - 10^{-10} M, and it was suggested that this level of $O_2^{\cdot-}$ is the driving force of the present-day spontaneous mutation rate [8].

Antioxidants are essential for the survival of all aerobic organisms. Anaerobic bacteria are killed by oxygen [819]. The enzymatic defenses are only effective over a limited range of oxygen concentrations. Normal air consists of 21% oxygen. Exposure of humans to pure (100%) oxygen for only a short time causes many deleterious effects and exposure over prolonged periods leads to alveolar damage. This effect of increased oxygen concentration varies considerably depending on the organism, age, physiological condition, and diet (initial conditions!). The damaging effects of high oxygen concentrations were first attributed to the formation of oxygen radicals by Gerschman et al [391] in 1954. Hyperoxia increased H_2O_2 release from lung mitochondria and microsomes [820]. The relation of free radical formation and hyperoxia has been reviewed [821]. The discovery of $O_2^{\cdot-}$ and SODs in all aerobic cells by McCord and Fridovich [83,84] gave rise to the superoxide theory of oxygen toxicity, which

262

I have already discussed in Chapter 3. Aerobic organisms have to defend themselves not only against oxygen radicals (Chapter 3), but also against singlet oxygen and electronically excited carbonyl compounds (Chapter 4).

The term antioxidant has become a new catch word in the popular literature. It usually refers to certain vitamins in our diet (Vitamins A, C, and E). In the scientific realm, however, the term "antioxidant" encompasses a much greater variety of compounds. Antioxidant has been defined by Halliwell and Gutteridge [822] as "any substance that, when present in low concentration compared to those of an oxidizable substrate, significantly delays or inhibits oxidation of that substrate".

As pointed out in Chapter 2, a radical reacts with a molecule by producing another radical, which is usually of lower reactivity (more stable). The secondary radical can, in turn, form still other radicals via combination with oxygen to give peroxyl radicals or via fragmentation to give smaller radicals. However, there are reactions of radicals, which give non-radical products. These reactions are one-electron oxidations and reductions:

$$O_2 \xleftarrow{-e^-} O_2^{\cdot-} \xrightarrow{+e^-} O_2^{2-}$$

$$H^+ + O_2 \xleftarrow{-e^-} HO_2^{\cdot} \xrightarrow{+e^-} HO_2^-$$

This redox property of $O_2^{\cdot-}/HO_2^{\cdot}$ is used by SODs to transform the superoxide radical anion to H_2O_2 (reduction) and O_2 (oxidation). I have pointed out the important differences between the superoxide radical anion and the hydroxyl radical. The hydroxyl radical is highly reactive, reacting with most molecules either via addition or abstraction at diffusion-controlled rates ($10^{10} M^{-1} s^{-1}$). Therefore most molecules can be considered scavengers of hydroxyl radicals especially electron-rich aromatics. It is therefore not surprising to find allopurinol and oxypurinol are hydroxyl radical scavengers [823] or that thiols (RSH) offer protection against radiation-induced damage [216], and that carnosine (β-alanyl-L-histidine) protects deoxyguanosine from oxidation *in vitro* [824]. These results are obvious to any chemist without doing an experiment. If we want to protect a biomolecule from hydroxyl radical attack, we have to add a radical scavenger, which can compete with the biomolecule:

$$\cdot OH \ + \ B \longrightarrow B\cdot$$

$$\cdot OH \ + \ S \longrightarrow S\cdot$$

The scavenger can compete only at high concentrations, which is physiologically impossible (theoretically it should be possible to scavenge all oxygen radicals by high concentrations of sugar!). In other words we cannot protect our cells and tissues against hydroxyl radicals. However we can prevent hydroxyl radical formation by eliminating the precursors of hydroxyl, namely the hydrogen peroxide and metal ions (Fe^{2+}, Cu^+). The hydrogen peroxide is eliminated by catalase and glutathione peroxidase and the metal ions can be sequestered by metal-complexing proteins like ferritin, transferrin, and ceruloplasmin. Since hydroxyl radical is highly reactive it can damage DNA only if it is formed in close proximity to the target. Metal ions are complexed to DNA and react with H_2O_2 in a site-specific manner. We can classify antioxidants into several groups: 1. Enzymes produced by our cells (SOD, CAT, GSH-Px) [7,9,103,104], 2. Uric acid [825], which is produced via the metabolism of purines, 3. Small molecules taken up in our diet (β-carotene, ascorbic acid, α-tocopherol), and 4. Proteins, which form complexes with transition metal ions. The antioxidant defenses are summarized in Fig. 1.

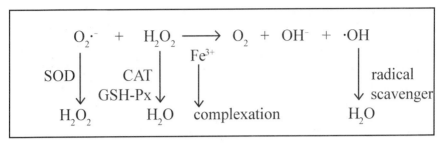

Fig. 1. Effect of SOD, CAT, complexing agent, and radical
scavengers on the Haber-Weiss reaction.

Whenever these pathways involving SOD, CAT, or complexing agent are not effective and hydroxyl radicals are formed (in a non-site-specific manner) the last line of defense are radical scavengers like Vitamins A, C, and E (β-carotene, ascorbic acid, α-tocopherol), and uric acid. The oxidation product of uric acid (allantoin) has been used as a measure for oxidative damage *in vivo* [826] (Chapter 6). *In vivo* it is difficult or impossible to protect biomolecules against the fast reactions of hydroxyl radicals. Other oxygen radicals ($O_2^{\cdot-}/HO_2\cdot$ or $LOO\cdot$) on the other hand, are far less reactive and ascorbic acid and α-tocopherol can scavenge these radicals and compete with other more damaging reactions. This is especially true for the α-tocopherol, which is incorporated into the membrane where the oxidative damage occurs.

The study of antioxidant defenses presents an excellent example of the pros and cons of interfering with a complex dynamic system. Oxygen is necessary for survival, but too much oxygen is toxic. A deleterious effect of too little oxygen was also observed. Fibroblasts in tissue culture intermittently exposed to an atmosphere of nitrogen (anaerobiosis) underwent neoplastic transformation [827]. A certain amount of ascorbic acid is necessary to avoid scorbut and as a cofactor for prolyl and lysyl hydroxylases [828]. These enzymes are important in the biosynthesis of collagen. Elevated doses may be beneficial by scavenging oxygen radicals. Ascorbic acid in moderate amounts (250 mg/day) protected DNA in human sperm cells from oxidation [829] (formation of 8-HO-G decreased considerably).

Ascorbic acid also inhibits carcinogenic N-nitrosamine formation in the stomach [681], but high doses can lead to many damaging effects, especially in presence of iron and copper ions [12-15]. This dual character of ascorbic acid can also be observed with the flavonoids. These compounds act at low concentrations as antioxidants, but as prooxidants in presence of metal ions. SOD is essential for survival in an oxygen atmosphere, but too much SOD is bad (SOD paradox).

B. SUPEROXIDE DISMUTASES

It has been clearly established that $O_2^{\cdot-}$ is formed in all aerobic cells and all aerobic cells contain superoxide dismutase (SOD) as a defense against superoxide toxicity [9]. The function of SODs was first illuminated by McCord and Fridovich [83,84]. The SODs come in 2 varieties: 1. SODs containing Cu(II) and Zn(II), and 2. SODs containing Mn(III) or Fe(III) at the catalytic center. These SODs have been extensively reviewed [9,103,104].

In biology the study of the abnormal helps us to understand the normal. *E. coli* mutants unable to produce the MnSOD (SodA) or the FeSOD (SodB) enzymes have shown a hypersensitivity toward oxygen, to agents that increase the production of $O_2^{\cdot-}$ (like paraquat), and toward H_2O_2 [108,109]. These *E. coli* mutants show enhanced oxygen-dependent mutagenesis (Chapter 3).

A disease in humans caused by a genetic defect in the SOD gene is Lou Gehrig's disease. This condition is associated with point mutations in the Cu,ZnSOD gene [285-287] and with decreased cytosolic SOD activity. The inhibition of cytosolic Cu,ZnSOD causes apoptosis of spinal neurons [830].

The dismutation of $O_2^{\cdot-}$ is an example of both the oxidizing and reducing ability of superoxide radicals. The dismutation by the CuZnSOD has been assumed to proceed via reduction and oxidation of the metal center [167,168]:

$$Cu^{2+}ZnSOD \ + \ O_2^{\cdot-} \ \longrightarrow \ Cu^+ZnSOD \ + \ O_2$$

$$Cu^+ZnSOD \ + \ O_2^{\cdot-} \ \overset{2\,H^+}{\rightleftharpoons} \ Cu^{2+}ZnSOD \ + \ H_2O_2$$

However, another hypothesis was put forward by Czapski et al.[186] These authors postulated that $O_2^{\cdot-}$ does not react with Cu(II) as a reducing agent, but as an oxidizing agent:

$$Cu(II) \ + \ O_2^{\cdot-} + 2\,H^+ \ \longrightarrow \ Cu(III) \ + \ H_2O_2$$

$$Cu(III) \ + \ O_2^{\cdot-} \ \longrightarrow \ Cu(II) \ + \ O_2$$

In this mechanism the electron shuttles between Cu(II) and Cu(III) instead of Cu(I) and Cu(II). It is interesting to note, that complexes of small peptides with Cu^{2+} show equal SOD activity as the native SOD [831a, 831b]. The rate constants were determined by pulse radiolysis [831b]. The structure of these Cu^{2+}-protein complexes has been examined by ^{13}C-NMR spectroscopy [831c].

Superoxide, unlike its dismutation product H_2O_2, does not easily cross cellular membranes (due to its negative charge). Therefore the $O_2^{\cdot-}$ must be detoxified within the cellular compartment within which it is produced. However, there are SODs present in the extracellular environment. This clearly demonstrates the need for defense against numerous extracellular sources of $O_2^{\cdot-}$. There are many extracellular sources of $O_2^{\cdot-}$. In an oxygen atmosphere the process of autoxidation is ever present. The membrane-associated NADPH oxidase (abundant in phagocytic cells) releases $O_2^{\cdot-}$ into the extracellular environment. [17,18,115].

The formation of hydroxyl via the metal-catalyzed reaction between $O_2^{\cdot-}$ and H_2O_2 (the Haber-Weiss reaction) is obviously inhibited by SOD, CAT, or metal chelating agents. Observations of such inhibitions are legions (see review by Fridovich [9]). These inhibitions have served as diagnostic tools for the involvement of $O_2^{\cdot-}$ or H_2O_2 or both in many pathological processes.

1. Superoxide/SOD balance and the SOD paradox

Our cells are carrying out a precarious balancing act between superoxide and superoxide dismutase. The SOD protects us from the damaging effects of superoxide, but it produces hydrogen peroxide, which in absence of catalase or GSH-Px or of low concentrations of these enzymes may accumulate and cause serious damage via formation of hydroxyl radicals.

$$2\ O_2{}^{\cdot-} + 2\ H^+ \xrightarrow{\ \ \text{SOD}\ \ } H_2O_2 + O_2$$

$$H_2O_2 \longrightarrow \cdot OH$$

This reaction sequence represents a highly simplified picture. Hydrogen peroxide is not only formed (via SOD), but it is also consumed by CAT. The H_2O_2 reaches a steady state concentration (where the rate of formation is equal to the rate of disappearance).

Superoxide and hydrogen peroxide are formed via numerous metabolic reactions [18,85,86,115,127] (Chapter 5), but they are rapidly consumed via many reactions. In order to understand the complex relationships between superoxide, hydrogen peroxide, SOD, CAT, GSH-Px, and other antioxidants we need quantitative information. There are many reactions taking place simultaneously and we have to know the concentrations of the species involved and the rate constants for their formation and disappearance. This knowledge allows us to calculate the steady state concentrations of these reactive oxygen metabolites. The steady state concentration of $O_2{}^{\cdot-}$ in *E. coli* in absence of SOD has been determined by Imlay and Fridovich [392] as 6.7×10^{-6} M, wheras in presence of SOD it is 2×10^{-10}M, a factor of 30,000 smaller! It is of course important to determine the minimum steady state level of $O_2{}^{\cdot}$, which is needed to damage various cell functions. As pointed out by Imlay and Fridovich [392], in order to establish a causal relationship between $O_2{}^{\cdot-}$, H_2O_2, $\cdot OH$, and $^1O_2$ and various cellular dysfunctions, we need quantitative information concerning the steady state of these reactive oxygen metabolites, the concentrations of antioxidant defenses, and their rate constants. Boveris and Cadenas [832] have recently reviewed these calculations and have compiled the rate constants and steady state concentrations of reactive oxygen metabolites. In the cytosol of most normal cells the ratio $[H_2O_2]_{ss}/[O_2 \cdot^-]_{ss}$ is approximately 10^3 [832].

Considering only the dismutation of superoxide radical anion we may conclude that SOD increases the H_2O_2 concentration. The dismutation of $O_2 \cdot^-$, either catalytic (with SOD) or non-catalytic, leads to the formation of ½ mole of H_2O_2 (pp. 50-51). However as I have already discussed (pp. 47- 48) the $O_2 \cdot^-$ in absence of SOD reacts *in vivo (E. coli)* with the dehydratases to yield 1 mole of H_2O_2 with the concomitant release of free Fe(II). It also reacts with NAD(P)H attached to lactate dehydrogenase via a chain reaction to generate high yields of H_2O_2 (p. 50). This means that SOD does not increase the yield of H_2O_2, but rather decreases it, by preventing $O_2 \cdot^-$ from undergoing these oxidations, as was pointed out by Liochev and Fridovich [157]. There is yet another way by which SOD can decrease the H_2O_2 burden borne by aerobic cells. We know that catalase [833] and peroxidases [834] are inactivated by $O_2 \cdot^-$. By removing $O_2 \cdot^-$, SOD protects CAT and therefore decreases the steady state of H_2O_2. It is also known that SODs (the CuZn and the FeSOD, but not the MnSOD) are deactivated by H_2O_2 [283,284]. H_2O_2 does not deactivate MnSOD, because M(II) does not participate in Fenton chemistry (Mn(II) + H_2O_2 $\rightarrow\!\!\!\times\!\!\!\rightarrow$ Mn(III) + OH^- + $\cdot OH$). The two enzymes SOD and CAT act synergistically. These interrelationships are summarized in Fig. 2.

In any reaction milieu, which produces both $O_2 \cdot^-$ and H_2O_2, the effectiveness of CAT is enhanced by SOD. The CuZnSOD and FeSOD can be deactivated by H_2O_2 and thus we have a reverse synergism in which CAT prevents inactivation of SOD. CAT and SOD represent a mutually protective set of antioxidant enzymes.

SOD mutants show the important role of SODs in protecting the cells against $O_2 \cdot^-$ (Chapter 3). We may be tempted to conclude: the

$$\begin{array}{cccc} & SOD & & CAT \\ O_2 \cdot^- & \longrightarrow & H_2O_2 & \longrightarrow & H_2O \end{array}$$

$$H_2O_2 + SOD \text{ (Fe, CuZn)} \longrightarrow \text{deactivation}$$

$$O_2 \cdot^- + CAT \longrightarrow \text{deactivation}$$

The H_2O_2 controls its own **rate of formation** and $O_2 \cdot^-$ controls the **rate of H_2O_2 disappearance**

Fig. 2. The interrelationship between $O_2 \cdot^-$, H_2O_2, SOD and CAT.

more SOD the better! However the study of SOD-rich *E. coli* bacteria led to a paradoxical increase in $O_2 \cdot^-$ toxicity. *E. coli* with multiple copies of the gene for bacterial FeSOD were studied using paraquat as the super-oxide-generating compound ($PQ^{+ \cdot} + O_2 \longrightarrow PQ^{2+} + O_2 \cdot^-$). These studies [16,835,836] led to the following observations: 1. SOD-rich bacteria were more readily killed by paraquat than the controls. 2. SOD-rich bacteria accumulated more H_2O_2. 3. High SOD bacteria show a greater decrease in GSH than the controls. 4. High SOD bacteria are more readily killed by hyperoxia. 5. SOD-rich bacteria and the controls showed no difference to killing by exogenous H_2O_2.

From these observations it was concluded that the increase in H_2O_2 accounts for the increased sensitivity of SOD-rich bacteria. The results show that the product of the reaction (H_2O_2) is as toxic as the substrate ($O_2 \cdot^-$). This result is not surprising, since we know that H_2O_2 is the precursor of the hydroxyl radical. The explanation therefore appears quite reasonable, but reasonable does not mean that it is correct.

The investigations of Liochev and Fridovich [836] seem to suggest a different explanation. They observed that the SOD-rich *E. coli* were not killed but only growth-inhibited. Superoxide induces the enzymes of *E. coli*, which constitute the *soxR* regulon. Overproduction of SOD lowers the steady state concentration of O_2·- and thus suppresses the induction of these enzymes. The induction of glucose-6-phosphate dehydrogenase, which is a member of the *soxR* regulon, was much greater in the controls than in the SOD overproducer. Glucose-6-phosphate dehydrogenase supplies the NADPH needed for the action of alkylhydroperoxide reductase and glutathione reductase. Glutathione reductase catalyzes the following reaction:

$$GSSG \longrightarrow 2\,GSH$$

SOD lowers the concentration of O_2·- and consequently decreases GSH, one of the important antioxidant defenses.

The controversies, contradictions and paradoxes of O_2·- have been reviewed by McCord [837] who asked the provocative question: "is superoxide radical good or bad?" Here we encounter the repetition of the nitric oxide story, which I discussed in chapter 7. Superoxide radical and nitric oxide are both good and bad, depending on the site, rate of formation, and rate of disappearance. Superoxide radical initiates lipid peroxidation and thus affects the structural and functional integrity of the membrane, followed by many deleterious consequences. However, as was shown by McCord and coworkers [838] in the attenuation of ischemia-reperfusion injury by SOD, O_2·- also can terminate the lipid peroxidation chain:

$$LO· + O_2·^- + H^+ \longrightarrow LOH + O_2$$

$$LOO· + O_2·^- + H^+ \longrightarrow LOOH + O_2$$

Superoxide, in addition to liberating Fe^{2+} from ferritin, and initiating lipid peroxidation, can also terminate the chain reaction. Therefore, too little or too much SOD will be bad. At some intermediate SOD concentration, initiation by superoxide would be largely suppressed, but termination would still be making a contribution and net lipid peroxidation would be at a minimum [838]. Plotting net lipid peroxidation and recovery of function versus SOD concentration yields the now famous bell-shaped curves as shown in Fig. 3. This reminds me of the words of Hermann Hesse: "straight lines evidently belong to geometry, not to nature and life" [839].

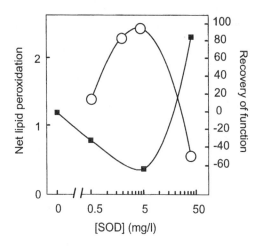

Fig. 3. Relationship of net lipid peroxidation and functional recovery of isolated rabbit hearts to administered dosages of superoxide dismutase (reprinted from [837] with permission).

The absence of SOD is responsible for some serious diseases (ALS), but too much SOD is also bad by lowering the antioxidant defenses. It is clearly a question of balance. Overproduction of SOD has been observed in Down's syndrome and in Parkinson's disease (Chapter 9).

2. SOD mimics

The superoxide radical is the first metabolite of oxygen. It can cause serious damage either by itself or via its secondary products. Superoxide radical plays an important role in autoxidation (lipids), inflammation, reperfusion injury, aging, and cancer, just to name a few. It should therefore be possible to counteract these processes by low molecular weight drugs, which mimic SOD activity. Many reports on SOD mimics have been made [9,840]. The most important ones are Mn(III) and Fe(III) complexes of substituted porphins. The Mn(III) complexes are of special interest since the possible release of Mn(II) does not lead to Fenton-type chemistry. Another group of SOD mimics are complexes of Mn(III) with cyclic polyamines. However, not surprisingly, (it took nature millions of years to develop the most efficient SODs) so far all SOD mimics are far less efficient than SODs.

C. REACTIONS OF OXYGEN AND OXYGEN RADICALS WITH ASCORBIC ACID

Ascorbic acid (AH_2) is a very important water soluble radical scavenger or antioxidant. Ascorbic acid has the following structure:

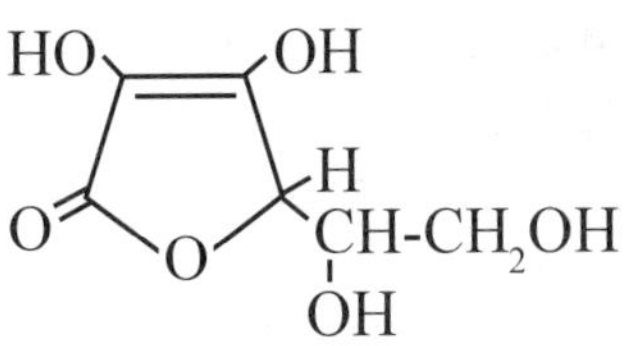

It is evident from the structure that ascorbic acid has two easily abstractable H-atoms (the 2 OH groups on the 5-membered ring). The most important property of ascorbic acid is its ability to be oxidized to dehydroascorbic acid (A). This transformation represents a two-electron oxidation, which proceeds in two one-electron steps :

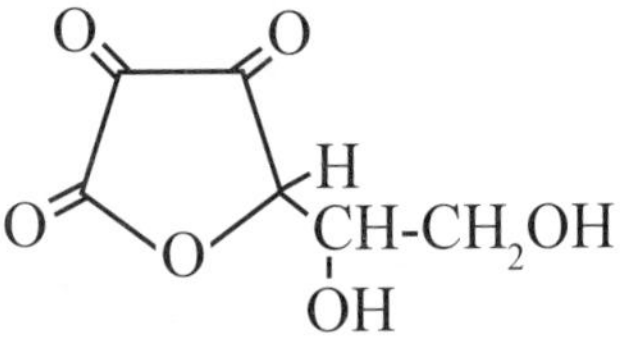

$$AH_2 \longrightarrow AH \cdot \longrightarrow A$$

The transformation of ascorbic acid to dehydroascorbic acid of course destroys its effectiveness as a vitamin and antioxidant. The mechanism of ascorbic acid oxidation by oxygen was therefore studied extensively. It is beyond the scope of this text to review all of these studies [841].

1. Non-catalytic oxidation of ascorbic acid by oxygen

As I have discussed in Chapter 3, oxygen can oxidize some organic compounds via one-electron transfer to yield the superoxide anion, via two-electron transfer to yield hydrogen peroxide, and via four-electron transfer to yield two moles of water. It has been suggested that oxygen oxidizes ascorbic acid via a two-electron transfer mechanism [842]:

$$AH_2 + O_2 \longrightarrow A + H_2O_2$$

However ascorbic acid (as the name implies) is an acid. In aqueous solutions it is dissociated to the ascorbate anion and a proton:

$$AH_2 \rightleftarrows AH^- + H^+ \qquad pk_a = 4.25$$

As the pK indicates, at physiological pH of 7.4, ascorbic acid exists mainly in the form of the ascorbate anion. It is therefore possible that the oxidation proceeds via two one-electron transfer steps [843]:

$$AH^- + O_2 \longrightarrow AH\cdot + O_2^{\cdot-}$$

$$AH\cdot + O_2 \longrightarrow A + O_2^{\cdot-} + H^+$$

However it was later shown that non-catalytic oxidation of AH$^-$ is practically absent in aqueous solutions [175,176].

2. Metal-catalyzed oxidation of ascorbate by oxygen

Metal ions can facilitate the transfer of an electron from ascorbate to oxygen [841]. These reactions involve the following steps:

$$AH^- + Fe^{2+} \rightleftarrows [AH^-Fe^{2+}]$$

$$[AH^-Fe^{2+}] + O_2 \rightleftarrows [AH^-Fe^{2+}O_2]$$

$$[AH^-Fe^{2+}O_2] \rightleftharpoons [AH^-Fe^{3+}O_2\cdot^-] \rightleftharpoons A\cdot^- + Fe^{2+} + HO_2\cdot$$

$$A\cdot^- + O_2 \longrightarrow A + O_2\cdot^-$$

The oxidation consists of two one-electron transfer steps. The first step is a transfer from Fe^{2+} to O_2 and the second step is a transfer from AH^- to Fe^{3+}. It has been shown by Buettner [271,272,844] that various chelates inhibit iron and copper-catalyzed autoxidation of ascorbate at pH 7.4. This method has been used by Buettner [845] for the analysis of trace amounts of metal ions in buffers.

In addition to oxidation with oxygen, ascorbic acid can be oxidized by Fe^{3+}, Cu^{2+}, $Fe(CN)_6^{3-}$, $S_2O_8^{2-}$, and quinones [846]. These oxidations involve the ascorbate anion:

$$AH^- + Fe(CN)_6^{3-} \longrightarrow A\cdot^- + Fe(CN)_6^{4-} + H^+$$

$$A\cdot^- + Fe(CN)_6^{3-} \longrightarrow A + Fe(CN)_6^{4-}$$

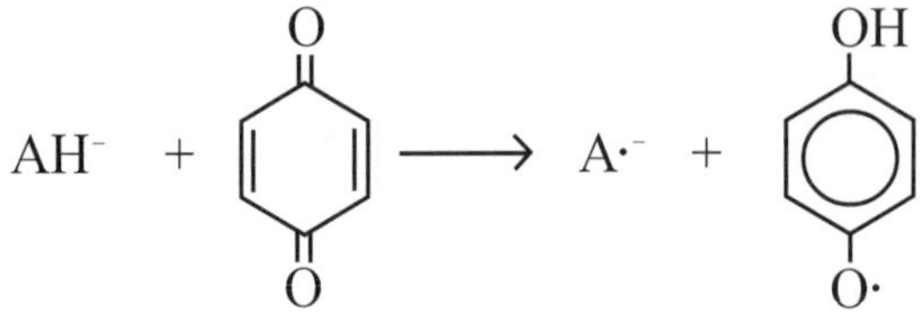

semiquinone radical

In the oxidation with Fe^{3+}, the one-electron transfer has been shown to involve an intermediate $[FeAH]^{2+}$ complex [847,848]:

$$Fe^{3+} + AH_2 \longrightarrow [FeAH]^{2+} + H^+$$

$$[FeAH]^{2+} \longrightarrow Fe^{2+} + AH\cdot$$

$$AH\cdot + Fe^{3+} \longrightarrow A + Fe^{2+} + H^+$$

$$\overline{2\,Fe^{3+} + AH_2 \longrightarrow 2\,Fe^{2+} + A + 2\,H^+}$$

3. Oxidation of ascorbic acid by superoxide/perhydroxyl radicals

It has been suggested that the reaction of ascorbic acid with $O_2^{-}/HO_2\cdot$ represents the real autoxidation reaction, which is initiated by $O_2^{-}/HO_2\cdot$. It was indeed observed that xanthine/xanthine oxidase-generated superoxide accelerated the autoxidation of ascorbic acid and that this autoxidation was completely suppressed by SOD [849]. On the basis of these results, Nishikimi proposed the following reaction:

$$O_2^{-} + AH_2 + H^+ \longrightarrow H_2O_2 + AH\cdot$$

However, as I have already discussed, the reactions involving O_2^{-} with compounds that have easily abstractable hydrogens start with a deprotonation step. We have to consider a number of different reaction steps involving the following species: AH_2, AH^-, O_2^{-}, $HO_2\cdot$. In aqueous solutions AH_2 ($pK_a = 4.25$) at physiological pH the ascorbic acid is completely dissociated to AH^-. These reactions have been studied by pulse radiolysis and stopped-flow EPR spectroscopy [143]. We can summarize the oxidation of ascorbic acid in aqueous solutions by $O_2^{-}/HO_2\cdot$ as follows:

$$AH_2 \rightleftharpoons H^+ + AH^-$$

$$O_2^{-} + H^+ \rightleftharpoons HO_2\cdot$$

$$AH^- + O_2 \xrightarrow{M^{n+}} complex \longrightarrow A^{-} + O_2^{-} + H^+ \quad \text{metal-catalyzed autoxidation}$$

$$O_2^{-} + AH^- \longrightarrow transient$$

$$O_2^{-} + A^{-} + H^+ \longrightarrow HO_2^- + A$$

$$O_2^{-} + HO_2\cdot \longrightarrow HO_2^- + O_2$$

$$HO_2\cdot + AH^- \longrightarrow H_2O_2 + A^{-}$$

$$HO_2\cdot + A^{-} \longrightarrow HO_2^- + A$$

In these oxidation reactions the ascorbate scavenges the O_2^{-}/ HO_2·, but produces H_2O_2, which in presence of iron or copper ions forms the hydroxyl radical. The ·OH is much more destructive than O_2^{-}/HO_2·. In presence of metal ions, ascorbic acid acts as a pro-oxidant.

4. Reaction of ascorbic acid with hydroxyl radicals

In addition to the oxidation of ascorbic acid by oxygen, superoxide, and perhydroxyl radicals, ascorbic acid is oxidized by hydroxyl radicals. The rate constant for this reaction is faster by several orders of magnitude than the aforementioned oxidations. The reaction is diffusion controlled ($k=1.2x10^{10}M^{-1}s^{-1}$) [99]:

$$AH_2 + ·OH \longrightarrow AH· + H_2O$$

$$AH^- + ·OH \longrightarrow A·^- + H_2O$$

At neutral pH the ascorbic acid ($pK_a = 4.4$) exists mainly in the anionic form (AH^-). The reaction of AH^- with ·OH has been studied by the pulse-radiolysis-in situ EPR technique by Schuler and coworkers [850]. The following structure for $A·^-$ was suggested:

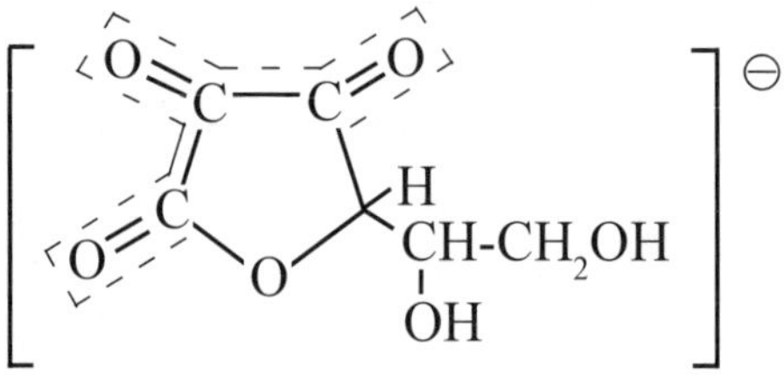

The unpaired electron is distributed over the 3 carbonyl groups. The following bicyclic structure for $A·^-$ has also been discussed [851]:

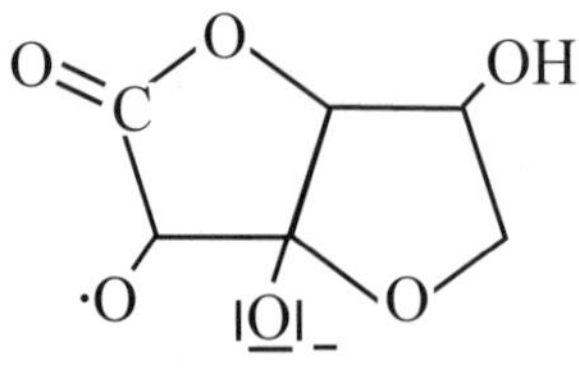

The ascorbyl radicals react in turn with oxygen:

$$AH\cdot \ + \ O_2 \longrightarrow A \ + \ HO_2\cdot$$

$$A\cdot^- \ + \ O_2 \longrightarrow A \ + \ O_2\cdot^-$$

We have transformed the highly reactive hydroxyl radical into the much less reactive superoxide/perhydroxyl radical. This is one of the important functions of ascorbic acid. It is a radical scavenger and a true antioxidant, i.e., it transfoms a highly oxidative radical to a less oxidative radical. However, further reactions of the $HO_2\cdot$ with AH_2 gives H_2O_2, the precursor of the hydroxyl radical.

5. The pros and cons of ascorbic acid

Vitamin C as a diet supplement gained prominence mainly through the work of Pauling and coworkers. Vitamin C has been recommended against the common cold [10] as well as for the prevention of cancer [11]. The recommendation by Pauling of mega-doses of Vitamin C has fallen in disrepute. We have seen that ascorbic acid reacts with oxygen radicals. However we should remember: a radical begets another radical. Further oxidations of the ascorbyl radical ($AH\cdot$) gives H_2O_2, which is the precursor of the hydroxyl radical. Ascorbic acid is an effective reducing agent for Cu^{2+} and Fe^{3+}, which are essential catalysts for Fenton-type reactions. It has been shown that ascorbic acid is mutagenic [12] and in presence of Cu^{2+} or Fe^{3+} can produce $\cdot OH$ *in vitro* [13-15] and *in vivo* [282]. The effect of ascorbic acid can be quite damaging in patients with iron-overload disease [852] or in trauma. Mechanical crushing of tissue releases the cell content into the extracellular space and can release free metal ions [853]. Trauma to the brain or via a stroke can release Fe ions and cause damage by radical reactions [854]. In all of these cases one should avoid taking Vitamin C. Vitamin C combined with mineral supplements should be avoided altogether [14,15]. Death from high doses of Vitamin C (by injection) in one patient with glucose-6-phosphate dehydrogenase deficiency has been reported [855].

The reducing ability of ascorbic acid may also have beneficial physiological effects. Ascorbic acid can regulate the blood supply in the brain. The brain requires a precise means of regulating oxygen delivery to the neurons. Too much or too little oxygen can cause serious damage. Nitric oxide is a vasodilator produced by many different cells including neurons. So the blood supply could possibly be regulated via nitric oxide. However in hypoxia the synthesis of ·NO from L-arginine is unlikely, since this synthesis requires oxygen. It has been suggested that ascorbate release from neurons generates ·NO via reduction of nitrite (NO_2^-) in the extracellular space [682]. This way neurons may regulate their own oxygen supply. The consequences of low ascorbate in the brain would be progressive damage from inaccurate oxygen delivery.

The same reaction that protects the brain from inadequate oxygen supply also inhibits N-nitrosamine formation in the stomach [681]. We have the following competing reactions:

$$R_2\text{-NH} + HONO \longrightarrow R_2\text{-N-NO}$$

$$AH_2 + HONO \longrightarrow AH\cdot + H_2O + \cdot NO$$

Nitric oxide does not react with secondary amine to give N-nitrosamines. We have a competition between secondary amines and ascorbic acid for nitric acid.

The prooxidant character of ascorbate however, has also been used to some advantage. It was observed that ascorbate mediated iron release from cellular ferritin in neuroblastoma cells and caused DNA damage [856]. Ascorbate is autoxidized to give $O_2^{\cdot-}$, which is know to cause iron release from ferritin [158,314]. It has been suggested that ascorbate could be a powerful enhancer of some cytostatic drugs in neuroblastoma therapy.

All these prooxidant effects should not distract us from the importance of ascorbate as a cofactor for some enzymes and as an antioxidant. Ascorbic acid is a cofactor for some hydroxylase enzymes, the lysyl and prolyl hydroxylases [828]. These hydroxylations play a role in the biosynthesis of collagen. Ascorbic acid accelerates hydroxylation by donating an electron to the metal-containing enzyme for which the metal ion is essential for optimal activity [828].

So the biochemically important role of ascorbate is to act as an electron donor. This same property is also responsible for the prooxidant activity of ascorbate by producing $O_2^{\cdot-}$ followed by dismutation and hydroxyl radical formation. *In vitro* ascorbate acts in the same way. In the Udenfriend reaction [130,131] ascorbate reduces Fe^{3+} to Fe^{2+} and oxygen to $O_2^{\cdot-}$ followed by $\cdot OH$ formation [51]. The $\cdot OH$ of course initiates aromatic hydroxylation. The property that makes ascorbic acid useful in biochemical synthesis also makes it a possibly dangerous drug via formation of $O_2^{\cdot-}, H_2O_2$ and $\cdot OH$.

Ames and coworkers [857] have studied the effectiveness of ascorbate in human plasma. Their results show that ascorbate protects plasma lipids against peroxidation by peroxyl radicals and that ascorbate was the only antioxidant that could offer this protection. At concentrations up to 5 mM no prooxidant effect was observed. This seems to indicate that the catalytic metal ions (Fe, Cu) are tightly bound to proteins and are not available to catalyze radical reactions. Normal concentrations of ascorbate in human plasma are ~27-51 μM. On the other hand, normal levels in tissue are much higher. The aqueous humor of the eye contains up to 1.5 mM. Brain, heart, liver, spleen, kidney, and pancreas contain ascorbate up to 0.8 mM [858]. As with other antioxidant defenses, such as SOD, CAT, GSH-Px we find higher concentrations of ascorbate in organs that are more exposed to oxidative stress due to higher metabolic rate and oxygen consumption. As I have pointed out above, an adequate oxygen supply is very important for the brain. The human brain makes up only about 2% of the total body weight, but uses about 18% of the body's total oxygen consumption [859]. The pros and cons of Vitamin C supplementation have been excellently reviewed by Halliwell [14], who states, and I quote: "Hence there is no clear evidence for any great benefit to be obtained by mega-doses of vitamin C, and we cannot yet prove that it is not harmful over a lifetime".

D. URIC ACID

The uric acid is formed together with the superoxide radical anion and hydrogen peroxide in the oxidation of xanthine by xanthine oxidase (Chapter 5). It is therefore not surprising that uric acid does not scavenge $O_2 \cdot^-$, otherwise we might not be able to observe it in the xanthine-xanthine oxidase reaction. Uric acid however is an effective scavenger of hydroxyl radicals, peroxyl radicals, and singlet oxygen [825,826]. The uric acid can also be oxidized by one-electron oxidizing agents like $KMnO_4$ to the urate radical anion [860], which was detected by EPR spectroscopy. The urate radical anion was found to react rapidly with ascorbic acid to reform urate.

Uric acid has also been shown to be an effective scavenger of peroxynitrite [861,862]:

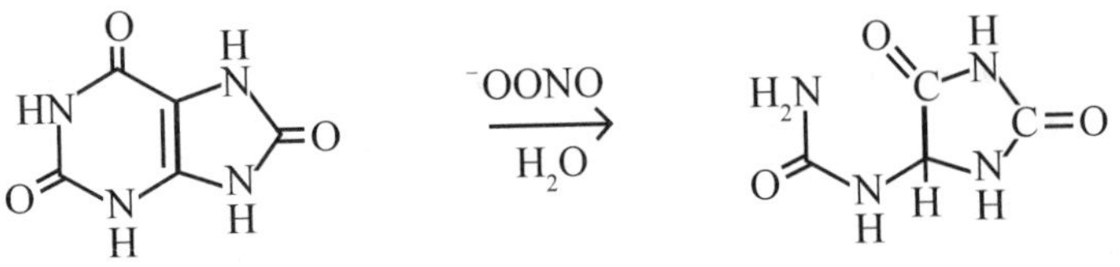

uric acid   allantoin

Uric acid does not react with ·NO [862] and therefore does not inhibit the many important physiological functions of ·NO. Uric acid only scavenges the highly damaging nitric oxide deactivation product, the peroxynitrite. Uric acid is therefore an evolutionary success story: it does not affect the physiological functions of ·NO, but only protects cells from its toxic effects.

In all of these scavenger reactions the uric acid is consumed. The main product of urate oxidation is allantoin [825,826] (Chapter 6), but many other minor products have been identified [863]. However, there is another way by which uric acid can act as an antioxidant, without being oxidized itself. This process is complexation of metal ions, such as Fe^{2+} or Fe^{3+}. Uric acid can inhibit the iron-catalyzed lipid peroxidation by complexation of the iron ions [864]. This complexing ability of uric acid is most likely responsible for the inhibition of ascorbate autoxidation in human serum [865].

Uric acid in presence of hydrogen peroxide is oxidized by a number of heme-containing proteins like hemoglobin, methemoglobin, catalase, myeloperoxidase, horseradish peroxidase etc.[866]. Uric acid is indeed the most important antioxidant in human plasma. The urate level is considerably higher (160~450 μM) than the ascorbate level (~50 μM) [825]. The increase in urate levels occurred during the course of 60 million years of evolution and coincided with a large increase in life span and brain size. At the same time, an enormous decrease in the age-specific cancer rate has occurred in humans compared to short-lived mammals [867,868].

The lower life expectancy of rats has been suggested to be due to the higher metabolic rate of rats and consequently, formation of higher concentrations of oxygen radicals. Rats excreted much more 8-OH-G (a measure for *in vivo* oxidative damage) in the urine than humans. However, as suggested by Ames et al [825] we may also surmise that the longer lifespan is due to better antioxidant defenses. The uric acid concentration in humans is considerably higher (5 mg/100 ml of blood) than in rats (≤0.5 mg/100 ml of blood). The longer life expectancy of humans is most likely due to a combination of these factors (less oxidative stress and better antioxidant defenses).

If we exercise we use more oxygen and produce more reactive oxygen metabolites, as shown by increased lipid peroxidation. This effect can be determined *in vivo* by measuring hydrocarbons in the exhaled breath [676,677]. At the same time, urate levels in the blood increase, possibly as a physiological response to this increased oxidative stress [825]. Exercise also increases the amount of carnosine, homocarnosine and anserine in the muscles of rowers and sprinters [869]. These defensive responses are unlikely to be perfect, and we

282

may therefore ask: Is exercise good or bad for you? This question is difficult to answer, since we are dealing with a complex dynamic system where the initial conditions are different for each individual.

Evolution has replaced ascorbic acid with uric acid [870], which is a more effective antioxidant with fewer of the damaging side effects of Vitamin C, which we have already discussed. It took evolution millions of years to adopt the best strategy against oxidative stress. In light of these considerations, the attempt to improve on this strategy, by recommending high doses of Vitamin C appears futile at best. Another evolutionary adaptation may be the formation of carnosine, anserine and homocarnosine in human skeletal muscle and brain [824]. The antioxidant effects of these dipeptides will be discussed later in this chapter.

E. MELATONIN

Melatonin (N-acetyl-5-methoxytryptamine) is a hormone, which for many years has been thought to be produced exclusively by the pineal gland. However, it has been shown that melatonin is produced in vertebrates in a number of cells and organs. Melatonin has the following structure:

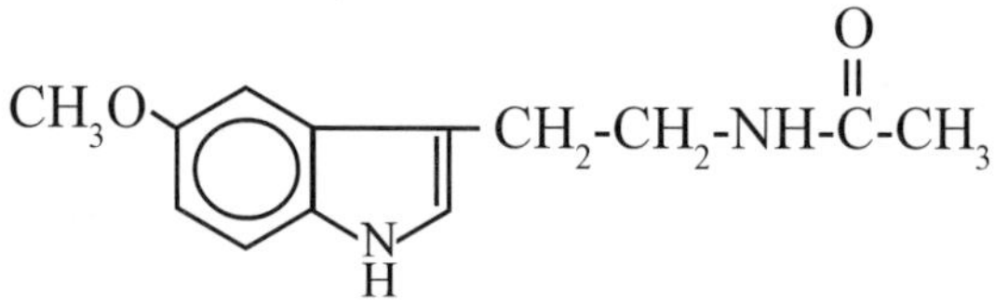

Melatonin has been shown to be an effective hydroxyl radical scavenger *in vitro* [871]. This is not surprising considering the electron-rich aromatic ring and the highly electrophilic hydroxyl radical (chapter 3). However, if melatonin indeed acts as an antioxidant *in vivo*, it must be able to penetrate cells and subcellular compartments where the ROMs are generated. Since melatonin is highly lipophilic, it meets this requirement admirably. It has been suggested by Reiter et al. [872] that melatonin was conserved during evolution because of its ability as a radical scavenger. The different aspects of cellular antioxidant defense mechanisms have been reviewed by Reiter et al. [873].

As I have discussed on previous occasions, it is diffficult to protect against OH-induced damage *in vivo*, since the hydroxyl radical is so highly reactive and is produced in a site-specific manner. A more successful strategy against OH induced damage is therefore the scavenging of the hydroxyl radical precursors, the $O_2\cdot^-$ (by SOD), H_2O_2 (by CAT or GSH-Px), Fe(II)/Fe(III) (by complexation with deferoxamine) or scavenging of HOCl. Melatonin has been shown to scavenge HOCl, *in vitro* [874,875] and may in this way protect cells against oxidative damage. Melatonin does not scavenge $O_2\cdot^-$ [874], but scavenges $^1O_2$ [876] and ·OH [871]. Melatonin in scavenging HOCl, also protects CAT from inactivation by HOCl [874]. Melatonin may offer protection in some neurodegenerative disorders, such as Parkinson's disease (PD) and Alzheimer's disease (AD). I shall discuss these diseases in Chapter 9. Catecholamine autoxidation is inhibited by melatonin *in vitro*. It has been suggested that this inhibition may be an important component of the brain's antioxidant defenses against catecholamine autoxidation and may protect against consequent dopaminergic neurodegeneration [877] as occurs during normal aging and in Parkinson's disease. A potential relationship between melatonin and one aspect of Alzheimer's disease was suggested by Pappolla et al. [878]. The amyloid beta protein, characteristic of AD, has neurotoxic properties by radical formation (see Chapter 9). Poppolla et al. demonstrated that melatonin was remarkably effective in preventing death of cultured neuroblastoma cells as well as oxidative damage and intracellular Ca^{2+} increases induced by a cytotoxic fragment of amyloid beta protein. Since secretion levels of melatonin are decreased in aging and even more reduced in AD, melatonin may be of therapeutic relevance in AD.

Protective effects of melatonin on oxidative damage *in vivo* has been demonstrated in several animal models. Paraquat-induced injury to lung and liver of rats was determined by measuring lipid peroxidation products (malonaldehyde and 4-hydroxyalkenals). This damage was significantly reduced by cotreatment with melatonin [879,880]. Safrole, an extract of sassafras oil, is widely used as a cancer initiator. Rats injected with safrole showed extensive damage to their liver DNA by adduct formation [881,882]. Cotreatment with melatonin significantly reduced the number of DNA adducts. In addition to the antioxidant effects, melatonin acts as a biological clock. Blood levels of melatonin are low during the day and high at night [883]. At high nocturnal levels of melatonin, safrole induced less DNA damage than at low melatonin levels during the day [872]. If melatonin was preserved during evolution as an antioxidant defense, then we may ask the following question: why are melatonin levels higher at night when oxygen consumption is low?

F. VITAMIN E

Next to ascorbic acid, Vitamin E represents the most important antioxidant [884,885]. Vitamins C and E complement each other. Ascorbic acid is water soluble and Vitamin E is lipid soluble and is incorporated into the lipid bilayer, where it is needed most. Unsaturated fatty acids undergo autoxidation (rancification) with formation of alkylhydroperoxides:

$$L\text{-}H + X\cdot \longrightarrow L\cdot + XH \qquad \text{initiation}$$

$$L\cdot + O_2 \longrightarrow LOO\cdot$$
$$\qquad\qquad\qquad\qquad\qquad\qquad \text{propagation}$$
$$LOO\cdot + L\text{-}H \longrightarrow LOOH + L\cdot$$

$$LOO\cdot + LOO\cdot \longrightarrow LOOOOL \qquad \text{termination}$$

The initiation step is due to radical attack ($O_2^{\cdot-}$/$HO_2\cdot$, $LOO\cdot$, $\cdot OH$) on the unsaturated lipid. The peroxyl radicals derived from lipids can react with tocopherols and flavonoids via H-abstraction and thus terminate the chain:

$$LOO\cdot \ + \ TOH \ \xrightarrow{\ k_t\ } \ LOOH \ + \ TO\cdot$$

Vitamin E consists of a number of tocopherols, of which the α-tocopherol is the most active radical scavenger. The rate constant for these reactions with tocopherols *in vitro* has been determined by Burton and Ingold [885,886]. In homogeneous lipid-like solvents k_t for α-TOH is greater than for any other lipid soluble antioxidant, i.e., α-TOH is the most powerful natural inhibitor of lipid peroxidation ($k_t = 23.5 \times 10^5 \ M^{-1}s^{-1}$) [885]:

$$LOO\cdot \ + \ \alpha\text{-}TOH \ \xrightarrow{\ k_t\ } \ LOOH \ + \ \alpha\text{-}TO\cdot$$

However, the rate constants of peroxyl radical trapping by α-TOH dispersed in SDS micelles and phospholipid liposomes are lower than the value in homogeneous systems [887-889]. Therefore, the true rate constant for the LOO$\cdot$ - α-TOH reaction in aqueous suspensions of lipid and in low-density lipoprotein (LDL) particles is unknown [887]. It has been determined by Ingold and coworkers [887] that the trapping of peroxyl radicals by α-TOH depended on the polarity of the solvent. The k_{inh} in polar solvents was considerably lower than in non-polar solvents. This effect is due to hydrogen bonding of α-TOH to the solvent [890].

The α-tocopherol is a compound with an aromatic hydroxyl group, which is an easily abstractable H-atom:

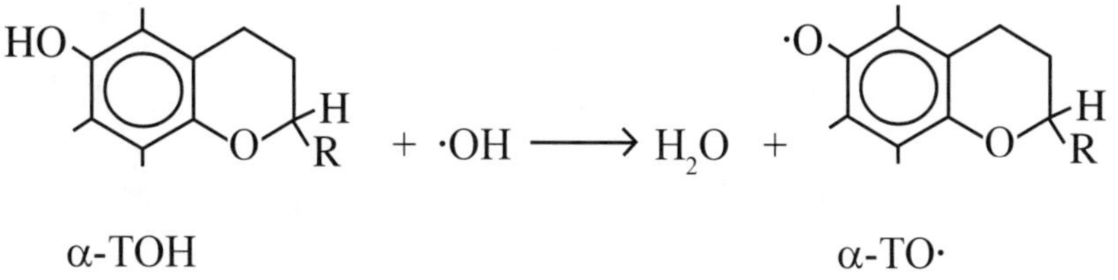

$$\alpha\text{-}TOH \qquad\qquad\qquad\qquad \alpha\text{-}TO\cdot$$

286

The α-tocopheroxyl radical (α-TO·) is relatively stable. This is another example in which a highly reactive radical is transformed to a less reactive one.

Since α-tocopherol is not water soluble many studies have been carried out using a water soluble model compound. In this compound (trolox) the side chain is replaced with a COOH group, which makes the molecule more water soluble. The xanthine/xanthine oxidase system that generates $O_2^{·-}$ oxidizes trolox [891]. This oxidation was completely inhibited by SOD, thus clearly implicating $O_2^{·-}$ in the oxidation. In aprotic solvents it was shown by spectroscopy that tocopherol reacts with $O_2^{·-}$ to immediately yield the tocopherol anion (TO^-), which then reacts with oxygen to form the tocopherol radical $(TO·)$. The reaction of TOH with $O_2^{·-}$ is therefore a multistep process (deprotonation followed by abstraction) [892]:

$$TOH + O_2^{·-} \longrightarrow TO^- + HO_2· \qquad k = 0.59 \pm 0.23 \ M^{-1}s^{-1}$$

$$TOH + HO_2· \longrightarrow TO· + H_2O_2 \qquad k = 2 \times 10^5 \ M^{-1}s^{-1}$$

$$TO^- + O_2 \longrightarrow TO· + O_2^{·-}$$

These reactions are analogous to the reactions with ascorbic acid. The reaction of superoxide with tocopherol is a very slow process, and it is therefore doubtful that tocopherol can function as a superoxide scavenger in biological systems. In addition, tocopherol anion and ascorbate anion react with oxygen to produce superoxide.

The α-TO· may under certain conditions (LDL peroxidation), continue the chain oxidation [890]:

$$\alpha\text{-}TO· + LH \xrightarrow{k_{init}} \alpha\text{-}TOH + L· \qquad (1)$$

$$L· + O_2 \longrightarrow LOO· \qquad (2)$$

$$\text{LOO·} + \text{α-TOH} \longrightarrow \text{LOOH} + \text{α-TO·} \qquad (3)$$

$$\text{α-TO·} + \text{X·} \xrightarrow{\quad k_t \quad} \text{non radical products} \qquad (4)$$

This sequence of reactions, the tocopherol-mediated peroxidation (TMP), endows tocopherol with a prooxidant activity. The TO· radical undergoes two competing reactions (reaction 1 and 4). Tocopherol acts as a prooxidant only if reaction (1) competes effectively with reaction (4). The complex kinetics involved in the α-TOH-mediated peroxidation (TMP) of low-density lipoprotein (LDL) particles have been investigated and reviewed recently by Bowry and Ingold [890] and will be briefly discussed in chapter 9 (atherosclerosis).

Another important reaction of the α-TO· is the regeneration of α–TOH by its reaction with coantioxidants [783,893,894], such as ascorbate or ubiquinol:

$$\text{α-TO·} + \text{AH}^- \xrightarrow{\quad k \quad} \text{α-TOH} + \text{A·}^- \quad k = 1.55 \times 10^6 \ \text{M}^{-1}\text{s}^{-1} \ [783]$$

Both Vitamin C [895-897] and Vitamin E [884,885,898] are chain-breaking antioxidants, and there is extensive and conclusive evidence that indicates there is a synergistic antioxidant interaction between these vitamins in a variety of *in vitro* systems (see Burton et al. [899] and 31 references cited therein). This synergism may be due to the regeneration of α-TOH by reduction of the α-TO· by ascorbate (AH⁻), but another mechanism involving enzymic reduction has been discussed by Burton et al.[899].

Somewhat less compelling evidence has been reported for the synergistic interaction between Vitamins C and E *in vivo* [899]. Dietary Vitamin C has been reported to enhance plasma levels of Vitamin E as well as decrease them. Harats et al. [900] were unable to demonstrate increased plasma Vitamin E levels in Vitamin C supplemented versus nonsupplemented smokers.

Model systems are never as good as the real thing. It has been demonstrated by Burton et al. [899] that even the most carefully modeled *in vitro* systems fail to reproduce the *in vivo* reality. Careful kinetic analysis of Vitamin C and E levels in male guinea pigs, showed no synergistic interaction between the two vitamins, i.e., Vitamin C does not spare Vitamin E *in vivo* in animals, that are not subjected to enhanced oxidative stress.

Although Vitamin C does not spare Vitamin E in plasma *in vivo*, it acts as a coantioxidant in the tocopherol-mediated peroxidation (TMP) of low-density lipoprotein particles and thus inhibits the first step in the development of atherosclerosis (Chapter 9).

Another way by which α-TOH may act as a prooxidant is by reduction of transition metal ions [901-905]:

$$\alpha\text{-TOH} \ + \ Cu^{2+} \longrightarrow \alpha\text{-TO}\cdot \ + H^+ \ + \ Cu^+$$

followed by :

$$LOOH \ + \ Cu^+ \longrightarrow LO\cdot \ + \ OH^- \ + \ Cu^{2+}$$

$$LO\cdot \ + \ LH \longrightarrow LOH \ + \ L\cdot$$

However, this prooxidant effect only manifests itself in the presence of lipid hydroperoxides (LOOH), but not in peroxide-free systems [904,906].

These reactions are analogous to the ascorbic acid case. We have again a pro and con situation. It has been suggested that Cu^+ reacts with traces of LOOH to give alkoxyl radicals, which initiate the chain oxidation. The alkoxyl radicals can also undergo β-cleavage to give a great variety of aldehydes, the most important of which are malonaldehyde (MA) and 4-hydroxy-2-nonenal (4-HNE) (Chapter 6). On the other hand Cu^+ can react with plain oxygen to give $\cdot OH$ radicals [51]. The combination of Cu^+- O_2, Cu^{2+}-ascorbic acid-O_2, Fe^{3+}-ascorbic acid-O_2 (Udenfriend's reaction) has been known for a long time to hydroxylate aromatic compounds [51].

Results by Lynch and Frei [903] show that LDL containing α-TOH specifically reduces Cu^{2+} but not Fe^{3+}. The iron-dependent oxidative modification of LDL requires an exogenous reductant, such as $O_2^{\cdot-}$ to facilitate reduction of Fe^{3+} to $Fe^{2+\cdot}$. The prooxidant/antioxidant activity of α-TOH and the role of Cu^{2+} in the oxidative modification of LDL, as the first step in atherogenesis, will be discussed in Chapter 9.

G. ANTIOXIDANT AND PROOXIDANT ACTIVITY OF FLAVONOIDS

The flavonoids are natural products present in fruits, vegetables, tea, and wine. These compounds contain phenolic and polyphenolic groups (easily abstractable H-atoms), which form a relatively stable phenoxy radical. Therefore, by just looking at their chemical structure one would predict that they are good antioxidants. Flavonoids have indeed been found to scavenge $\cdot$OH [907], $O_2^{\cdot-}$ [908], ROO$\cdot$ [909], and $^1O_2$ [910,911]. The observation that flavonoids scavenge $\cdot$OH is of course no surprise, nor is their reaction with superoxide radical anion, since these reactions have been known for decades [178,179]. Flavonoids (quercetin, gossypol, myricetin, silymarin) act as inhibitors of lipid peroxidation via the same mechanism as α-tocopherol. The chain carrying peroxyl radical abstracts a H-atom from the phenolic -OH group, producing a stable phenoxy radical. This inhibition of chain autoxidation was observed in purely chemical systems [912,913] as well as in rat liver mitochondria and microsomes [914], and in myocardial phospholipids [915].

The autoxidation of linoleic acid gives two isomeric hydroperoxides, the trans, cis and the trans, trans configuration. In presence of flavonoids the trans, cis isomer increased and the trans, trans isomer [909] decreased. This clearly indicates that the intermediate peroxyl radical abstracts a H-atom from the flavonoid, thus terminating the chain.

However the flavonoids also act as prooxidants in presence of H_2O_2 and metal ions [913,916-918]. A number of flavonoids (quercetin, myricetin, gossypol) increased formation of hydroxyl radical (as measured by deoxyribose degradation) in presence of Fe^{3+}-EDTA at pH 7.4 [913]. The increased hydroxyl radical production was inhibited both by SOD and CAT, thus clearly indicating that both $O_2\cdot^-$ and H_2O_2 are involved in the formation of hydroxyl radicals. Remember: deoxyribose is not a reliable probe for the $\cdot OH$ radical. The results nevertheless indicate that some oxidizing species is formed and this species can damage deoxyribose and other biomolecules as well.

As pointed out in Chapter 3, polyphenolics react with oxygen to give the superoxide radical anion. The $O_2\cdot^-$ dismutates to H_2O_2 and O_2, and the H_2O_2 gives in presence of metal ions the $\cdot OH$ radical. Polyphenolics also react with $O_2\cdot^-$ to give H_2O_2 and semiquinones, which in turn react with O_2 to give more $O_2\cdot^-$ [132,133,178,179]. Remember: the important ingredient in these reactions are catalytic metal ions (Fe^{3+}, Cu^{2+}). Therefore *in vivo* the presence of free metal ions must be carefully controlled. The metal ions are complexed to proteins (ferritin, albumin, ceruloplasmin etc.). Polyphenolics not only produce $O_2\cdot^-$, but also cause the release of free iron from ferritin [316]. The superoxide also releases iron from ferritin [158,314]. The reduction of Fe(III) to Fe(II) by $O_2\cdot^-$ was suggested as the most likely mechanism for iron release *in vivo* [158]. The polyphenolics could therefore, *a priori* cause iron release either directly or via intermediate formation of $O_2\cdot^-$. In the work of Boyer et al. [316] superoxide dismutase had no effect on the rate of iron release. The released Fe(II) provides the essential catalyst for the Haber-Weiss reaction.

These flavonoids behave like ascorbic acid, acting as antioxidants, but as prooxidants in presence of metal ions *in vitro*. In many bacterial tests quercetin has been shown to be mutagenic [912]. Gossypol caused degradation of rat liver DNA incubated *in vitro* even in absence of metal ions and reducing agents [916]. This is not surprising since these polyphenolics are easily oxidized to give $O_2\cdot^-$ and H_2O_2, the precursors of $\cdot OH$.

Quercetin in presence of O_2 and Cu (II) induces strand scission of DNA *in vitro* [918]. Another natural polyphenolic is tannic acid, a derivative of glucose. Tannic acid is frequently used as a food additive. The combination tannic acid, oxygen, and Cu (II) has been found to cause strand scission of DNA *in vitro* [917]. Oxygen and Cu(II) were absolutely essential for DNA damage in both of these investigations. Since quercetin and tannic acid reduce Cu(II) to Cu(I) and O_2 to $O_2\cdot^-$, it has been suggested that the hydroxyl radical is the reactive agent. Well-known hydroxyl radical scavengers like mannitol, formate, and iodide as well as SOD and CAT considerably inhibited the DNA strand breakage [918]. These results implicate the hydroxyl radical formed in a Fenton-type reaction [202]:

$$Cu(I) + H_2O_2 \longrightarrow Cu(II) + OH^- + \cdot OH$$

Hydroxylation of aromatic compounds by Cu^+-O_2 has been known for a long time. The evidence for the involvement of hydroxyl radical in these hydroxylations has been reviewed [51]. It is therefore not surprising that the combination tannic acid-Cu(II)-O_2 and quercetin-Cu(II)-O_2 causes DNA strand breakage *in vitro*.

There are many other phenolics and polyphenolics present naturally in foods or added as an additive to prevent spoilage (inhibition of lipid peroxidation). These compounds can act as anticarcinogens, but can also cause mutations and act as carcinogens and cocarcinogens. The dual character of these compounds has been reviewed [919,920].

H. CAROTENOIDS

The carotenoids comprise a group of natural products present in many fruits and vegetables. Their common chemical characteristic is a long hydrocarbon chain with conjugated double bonds:

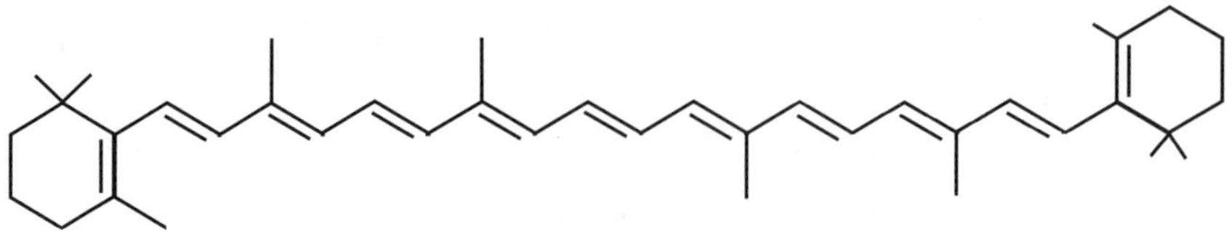

β-carotene

These conjugated chains give the carotenoids their characteristic color (yellow-red). The carotenoids are best known for their ability to quench singlet oxygen. The most effective singlet oxygen quencher among the carotenoids is lycopene (k = 3.1 x 10^{10} $M^{-1}s^{-1}$) [921]. The quenching rate constant of β-carotene is 1.4 x 10^{10} $M^{-1}s^{-1}$ [921]. Other antioxidants, like ascorbic acid, tocopherols, and thiols also can quench singlet oxygen, but at rates many orders of magnitude smaller than the carotenoids [922].

There are many examples that demonstrate the quenching efficiency of the carotenoids. Erythropoietic protoporphyria is a human photosensitivity disease. The skin of these patients accumulates high concentrations of porphyrins, which react with oxygen to give singlet oxygen. The symptoms of these patients can be relieved by β-carotene [462,463]. This result indicates that singlet oxygen is involved in causing the symptoms of this disease, but not in causing the underlying pathology.

However, evidence has been accumulating indicating that β-carotene can also act as a radical scavenger. Krinsky and Deneke [923] demonstrated that carotenoids inhibited free radical-induced oxidation of liposomal lipids. β-carotene was also found to decrease the rate of formation of linoleate hydroperoxides [924]. The carotenoids are effective in protecting cells not only against singlet oxygen, but against other oxygen radicals as well (ROO·, ·OH, HO_2·/O_2·⁻). The antioxidant functions of carotenoids have been reviewed by Krinsky [925].

Another interesting effect was ascribed to the scavenging of singlet oxygen. Human polymorphonuclear leukocytes killed a colorless mutant of *Sarcina lutea* much more readily than the carotenoid-containing strain [458]. Since carotenoids are singlet oxygen quenchers, the protective effect was considered as evidence of singlet oxygen formation in the bactericidal action of leukocytes. However, as the later results of Krinsky and Deneke [923], Terrao [924], and Packer et al. [926] indicate, carotenoids can also be effective radical scavengers. The results with *Sarcina lutea* show that although the conclusion (PMNs produce $^1O_2$) is correct, the conclusion has been reached for the wrong reason.

It has been suggested that β-carotene is a very special antioxidant, acting as an antioxidant at low oxygen concentrations (as usually present in normal tissue), but as a prooxidant at high oxygen tension [927]. However, computer modeling of β-carotene consumption and oxygen uptake led to the conclusion that β-carotene behaves as an interceptor of radicals regardless of oxygen tension [928]. It has been shown that β-carotene efficiently traps both the carbon-centered radicals of lipids as well as the peroxyl radicals (LOO·). It is thus conceivable that β-carotene intercepts radicals both in absence or presence of oxygen.

Singlet oxygen can be produced *in vivo* via many reactions. I have already discussed in chapters 3, 4, and 5 (phagocytosis) the many possible reactions by which singlet oxygen can be produced *in vivo*. A reaction important in membranes is the dismutation of lipid-peroxyl radicals (the Russell mechanism).

I. PROTEINS THAT FORM COMPLEXES WITH TRANSITION METAL IONS

The formation of oxygen radicals like ·OH (from H_2O_2) and RO· (from ROOH) requires transition metal ions (Fe^{2+}, Cu^+). Therefore any protein that complexes metal ions also provides protection by sequestering the metal ion. The potent antioxidant activity of human plasma is mainly dependent on naturally present complexing proteins like transferrin, ferritin, ceruloplasmin, albumin, and metallothionein. These proteins are present intracellularly as well as in extracellular fluids. The major extracellular metal binding protein is transferrin with a high affinity for iron, and ceruloplasmin, which binds Cu^{2+}. Extracellular binding is also provided by albumin. The intracellular proteins that bind metal ions are ferritin (binds Fe) and metallothionein (which binds Cu). The role of transition metal ions in diverse disease processes has been periodically reviewed [258,277,929].

Ceruloplasmin binds Cu^{2+} and Fe^{3+} and can therefore inhibit copper and iron ion-stimulated formation of hydroxyl radicals and lipid peroxidation [930]. The iron-binding protein lactoferrin was also shown to inhibit lipid peroxidation [931]. Autoxidation of ascorbic acid is catalyzed by metal ions. This autoxidation is inhibited by copper-binding ceruloplasmin, albumin, and apotransferrin [932]. The formation of hydroxyl radicals from $O_2^{\cdot-} / H_2O_2$ in the presence of iron was inhibited by apo-lactoferrin and apo-transferrin [933].

During phagocytosis (Chapter 5) $O_2^{\cdot-}$ and H_2O_2 are formed in huge quantities. These reactive oxygen metabolites in presence of iron or copper salts form the reactive hydroxyl radical (Chapter 3), which can attack almost any molecule it encounters. In patients with rheumatoid arthritis, large numbers of activated neutrophils enter the joints and the hydroxyl radicals initiate lipid peroxidation (Chapter 6). This process leads to the destruction of membranes and release of proteolytic enzymes, which further aggravate the problem. In patients with rheumatoid arthritis the naturally present antioxidant defenses are defective. An *in vitro* study of lipid peroxidation (as measured by the thiobarbituric acid test) showed that serum from normal subjects inhibited lipid peroxidation considerably more than serum and synovial fluid from patients with rheumatoid arthritis [934]. The addition of ceruloplasmin at physiological concentrations offered protection against the copper-induced lipid peroxidation.

The unnatural occuring complexing agent deferoxamine, which chelates Fe has been used as a probe for many radical-induced processes, such as lipid peroxidation [935]. Asbestos-induced lung injury [304] was studied. Lung fibroblasts exposed to asbestos showed longer survival rates when the cells were pretreated with defer-oxamine. This result supports the hypothesis that metal ions on the surface of asbestos fibers are responsible for hydroxyl radical formation [300,301].

J. PEPTIDES AS ANTIOXIDANTS

Some dipeptides with antioxidant properties have been found in skeletal muscle and brain. These peptides are carnosine (β-alanyl-L-histidine), anserine (β-alanyl-3-methyl-L-histidine), and homo-arnosine (γ-aminobutyryl-L-histidine). Carnosine has the following structure:

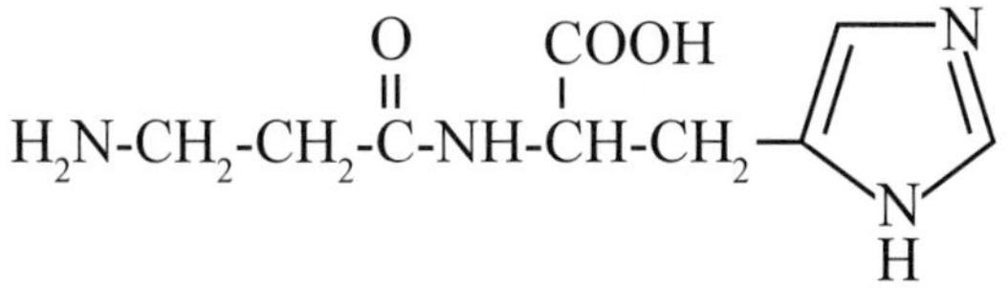

$$H_2N-CH_2-CH_2-\overset{O}{\overset{\|}{C}}-NH-\overset{COOH}{\overset{|}{CH}}-CH_2-\text{(imidazole ring)}$$

From the antioxidant point of view, the important part of this peptide is the imidazole ring. Aromatic rings are excellent scavengers of ·OH radicals (benzoic acid, phenylalanine, guanine). It is therefore no surprise to find that carnosine inhibits lipid peroxidation and deoxyguanine hydroxylation *in vitro* [824]. The hydroxylation of the deoxyguanosine was studied with the Cu^{2+}-AH_2 - O_2 system, which has been shown to produce ·OH radicals [51]. Imidazole itself has been shown to inhibit lipid peroxidation [824].

The reaction of singlet oxygen ($^1O_2$) with imidazoles has been studied extensively [936]. Many enzymes containing the imidazole ring are destroyed by oxygen and light. This effect has become known as the "photodynamic effect" and is due to the reaction of the imidazole moiety with $^1O_2$.

In addition to reacting with oxygen radicals and $^1O_2$, the carnosine has been shown to be an efficient copper chelating agent [937]. This is an important property since skeletal muscle contains about one-third of the total copper content in the body (20-47 mmol/kg).

The presence of carnosine in skeletal muscle could indicate its physiological role as a chelator of Cu-ions.

The amount of these dipeptides in humans is quite high (up to 5 μ moles/g in muscle). An increase in carnosine was noted in sprinters and rowers [869]. This has been interpreted as a physiological adaptation to increased oxidative stress. This observation is analogous to the increased pentane and ethane formation during exercise and the simultaneous increase in plasma levels of uric acid [825,938]. Human cerebrospinal fluid contains up to 50 μM of homocarnosine. Brain tissue has been found to contain carnosine, homocarnosine, and anserine in the 0.3 - 5 mM range [824]. Other antioxidants in human cerebrospinal fluid are ascorbic acid (100 μM), and urate (18 μM) [825].

I have already pointed out the high levels of ascorbic acid in the brain. The human brain is the most likely place for oxidative stress to occur. The human brain consumes about 18 % of the total oxygen [859], but represents only 2% of the total body mass. The brain also contains many unsaturated lipids, which can easily become autoxidized. The presence of carnosine, anserine, and homocarnosine in the brain may represent an evolutionary adaptation to cope with increased oxidative stress. These small dipeptides are produced endogenously and do not depend on diet like ascorbic acid.

K. DO ANTIOXIDANTS PREVENT DISEASES *IN VIVO*?

After we have some understanding of the chemistry of antioxidant action *in vitro*, we can ask whether these compounds in our diet offer protection against diseases caused by oxidative stress. There are epidemiological studies examining Vitamin C, Vitamin E, flavonoids, and β-carotene. There are numerous reviews on the subject. A list of these reviews can be found in the review by Byers and Perry [939]. However, *in vivo*, it is difficult to study the effects of antioxidants in isolation from many other effects. The antioxidants in our diet are not there in splendid isolation, but together with many other natural products, which may be beneficial or deleterious to our health. We have to realize that our daily diet not only contains antioxidants and anticarcinogens, but also mutagens and carcinogens (hydrazine, safrole, quercetin and similiar flavonoids, nitrite, nitrate and nitrosamines, and alcohol) [919,920]. In addition, many antioxidants can also act as prooxidants. It is therefore not surprising to find much confusion in the literature. Articles written by honest scientists are therefore full of phrases like: evidence is fairly good, perhaps, possibly, maybe, so far, but etc.

In order for our cells to synthesize these enzymatic defenses we need certain compounds for their synthesis. These compounds include selenium (Se) for GSH-Px, Zn, Cu, and Mn for SOD and CAT. A deficiency in these nutrients has deleterious effects on our health.

Ecologic studies in the U.S. have shown an inverse correlation between cancer mortality rates and Se in locally grown food crops [940] as well as with dietary Se intake [941-943]. There is increasing epidemiologic evidence indicating that Se plays a prominent role in cancer prevention [943-945]. Some interesting epidemiological studies have been carried out in China. People in certain regions of China have the lowest Se intake in the world and at the same time have very high rates of lung, esophagus, stomach, and liver cancers [946-948].

Selenium is needed for the synthesis of many biologically active selenoproteins. Among almost 30 known selenoproteins, the GSH-Px is the most extensively characterized and studied. All of these peroxidases catalyze the reduction of peroxides (ROOH, H_2O_2) with GSH as the reducing agent:

$$H_2O_2 + 2\,GSH \xrightarrow{\text{GSH-Px}} GSSG + 2\,H_2O$$

I have already discussed the greater effectiveness of GSH-Px compared to CAT in removing H_2O_2 (Chapter 5, phagocytosis).

Any deficiency in Se will therefore cause a decrease in the GSH-Px concentration and thus an increase in the steady state concentrations of ROOH and H_2O_2. This changes the prooxidant/antioxidant balance and leads to pathological changes.

The antioxidant defenses are not evenly distributed throughout the body. We again encounter the principle of economy of design. The antioxidant enzymes are produced where they are needed most. Cells do not waste their energy producing something they do not need. GSH-Px and GSH concentrations are higher in organs like the liver and lung, which have a high flux of ROMs [949-953]. SOD concentrations are higher in the heart compared to the liver. The heart has a greater need for $O_2\cdot^-$ destruction (has a higher O_2 concentration than the liver). The steady state concentration of $O_2\cdot^-$ in heart mitochondria is about twice as high as in liver mitochondria [832].

Intracellular formation of reactive oxygen metabolites induces the synthesis of enzymes that participate in the destruction of these ROMs. Some recent work [954] on Se concentrations in human plasma and in malignant tissue of lung cancer patients, has shown a considerably lower plasma level of Se, but significantly higher Se concentrations and elevated GSH-Px levels in cancerous lung tissue than in the surrounding cancer-free tissue. The Se accumulated in the tissue induces the synthesis of GSH-Px. In addition to Se, other trace elements such as Zn, Cu, and Mn, which are incorporated into superoxide dismutase (SOD) are significantly elevated in malignant tissue compared with tumor-free tissue. It has been suggested [954] that the increased activity of GSH-Px in cancerous lung tissue may be a consequence of the increased accumulation of ROMs in the tissue.

In a recent study of cancer prevention by Se supplementation [955] it was shown that Se did not lower the risk of skin cancer (basal cell and squamous cell carcinomas), but showed profound effects on the incidence of lung cancer, colorectal cancer, and prostate cancer. Antioxidant concentrations are higher in organs exposed to high oxidative stress, especially the lung. Selenium and the enzyme glutathione peroxidase derived from Se are therefore expected to have a more profound effect on lung tissue than on skin.

Epidemiology is a primitive first step in the scientific understanding of the etiology of disease. We observe that a certain kind of food (from geographic data) increases or lowers the incidence of a disease. This observation does not tell us which component of the food is responsible and how this component is metabolized and transformed into the final damaging agent.

An interesting example of the complexities encountered in epidemiological studies is the case of prostate cancer. Ecological studies established high milk and cheese consumption as major risk factors for prostate cancer [956a]. Many of these studies have been reviewed by Giovannucci [956a]. Laboratory and clinical data showed an antitumor effect of $1,25 \, (OH)_2$-vitamin D $(1,25 \, (OH)_2 D)$ on prostate cancer [956b]. Since milk is a major source of dietary Vitamin D, these results seem paradoxical. Since milk and dairy products are high in fat content and since unsaturated fatty acids are easily autoxidized to peroxyl radicals and hydroperoxides, and subsequently many mutagenic degradation products (aldehydes) the milk-prostate cancer connection may be due to the fat content of milk [956a]. However this ain't necessarily so.

The $1,25 \, (OH)_2 D$ has a well-established role in Ca^{2+} and PO_4^{3-} homeostasis, but has many other functions [957]. A reduction in plasma phosphate levels increases the level of $1,25 \, (OH)_2 D$ considerably and thus inhibits the development of prostate cancer. Dietary fructose lowers plasma phospate levels by 30 - 50% (fructose is phosphorylated), thus increasing the level of $1,25 \, (OH)_2 D$ and decreasing the risk for prostate cancer [956b]. On the other hand, high plasma levels of Ca^{2+} decrease the formation of $1,25 \, (OH)_2 D$, thus favoring the development of prostate cancer [956b]. The increased Ca^{2+} intake through dairy products and not the more obvious high fat intake, is responsible for the increased risk for prostate cancer. Low dairy product consumption and high fruit consumption may be the right prescription for the prevention of prostate cancer.

I have already pointed out the pros and cons of ascorbate and multivitamin-mineral combinations. Ascorbic acid, α-tocopherol, and the phenolic flavonoids can act as antioxidants or as prooxidants depending on the presence or absence of transition metal ions.

Epidemiologic studies on the effects of Vitamin E on coronary heart disease has been carried out [958-960]. The authors seem convinced that their data support the hypothesis that vitamin E supplements have a beneficial effect in preventing coronary heart disease. In these studies the benefits of Vitamin E were, however, confined to the subgroup of subjects taking large amounts of Vitamin E supplements. Vitamin E intake through diet alone afforded no protection, even when supplemented with multivitamin tablets. A critical analysis of these studies is given by Byers [961] and Steinberg [962]. Nothing is known about the long-term toxic effects (if any) of high Vitamin E intake. Therefore the warning by Steinberg: hold the Vitamin E!

It has long been suggested that carotenoids can function to reduce human cancer rates [963]. Although this hypothesis has been scrutinized extensively, it is not yet proven right or wrong. The hypothesis rests on epidemiological evidence in humans as well as experimental evidence in cellular systems and in animals [963-966]. The evidence suggests that carotenoids inhibit mutagenesis, malignant transformation, tumor formation, and immunoenhancement. These results have been reviewed by Krinsky [966] and Ziegler [967]. In UV-B-induced skin tumors on hairless mice, however, a protective effect of carotenoids was noted only at very high doses (0.7 - 7 g. carotenoids/kg.), while at doses of 70 mg/kg the carotenoids offered no protection [968,969].

Vitamin E and β-carotene were examined (over a period of 5-8 years) as to their effectiveness on the incidence of lung cancer on male smokers [970]. Vitamin E showed no reduction in lung cancer incidence, but β-carotene showed a small increase (18%). On the other hand fewer cases of prostate cancer were diagnosed among those patients receiving α-tocopherol (99 versus 152) than among those who did not. β-carotene on the other hand had no effect on prostate cancer. Conclusion: Vitamin E does not prevent lung cancer and β-carotene may even be harmful. However, in another study a relationship between high plasma levels of β-carotene and lower lung cancer incidence was established [964]. A large scale nutritional study, involving 3968 men and 6100 women, aged 25-74 years over a median follow-up period of 19 years, examined the intake of Vitamins E, C, and A and the risk of lung cancer [971]. Conclusion: there was no additional protective effect of supplements of Vitamins E, C, and A beyond that provided through dietary intake. The moral of the story: an ounce of prevention (do not smoke) is better than an ounce of antioxidants (the multivitamin supplements). The controversies, contradictions, and paradoxes have been briefly discussed by Gonzalez et al. [972].

Exercise increases oxygen consumption and increased lipid peroxidation. The amount of pentane in the exhaled breath of human subjects increased and this increase was lowered in subjects whose daily diet was supplemented with 1200 IU dl-α-tocopherol for two weeks [938]. This result clearly indicates protection from oxidative damage by Vitamin E *in vivo*. The antioxidant defenses are, however, far from perfect, otherwise we wouldn't age nor die. Epidemiological studies are hailed as medical breakthroughs if a certain vitamin or antioxidant lowers the risk of a disease by 20 - 30%.

The tocopherols and flavonoids may offer some protection against the damaging effects of lipid peroxidation, however, after several glasses of wine the negative effects of alcohol may outweigh the protective effect of flavonoids. In the meantime, the multivitamin industry is making billions selling something which at best does no harm, but at worst can be hazardous to our health.

Chapter 9

PATHOLOGICAL PROCESSES INVOLVING
REACTIVE OXYGEN METABOLITES

The spirit of medicine is easy to know:
Through the macro-and microcosm you breeze,
And in the end you let it go
As God may please.
Mephisto in Goethe's Faust, Part 1

A. INTRODUCTION

1. General considerations

We live in a chemical world [973]. Our bodies are made of chemicals, and these chemicals are constantly oxidized (metabolized) and converted to other chemicals, which may be inocuous or damaging to our health. Even in absence of all the chemicals in our external environment would we observe mutations, carcinogenesis, aging and death. These type of changes (absence of external sources) in our DNA are referred to as spontaneous mutation and spontaneous carcinogenesis [974]. Spontaneous mutations are essential for evolution. As pointed out by Raold Hoffmann [21] we are constantly bombarded with high energy radiation from the small amounts of tritium in normal water. This mutagenic effect of radiation may have played an important role in our evolution. Spontaneous mutation has been defined as "the net result of all that can go wrong with DNA during the life cycle of an organism" [975]. It has been demonstrated that the types and frequencies of spontaneous mutation change considerably with subtle changes in conditions. This of course always happens with complex non-linear dynamic systems, where the initial conditions are not clearly defined. Many types of mutations are produced spontaneously, i.e., base eliminations, frame shifts, insertion, and deletion. The intrinsic instability of DNA can lead to mutations via errors made in replication, repair, and recombination.

304

Life began in the absence of oxygen. However due to photosynthetic activity (by splitting of water) oxygen became present in the atmosphere. Due to the increasing efficiency of energy storage by ATP, most organisms developed and depended on oxygen as the final electron acceptor. Unfortunately oxygen is also responsible for the metabolic formation of some highly toxic and reactive by-products, namely the formation of $O_2^-\cdot/HO_2\cdot$, H_2O_2, $\cdot OH$, and $^1O_2$. Any advantage gained has to be paid for by some disadvantage, in other words "there is no such thing as a free lunch". As suggested by Sies [393], biological systems are in a state of approximate equilibrium between prooxidant forces and the antioxidant capacity of most biological systems. Our cells, tissues and orgens function within very narrow limits of oxygen tension, pH, Fe (III), Cu(II), Ca(II), thiols, and enzymes, such as SOD and external antioxidants (ascorbic acid, tocopherols, and selenium). Among the radicals $O_2\cdot^-$ and $\cdot NO$ we encounter the same situation: too little or too much may have deleterious consequences. Biological oxygen metabolizing systems are maintaining a precarious balance between life and death. The prooxidant forces are likely to be predominant, otherwise we may not age nor die. A well-known measure for oxidative damage is the accumulation of lipofucsin during the lifespan of an organism. Lipofucsin is a PUFA derived colored oxidation product. Oxidative stress causes extensive damage, including lipid peroxidation of membranes, oxidation of proteins, and damage to DNA and RNA. I have discussed the basic chemistry of ROMs in previous chapters. All we have to do now is apply this knowledge to the interpretation of some pathological processes.

Since DNA and RNA are the single most important molecules for the survival of the species and the proper functioning of metabolism, these molecules are well protected and far removed from the site of oxygen radical and oxygen metabolite formation. DNA is surrounded by a protein sheath and enclosed in a nuclear membrane.

Among the immediate oxygen metabolites ($O_2\cdot^-$ and H_2O_2) only H_2O_2 (because it is a neutral molecule) can penetrate the membrane.

Superoxide can only pass through special ion channels, like in erythrocytes. Since erythrocytes are high in SOD and CAT, erythrocytes can act as a superoxide radical sink [976]. This protective role of erythrocytes is important in inflammatory lung disease.

Hydroxyl radicals are far too reactive to travel very far from their site of formation. Hydroxyl radicals react with most organic molecules at close to diffusion-controlled rates, which means they react at each collision. Therefore in order for hydroxyl radicals to damage DNA, the hydroxyl radical has to be formed close to the target, i.e., it has to react in a site-specific manner. In this case hydroxyl radical scavengers cannot interfere with the reaction. This type of observation has let to many wrong conclusions concerning the nature of the attacking oxygen species (crypto hydroxyl radical).

ROMs are formed in normal metabolism via the mitochondrial respiratory chain, microsomal enzymes, phagocytosis, endothelium and neurons (Chapter 5) and they are consumed by many reactions and are, therefore, present in a steady state (the rate of formation equals the rate of disappearance). This steady state is normally low enough and causes no major damage. The antioxidant defenses are sufficient to keep us from disintegrating. Any change in this delicate prooxidant/antioxidant balance will affect the operation of a cell. This change can be brought about by some abnormality in the metabolism, which can have many causes, such as genetic defects (ALS, Down's syndrome, multiple sclerosis) or can be brought about by external factors. Some of these factors can be exposure to viruses, toxic chemicals, or improper lifestyle (smoking, alcohol consumption, exercise, diet).

Whenever this balance is disturbed we have a situation that has been called "oxidative stress" [393]. Continued oxidative stress leads to cellular damage and pathological changes. The major factor in pathological changes are the attack on lipids, enzymes and DNA (Chapter 6). The term "oxidative stress" has been criticized [977]. It was argued that O_2^- is not an oxidizing agent, but a reducing agent. This is, however, only partly true. The oxidation-reduction property of O_2^- depends on the pH (Chapter 3). At a pH of 4.5 as prevalent in the lipid bilayer the $O_2^{\cdot-}$ is present to a considerable extent as $HO_2\cdot$, which is an oxidizing agent:

$$O_2 \xleftarrow{\text{oxid..}} O_2^{\cdot-} \rightleftarrows HO_2\cdot \xrightarrow{\text{red.}} HO_2^-$$

The $HO_2\cdot$ oxidizes lipids. In addition, oxidative stress does not only involve an increase in $O_2^{\cdot-}/HO_2\cdot$ but also of H_2O_2, $\cdot OH$, $^1O_2$, $HOCl$, $\cdot NO$, ^-OONO, and excited carbonyl compounds. All of these species are oxidizing species and the term "oxidative stress" is therefore quite appropiate [393, 565].

Words are of course never as precise as formulas and equations, but remember: a rose by any other name smells just as sweet (Shakespeare)! The concept of oxidative stress gives us an answer to the following question: what do aging, carcinogenesis, amyotrophic lateral sclerosis, Alzheimer's disease, Down's syndrome, Parkinson's disease, multiple sclerosis, strenuous exercise, and psychological stress have in common?

2. Cause and effect

The most frequently used marker for tissue damage has been lipid peroxidation. However we have to ask: is lipid peroxidation the cause of cell death or a consequence of it? Halliwell and Gutteridge have called attention to this problem and have argued in favor of the second possibility [978]. It has been well known that damaged tissues are more easily autoxidized [979-981]. Dead cells release metal ions from their storage proteins and catalyze autoxidations. Metal ions are essential for the formation of hydroxyl radicals (Chapter 3).

The first radical-induced pathological condition for which a cause and effect relationship has been established is the liver toxicity of CCl_4. The carbon tetrachloride is metabolized to the $\cdot CCl_3$, which initiates the damage via lipid peroxidation [595].

There are over 100 diseases documented to involve reactive oxygen metabolites [4]. I have no intention of discussing all of these in this introductory text. I am going to discuss the evidence for the involvement of ROMs in aging, cancer, and diseases in a number of important organs (central nervous system, cardiovascular system, pulmonary system). The mechanisms of ROM-induced damage are basically all the same. I have discussed the reactions of these species with many biomolecules in Chapters 6 - 8.

We have made tremendous progress in our endeavor to reduce the many diseases with a great variety of clinical manifestations to the simplest common denominator. We have arrived at just a handful of simple chemical species, like $O_2 \cdot^-$, $\cdot NO$, ^-OONO, $HOCl$, $\cdot OH$, and $^1O_2$. These species are so simple that they most likely have been around since the beginning of life on earth and they are responsible for why we are what we are.

Among the three primary oxygen metabolites, $O_2 \cdot^-$, H_2O_2, and $\cdot NO$ only two are radicals. These two radicals are not only damaging but also fulfill regulatory functions [157]. The secondary products derived from these primary metabolites ^-OONO, $\cdot OH$, and $^1O_2$ are the reactive species responsible for oxidative damage. The combination $O_2 \cdot^-$ and $\cdot NO$ requires high concentrations of both (Chapter 7) and is therefore restricted to phagocytic cells, which produce both high concentrations of $O_2 \cdot^-$ and $\cdot NO$ and where the formation of ^-OONO is part of the antimicrobial defense mechanism. Without this restriction we would not exist.

The observation of reactive oxygen species, either directly by spin trapping, indirectly by quenching, or by amelioration of the symptoms by antioxidant therapy, is no proof that the ROMs are causing the pathological abnormality. The cause most likely is some genetic defect such as the over- or underexpression of some enzymes, which changes the prooxidant-antioxidant balance.

In order to establish a causal relationship between ROMs and disease it is important to know their rate of formation and their rate of disappearance as well as the concentrations of the antioxidant

308

defenses present in the tissue under study. Remember: the steady state concentrations of prooxidants and antioxidants is different in different tissues (chapter 8). The knowledge of the steady state under normal conditions will give us a yard stick to determine the minimum steady state concentration required to initiate pathological changes. The important aspect is the proxidant/antioxidant balance:

$$[\text{prooxidant}]_{ss}/[\text{antioxidants}]_{ss}$$

Whenever the ratio is lower or higher than normal we can expect pathological consequences.

High concentrations of H_2O_2, $O_2^{\cdot-}$, $\cdot NO$, and ^-OONO are not the causes of amyotrophic lateral sclerosis (ALS), Alzheimer's disease (AD), Parkinson's disease (PD), Down's syndrome (DS) or multiple sclerosis (MS), but arise as a consequence of some metabolic imbalance. The metabolic imbalance (genetic defect, hormonal imbalance, loss of homeostatic control, infection, toxic chemicals etc.) may have many causes and these causes may have other causes all the way down the line to the first cause. Likely candidates for the first cause are radicals that are produced either internally in our bodies (spontaneous mutation, spontaneous carcinogenesis) or can be produced by some toxin (via metabolism to radicals) or some other outside agent like bacteria or viruses. It has been suggested by Hassan [8] that the steady state concentration of $O_2^{\cdot-}$ of $10^{-12}-10^{-10}$ M is responsible for the present day spontaneous mutation rate. Hydroxyl radicals are formed from the radiation of trace amounts of tritiated water contained in the water of our bodies. The tritiated water was formed in the biggest explosion ever, the 'BIG BANG', and that's why we are here.

The present text is dealing with reactive oxygen metabolites and it is therefore beyond the scope of this text to discuss the causes of the metabolic imbalances. I am emphasizing the "oxidative stress hypothesis" i.e., the disturbance of the prooxidant/antioxidant balance. Any over- or underproduction of the enzymes involved in oxygen metabolisms (SOD, CAT, GSH-Px, NOS, iNOS, NADPH oxidase, MAO, XO, etc.)

is leading to pathological changes. Some of the antioxidant defenses are produced endogenously (SOD, CAT, GSH-Px, peptides, uric acid, melatonin, metal ion chelating proteins) while others are supplied in our diet (Vitamins A, C, E, and selenium).

3. Lipofuscin

The accumulation of lipofuscin is the only age-related morphological change detected so far. Lipofuscin has been known for over a century, long before the importance of reactive oxygen metabolites was recognized as an important part in many pathological processes. References to numerous older reviews can be found in the review by Sohal and Wolfe [982].

Unlike the other biological markers for oxidative damage (Chapter 6), the lipofuscin is chemically not well defined. It is a complex polymeric mixture, that contains lipids and proteins, exhibits auto-fluorescence, and accumulates in the cytoplasm with age under normal physiological conditions. The fluoroescence makes lipofuscin easy to detect in biological samples. The formation of lipofuscin according to a scheme proposed by Tappel and coworkers [983,984] involves formation of malonaldehyde or other aldehydes formed by lipid peroxidation (Chapter 6). The malonaldehyde interacts with amino-groups of proteins to form a Schiff base leading to a complex polymeric material. Proteins and lipids have been shown to be the main constituents of lipofuscin. Since proteins are complexing agents for metal ions (Chapter 8) it is not surprising to find high levels of Cu^{2+} and Fe^{3+} present in lipofuscin [985,986]. The autofluorescence has been suggested to involve degradation of the lipids. The oxidation of Schiff bases to give electronically excited states has already been discussed in Chapter 5. The chemical composition of lipofuscin depends on the cell's type and its physiological status. The neurons of substantia nigra contain dopa and dopamine, which are easily oxidizable polyhydroxy-aromatic compounds (Chapter 3). Dopamine and its oxidation products are incorporated into the lipofuscin [987,988]. This type of modified lipofuscin has been called "waste pigment" in PD patients. Lipofuscin has been characterized as a waste basket for cellular wastes primarily consisting of intracellular membranes.

The relationship between oxygen metabolites and lipofuscin accumulation has been demonstrated in cultured human glial cells [989]. Glial cells are the macrophages of the CNS. These phagocytic cells are the source of copious amounts of reactive oxygen metabolites (Chapter 5). Glial cells were grown in 5%, 10%, 20%, and 40% oxygen. The rate of lipofuscin accumulation increased with increasing oxygen concentration. The presence of the prooxidant combination ascorbic acid - Fe^{3+} increased and the presence of antioxidants such as vitamin E, Se, GSH, and DMSO decreased the accumulation of lipofuscin.

From these results it follows that lipofuscin formation is influenced by diet and pathological conditions. The increased formation of lipofuscin in some diseases (Alzheimer's, Parkinson's , amyotrophic lateral sclerosis, Down's syndrome, and multiple sclerosis) will be discussed later in this chapter. Lipofuscin in absence of any pathological condition, is a useful marker for physiological aging. There is no evidence that lipofuscin in the cytoplasm hinders cellular functions. In other words: lipofuscin is a marker for aging, but does not cause aging just as the wrinkles in our face are signs of aging, but not the cause of it.

4. On men, mice, and flies

If reactive oxygen metabolites are indeed responsible for aging and many degenerative diseases of old age, then there could be a possible relationship between metabolic rate and life expectancy. The more ROMs that are produced by an organism the more damage accumulates and the life expectancy drops, unless of course the organism in question also has high levels of antioxidant defenses. I have already discussed many biological markers for oxidative damage in Chapter 6. Using the 8-OHdG as a marker, Ames and coworkers [867] have carried out extensive studies on men and mice. The relationship between metabolic rate and aging has been reviewed by Sohal and Allen [990]. Mice have a much higher metabolic rate than men and excrete higher levels of 8-OHdG in their urine. Interestingly enough mice do not have higher levels of antioxidants, but lower levels than men. The endogenously produced antioxidant uric acid is many times higher in the plasma of humans than in mice. The combination of high metabolic rate and low antioxidant defenses in mice is the reason why there are "too many rodent carcinogens"[868].

Lipofuscin accumulation was studied by Sohal and coworkers in flies [982]. Flies (*Musca domestica*) whose flying ability was restricted (by keeping them in a small jar) accumulated less lipofuscin than flies who were flying more. The more active flies had a 60-100-fold higher oxygen consumption [991] and exhaled more pentane [992] (a marker of lipid peroxidation) than the non active flies. At the same time the life expectancy of the active flies dropped. Three cheers for a siesta! Studies on house flies have indicated a decrease of antioxidant defenses and an increase in exhaled pentane with age [992].

A long time ago it was reported that lipofuscin accumulation was absent in muscles of human stroke victims [993,994]. Similiar results were obtained in studies on Turkish hamsters [995,996]. The hibernating hamsters accumulated less lipofuscin and had a longer life expectancy than the non hibernating kind. Lipofuscin is not a marker of physical time but of physiological time. However, some exceptions (how could it be otherwise) to this activity-lipofuscin relationship have been reported. The flight muscles of the housefly *Musca domestica* rarely showed detectable lipofuscin structures [997]. This exception to the rule could be due to the fact that cells with a higher metabolic rate also have higher levels of antioxidant defenses.

In Chapter 8 I asked the question: is exercise good or bad ? Based on the preceeding results and arguments the answer appears to be: it is bad for you, it lowers your life expectancy. However, I have also pointed out that the endogenously produced antioxidant uric acid [825] and carnosine, anserine, and homocarnosine increase with exercise [869]. Whether the increase in antioxidants is sufficient to compensate for the additional ROM-induced damage is doubtful. It is a question of balance. Another factor that increases life expectancy in men and mice is restriction of caloric intake [998]. Moderate physical activity combined with moderation in everything else (drinking, eating) may be the prescription for a longer life. These life style factors will be discussed in detail later in this chapter.

B. AGING AND CANCER

A deleterious biological process obvious even to the most casual observer is the process of aging. This process is a direct consequence of living in an oxygen atmosphere. Oxygen is essential for the operation of a cell, but as a side-product (about 1 - 5 % of total oxygen consumption) [7,85,127, 832] it also produces highly toxic and reactive oxygen radicals ($O_2 \cdot^- /HO_2 \cdot$ and $\cdot OH$), oxygen containing metabolites (H_2O_2, hydroperoxides, epoxides, 1,2-dioxetanes), singlet oxygen, and excited carbonyl compounds. These radicals and reactive oxygen metabolites are formed by normal metabolism as well as by metabolism of xenobiotics. In addition, they are formed by stimulated phagocytes in response to invading microorganisms and other foreign bodies (asbestos). All of these processes contribute to DNA damage, mutations, and cancer. DNA damage induced in somatic cells will accumulate over time and contribute to age-related pathologies and cancer [999-1001]. The different theories of aging and the factors that affect longevity, like diet, antioxidants and lifestyle (exercise, smoking, alcohol consumption, and psychological stress) have recently been discussed in detail by Marie-Françoise Schulz-Allen [1002].

Endogenous processes leading to DNA damage are oxidations, methylation, deamination, and depurination (base loss). The DNA damage produced by oxidation appears to be the most significant factor in DNA damage. In order to prove the theory that oxidative endogenous damage to DNA is responsible for aging and age-related degenerative pathologies such as cancer and heart disease, we need a specific marker to measure this damage and a sensitive analytical technique to detect this marker. Thymine glycol and 5-hydroxymethyl-uracil are two specific products of DNA oxidation [394]. These products have been shown to be produced via γ-radiolysis of aqueous solutions of thymine [599,600]. An indication that these products are important products of oxidative damage *in vivo* is the fact that specific repair enzymes that repair this damage are formed *in vivo* [600]. A cell would not waste its energy in producing something it does not need.

More recently, the formation of 8-OHdG has been used as a marker for studying oxidative damage *in vitro* and *in vivo*. The popularity of 8-OHdG as a marker is due to the availability of a sensitive, specific probe for 8-OHdG (HPLC-EC) developed by Floyd and coworkers [120,121]. I have already discussed the formation of 8-OHdG by various reactive oxygen metabolites (Chapter 6).

The 8-OHdG causes mutations by G$\longrightarrow$T and A$\longrightarrow$C transmutations. These transmutations have been implicated in lung and liver carcinomas [647]. The detection of 8-hydroxydeoxyguanosine has provided strong support for the involvement of oxygen radicals in carcinogenesis [122,1003-1005]. In a number of cancer causing protocols it was found that the yield of 8-hydroxyguanine is increasing substantially. The exposure of human granulocytes to the tumor promoter tetradeconylphorbolacetate (TPA) resulted in strand breaks and in the accumulation of 8-hydroxyguanine in the DNA of the treated cells [1006]. Superoxide dismutase (SOD) and catalase (CAT) caused a marked decrease in the amount of 8-hydroxyguanine formed. This effect of SOD and CAT, of course, implies the intermediate formation of $O_2^{\cdot-}$ and H_2O_2, which in turn form hydroxyl radicals via the Fe-catalyzed Haber-Weiss reaction. Some other *in vitro* and *in vivo* examples are: DNA incubated with H_2O_2 and asbestos fibers caused a large increase in 8-OHdG [302]. The formation of hydroxyl radicals via asbestos-H_2O_2 [301] has already been discussed (Chapter 3). Potassium bromate ($KBrO_3$) in rats increased 8-OHdG about four-fold in kidney (the target organ) DNA [1007]. 2-nitropropane treatment of rats gave a two-fold increase of 8-OHdG in liver DNA [1008].

Increased levels of 8-OHdG were also observed in H_2O_2-treated *Salmonella typhimurium* cells [1009]. Mammalian cells treated with the carcinogen 4-nitroquinoline-1-oxide also resulted in 8-OHdG formation [1010]. The accumulation was also observed in naked DNA when exposed to carcinogens or agents that generate oxygen radicals [637,638,649-651] (see Chapter 6).

Epidemiology provides a compelling demonstration that human cancers increase with age and that cancer development is a multi-step process. Statistical analysis has shown that the risk of contracting, for example, colon cancer increases approximately as the fifth power of elapsed time (age) [398]. This implies a succesion of five distinct events. Similiar behaviour was observed with many tumors. Cancer is, therefore, one of the degenerative diseases of old age, although exogenous factors may increase (lifestyle) or decrease (low caloric intake, proper nutrition) the cancer incidence. Since reactive oxygen metabolites are formed by normal metabolism, there should be a relationship between basal metabolic rate, life span, and cancer incidence. A higher metabolic rate would increase the level of endogenous mutagens produced. The basic metabolic rate in man is much lower than in rodents, which may be an important factor in their different longevity [867,868]. It has been suggested that the higher metabolic rate in rodents (compared to humans) is responsible for the fact that over half of all chemicals are carcinogenic in rodents [868].

Ames and coworkers [395-398] studied the amounts of 8-OHdG excreted in the urine of humans, rats, and mice. They observed that mice excreted about 3.3 times as much 8-OHdG as humans (582 versus 178 residues / cell/ day). This result supports the theory that longevity is related to lower metabolic rate and caloric intake [396,867,998].

The observation of 8-OHdG in the urine does not tell us where this product comes from (nuclear or mitochondrial DNA). DNA damage was assesssed in nuclear DNA and in mitochondria exposed to oxidative insult. The amount of 8-OHdG in m-DNA was about 16 times higher than in n-DNA [666]. This may be due to several factors. The mitochondria consume about 90% of the cell's oxygen and produce, via the respiratory chain, a continuous flux of reactive oxygen species. In addition, m-DNA is not protected by a sheath of proteins (histones) and is, therefore, more susceptible to attack. The theory of aging, proposed by Harman [558], claimed that accumulated damage to nuclear DNA is responsible for aging and degenerative diseases associated with old age. However, recent work by Yakes and van Houten [559] indicates that damage to the nuclear DNA is repaired more efficiently than damage to mitochondrial DNA. This could mean that damage to m-DNA is the more important factor in aging and age-related diseases.

Another important aspect in aging is the damage to proteins. Stadtman and coworkers [804,1011] studied accumulation of oxidatively damaged proteins and inactivation of metabolic enzymes during aging. They found that during aging, many critical enzymes inactivated by mixed-function oxidation systems, accumulated as inactive forms. The accumulation of these oxidized proteins plays an important role in the disruption of normal cellular functioning and promotes early aging.

C. INFLAMMATION, CELL DIVISION, AND CANCER

A variety of epithelial malignancies have been associated with chronic inflammation for a long time by clinicians [1012]. Many malignant tumors have been described as arising after long periods of chronic inflammation. Bowel cancer after ulcerative colitis or Crohn's disease [1013], bladder cancer after schistosomiasis [1014], and gastric cancer after atrophic gastritis [1015]. In controlled animal experiments it was demonstrated that inflammation contributed to carcinogenesis in presence of a known carcinogenic agent. Tumors induced in rodents by a systemically given carcinogen appear preferentially at sites of wounding and inflammation [1016-1018].

Tumors induced by the Rous sarcoma virus appears preferentially at the site of injury or inflammation [1018]. Inflammation increased tumor formation in the colon, bladder, and skin [1019-1022]. Oxygen radicals produced by chemical carcinogens have been related to environmental carcinogenesis [1004].

Stimulated phagocytes produce abundant amounts of O_2^{-}/HO_2·, H_2O_2, ·OH, $^1O_2$, ·NO, NO_2, N_2O_3, and ^-OONO. All of these reactive metabolites can cause damage to important biomolecules including DNA and produce 8-OHdG, which is highly mutagenic [647]. This observation provides us, therefore, with a rationale for, why inflammation and carcinogenesis might be related. Activated phagocytes cause mutations, DNA strand breaks, malignant transformations and activation of xenobiotics occurs. Human neutrophils and macrophages produce mutations in bacteria and mamallian cells [108,1023-1025]. Neutrophils from patients with chronic granulomatous disease (which do not produce reactive oxygen metabolites) are not mutagenic [1023], thus clearly implicating these reactive oxygen metabolites as the mutagenic agent.

It is well known that many chemical carcinogens are not carcinogens per se, but have to be activated via mixed function oxidases (cytochrome P450) to reactive intermediates, which, in turn, react with DNA. A most famous example of this type of activation is benzo(a)pyrene , which is converted to the 7,8-dihydroxy-7,8-dihydro-9,10-epoxybenzo[a]pyrene. It has been shown that activated phagocytes can replace the mixed function oxidase in activating this carcinogen [1026,1027]. .

The involvement of the hydroxyl radical in these DNA base modifications is quite straightforward in solution (Chapter 6). In intact cells, however, we have a different situation. The question arises: how does the highly reactive hydroxyl radical reach the DNA ? The DNA is well protected by a nuclear membrane and a sheath of histones. Among the oxygen metabolites, only H_2O_2 can penetrate membranes and the formation of hydroxyl radical must occur in close proximity to DNA involving one-electron transfer from a DNA associated metal ion. I have already discussed this site-specific formation of hydroxyl radicals (Chapter 3). However, even if H_2O_2 is converted to hydroxyl extracellularly, it still can cause damage to DNA by oxidation of membrane lipids leading to toxic products (aldehydes), which I have already discussed (Chapter 6).

The oxidants produced by activated phagocytes are signals for mitogenesis (promotion of wound healing) [1028]. Cell division is critical for mutagenesis and carcinogenesis [868,1029,1030]. During cell division, single stranded DNA is without base pairing or histones and is therefore more sensitive to damage than double stranded DNA. Therefore endogenous or exogenous damage is increased if cells are proliferating. Some of this damage is repaired. However, the amount of unrepairable DNA damage is a function of the level of cell division, because division must occur for any DNA error to be propagated. Non-dividing cells, such as adult nerve cells never develop tumors.

Epidemiology clearly shows that increased cell division induced by external or internal stimuli is a common factor in the pathogenesis of many human tumors. Agents that can lead to increased cell division include a wide variety of chemical, physical, and infectious agents. Increased cell division can be caused by internal stimuli, like hormones (estrogen, testosterone) [1029] or external stimuli like drugs, chemicals, bacteria, viruses, such as hepatitis B and C in liver cancer [1031] or heliobacter pylori in stomach cancer [1032,1033], physical and mechanical trauma and chronic inflammation.

Infectious agents and prolonged irritation by chemical and physical agents cause cell death. The subsequent cell division to repair the damaged tissue increases the risk of cancer at the damaged site. Mechanical abrasions of epithelial cells initiates cell proliferation. This type of abrasion has been suggested as causing cancer of the stomach (by rough or salty foods), and gallbladder [1034] (stones damage the walls of the gallbladder). Physical trauma caused by asbestos in lung epithelial cells contributes to increased lung cancer rates [301,1026]. Asbestos may also get caught in the intestinal tract and cause chronic inflammation, thus contributing to bowel cancer.

Another factor in the etiology of bowel cancer is the formation of nitrogen oxides by activated phagocytes (Chapter 5). Intestinal bacteria produce many different amines [696], which can react with N_2O_3 to give carcinogenic N-nitrosamines (Chapter 7). Peroxynitrate activates cycloxygenase, a key enzyme in the production of inflammatory molecules, the prostaglandins [1035]. Nitric oxide produced in excess may be functioning as a inflammatory mediator [1036]. In patients with chronic inflammation of the colon this mechanism may contribute to the high incidence of colon cancer in these patients.

1. Bacterial and viral infection

The damage caused by activated phagocytes depends on the nature of the invading pathogen. The different effects of bacteria versus viruses has recently been reviewed by Maeda and coworkers [1037]. The simplest host defense response is physical containment of the pathogen in a confined area of septic foci. The containment of the pathogen is characterized by abscess or granuloma formation in the tissue. Liver infected with *S. typhimurium* showed multiple microabscesses [1038,1039]. The bacteria are contained by the phagocytes in the localized septic lesions. The reactive oxygen metabolites affect the invading pathogen only in the confined area and mostly intracellularly, and therefore cause limited damage to the surrounding area. In contrast, viruses attack tissues indiscriminately, although a specific tropism is well recognized [1040]. Many types of viruses propagate and spread in the organ not only from cell to cell, but also by free diffusion, like "flying sparks". Therefore the physical containment strategy does not work well in viral infections, and the reactive oxygen metabolites produced by the phagocytes will attack both normal cells and the virus infected cells. The host response to virus infection clearly represents a double-edged sword. Since the mutagenic potential of radicals and ⁻OONO is clearly established, we have a link between viral infections and carcinogenesis [1041-1044]. I have already mentioned the relationship between hepatitis virus and liver cancer [1031]. A cause and effect relationship between *helicobacter pylori* infection and gastric cancer has recently been extensively investigated [1045-1048]. The virus-cancer link was first observed in the Rous sarcoma virus. Like all revolutionary ideas it took a long time before the virus-cancer link was accepted by the scientific community. In 1966 P. Rous received the Nobel Prize for his discovery of tumor-inducing viruses.

D. NEURODEGENERATIVE DISORDERS

1. General considerations

The CNS is particularly vulnerable to oxidative stress for several reasons. The brain consumes about 18% of total oxygen, but represents only about 2% of the total body weight [859]. This indicates increased oxygen metabolism (higher ROM levels), but at the same time the CNS is low in antioxidants (SOD, CAT, GSH-Px) and high in unsaturated fatty acids. Antioxidants like GSH are found to decrease with age in the rat brain [1049]. Since the brain has a high demand for oxygen, any interruption of the oxygen supply (trauma, atherosclerosis, stroke) has serious consequences. It was observed that hypoxia caused the release of dopamine from dopaminergic neurons [1050]. Dopamine is oxidized by monoamine oxidase (MAO) to give H_2O_2. The formation of H_2O_2 from dopamine is an important step in Parkinson's disease. Specific areas of the human brain (substantia nigra, globus pallidus) are rich in iron and the cerebrospinal fluid has a low capacity for iron complexation [1051]. Therefore, any injury to the brain by trauma or stroke will release iron ions and lead to formation of hydroxyl radicals [852-854]. The topic of neurode-generative diseases and its relation to reactive oxygen metabolites has been periodically reviewed [1051-1057].

The CNS produces ROMs via the action of enzymes, like monoamine oxidase (MAO), xanthine oxidase (XO), and NO synthases (Chapter 5). In addition, the CNS contains microglia, which are the ontogenetic and functional equivalents of mononuclear phagocytes in somatic tissue [1058,1059]. Microglia are an abundant source of ROMs in the central nervous system. Appropriately stimulated, microglia produce superoxide [1060,1061] as well as ·NO [1061,1062] and cause cell injury [1063,1064]. I have already discussed the formation of these ROMs in macrophages (Chapter 5). In stimulated, microglia we therefore have the right conditions for peroxynitrite formation (high concentrations of $O_2^{·-}$ and ·NO), which is the most destructive metabolite of ·NO (Chapter 7). Peroxynitrite reacts with any molecule it encounters (lipids, AH_2, TOH, uric acid, proteins with tyrosine or SH groups, and DNA).

Several neurodegenerative diseases like AD and PD [1065] and MS [1064] have been thought to be associated with increased microglia activity. During the last decade hundreds of reviews have been written about NO and various diseases. A whole issue of the *Annals of Medicine* has been devoted to nitric oxide and medicine [553]. Everybody is jumping on the bandwagon. The topic has even entered the realm of science-in-fiction. Carl Djerassi's recent book is entitled *NO* [1066]. Faced with this tremendous flood of information it is no wonder that ordinary mortals are stunned and confused. But do not worry! All you have to remember is the prooxidant/antioxidant balance, a few basic reactions, and all the possible sources for ROM formation, which I have discussed in previous chapters. All the rest is filling in the details.

Nitric oxide has three essential functions: 1. in the vascular system it relaxes vascular smooth muscle cells, 2. in phagocytes it mediates bactericidal activity and 3. in the nervous system it acts as a neurotransmitter. In addition to these physiological effects ·NO has many toxic effects whose chemistry I have discussed (Chapter 7). Any change in the prooxidant/antioxidant status of these systems will lead to pathological conditions.

The neurodegenerative diseases are excellent examples of the disturbance of the prooxidant/antioxidant balance. Cases of over- or underproduction of antioxidant-regulatory enzymes are well documented in ALS, AD, DS, PD, and MS. In most cases the etiology of these diseases is unknown. However reactive oxygen metabolites are involved in disease progression and make the condition worse. Strong support for ROMs comes from results with antioxidants, which ameliorate the symptoms. It is interesting to note that the antioxidant defenses remove hydrogen peroxide to produce the non-toxic H_2O molecule, whereas the deactivation of $\cdot NO$ leads to more reactive and toxic products (NO_2, N_2O_3, ^-OONO). Since $\cdot NO$ is an intercellular messenger, the neurodegenerative diseases are expected to involve $\cdot NO$ or its downstream metabolites NO_2, N_2O_3, and ^-OONO.

In the sequence $O_2 \rightarrow O_2\cdot^- \rightarrow H_2O_2 \rightarrow \cdot OH \rightarrow H_2O$ we have several enzymes involved in the deactivation and regulation of the intermediates (SOD,CAT,GSH-Px). No such deactivation enzymes are involved in the deactivation of $\cdot NO$. Nitric oxide is of low reactivity and exerts its damaging effect via its metabolites. The same can be said of $O_2\cdot^-$ and H_2O_2, whose damaging effects are mainly mediated by $\cdot OH$ or $^1O_2$ (Chapters 3-6). The deactivation of $\cdot NO$ leads to more reactive species. The reaction of $\cdot NO$ with O_2 or $O_2\cdot^-$ can be considered a deactivation if we are talking about the physiological role of $\cdot NO$, but an activating reaction if we are talking about the toxic effects of nitric oxide.

In cases where we have an overproduction of SOD, the steady state concentration of $O_2\cdot^-$ is low, too low for the formation of ^-OONO. High SOD levels together with low levels of CAT and GSH-Px leads to high steady state levels of H_2O_2, which is a relatively stable and easily diffusible molecule. Since the brain is high in iron concentration [1051], we have the perfect prescription for the formation of $\cdot OH$ by the Fenton reaction. High H_2O_2 concentrations also deactivate CuZnSOD by release of free Cu^{2+} and subsequent formation of hydroxyl radicals (Chapters 3 and 8).

The H_2O_2 only deactivates CuZnSOD and FeSOD, but not the MnSOD [283,284], because Mn(II) does not participate in Fenton-type chemistry. In the literature there are still many references to the formation of $\cdot OH$ from H_2O_2 involving Mn. No one ever specifies Mn-ion or gives any references [1056]. The only biologically relevant metal ions that produce $\cdot OH$ from H_2O_2 are Fe^{2+} and Cu^+ (Chapter 3).

I have already discussed the favorite markers for oxidative damage in Chapter 6. Some of these markers are 8-OHdG for DNA damage, 3-nitrotyrosine for protein damage, allantoin from uric acid as peroxynitrite-induced damage, pentane exhalation for lipid peroxidation, the GSH/GSSG ratio, and lipofuscin and ceroid (age pigments) as a measure of overall oxidative damage. Whenever we observe an increase in any of these biological markers we can be sure that in the pathological process under study, reactive oxygen metabolites are involved. These complex relationships in amyotrophic lateral sclerosis (ALS), Alzheimer's disease (AD), Down's syndrome (DS), Parkinson's disease (PD), and multiple sclerosis (MS) are briefly and most likely oversimplified as summarized in Fig. 1.

2. Amyotrophic lateral sclerosis

Amyotrophic lateral sclerosis, better known as Lou Gehrig's disease is a highly debilitating disease with lethal consequences. It affects adults in midlife and is characterized by degeneration of motor neurons with progressive paralysis [1067].

Recent studies have shown that only about 10% of ALS cases are inherited (familial ALS). Familial ALS is dominantly inherited, i.e., only one copy of the abnormal gene is required to cause the disease. The genetic defect causes underproduction of CuZnSOD. Genetic disorders that produce too little SOD may cause the damaging effect because the lower SOD increases the level of ROMs. However, it may also result from an altered protein (produced by the altered gene), which is toxic to the neurons [285,1068].

Experiments with transgenic mice, producing extra copies of normal human CuZnSOD or containing extra copies of the ALS gene, have given support to the latter hypothesis [286]. In one line of mice the overproduction of human CuZnSOD (with ALS mutation) a disorder closely resembling human ALS developed [286]. It was reported that some cases of inherited ALS arise because of mutations in the gene encoding the cytosolic form of CuZnSOD, which catalyzes the dismutation of $O_2^{\cdot-}$ to H_2O_2 , which is the precursor of the highly reactive ·OH radical. Familial ALS patients (heterozygous for SOD mutation) have less than 50% of normal SOD activity in their erythrocytes and brain [285]. The different SOD gene is located on chromosome 21, whereas the gene for mitochondrial SOD is located on chromosome 6 [1069].

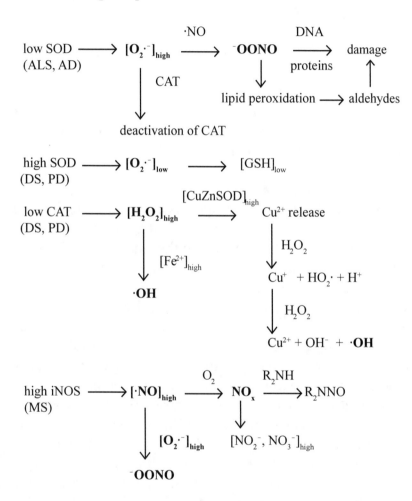

Fig. 1. Effect of over- or underexpression of some ROM-producing and regulating enzymes in some neurodegenerative diseases.

It has been suggested [281,288] that the mutant enzyme binds copper ions less tightly and they are, therefore, released from the enzyme. Copper ions can then react with H_2O_2 to produce $\cdot OH$ radicals [202]:

$$Cu^+ + H_2O_2 \longrightarrow Cu^{2+} + OH^- + \cdot OH$$

The hydroxyl radical then damages the DNA [289] as detected by the formation of 8-OHdG.

Another toxic agent is peroxynitrite. Low levels of SOD (high $[O_2^{\cdot-}]$) favor formation of peroxynitrite, which damages proteins via nitration of tyrosine residues [765], but which also reacts with almost any molecule in sight (Chapter 7).

3. Alzheimer's disease

Alzheimer's disease is an age-related disease affecting the central nervous system. It has been estimated that if the average life expectancy reaches 100, about one-third of the population will be affected by Alzheimer's disease (AD) [1070]. In addition to aging, AD has a genetic factor.

Lipofuscin and ceroid accumulate in liver, heart, and brain as aging progresses and these pigments also increase in Alzheimer's dementia [1071]. I have already discussed the formation of lipofuscin as a marker for aging (physiological time) and oxidative damage. Studies with cultured human glial cells showed that lipofuscin formation is accelerated by iron ions, but slowed down by antioxidants, like Vitamin E, GSH, and Se [989,1053]. Vitamin E may not offer protection against lung cancer [971] (Chapter 8), but it may slow the aging process and Alzheimer's disease. It is interesting to note that Se, which has been shown to substantially lower the risk for prostate cancer [955] another age-related disease (Chapter 8) may also be effective against AD.

Some studies have suggested that AD represents an acceleration of the aging process [1072]. However, other investigators [1073] concluded that the neurodegenerative processes associated with AD are different from those of normal aging. The gene defect in familial Alzheimer's disease is located on chromosome 21, where the CuZnSOD gene is located [1074]. In addition, chromosome 21 encodes for amyloid protein. A decrease in CuZnSODmRNA in AD patients was observed [1075]. This decrease in turn leads to low levels of CuZnSOD and high steady state concentrations of $O_2 \cdot^-$. High concentrations of $O_2 \cdot^-$ favor combination with $\cdot NO$ to form the highly destructive ^-OONO.

A neuropathological marker for Alzheimer's is the deposition of extracellular aggregates of β-amyloid peptides. It has been observed that β-amyloids are toxic to neurons *in vitro*, and it was therefore suggested that neuronal degeneration in AD is due to these β-amyloids [1076,1077]. Several observations led to the hypothesis that oxidative stress is involved in the toxicity of β-amyloids. After exposure of cells to β-amyloids, an increase in H_2O_2 was detected and vitamin E and CAT protected the cells [1078,1079]. NADPH oxidase (produces $O_2 \cdot^-$) inhibitors were also protective. Another effect of β-amyloids is the release of NO from microglial cells [1080], which are present in many neurodegenerative disorders. The β-amyloids were also found to break apart into smaller peptide radicals in aqueous solution [1081], and may therefore be toxic by themselves via inactivation of some enzymes (glutamine synthetase, creatine kinase).

Other indications for ROM-induced damage comes from observations of increased protein [1082] and lipid peroxidation [1083,1084]. Jeandel and coworkers [1072] studied lipid peroxidation and antioxidant defenses in the serum of AD patients and in controls. They found a significant decrease in serum levels of GSH-Px, Vitamins A, C, and E, zinc ions, transferrin, and albumin (metal complexing proteins) in AD patients. The 8-OHdG, another marker for oxidative damage (Chapter 6), has been found to be three-fold higher in mitochondria of AD patients compared to controls [1085].

Another marker for AD are the neurofibrillatory tangles (NFTs). The presence of 3-nitrotyrosine in NFTs of Alzheimer's patients has been reported [1086]. It is indeed amazing how difficult it is to erase an erroneous hypothesis from the scientific subconscious mind. The hypothesis that peroxynitrite decomposes to give hydroxyl radical has had its supporters and detractors (Chapter 7), although at present it appears highly unlikely that free hydroxyl radicals are formed. Even in the latest reviews on ROMs and disease one still finds the assertion that peroxynitrite decomposes to give the highly destructive hydroxyl radical [1056,1086,1087]. The latest results by Koppenol and Kissner [775] and by Douki and Cadet [223] show that the decomposition of peroxynitrous acid does not produce $\cdot OH$.

Several neurodegenerative diseases like AD, PD [1065,1088] and MS [1064] have been thought to be associated with increased microglia activity. Therefore, an increase in [$\cdot NO$] could possibly arise via the iNOS in microglia [1062,1063]. This increase in [$\cdot NO$] should lead to an increase in $NO_2^- + NO_3^-$. However, instead of an increase, a decrease in NO_3^- in CSF was observed in patients with PD, AD, and MSA [1089].

The iNOS is present in macrophages, neutrophils, and microglial cells and is expressed in these phagocytic cells by a variety of stimuli, like bacteria, endotoxins, and cytokines. However NOS depends for NO synthesis on L-arginine and tetrahydrobiopterin (BH_4) (Chapter 5). A deficiency in BH_4 was observed in AD [1090], in PD [1091] and in multiple system atrophy [1089]. In these cases hydrogen peroxide and oxygen radicals can be synthesized by NOS [1092]. The BH_4 -deficiency leads to low [$\cdot NO$] and low levels of NO_3^- in CSF of AD, PD, and MSA patients. Since in AD patients the antioxidant defenses (CAT, GSH-Px, GSH, Vitamins C and E, uric acid, selenium) are low, the reactive species ^-OONO and $\cdot OH$ are not scavenged, but react with lipids, proteins, and DNA and cause extensive damage. The formation and reactions of ROMs in AD are summarized in Fig. 2.

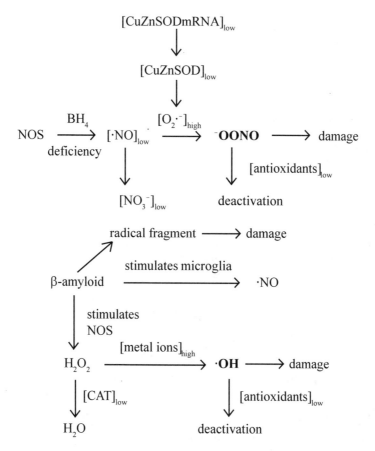

Fig. 2. Formation and reactions of ROMs in Alzheimer's patients.

AD like ALS is an underproducer of SOD leading to high $[O_2^{\cdot-}]$. The complexities involved in radical-radical combinations ($\cdot$NO + $O_2^{\cdot-}$) have been discussed (Chapter 7). Both low [$\cdot$NO] and low $[O_2^{\cdot-}]$ do not give $^-$OONO, but low [$\cdot$NO] and high $[O_2^{\cdot-}]$ does. This latter result is due to the fact that the competing reaction, the oxidation of $\cdot$NO by oxygen is second order in [$\cdot$NO] and is therefore unfavorable at low [$\cdot$NO].

In AD the damaging agents are $^-$OONO and $\cdot$OH. Since uric acid is an effective scavenger of both, hyperuricemia and AD should be mutually exclusive. In multiple sclerosis such a relationship was indeed observed [861].

4. Down's syndrome

The pathological symptoms of DS are mental retardation and ac-celerated aging. DS is characterized by an extra chromosome 21 [1093]. The assignment of the CuZnSOD gene to chromosome 21 and the detection of increased expression of SOD activity in red blood cells of DS patients, provided the rationale for the hypothesis that oxygen radicals are involved in the progression of the disease. In Down's syndrome we have an overexpression of the CuZnSOD gene. The overproduction lowers the steady state of $O_2 \cdot^-$ and leads to an increase in H_2O_2. SOD, CAT, and GSH-Px are not only protective enzymes but also regulatory enzymes. Increased damage by excess SOD (low $O_2 \cdot^-$) in SOD-rich *E. coli* mutants has become known as the SOD paradox, which I have discussed in Chapter 8. McCord and coworkers [838] have shown that $O_2 \cdot^-$ is not only an initiator of lipid peroxidation, but may also act as a chain terminator. I have also dis-cussed the results of Liochev and Fridovich [836], who showed that the superoxide radical induces the formation of some enzyme, the glucose-6-phosphate dehydrogenase. This enzyme is required for the action of glutathione reductase (GSSG $\rightarrow$ 2GSH). A decrease in this enzymatic activity, therefore, lowers the antioxidant defenses. Al-though too much SOD lowers the prooxidant $O_2 \cdot^-$, it also lowers the level of the antioxidant defenses. High levels of SOD combined with low levels of CAT leads to accumulation of H_2O_2, which in turn de-activates CuZnSOD via release of free Cu^{2+} and eventual formation of $\cdot OH$. Therefore it has been suggested that antioxidants may slow the progress of the disease [1094]. The complex interrelationship between $O_2 \cdot^-$, SOD, H_2O_2 and CAT has been discussed in Chapter 8 and is summarized in Fig. 2 of that chapter.

5. Multiple sclerosis

Chronic inflammation is a major factor in disease progression not only in the CNS, but also in other organs like the gastrointestinal tract, liver, and lung. As I discussed in Chapter 5 (phagocytosis) phago-cytes are the most abundant source of $O_2 \cdot^-$, H_2O_2, $^1O_2$, $\cdot OH$, HOCl,

·NO, and ⁻OONO. The phagocytes are stimulated by bacteria, viruses, endotoxins, and cytokines. Therefore, whenever we have inflammation we have a change in the prooxidant/antioxidant balance with damaging consequences.

Chronic inflammation is involved in bowel cancer as well as in multiple sclerosis. Microglial cells, which are the macrophages of the CNS, are activated in MS lesions [1064,1095]. Activated microglial cells produce more NO. Metabolic balance studies have shown excess formation of NO_2^- and NO_3^- in patients with fever long before the importance of ·NO in inflammation was recognized. Cytokines are associated with occurrence of disease and have been shown to inhibit or kill oligodendrocytes *in vitro* [1064]. CNS inflammation is observed in rats with experimental allergic encephalomyelitis (EAE) and in humans with MS (increased formation of NO_2^-, NO_3^- in CSF and serum). EAE in rodents has been used as an animal model for human MS. Direct evidence for ·NO in EAE in mice has been obtained. EPR spectroscopy showed the ·NO-radical complex to iron-sulfur protein [1096]. The reaction with Fe-S centers has been proposed by Ignarro and coworkers [1097] as one pathway for neuronal damage in MS. MS patients are overproducers of ·NO synthesized via the inducible NO synthase (iNOS) in microglial cells [1098]. Since the deactivation products, of ·NO are NO_2^- and NO_3^-, it is not surprising that MS patients have higher concentrations of these anions in the cerebrospinal fluid and in serum [1099]. Cerebrospinal fluid of MS patients stimulates microglial cells to produce more ·NO [1100,1101]. The ·NO production correlated positively with lactate dehydrogenase (LDH) release. The LDH release is a measure for neuronal damage.

Risk factors for MS include genetic susceptibility [1102,1103], higher concentrations of nerve growth factor in certain tissues during development [1104], and environmental factors, such as viral infections. Tissue extracts from MS brains contained increased levels of iNOSmRNA compared to normal subjects [1098]. MS -CNS tissues produce more NO. The NO and downstream metabolites

(NO_2^-, NO_3^-, ^-OONO) are not the cause of the disease. Nitric oxide production is a consequence of increased expression of iNOSmRNA. We have the following chain of events:

First cause (radicals)? $\longrightarrow$ .... $\longrightarrow$ increase in iNOSmRNA $\longrightarrow$

increase in iNOS $\longrightarrow$ increase in $\cdot NO$ $\longrightarrow$

increased formation of NO_2, N_2O_3, ^-OONO $\longrightarrow$

NO_2^-, NO_3^-, 3-nitrotyrosine and allantoin (final stable products).

First we observe an increase in mRNA which expresses the NO synthase, then increased formation of iNOS, which leads to increased formation of $\cdot NO$. Nitric oxide and its downstream metabolites are responsible for disease progression, but are not the cause. A knowledge of the above sequence of events shows us how we can possibly intervene therapeutically at different stages of disease progression. We can inhibit NOS (NOS inhibitors like L-arginine analogs). We can also scavenge the peroxynitrite (^-OONO). Aminoguanidine, an inhibitor of inducible NOS was indeed found to ameleriorate experimental autoimmmune encephalomyelitis (EAE) in mice [1105].

In MS the symptoms are demyelination of the myelin sheath of neurons. This demyelination is caused by ^-OONO. The presence of 3-nitrotyrosine residues, a biological marker for ^-OONO, as discussed in Chapter 6, was observed in higher concentration in brain tissues of MS patients [1098]. We should therefore be able to ameliorate MS symptoms by scavenging ^-OONO. I have already discussed the role of uric acid as such a scavenger in Chapter 8. Experiments with experimental allergic encephalomyelitis (EAE) in mice, which is the animal model for MS, clearly demonstrated the protective role of uric acid [861].

Patients with MS had significantly lower levels of serum uric acid than controls. Statistical evaluation of over 20 million patient records for the incidence of MS and gout (hyperuricemia) revealed that the two diseases are almost mutually exclusive, supporting the peroxynitrite scavenging by uric acid [861]. The excess formation of $^-$OONO also implies increased lipid peroxidation (Chapter 7) as evidenced by increased exhalation of ethane and pentane by MS patients [1106].

Another interesting aspect of MS is the increased deposition of homosiderin (an iron-complexing protein) in brain of MS patients [1107]. Homosiderin is believed to be a large aggregate of ferritin molecules with a much higher content of iron. The involvement of homosiderin in the formation of hydroxyl radicals is another possibility for oxidative stress.

6. Parkinson's disease

Dopamine, a metabolite of tyrosine, is a dihydroxy-aromatic compound, that has long been recognized as a source of ROMs. Dopamine, like other polyhydroxy-aromatics is easily autoxidized non-enzymatically to yield superoxide radical anion (Chapter 3). In dopaminergic neurons, dopamine is stored in vescicles where it is non-reactive. However, when dopamine is released into the cytosol it is metabolized in the mitochondria by monoamine oxidase type B enzyme (MAO-B) to yield H_2O_2 [1056]:

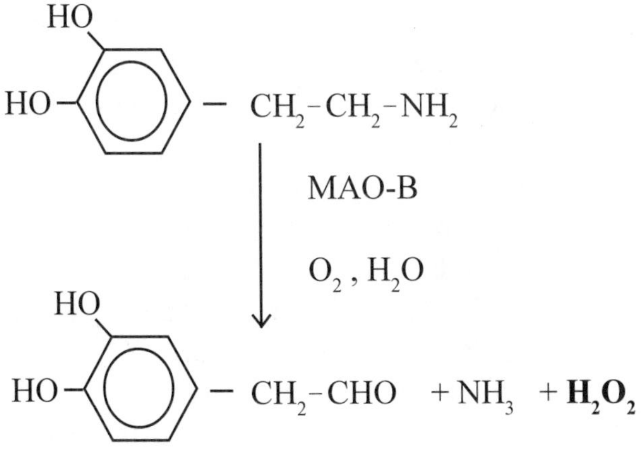

332

It has been observed that hypoxia induces the release of dopamine from the storage vesicles and thus can cause PD [1050]. Dopamine depletion protected striatal neurons from ischemia-induced cell death [1108]. The H_2O_2 formed in this reaction may react with Fe^{2+} or Cu^+ or with CAT:

$$H_2O \overset{CAT}{\longleftarrow} H_2O_2 \overset{Fe^{2+} \text{ or } Cu^+}{\longrightarrow} \cdot OH$$

We have a competition between CAT and metal ions for H_2O_2. Since in Parkinson's disease CAT levels are low and iron levels are high, the winner in this competition is obvious. Dopamine or L-DOPA the metabolic precursor of dopamine are neurotoxic via the above sequence of reactions.

The development of PD involves several factors: a genetic component, aging, and environmental toxins [1109]. Biochemically, Parkinson's disease is characterized by the following abnormalities: An increase in CuZnSOD, which is preferentially expressed in the neuromelanin-containing neurons of substantia nigra [1110-1112], leads to low $[O_2\cdot^-]_{ss}$. This subset of neurons is known to be particularly vulnerable to degeneration in PD [1113]. The observation that an increase in SOD (an antioxidant) and a decrease in $[O_2\cdot^-]$ (a prooxidant) causes oxidative damage has become known as the SOD paradox and has been discussed in Chapter 8. SOD also fulfills a regulatory function and low $[O_2\cdot^-]$ leads to lower antioxidant defenses like GSH. In PD patients, a decrease in GSH was indeed observed [1114].

In addition to an increase in CuZnSOD, decreased levels of CAT and GSH-Px have been reported [1051,1115-1118]. High SOD combined with low CAT and GSH-Px, of course, yield high steady state concentrations of H_2O_2, the precursor of $\cdot OH$. PD patients have low levels of CAT, GSH-Px and low levels of GSH. GSH in the substantia nigra was lower by 40% compared to controls (lower GSH/GSSG ratios) [1117,1119]. I have already discussed the neurotoxicity of GSSG (Chapter 6).

An increased dopamine turnover results in increased concentrations of H_2O_2. The dopaminergic neurons contain neuromelanin, a complex mixture of lipofuscin and polymer derived from oxidation of catecholamines [1120,1121]. This so called "waste pigment" is derived from the oxidation of dopamine and other catecholamines [1120] and is a good complexing agent for metal ions. In PD we have increased dopamine turnover (high $[H_2O_2]$) together with high [Fe-ions and Cu-ions] in the waste pigment of substantia nigra. This is the perfect prescription for hydroxyl radical formation. Since the ·OH radical is highly reactive and reacts in close proximity of its site of formation (site-specific), the damage occurs preferentially in these neuromelanin-containing neurons of substantia nigra. Due to the low levels of antioxidant defenses (CAT, GSH-Px) the H_2O_2 accumulates and reacts with the excess of iron ions (40% higher in PD patients) [1122,1123] to give ·OH, which is the ultimate damaging radical. A change of the Fe(II)/Fe(III) ratio from 2:1 in normal subjects to 1:2 in PD patients [1122-1124] lends support to the oxidation of Fe(II) by H_2O_2 (Fenton reaction). Additional support for ROM-induced damage in PD comes from a decrease in polyunsaturated fatty acids (PUFAs) and an increase in lipid peroxides [1125].

Nitric oxide does not play an important role in Parkinson's disease. The plasma levels of NO_3^- did not differ significantly between PD patients and controls [1126]. Another group of investigators found a decrease in NO_3^- in CSF of PD, AD, and MSA patients [1089]. The formation of low levels of ·NO is consistent with the observed BF_4-deficiency in PD [1091]. However, other investigators have suggested nitric oxide involvement [1127].

Summary. Overproduction of CuZnSOD and underproduction of CAT and GSH-Px leads to low $[O_2^-]$ and high $[H_2O_2]$. Low $[O_2^-]$ also leads to low [GSH]. Since PD patients have high concentrations of iron ions in the neuromelanin-containing neurons of substantia nigra, the high levels of H_2O_2 lead to hydroxyl radicals. The hydroxyl radical may be formed either by reactions with already available Fe(II) or Cu(I) or by H_2O_2- induced damage to CuZnSOD (Chapter 8). PD patients also have low levels of ·NO, due to a deficiency in BH_4. These reactions are summarized in Fig. 3.

Based on this mechanism of action, we can envision several possibilities for intervention. These studies have been recently reviewed [1128]. We can inhibit MAO to decrease the concentration of H_2O_2 or we can scavenge the ·OH radical. The first strategy is more effective and the MAO inhibitor selegiline has been used for many years and has been very effective in ameliorating the symptoms of Parkinson's disease. However, in order to scavenge ·OH the scavenger has to be located close to the site of ·OH formation. If we could devise a scavenger that complexes to CuZnSOD we might have an effective scavenger. α-Tocopherol has, not surprisingly, been shown to be ineffective [1128].

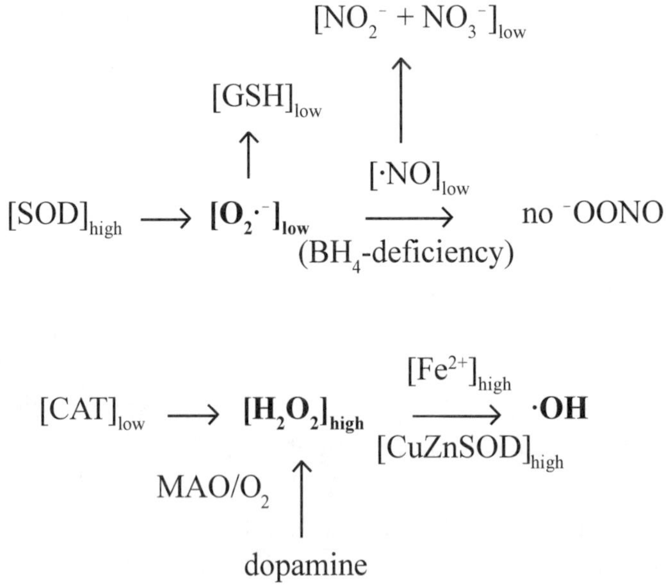

$$[NO_2^- + NO_3^-]_{low}$$

$$[GSH]_{low} \qquad \uparrow$$

$$\uparrow \qquad [·NO]_{low}$$

$$[SOD]_{high} \longrightarrow [O_2^{·-}]_{low} \xrightarrow{\quad\quad} \text{no } ^-OONO$$
$$(BH_4\text{-deficiency})$$

$$[Fe^{2+}]_{high}$$
$$[CAT]_{low} \longrightarrow [H_2O_2]_{high} \xrightarrow{\quad\quad} ·OH$$
$$[CuZnSOD]_{high}$$

$$MAO/O_2 \uparrow$$

$$\text{dopamine}$$

Fig. 3. Formation of reactive oxygen metabolites in PD.

E. LUNG DISEASES

1. Introduction

Since the main entry of oxygen into the body is through the lungs, the relationship between oxygen consumption, lung diseases, and the antioxidant defenses has been studied extensively and many reviews are available [307,821,952,1129-1131].

The lung is exposed to bacteria, viruses, inorganic particles (asbestos, silica), toxins like paraquat, CCl_4, bleomycin, anthracyclin, and toxins that are part of air pollution, like NO_2, N_2O_3, O_3, and airborne particulates (B(a)P). The response of the lungs to these toxins has been reviewed [1129]. In addition, lungs of active and passive smokers are exposed to numerous chemicals including the carcinogenic polycyclic aromatic hydrocarbons. Tobacco smoke contains aldehydes, epoxides, peroxides, nitric oxides, which all have long lifetimes to reach the alveolar space and cause serious damage [1132,1133]. The numerous reactions of these compounds with many biomolecules has been discussed in previous chapters. Oxidants in tobacco smoke also deplete antioxidant defenses *in vitro* [1134]. Even if certain pollutants in air are present in low concentrations, due to the huge volume of air passing through the lung and the large surface area , the lung is undergoing considerable oxidative stress. Due to this exposure to oxidative stress, the lungs have developed a sophisticated armamentarium of antioxidant defenses [952].

As I have pointed out before, the antioxidant defenses are not evenly distributed throughout the body and they are compartmentalized (we are not dealing with a test tube situation).The antioxidants are located where they are needed most. Normal alveolar epithelial lung fluid contains high levels of glutathione [1135,1136]. The different types of antioxidants and their reactions with ROMs have been discussed in Chapter 8. In defense against oxidative stress, the lung uses the whole gamut of available defenses. These defenses include enzymes (SOD,CAT,GSH-Px), high molecular weight proteins (metal complexing agents), uric acid, taurine [1137], and antioxidant supplied in the diet (Vitamins A,C, E, and Se).

Another aspect of antioxidant defense, which I have not yet discussed, is the recruitment of cells, that can act as antioxidants. Some of these cells in the lung are erythrocytes, platelets, and phagocytes (monocytes, macrophages, and neutrophils). How do these cells exert their antioxidant effect? Erythrocytes are rich in SOD, CAT, and compounds of the GSH-redox cycle [1138,1139]. In conditions of high oxidative stress (when ROMs overwhelm the antioxidant defenses of pulmonary resident cells, the lung can recruit additional antioxidant defenses from circulating blood cells (erythrocytes and platelets). Erythrocytes also have anion channels, which permit the entry of $O_2 \cdot^-$ into the cells where it can be detoxified via SOD and CAT to H_2O, thus preventing its conversion to the more toxic $\cdot OH$ and HOCl [976]. Erythrocytes exposed to oxidizing agents preferentially adhere to endothelial cells, most likely through a change in the erythrocyte membrane [1140]. This greater adherence to ROM-producing cells allows the erythrocytes to act as a sink for $O_2 \cdot^-$ and H_2O_2 [976].

I have already discussed the SOD paradox (Chapter 8). Liochev and Fridovich [836] have shown that $O_2 \cdot^-$ induces the formation of antioxidant defenses. Reactive oxygen metabolites induce in erythrocytes the formation of more GSH and CAT [1141]. It was observed that erythrocytes of smokers contained more GSH and CAT than those of non-smokers. The erythrocytes of smokers protected endo-thelial cells in culture from H_2O_2 more effectively than erythrocytes from non-smokers [1141]. This example shows that the lungs are not helpless against the constant onslaught by ROMs, and that an occasional cigar won't kill us!

Erythrocytes also protect ischemic lung from injury during reperfusion [1142]. Erythrocytes are not only carriers of oxygen, but also fulfill important defensive functions during oxidative stress. However, erythrocytes can also have a prooxidant effect. ROMs cause lysis of erythrocytes [1143], thus releasing not only their antioxidant enzymes into the microenvironment, but also hemoglobin-bound iron, which may serve as a catalyst for $\cdot OH$ formation.

Platelets are also rich in CAT and compounds of the GSH redox cycle [1144]. Platelets have several times more GSH than erythrocytes [1145]. Hydrogen peroxide causes aggregation of platelets and may serve, therefore, as a signal to recruit platelets to the site of high oxidative stress [1144]. The platelets become closely associated with alveolar neutrophils on the endothelial cell surface [1146]. This close association allows platelets to act as a protector against ROMs generated by the neutrophils. In addition to *in vitro* evidence for a protective role of platelets, there is also evidence *in vivo*. It has been shown that in rats, lung toxicity induced by α-naphthylthiourea is potentiated in absence of circulating platelets [1147]. Platelets also synthesize ·NO, which depending on the special circumstances, can react via its metabolites (NO_2, N_2O_3, ^-OONO) as an oxidizing agent or as a radical scavenger (Chapter 7). The radical scavenging ability may contribute to the antioxidant effect of platelets.

Another mechanism by which platelets may exert an antioxidant effect is via modulation of ROM release from activated neutrophils. It was observed that platelet lysate inhibited $O_2\cdot^-$ release from activated neutrophils [1148]. However, other results showed that platelets and platelet lysate increased $O_2\cdot^-$ release from activated neutrophils [1149]. These conflicting results may indicate that platelets have different functions depending on slight differences in conditions and their functional state.

In addition to erythrocytes and platelets, it has been suggested that phagocytes contribute to pulmonary antioxidant defense [949]. The possibility that phagocytes, the most abundant source of ROMs (Chapter 5) can act as an antioxidant seems strange indeed, but truth sometimes is stranger than fiction. Phagocytes contain considerable amounts of antioxidants in order to protect themselves against their own ROMs. Phagocytes have indeed been shown to decrease H_2O_2 *in vitro* [1150]. Phagocytes, attracted to areas of high oxidative stress (chemotaxis), may lyse and loose their ability to form ROMs while releasing their antioxidants into the microenvironment.

Platelets and phagocytes may have both a prooxidant and a anti-oxidant role. These cells may have the capacity to either promote or attenuate lung injury depending on their functional state. Remember: small changes in a complex dynamic system can have dramatic consequences. We have a pro and con situation, which we have encountered on previous occasions: the ascorbic acid problem, the SOD paradox, exercise (too much or too little), oxygen pressure (hypoxia or hyperoxia), erythrocytes, and platelets.

The lungs are the main port of entry for bacteria and viruses. Bacteria and viruses are attacked by phagocytes, which produce copious amounts of ROMs (Chapter 5). While in bacterial infection the damage inflicted on the surrounding tissue is minimal, it is quite substantial in viral infection (see previous section in this chapter). Viruses leave many battle scars, which affect the viability of the lung. On the other hand, inorganic particles, like asbestos fibers or silica dust undergo failed phagocytosis. In these cases the particles cannot be destroyed and the phagocytes release ROMs into the microenvironment, providing a continuous source of damaging ROMs. Asbestos fibers can be considered as acting like an implanted radiation source, producing hydroxyl radicals over the lifetime of the organism.

Most lung injuries are caused by external factors, such as smoking, air pollution (nitrogen and sulfur oxides, ozone, polycyclic aromatics, metal dusts, asbestos, silicate), bacteria, and viruses. I am now going to discuss some examples of pathological effects of ROMs on the lung.

2. Hyperoxia

Since oxygen is metabolized to a number of highly reactive metabolites, hyperbaric oxygen causes tissue damage and alveolar dysfunction. Long before the discovery of O_2^- by Fridovich and McCord [83] the damaging effect of hyperbaric oxygen was noted. Gershman et al. [391] established a connection between hyperbaric oxygen damage and X-irradiation. Mengel et al. [1143] showed that hyperbaric oxygen caused haemolysis in mice. Gershman et al. proposed that the damage is due to increased formation of oxygen radicals at a rate in excess of the antioxidant capabilities. This proposal has been supported by numerous studies, which have shown that with increased oxygen tension the oxygen consumption increases, and the fraction of oxygen converted to ROS is also increasing [1151-1154]. Antioxidant depletion decreases survival of animals exposed to hyperoxia [1155,1156]. However sublethal hyperoxia induces antioxidant defenses and increases survival [1157-1159]. Remember: small amounts of $O_2 \cdot^-$ induces formation of antioxidants (GSH) as discussed in chapter 8 under SOD paradox.

There is evidence that in hyperoxia-induced damage all the different sources of ROMs (Chapter 5) are involved, such as mitochondria [1151,1154], microsomes [1152], nuclear membrane [1160], xanthine oxidase [1161], and activated neutrophils [1162-1164]. The evidence for neutrophil participation in hyperoxia-induced damage rests on the following observations: 1. Shasby et al. [1164] observed a reduction of the edema by granulocyte depletion, and 2. Parrish et al. [1165] found decreased injury and neutrophil accumulation in lungs of mice deficient in C5, a potent neutrophil chemoattractant, after exposure to hyperoxia. The study of the abnormal is an important tool in the study of the normal. The primary target of hyperoxia-generated ROMs are lipids [821]. This results in increased membrane permeability (Chapter 6). Additional targets are SH-containing enzymes [1166,1167] and cytoskeletal structures of endothelial cells [1168].

However, the damage caused by hyperoxia may also be put to good use. In radiation therapy, increased oxygen tension has been extensively used to increase the killing of tumor cells.

3. Smoking

There are numerous studies which show that smoking increases the risk of coronary heart disease [1169], emphysema [1170], asthma [1171], and lung cancer [1172,1173]. How does smoking bring about these changes? It has been shown by Church and Pryor [1133] that cigarette smoke contains reactive oxygen species, as well as many toxic substances like aldehydes, epoxides, endoperoxides, B(a)P, hydrogen peroxide, and nitric oxides. ROMs lead to an increase in lipid peroxidation (as was found in plasma of smokers) [1174,1175] and to an increase in urinary excretion of 8-OHdG [1176,1177]. The first demonstration that cigarette smoke causes DNA damage has been given by Nakayama et al. [1178]. These authors showed that cigarette smoke induces a considerable number of DNA single strand breaks in cultured human cells. I have already discussed the formation of 8-OHdG by ROM-producing systems (Chapter 6). Cigarette smoke was found to considerably increase the yield of 8-OHdG [123]. As I have discussed in Chapter 6, the reactive oxygen metabolites react with many biomolecules and with the dietary antioxidants, vitamins A, C and E. It is therefore not surprising that Vitamin E, which is a good protector against lipid peroxidation, was found to decrease in the alveolar fluid of smokers [1179]. However, surprisingly, the content of Vitamin C increased [1180]. These authors compared smokers with non-smokers given the same amount of Vitamin C supplementation. The smokers concentrated more Vitamin C (by a factor of two) in the alveolar fluid and in alveolar macrophages than non-smokers, possibly as a response to the increased oxidative stress. Vitamin E supplementation was found to decrease lipid peroxidation [1175]. This protective effect against lipid peroxidation, however, does not translate into a decreased risk for lung cancer in smokers. In a Finnish study on male smokers [970], Vitamin E did not prevent lung cancer and Vitamin A may even be harmful (Chapter 8).

4. Asbestosis

Due to its industrial importance the effect of asbestos on lung disease has been extensively investigated and reviewed [304]. *In vitro* exposure of red blood cells to different asbestos fibers resulted in increased lipid peroxidation (a marker for ROMs) as shown by the formation of thiobarbituric acid reactive products [1181] (Chapter 6).

When small asbestos fibers, are inhaled, the immune system reacts to deal with the foreign intruder. Macrophages try to digest the particles, but because asbestos is an inorganic fiber, this cannot be accomplished. Instead we have a process known as "failed phago-cytosis". The phagocytes produce ROMs, which are released into the microenvironment. The asbestos fibers react with H_2O_2 to give ·OH. This process has been demonstrated in cell-free solutions via DMPO spin trapping [300] and via the DMSO probe [301]. Since hydroxyl radicals are formed, it is to be expected that asbestos fibers- H_2O_2 - DNA *in vitro* cause DNA strand breaks and formation of 8-OHdG [302]. In the reaction of different asbestos fibers and other silicates with H_2O_2, a group of investigators [303] established a relationship between ·OH formation and the ability of these silicates to cause pneumoconiosis.

The reaction involves Fe- or other metal ion sites on the asbestos surface. It has been shown that pretreatment of the fibers with metal complexing agents (like deferoxamine) inhibit ·OH formation [300]. Failed phagocytosis provides a continuous source of ·OH radicals, which damage the tissues. The relationship between chronic inflam-mation (production of ROMs) and cancer has already been discussed (previous section, this chapter). It has been demonstrated that B(a)P is activated by asbestos fibers and initiates carcinogenesis [301,1026]. Asbestos fibers lodged in the airways or in the intestine provide a continuous source of radicals (like an implanted radiation source), and contribute to lung and intestinal cancers. In the cases of asbesto-sis, pneumoconiosis, and B(a)P activation we have indeed a causal relationship between formation of ·OH and the pathology.

5. Silicosis

Metal ions on the surface of the asbestos fibers react with H_2O_2 to give ·OH. Contrary to asbestos, the silicate SiO_2 has been shown to produce radicals via mechanical fracture of the crystals [1182]. This radical formation has been postulated to be the cause of silico-sis, which is common in silicate miners. It was suggested that the silicate radicals react with water to produce hydroxyl radicals:

342

$$SiO\cdot \ + \ H_2O \ \longrightarrow \ SiOH \ + \ \cdot OH$$

Although the hydroxyl radical in the above reaction was identified by DMPO spin trapping (the DMPO/·OH signal), this proposed mechanism appears to be doubtful. Due to the high dissociation energy of the O-H bond in H_2O (119 Kcal/mole) there are only few reactions that produce ·OH from H_2O: 1. radiolysis of H_2O [25], and 2. reaction of some aromatic radical cations with H_2O [44] (see Chapter 3).

6. Emphysema

Emphysema is characterized by a permanent enlargement of the air spaces distal to the terminal bronchioles accompanied by destruction of their walls. Emphysema is a consequence of cigarette smoking [1170]. I have already pointed out the many toxic substances in cigarette smoke, and it is, therefore, to be expected that these reactive species react with cellular walls. In addition to attack on the walls, ROMs in the case of emphysema, attack proteins.This protein is the alpha-1-proteinase inhibitor [1132]. It has been shown that the alpha-1-proteinase inhibitor recovered from lungs of cigarette smokers contained oxidized methionine residues and has decreased elastase inhibitory activity [1183].

7. Asthma

Asthma is a common disease in young people under 20 years of age. Statistics show a considerable increase in diagnosed asthma cases during the last decades [1171]. Like the previously discussed diseases of the lung, asthma is caused by external factors, like smoking, second-hand smoking, and air pollution. Cigarette smoke as well as polluted air contain numerous reactive compounds including a variety of nitrogen oxides [1133]. All of these species can react with lipids, proteins, and DNA as discussed in previous chapters. Epidemiological evidence has shown an increase in asthma in children of smokers [1171].

Cigarette smoke causes lipid peroxidation [1174,1175], DNA strand breaks [123,1178], and rapid oxidation of Vitamin C, and to a lesser degree, of Vitamin E. After all Vitamin C is consumed, cigarette smoke initiates lipid peroxidation. Low dietary intake of Vitamin C has been consistently associated with increased asthmatic symptoms [1184,1185]. As pointed out before, the amelioration of symptoms by antioxidants is an indication that ROMs are involved in the disease process. The Vitamin C content of plasma and blood leukocytes was examined in asthmatic and normal patients. The asthmatic patients had 35% less Vitamin C in their blood leukocytes and 50% less in their plasma [1186].

Environmental pollutants can react directly with vital targets, but they also can stimulate (like in asbestosis and silicosis) leukocytes to produce ROMs. It was proposed [1187] that asthma involves an overproduction of ROMs by leukocytes. Eosinophils, alveolar macrophages, and neutrophils from asthmatic patients produce more ROMs compared to normal patients [1188-1191]. ROMs contract airway smooth muscles, and this effect is enhanced when the epithelium is damaged or removed [1187]. Epithelial cells, of course, produce nitric oxide, a vasorelaxant. The importance of activated leukocytes in asthma is also evident from the observation that asthmatic patients exhale increased amounts of nitric oxide [1192], a product of activated leukocytes (Chapter 5).

F. ATHEROSCLEROSIS

1. Introduction

Atherosclerosis is the main cause of death in the U.S. and Western Europe. Narrowing of the arteries by atherosclerotic plaques causes myocardial infarction or stroke. The study of atherogenesis is therefore of great medical and economic importance. Research into the causes of atherosclerosis and possible strategies against it has accelerated during the last decades and numerous reviews have been written on the subject [977,1193-1202].

The classical studies of Brown and Goldstein [1193,1203] established the correlation between atherosclerosis and cholesterol metabolism and the critical importance of the low density lipoprotein (LDL) receptor in the cellular uptake of LDL. Cholesterol as a major constituent of membranes is vital for cell growth and survival [1203]. However, excessive amounts of cholesterol can be lethal, as evidenced by the cholesterol deposits in arterial walls and in macrophages.

The main morphological features of atherosclerosis are formation of the atherosclerotic plaque and accumulation of macrophages loaded with cholesterol esters (foam cells). The atherosclerotic plaque is a complex structure and has been studied in detail. Early on it was recognized that atherosclerotic plaques contain fibroblast cells, proliferating smooth muscle cells and esterified cholesterol [1204]. Later on the presence of cholesterol filled macrophages was detected [1205]. Since macrophages are the most prolific producers of ROMs, the presence of macrophages in atherosclerotic plaques immediately raises the possibility of ROM involvement in atherogenesis. In addition atherosclerotic lesions contain ceroid pigment, a complex of oxidized lipids and proteins [1206], which is an *in vivo* indicator of ROM-induced damage.

The major targets of ROMs in biological systems are lipids, proteins, and DNA (Chapter 6). In the case of atherogenesis, the initiating event is attack on the low density lipoprotein (LDL). The low density lipoprotein consists of a group of globular particles, each of which consists of a core of neutral lipids (mainly triglycerides or cholesterol esters) surrounded by phospholipids and proteins. LDL consists of 75% lipids and 25% protein (the apolipoprotein B) [1193]. The apolipoprotein B is the important site of LDL modification. The first modified LDL was the acetyl-LDL, which has an acetyl group at the ε-amino group of lysine [1207]. LDL is the most important transport protein of cholesterol. It provides the cells with the cholesterol needed for the synthesis of membranes and steroid hormones. The bulk of the cholesterol in LDL is in the form of cholesteryl ester, which, after receptor-mediated endocytosis, is hydrolyzed (in the ly-

sosomes) to free cholesterol [1195]. In addition, LDL contains varying amounts of antioxidants, like ubiquinol, tocopherols, lycopene,
etc. However, these antioxidants are present in low concentrations
and offer little protection against LDL oxidation [1201]. Full-blown
atherosclerosis can be produced in experimental animals simply by
feeding them a diet rich in cholesterol [1193].

2. The LDL-macrophage paradox

In vivo tissue macrophages take up large amounts of LDL cholesterol and are transformed into foam cells. On the other hand, *in
vitro* tissue macrophages take up native LDL at extremely slow rates
and do not accumulate excessive amounts of cholesterol even when
exposed to high concentrations of LDL over prolonged periods of
time. This paradoxical result started the search for modified LDL,
which could be taken up by macrophages at a more rapid rate. Paradoxes arise as a consequence of incomplete knowledge and their resolution advances our scientific understanding. The resolution of the
EDRF paradox (Chapter 5) started the explosion of research on nitric oxide, which was rewarded with a Nobel Prize in 1998. The resolution of the LDL-macrophage paradox by Brown and Goldstein
[1203] and Goldstein et al. [1207] led to the oxidative modification
hypothesis of atherogenesis [1196,1199,1208]. Brown and Goldstein
received the Nobel Prize in 1985 for their work on cholesterol metabolism.

The cholesteryl esters cannot cross cellular membranes. They require special receptors on the cellular surface. LDL receptors are
present on a variety of cells, like endothelial cells, smooth muscle
cells, and macrophages. The LDL receptors mediate the uptake of
LDL by cells, and thus play an important role in the control of plasma
cholesterol levels. Patients with homozygous familial hypercholesterolemia have a deficiency or total lack of LDL receptors and plasma
cholesterol therefore rises to high levels [1209].

Before LDL can be taken up by macrophages it has to be oxidatively modified. The key event in atherogenesis is oxidative modifi-

cation of LDL and endocytosis by macrophages to form foam cells
[1210]. The oxidation of LDL by the different cell types present in
atherosclerotic plaques has been examined in culture. It has been
shown that LDL is oxidatively modified by incubation with endothelial cells (EC) [1211-1214], smooth muscle cells (SMC) [1213,1215],
monocytes-macrophages (MOC/MPH) [1216-1219], and Cu^{2+} ions
[901,1214,1220-1222]. The first modified LDL that was shown to be
taken up by macrophages was acetyl-LDL [1195,1207], obtained via
a reaction of LDL with acetic anhydride:

$$R-NH_2 + (CH_3CO)_2O \longrightarrow R-NH-COCH_3$$

The acetylation occurs at the ε-amino group of lysine [1195]. The ac-
LDL was shown to be incorporated into macrophages with high efficiency, leading to cholesterol accumulation and foam cell formation.
In addition to acetylation, many other modifications are possible. The
different aldehydes formed via lipid peroxidation (Chapter 6) can
react with amino groups to form Schiff bases or react with SH containing proteins via Michael addition or crosslinking of amino groups
by MA:

$$R-NH_2 + L-CHO \longrightarrow R-N=CH-L \quad \text{Schiff base}$$

$$R-CH(OH)-CH=CH-CHO + RSH \longrightarrow$$

$$R-CH(OH)-\underset{SR}{CH}-CH_2-CHO \qquad \text{Michael addition}$$

$$R-NH_2 + O=CH-CH_2-CH=O + R'-NH_2 \longrightarrow$$

$$R-N=CH-CH_2-CH=N-R' \qquad \text{crosslinking}$$

MA (malonaldehyde) alteration of LDL leads to cholesteryl ester accumulation in human monocyte-macrophages [1223].

Cultured EC and SMC modified LDL to a form, that permitted rapid uptake by macrophages via the acetyl-LDL receptor [1211,1224,1225]. The interaction of aldehydes with lysine of the apolipoprotein B in LDL changes the electrophoretic properties of LDL and changes LDL to a form, that can be endocytosed by cells. The reason that ox-LDL is recognized by the acetyl receptor of macrophages is not due to lipid peroxidation per se, but due to derivatization of lysine amino groups of apolipoprotein B [1211].

Since oxidative modification is essential for LDL-cholesterol uptake by cells, and these modifications involve aldehydes derived from PUFA peroxidation, it follows that the oxidation of PUFAs is the first step in atherogenesis. I am now going to discuss how PUFAs in LDL are oxidized.

3. Oxidative modification of LDL

The two primary oxygen metabolites $O_2 \cdot^-$ and H_2O_2, are not very reactive. This low reactivity of $O_2 \cdot^-$ was one of the reasons why the need for SOD was so hotly debated (Chapter 3). It has been suggested that $O_2 \cdot^-$ and H_2O_2 react to give the more reactive hydroxyl radical (the Haber-Weiss reaction). This reaction is catalyzed by metal ions (Chapter 3). I have discussed the requirement for metal ions in ROM induced reactions in Chapter 3. The concentrations of these metal ions (Fe^{3+}, Cu^{2+}) in vivo has to be carefully controlled. The metal ions are complexed to a variety of proteins. In addition, the metal ions are usually present in their higher oxidation states and have to be converted to their lower oxidation state by reducing agents ($O_2 \cdot^-$, AH_2, TOH, RSH, flavonoids). Radicals like $O_2 \cdot^-$ or $\cdot NO$ are known to release free Fe ions from ferritin (Chapter 3). The increased formation of these radicals by endothelial cells could cause the release of Fe ions.

The importance of transition metal ions in the modification of LDL by cultured cells was recognized in EC [1211,1224-1226], SMC

[1227], MPH/MOC [1217,1228] by several groups. Oxidation of LDL by rabbit endothelial cells was increased in presence of metal ions [1217] and Cu^{2+} and Fe^{3+} enhanced LDL oxidation by human smooth muscle cells [1227].

Endothelial cells are known to generate superoxide radical anion [1229]. Since $O_2\cdot^-$ is the primary oxygen metabolite, it is a prime candidate for *in vivo* oxidative modification of LDL. SOD inhibition of LDL oxidation was indeed observed by several investigators [1215,1226], but not by others [1214]. Hypercholesterolemia increased endothelial superoxide anion production via the xanthine oxidase pathway [801].

However, in addition to xanthine oxidase, endothelial cells have an efficient prostacyclin synthesizing system, which is based on cyclooxygenase and lipoxygenase action. Lipoxygenase could act endogenously with cellular lipids, producing peroxyl radicals and peroxides, which are then transferred to the LDL by lipid exchange [1230]. Aspirin, an inhibitor of cyclooxygenase, had no effect on EC-induced modification of LDL [1230], indicating that cyclooxygenase is not involved, a conclusion also reached by van Hinsberg et al. [1214]. Lipid peroxides produced via lipoxygenase action on arachidonic acid or linoleic acid are essential in the Cu^{2+} -induced oxidation of LDL [1231].

Studies by van Hinsberg et al. [1214] on the Cu^{2+}-induced LDL oxidation showed similar physicochemical alterations of LDL as was found with endothelial cells, and was inhibited by SOD, indicating $O_2\cdot^-$ as an intermediate. On the other hand, these authors found no effect of SOD on the LDL modification by endothelial cells contrary to results by Steinbrecher [1226].

PUFAs are easily autoxidized in an oxygen atmosphere and accumulate lipid hydroperoxides (rancification). The amount of Cu^{2+} -catalyzed peroxidation, as determined by the TBA test, increased about 40-fold in 5-month-old erythrocyte membranes (kept at -20° C) compared to freshly prepared membranes [1232]. This clearly indicates accumulation of lipid hydroperoxides. The LDL particle contains endogenously produced trace amounts of hydroperoxides [1233]. These peroxides in presence of Cu^{2+}/Cu^{+} initiate further lipid peroxidation and provide an explanation for the observation that high plasma Cu^{2+} levels increase the risk of CAD [1234].

A detailed study of LDL modification by EC was carried out by Steinbrecher [1226]. Modification of LDL by EC required the presence of O_2 and was found to be inhibited by SOD, but not by catalase and hydroxyl radical scavengers [1226]. In addition, the results of Steinbrecher showed that the modification of LDL by different cell types correlated well with their ability to generate O_2^{-}. As I have pointed out on previous occasions, the radiolysis of water provides us with a clean and quantitative source of O_2^{-}, $HO_2\cdot$, $\cdot OH$. Radiolysis allows us to study the effect of one radical at a time. This approach was used by Bedwell et al. [1235] in a study of oxidative modification of LDL. Both $HO_2\cdot$ and $\cdot OH$ caused depletion of α-tocopherol and accumulation of hydroperoxides:

$$\cdot OH (\text{or } HO_2\cdot) + LH \longrightarrow L\cdot + H_2O \text{ (or } H_2O_2) \quad \text{initiation}$$

$$L\cdot + O_2 \longrightarrow LOO\cdot$$

$$\text{propagation}$$
$$LOO\cdot + LH \longrightarrow LOOH + L\cdot$$

O_2^{-} produced only limited oxidation, however it was found to potentiate oxidation stimulated by Cu^{2+}.

These investigations, tell us something we already know: Superoxide is not very reactive, but $HO_2\cdot$ is [1236], and O_2^{-} can reduce Cu^{2+} to Cu^{+}. The Cu^{+} in turn, can react with lipid hydroperoxides :

$$Cu^{+} + LOOH \longrightarrow Cu^{2+} + LO\cdot + OH^{-}$$

Cu^{+} can also react with O_2 to produce a reactive species (most likely $\cdot OH$) [51], which is known to hydroxylate aromatic compounds and can initiate lipid peroxidation.

Hydroxyl radical did modify LDL, but not to a form that could be endocytosed by MPH *in vitro*. The results suggest, according to the authors, that it is therefore unlikely that *in vivo* $O_2\cdot^-$ is converted to the more reactive $\cdot OH$ [1235]. However, this argument does not hold up under scrutiny. It would be indeed amazing if radiolytically produced radicals would react with LDL to cause modification, which would be recognized by the acetyl-LDL receptor of macrophages. Radiolytically produced $\cdot OH$ is distributed uniformly throughout the solution, whereas $\cdot OH$ produced *in vivo* is produced in a site-specific manner and cannot therefore be scavenged.

Samples exposed to $O_2\cdot^-$, treated with CAT (to remove any radiation produced H_2O_2) and then incubated with Cu^{2+} showed a dose-dependent oxidation of LDL [1235]. Although $O_2\cdot^-$ is not very reactive, the reaction with LDL produces small amounts of hydroperoxides (LOOH) [1235]. LOOH in presence of Cu^{2+} can initiate lipid peroxidation:

$$LOOH + Cu^{2+} \longrightarrow LOO\cdot + H^+ + Cu^+$$

$$LOOH + Cu^+ \longrightarrow LO\cdot + OH^- + Cu^{2+}$$

The reduction of Cu^{2+} to Cu^+ can also be accomplished by AH_2, α-TOH, or $O_2\cdot^-$. We have the following competing reactions:

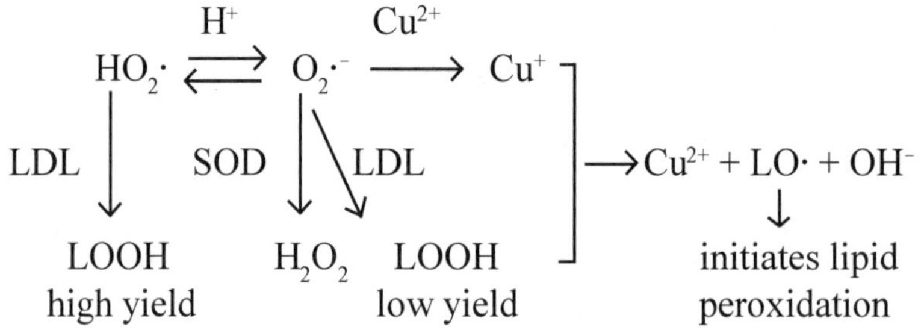

SOD protects LDL from oxidation [1215,1226], clearly indicating the involvement of $O_2 \cdot^-$ as an intermediate. The observation by Steinbrecher [1226] that oxidation of LDL by EC in culture is not affected by $\cdot$OH scavengers, does not necessarily mean that no $\cdot$OH is formed. This was indeed recognized and discussed by Steinbrecher. *In vivo* the reaction of Cu^+- H_2O_2 may lead to a site-specific formation of $\cdot$OH, which cannot be scavenged. The binding of Cu^{2+} by lysine or histidine residues in apolipoprotein B [831c] could catalyze the formation of $\cdot$OH in close proximity to the target (site-specific formation of $\cdot$OH).

In the oxidation of LDL by Cu^{2+} or Fe^{3+} in phosphate buffer, the Fe^{3+} was the far less effective catalyst [1221,1237]. Kuzuya et al. [1221] found a marked difference between Cu^{2+} and Fe^{3+} in the formation of complexes with LDL in phosphate buffers.

Oxidative modification of LDL was also accomplished with activated macrophages or neutrophils [1216-1219]. These phagocytic cells contain myeloperoxidase, an enzyme that converts H_2O_2 and Cl^- to HOCl, $^1O_2$, and $\cdot$OH (Chapter 5, phagocytosis). All of these species can initiate lipid peroxidation and it is therefore not surprising to find that MPO is involved in LDL oxidation [1238,1239]. The reaction between HOCl and $O_2 \cdot^-$ to give $\cdot$OH (the Long-Bielski reaction) is the only OH-forming reaction not requiring metal ions (see Chapters 3 and 5). The MPO-induced oxidation of LDL was indeed found to be independent of metal ions [1238]. MPO was found to be present in atherosclerotic plaques [1239]. Exposure of LDL to L-tyrosine and activated human neutrophils caused peroxidation of LDL. Incubation of LDL, L-tyrosine, MPO, and H_2O_2 caused lipid peroxidation, which was inhibited by heme poisons and by catalase. Replacement of L-tyrosine with O-methyl-tyrosine inhibited LDL oxidation [1238]. This result seems to indicate the important role of the tyrosyl radical in the initiation of LDL oxidation:

$$\text{tyr-OH} \xrightarrow[\text{H}_2\text{O}_2]{\text{MPO}} \text{tyr-O} \cdot$$

352

$$\text{tyr-O·} + \text{L-H} \longrightarrow \text{tyr-OH} + \text{L·}$$

It has therefore been suggested that the tyrosyl radical is a physiological catalyst for LDL oxidation [1238]. There is also increasing evidence that 15-lipoxygenase is involved in LDL oxidation *in vivo* [1240-1242].

Another oxidizing species is peroxynitrite. Activated macrophages and neutrophils produce $^-$OONO. Peroxynitrite oxidizes LDL [1243-1245] and also nitrates tyrosine to 3-nitrotyrosine [769], which has been detected in atherosclerotic plaques [1246].

4. Prooxidant/antioxidant balance

Atherogenesis is another textbook example of the disturbance of the prooxidant/antioxidant balance. Increased formation of prooxidants (·NO, $^-$OONO, $O_2^{·-}$/$HO_2·$, H_2O_2, ·OH, $^1O_2$, and HOCl) by a variety of cell types, EC , SMC , MOC, and MPH leads to increased oxidative modification of LDL. Ox-LDL is internalized by macrophages (located in the subendothelial space), which are converted to foam cells.

Once the LDL is minimally oxidized, it interacts with endothelial cells and smooth muscle cells to produce monocyte-chemotactic protein 1 (MCP-1) [1247], which attracts more monocytes. Minimally oxidized LDL stimulates monocyte-endothelial cell interaction [1248,1249]. The monocytes are differentiated to macrophages, which in turn produce more ROMs, which all can oxidize more LDL.

All the cells encountered in the atherosclerotic plaques produce ROMs (Chapter 5). There is substantial evidence that in atherosclerosis the formation of $O_2^{·-}$ by endothelial cells (EC) is increased [801,1226]. In the subendothelial space we have present·NO, $^-$OONO, $O_2^{·-}$/$HO_2·$ H_2O_2, ·OH, $^1O_2$, and HOCl. All of these species oxidize LDL as discussed in previous chapters.

Another way to increase oxidative stress is a decrease in antioxidants. Important naturally present antioxidants are SH compounds, like cysteine and glutathione. GSH decreases the prooxidant LOOH, but also decreases the GSH/GSSG ratio:

$$2 \, GSH \; + LOOH \; \longrightarrow \; GSSG \; + \; LOH \; + H_2O$$

Thiol compounds are important stabilizers of $\cdot NO$ via formation of nitrosothiols, thus preventing its deactivation by $O_2\cdot$ (chapter 7). The effect of a low GSH/GSSG ratio has already been discussed (Chapter 6). We should be able to counteract an increase in [prooxidant]$_{ss}$ with an increase in [antioxidant]$_{ss}$ by dietary supplementation (Vitamins A, C, E). Dietary supplementation with Vitamin A (β-carotene) *in vitro* and *in vivo* does not inhibit low-density lipoprotein (LDL) oxidation [1250-1252]. On the other hand, Vitamin E supplementation protected against LDL peroxidation *in vitro* [1251] and in human subjects [1252]. Epidemiological studies have shown that Vitamin E is useful in preventing atherosclerosis and coronary artery disease [958,959]. The important reactions in atherogenesis are briefly summarized in Fig. 4.

5. Antioxidants and atherogenesis

If the oxidation of LDL is indeed the initiating event in atherogenesis, then it should be possible to inhibit the process by antioxidants. There is no doubt that low molecular weight antioxidants (especially Vitamins C and E) protect LDL against oxidative modification*in vitro* [1197,1198,1253,1254]. However, *in vivo,* both Vitamin C and E may represent a double-edged sword, since these vitamins may act under certain conditions as prooxidants as well (Chapter 8). Although the effects and the mechanism of action of antioxidants on atherogenesis is still controversial, there is mounting evidence of the beneficial effects of antioxidant supplementation (especially Vitamin E) on cardiovascular events and mortalty in humans [958-961].

a. Effect of antioxidants in animal studies

Most extensive studies have been carried out with Watanabe heritable hyperlipidemic (WHHL) rabbits. These studies have shown a significant reduction in the extent of atherosclerosis and reduction in

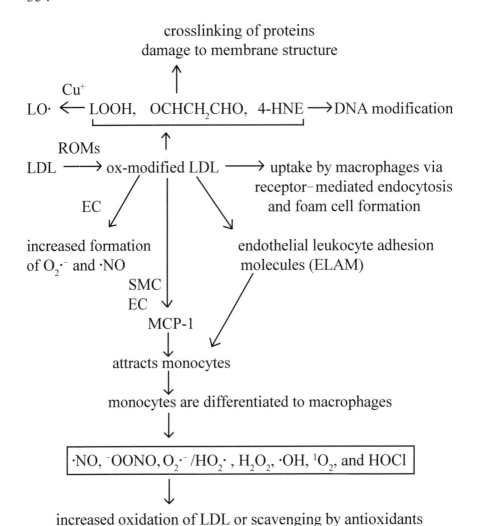

Fig. 4. Some important reactions in atherogenesis.

serum cholesterol with dietary supplementation of Vitamin E. In other animal models (rabbit, quail, and monkey) involving animals fed a high cholesterol diet, a significant inhibition of atherosclerosis with Vitamin E treatment was observed. In addition, decreased lipid peroxidation products (malonaldehyde) in blood of vitamin E treated animals was found [977,1255,1256]. In rabbits, a combination of Se and Vitamin E resulted in a reduction in atherosclerotic plaque formation. This combination of antioxidants lowers total serum lipids and increases HDL cholesterol (the good cholesterol) [1256].

An interesting study of Vitamin E was reported by Keaney et al. [1257]. These authors found that in hypercholesterolemic animals, low doses of Vitamin E restored endothelium-dependent vasorelaxation whereas high doses actually made the condition worse. We again have a pro and con situation. LDL isolated from rabbits treated with Vitamin E was found to be resistant against Cu^{2+}-induced oxidation [1258].

The most water-soluble antioxidant is Vitamin C. It is profoundly effectve in protecting LDL against oxidation [1253,1259]. Vitamin C deficient guinea pigs fed an atherogenic diet developed atherosclerosis and this effect was prevented by Vitamin C supplementation [1260]. In genetically scorbutic rabbits, Vitamin C decreased lipid peroxidation in plasma and LDL, suggesting that Vitamin C limits peroxidation *in vivo* [1261].

Another drug that has been used in the prevention of atherosclerosis is probucol, which has the following structure:

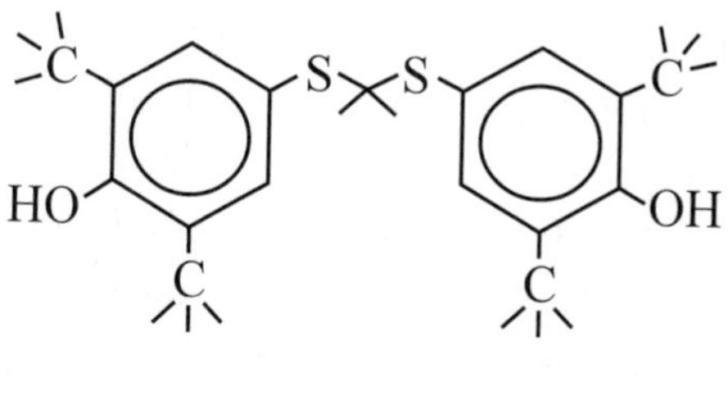

Probucol is a cholesterol-lowering drug with antioxidant properties. It is not always obvious which of these properties is causing a certain effect. The antioxidant property of probucol is quite evident to a chemist, since it has two easily abstractable H atoms (the two aromatic-OH groups). Probucol may even be an effective complexing agent for metal ions. Many investigations have shown that the beneficial effects of probucol is mainly due to its antioxidant property [977,1262].

b. Human trials with antioxidants

Large scale epidmiological studies (39, 910 men over a period of four years) [959] have shown a considerable reduction in the clinical expression of coronary artery disease (CAD) and atherosclerosis in patients treated with lipid soluble dietary antioxidants. Intake of Vitamin E was independently associated with lower risk of CAD, non-fatal myocardial infarction, coronary bypass grafting, and coronary angioplasty. In this study, carotene intake was inversely associated to risk only in smokers, and Vitamin C had no effect on CAD. In a similiar study in women Vitamin E lowered the risk of major CAD by 34% [958]. These results are in agreement with other observations, showing that plasma levels of Vitamins A and E are inversely associated with ischemic heart disease mortality [1263]. In conclusion, animal studies as well as epidemioplogical studies in humans support the hypothesis that antioxidants reduce the clinical expression of atherosclerosis and that ROMs must be involved in its progression.

Vitamin C-deficient guinea pigs fed an atherogenic diet developed atherosclerosis and this was prevented by Vitamin C supplementation [1260]. In human trials Vitamin C had no effect on the risk of CAD [959]. This result in no way contradicts the animal studies, since the guinea pigs were Vitamin C-deficient, whereas the human subjects were not. This result seems to indicate that normal levels of Vitamin C are enough to protect against LDL oxidation and that additional supplementation does no good and may actually be harmful, acting as a reductant for Cu^{2+} or Fe^{3+} (Chapter 8).

Further support for lipid peroxidation as a causative factor in atherogenesis comes from the following studies. LDL isolated from patients who have been treated for hypercholesterolemia with probucol or Vitamin E was found to be more resistant to oxidative modification induced by metal ions [1264,1265]. The same effect was observed by Williams et al. [1258] in rabbits.

When human LDL is isolated and exposed to Cu^{2+}, peroxidation occurs, first at a slow rate (lag period) and then at an accelerated rate. During the lag phase, the α-tocopherol is lost from LDL (sacrificial protection) [1266]. The concentration of α-tocopherol in the LDL of normal human subjects not consuming supplements is quite low, between 3-15 molecules /LDL particle [1201]. If LDL was isolated from subjects who had received Vitamin E supplementation, the length of the lag period increased significantly [1197,1198,1267]. In addition to α-tocopherol, LDL contains a number of other naturally occuring antioxidants, like ubiquinol, β-carotene, lycopene, and lutein. The amounts of these antioxidants depends on the diet, but they are usually present in very low concentrations (less than 1 molecule/LDL particle) [1201].

Further support for oxidation of LDL in atherogenesis comes from studies by Regnstrom et al. [1268]. These authors demonstrated a relationship between the susceptibility of LDL towards *in vitro* oxidation by Cu^{2+} and coronary atherosclerosis in men.

In Chapter 8 I have discussed another group of antioxidants, the flavonoids. Flavonoids are contained in red wine, and the positive effect of red wine on atherogenesis has been suggested [1269]. However, remember: the flavonoids, like Vitamins C and E can also act as prooxidants (Chapter 8).

In addition to Vitamin E there are other protective mechanisms involved in LDL oxidation. High density lipoprotein (HDL) has been shown to protect LDL [1270]. Epidemiology established an inverse correlation between plasma HDL levels and the risk of coronary heart disease [1271,1272]. On the other hand, elevated levels of plasma LDL correlated with increased risk of coronary heart disease [1272]. The HDL cholesterol has therefore become popularly known as the "good cholesterol" and LDL-cholesterol as the "bad cholesterol". The mechanism by which HDL protects LDL is not clear. It has been suggested by Parthasarathy et al. [1270] that the mechanism may involve exchange of lipid peroxidation products (aldehydes) between the lipoproteins.

β-carotene has been found to inhibit lipid peroxidation in several *in vitro* studies [1273]. However enrichment of LDL with β-carotene does not increase its resistance to peroxidation [1250].

6. Antioxidant/prooxidant effect of Vitamin C

As pointed out previously (Chapter 3), radicals are e^--acceptors or e^--donors. Ascorbic acid and α-tocopherol react with ·OH with very high rate constants [99]:

$$AH_2 \; + \; \cdot OH \longrightarrow AH\cdot \; + \; H_2O$$

$$\alpha\text{-}TOH \; + \; \cdot OH \longrightarrow \alpha\text{-}TO\cdot \; + \; H_2O$$

In these reactions, the ·OH acts as an oxidizing agent (an e^--or hydrogen atom acceptor) and the ascorbic acid and α-tocopherol act as reducing agents (donate an e^- or H·). If the AH_2 or TOH donate an electron or H-atom to a radical like ·OH they act as **antioxidants**. If however they donate an electron to a Cu^{2+} or Fe^{3+}, they act as **prooxidants**, because the reduced metal ions are the essential catalysts for oxygen radical formation (Chapter 3).

Oxidative modification of LDL *in vitro*, either Cu^{2+}- induced or mediated by endothelial cells in culture, is strongly inhibited by Vitamin C [1259]. However, it is well established that AH_2 in presence of metal ions (Cu^{2+} or Fe^{3+}) acts as a prooxidant, since it reduces these metal ions to their lower oxidation state. These metal ions are essential catalysts for ·OH formation (Chapter 3). The observation by Retsky et al. [1259] on the strong inhibition of Cu^{2+}- induced LDL oxidation by Vitamin C is therefore paradoxical. The study of Retsky et al. showed that AH_2 is rapidly oxidized when incubated with Cu^{2+} and LDL, leading to the transient formation of dehydroascorbic acid (A). AH_2 without oxidation to A offered no protection. When LDL is incubated with A, followed by gel filtration, LDL remains protected against subsequent Cu^{2+}- induced oxidative modification. This result suggests that LDL forms a stable modification in presence of A, which is resistant to lipid peroxidation. While A in contrast to AH_2 prevents metal ion dependent formation of ox-LDL, the situation is reversed under a metal ion-independent oxidation. It was suggested by the authors that the protection of LDL by A in Cu^{2+}- induced oxidation may be due to decreased binding of Cu^{2+} to LDL. Cu^{2+} bound to apolipoprotein B of LDL (not free Cu^{2+}) initiates lipid peroxidation in LDL [1222]. Decreased binding of Cu^{2+} to LDL may therefore result in less oxidative damage. However, other investigators [903] concluded that the Cu^{2+}- induced oxidation of LDL is independent of SH groups associated with apolipoprotein B.

7. Prooxidant effect of α-tocopherol

Recent studies [890,906] have shown that α-TOH promotes the peroxidation of low-density lipoprotein (LDL) exposed to a variety of non-enzymatic and enzymatic radical sources. α-TOH depleted samples were remarkably resistant to the different oxidizing conditions. Several studies [1274,1275] have established that enrichment of LDL with α-TOH increases the oxidizability of lipoproteins toward mild fluxes of ROO· and the transition metal ion-containing Ham's F-10 medium. How do we manipulate the levels of α-TOH in LDL?

The study of the abnormal helps us understand the normal. Such an abnormality is familial-isolated Vitamin E (FIVE) deficiency [906]. Plasma prepared from a FIVE patient before Vitamin E supplementation contained 4.4 μM α-TOH, whereas after Vitamin E supplementation the levels rose to 77.8 μM [906]. The two plasma samples were oxidized with 10 mM AAPH (a water soluble radical forming azo-compound), which decomposes to give peroxyl radicals:

$$R\text{-}N\text{=}N\text{-}R \xrightarrow{\Delta} N_2 + 2\,R\cdot \xrightarrow{2\,O_2} 2\,ROO\cdot$$

The oxidized samples showed formation of cholesteryl ester hydroperoxides (CE-OOH) in the Vitamin E supplemented, but not in the Vitamin E depleted plasma. Oxidation of LDL isolated from the FIVE patient before and after Vitamin E supplementation confirmed that α-TOH made the lipoprotein more susceptible to oxidation by aqueous and lipophilic ROO· radicals, soybean lipooxygenase (SLO) or Cu^{2+} ions [906].

Several studies [902-905,1275] have suggested that the reduction of Cu^{2+} by α-TOH to yield Cu^+ and α-TO·, is essential for the initiation of LDL lipid peroxidation in peroxide-free lipoproteins. In absence of peroxides, Cu^+ cannot initiate chain oxidation, and therefore, the α-TO· must be the chain-carrying radical. This could explain why α-TOH-depleted LDL is not oxidized even in the presence of high Cu^{2+} concentrations. LDL isolated from a FIVE patient (depleted in α-TOH) reduced Cu^{2+} at a much slower rate compared to LDL isolated from a normal donor [904].

Many studies [1197,1198,1266,1267], discussed on preceding pages have shown that Vitamin E acts as a strong antioxidant and chain terminator in LDL peroxidation. How can these results be reconciled with the prooxidant and chain transfer activities of α-TOH in LDL peroxidation?

The α-TOH-mediated peroxidation (TMP) may be presented by the following sequence of reactions [890]:

$$\alpha\text{-TOH} + \text{ROO·} \xrightarrow{k_{init}} \alpha\text{-TO·} + \text{ROOH} \qquad (1)$$

$$\alpha\text{-TO·} + \text{LH} \xrightarrow{k_{TMP}} \alpha\text{-TOH} + \text{L·} \qquad (2)$$

$$\text{L·} + \text{O}_2 \longrightarrow \text{LOO·} \qquad (3)$$

$$\text{LOO·} + \alpha\text{-TOH} \xrightarrow{k_{inh}} \text{LOOH} + \alpha\text{-TO·} \qquad (4)$$

$$\alpha\text{-TO·} + \text{ROO·} \xrightarrow{k_t} \text{non-radical products} \qquad (5)$$

The LOO· radicals react ~200 fold more likely with α-TOH than with LH [890], and thus regenerates the α-TO· radical. The sequence of Reactions (2) - (4) constitutes a chain reaction, which, however, only occurs if Reaction (2) can compete with Reaction (5). This appears at first glance unlikely, since an α-TO· radical would have to wait around for ~30 s between propagation steps (k_{TMP} [LH] 0.03 s^{-1}). The chain termination (Reaction 5) proceeds with a rate constant of 3×10^8 M^{-1}s^{-1} in a homogeneous system. However, as pointed out by Bowry and Ingold [890] an LDL particle is not a homogeneous solution. Diffusion of lipid soluble compounds between LDL particles is known to be slow [1276]. Therefore, it was suggested by Bowry and Ingold [890], that once an α-TO· radical was formed within an LDL particle, it might remain trapped within the particle and therefore be forced to "wait around" for a relatively long time (in the order of min.) for a second radical (ROO·) to strike the LDL particle. This scenario provides ample time for α-TO· to abstract a hydrogen from PUFAs within the LDL particle, especially if the radical flux of initiating ROO· radicals is low.

It has been pointed out by Stocker and coworkers [906] that the majority of previous studies on LDL oxidation, which showed an antioxidant effect of α-TOH [1198], used LDL without prior removal of lipid hydroperoxides, which may be formed during the isolation procedure. In addition, many studies exposed LDL to strongly oxidizing conditions, i.e., a Cu^{2+}/LDL particle ratio of 16 [1277]. Under these conditions the consumption of α-TOH is rapid and precedes lipid oxidation, and enrichment of the LDL with α-TOH increases the resistance towards oxidation [1277]. These findings do not contradict the prooxidant activity of α-TOH, but rather support the TMP model. Under conditions of high radical flux , the α-TO· will preferentially engage in radical-radical termination reactions (Reaction 5) rather than chain transfer (Reaction 2) The antioxidant or prooxidant activity of α-TOH towards LDL depends, in part, on the radical flux the LDL particle is exposed to [906]. An additional factor affecting the prooxidant /antioxidant activity of α-TOH toward LDL, is the reactivity of the oxidant. Highly reactive radicals, such as $\cdot$OH compared to $ROO\cdot$, require a lower radical flux to achieve a prooxidant activity [906].

A third factor that determines whether α-TOH acts as a pro-oxidant or antioxidant depends on the presence of coantioxidants. In the presence of coantioxidants, the chain transfer activity of α-TOH is prevented or attenuated through scavenging of the peroxidation chain-carrying α-TO$\cdot$:

$$\alpha\text{-TO}\cdot \;+\; \text{X-H} \;\longrightarrow\; \alpha\text{-TOH} \;+\; \text{X}\cdot$$

where X-H may be ubiquinol-10 or ascorbate [783]. I have already discussed this reaction in Chapter 8. In human plasma exposed to $ROO\cdot$, lipid peroxidation is prevented as long as the coantioxidants ubiquinol-10 or ascorbate are present, and this is independent of the concentration and/or presence of α-TOH [906]. From this model of LDL oxidation we would predict that the more α-TOH an LDL particle contains, the greater the requirement for coantioxidants to efficiently inhibit peroxidation under a given flux of radical oxidants.

However, the TMP model is difficult to reconcile with the observation of ascorbate in homogenates of advanced human atherosclerotic lesions [906,1278].

8. Impairment of EDRF action in atherosclerosis

In addition to producing fatty streak lesions, ox-LDL in the vascular wall leads to impaired EDRF action. It has been well documented that NO released from endothelium by certain stimuli causes vasorelaxation and inhibits platelet aggregation (Chapter 5). Endothelium derived relaxation is impaired in animals with atherosclerosis [1279] and in isolated atherosclerotic human arteries [1280]. In the intact human circulation, Ludmer et al. [1281] showed paradoxical vasoconstriction in response to acetylcholine in atherosclerotic coronary arteries.

Inhibition of the vasorelaxant effect means increased platelet aggregation and coronary artery disease. Platelet aggregation can also be the result of prostaglandin synthesis. Ox-LDL but not native LDL incubated with EC stimulated prostaglandin synthesis [1282].

There are several possible mechanisms for the impairment of EDRF action.

1. Impairment of the vasorelaxant effect in response to acetylcholine could be due to a decreased NO production via damage to the NO synthases. This was found not to be the case. Diet induced atherosclerosis actually increased the release of nitrogen oxides from rabbit aorta [1283].

2. Increased formation of $O_2\cdot^-$ can deactivate $\cdot$NO via combination [93,728,729] (Chapter 7). Strong evidence in favor of this hypothesis was presented by Mügge et al. [800]. These authors showed that SOD inhibited the deactivation in cholesterol-fed rabbits.

3. Increased production of ROMs leads to a decrease in thiols, which in turn lowers the half-life of $\cdot$NO. Remember: $\cdot$NO is stabilized via formation of S-nitrosothiols (Chapter 7).

4. LDL damages endothelial cells [1225,1284]. EC modifies LDL to ox-LDL, which is degraded to cytotoxic products like hydroperoxides, malonaldehyde, and 4-HNE. MA affects the structural integrity of the phospholipid bilayer of membranes [1285].

I have already discussed the complexities of NO deactivation in Chapter 7. I just like to remind the reader of the antioxidant properties of NO. Nitric oxide can scavenge other radicals, especially the lipid peroxyl radicals, which are the chain carriers in lipid peroxidation.

$$LOO\cdot \ + \ LH \ \longrightarrow \ LOOH \ + \ L\cdot \quad propagation$$

$$LOO\cdot \ + \ \cdot NO \ \longrightarrow LOONO \qquad termination \ [798]$$

Nitric oxide can therefore protect LDL from oxidative modification. A diet supplemented in L-arginine has indeed been shown to lower the risk of atherosclerosis in hypercholesterolemic rabbits [1286]. On the other hand, NO deactivation products like NO_2 or ^-OONO can initiate lipid peroxidation (Chapter 7). The answer to the question: is NO good or bad?, depends on the specific circumstances (Chapter 7).

Since the inhibition of EDRF action involves ROMs it should be possible to protect against this effect with Vitamin E supplementation. Here again we encounter a pro and con situation. Results by Keaney et al. [1257] on cholesterol-fed rabbits showed that low doses of Vitamin E improved, but high doses worsened endothelial vasodilation.

An interesting consequence of the damaged EDRF response in atherosclerosis is the observation that mental stress induces vasoconstriction [1287]. It has been shown that mental stress does not have any or only limited effect on the vasomotor response of normal epicardial coronary arteries. However, in patients with atherosclerosis, mental stress causes a pronounced and paradoxical vasoconstriction. In regions of atherosclerotic stenosis this constriction probably causes the decrease in blood flow and together with an increase in metabolic demand, leads to myocardial ischemia during mental stress [1288].

Summary. The best prescription for prevention of atherosclerosis is: low fat intake, (especially low in PUFAs), low cholesterol intake, and sufficient (whatever that means) intake of Vitamins E, C and selenium. Remember: too little or too much Vitamin E can be damaging [1257]. However as pointed out by Steinberg et al. [1208], no matter how successful, we deal with hypercholesterolemia, coronary heart disease will not disappear, because a high cholesterol level is by no means the only causative factor. In 1981 Hopkins and Williams [1272] compiled 246 suggested risk factors for coronary heart disease. Paying attention to all these risk factors would hardly leave one any time for living.

G. ISCHEMIA-REPERFUSION

1. Introduction
Ischemia causes tissue necrosis in many organs, such as heart, intestine, brain, kidney, and lung. Permanent deprivation of oxygen supply (blood flow) is lethal to any tissue or organ. Any interruption of the oxygen supply should therefore be immediately treated with reperfusion. This is especially true for the brain, which consumes about 18% of the total oxygen. Here again we have another pro and con situation. Reoxygenation is necessary, but causes additional damage. This observation has become known as the "oxygen paradox". Reperfusion causes additional tissue injury due to formation of reactive oxygen metabolites. This theory of reperfusion is based on the fact that antioxidants have been found to diminish injury in some model systems [1289-1294].

Radicals have also been demonstrated directly during reperfusion using EPR spectroscopy and the spin trapping technique. The DMPO/·OH adduct was identified in several studies [1295-1298]. Hydroxyl radicals are very reactive and may indeed kill any cell, by damaging its DNA or other important biomolecule. This, however, should not tempt us to believe that whenever cells are killed, hydroxyl radicals are involved. The high reactivity of hydroxyl radical is responsible why in many spin-trapping experiments carbon-centered rather than the originally formed hydroxyl radicals have been observed [1297-1300].

2. Pathogenesis of ischemic injury
There are many different causes for ischemia. It can result from atherosclerosis (narrowing of the arteries), thromboembolism, or external pressure on blood vessels (as in the case of tumors). Ischemia can also be iatrogenic. It can occur during surgery, when the blood flow to an organ must be interrupted. Ischemia is a lack of oxygen and substrate and therefore a lack of aerobic energy production. The ATP (energy storing molecule) content of the tissue falls rapidly. Since

ATP is essential for vital metabolic processes, a decrease in ATP initiates a cascade of damaging effects. Ca^{2+}-influx occurs during ischemia [1301,1302], but is more pronounced upon reoxygenation [1301,1303]. This influx is followed by the activation of Ca^{2+}-dependent proteases and phospholipases, membrane disruption, swelling of mitochondria, rupture of lysosomes, and release of proteolytic enzymes [1293,1304-1307]. Proteases lead to degradation of cytosolic proteins and degradation of vital enzymes and the cytoskeleton of the cell. Phospholipases damage membranes with the release of free fatty acids such as arachidonic acid (AA). AA is metabolized to epoxides, endoperoxides, hydroperoxides, and aldehydes. The ischemic tissue attracts neutrophils (chemotaxis), which causes further blockage of capillary arteries. The accumulating neutrophils, upon reperfusion, cause additional damage via the formation of oxygen radicals ($O_2^{·-}$ and $·OH$), singlet oxygen, HOCl, NO and ^-OONO. Ischemia causes large increases in the permeability of the endothelium to macromolecules, thus leading to organ dysfunction. The total injury depends on all these mentioned processes and there is no single event which can be said to be responsible for cell death.

Ischemia-reperfusion (I/R) injury may occur in any organ of the human body. From a medical point of view, however, the most significant organs are the heart and the brain. Heart attacks and strokes are among the leading causes of death in the U.S. and other Western countries. Oxygen radicals are definitely not a major contributor to cell killing, because oxygen is unavailable in ischemia. ROMs are only produced after reperfusion. Spin trapping has shown oxygen radicals and carbon-centered radicals in reperfused tissue, especially heart [1295-1300]. A dramatic decrease in intracellular ATP production correlates best with irreversible injury [1308].

Ischemia-reperfusion (I/R) represents two sides of the same coin. Reoxygenation damage depends on previous ischemia. The mechanisms of tissue damage by ischemia and reperfusion are quite distinct. The main events in these two phases are summarized as follows:

Ischemic period

1. Ca^{2+}- influx $\longrightarrow$ activation of Ca^{2+}-dependent proteases and phospholipases $\longrightarrow$ damage to proteins and membranes $\longrightarrow$ release of free fatty acids, such as AA.

2. ATP $\longrightarrow$ AMP $\longrightarrow$ adenosine $\longrightarrow$ inosine $\longrightarrow$ hypoxanthine.

3. Polymorphonuclear leukocytes (PMNs) infiltrate the ischemic tissue and adhere to vascular endothelium (causes capillary plugging).

These events give us an idea for possible sources of ROMs upon reoxygenation: AA metabolism, hypoxanthine metabolism, and PMN activation.

Reoxygenation period

1. Increased Ca^{2+} -influx [1301,1303,1309].

2. AA is metabolized by cyclooxygenase and 5- and 12-lipoxygenase to prostaglandins, thromboxanes, and leukotrienes via epoxides, hydroperoxides, and endoperoxides, which may give oxygen radical, as well as numerous cytotoxic aldehydes (Chapter 6).

3. Hypoxanthine leads to the formation of $O_2{\cdot}^-$ and $\cdot OH$ via the following sequence of reactions:

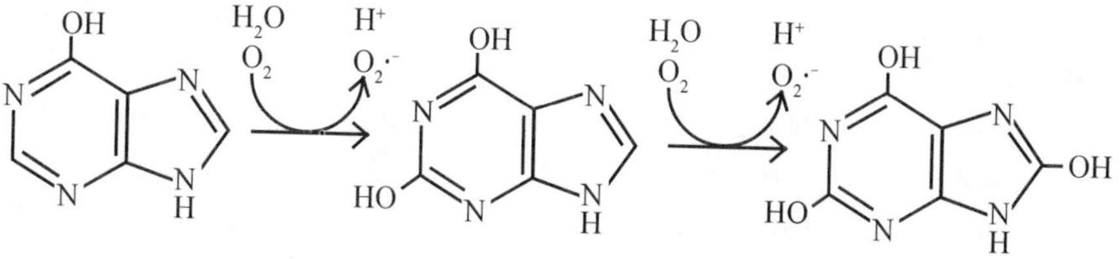

hypoxanthine   xanthine   uric acid

4. PMNs produce $O_2{\cdot}^-$, H_2O_2, $\cdot OH$, $^1O_2$, HOCl, $\cdot NO$, and ^-OONO (Chapter 5, phagocytosis). PMNs contain active lipoxygenase enzymes capable of metabolizing AA.

Based on the above sources of ROM formation, there are many ways in which we can therapeutically intervene to attenuate I/R injury:

1. Inhibition of XO, 2. Inhibition of COX, 5-LOX and 12-LOX, 3. Scavenging of $O_2^{\cdot-}$ by SOD, 4. Scavenging of H_2O_2 by CAT, 5. Complexation of iron ions by deferoxamine, 6. Scavenging of $\cdot OH$ (mannitol, α-tocopherol), 7. Scavenging of $^1O_2$, 8. Scavenging of HOCl (by the β-aminoacid taurine), 9. Inhibition of NO-synthase (by L-arginine analogs).

3. EPR and spin-trapping studies

The technique of spin trapping is a very sensitive method for the identification of radicals, although the method has been shown to produce some artifacts (Chapter 3). Therefore, in addition to spin trapping, direct EPR measurements at 77° K have been used. Zweier et al. [1310] observed 3 signals in reperfused rabbit heart. These signals were assigned to a semiquinone type radical, a ROO· or $O_2^{\cdot-}$, and a nitrogen-centered radical. However as pointed out by Nakazawa et al. [1311], the signal assigned to $O_2^{\cdot-}$ was an artifact produced by mechanical crushing of the frozen sample.

Subsequent studies by Zweier et al. [1298] examined the time course of EPR signal appearance and DMPO spin trapping. EPR signal intensity occurs after 15 s of reflow following 30 min. of ischemia. Measurements with the DMPO spin trap gave the DMPO/·OH and DMPO/·R signal peaking between 10-20 s of reflow. The DMPO/ adduct signals were quenched by SOD and deferoxamine. These results show that $O_2^{\cdot-}$-derived ·OH, R· and ROO· are generated in post-ischemic myocardium. Hearts reperfused with active recombinant human superoxide dismutase exhibited improved contractile function in parallel with a marked reduction in radicals [1297].

Another spin trap, the alpha-phenyl-N-tert-butyl nitrone (PBN), has been used by Bolli et al. [1300] to study reperfusion in dogs. As I have discussed in Chapter 5 (phagocytosis) this trap reacts an order of magnitude faster with radicals than DMPO and the adducts are more stable. The PBN trap showed secondary oxygen-and carbon-centered radicals, such as alkoxy and alkyl radicals. These radicals are formed by reactions of the primary oxygen radicals with membrane lipids [1300]. PBN was also used by Garlick et al. [1299] on reperfused rat hearts. After 15 min. of total ischemia, aerobic reperfusion resulted in a sudden burst of radical formation, peaking after 4 min. When hearts were perfused with anoxic buffer, no dramatic increase in radical production was observed. Subsequent introduction of oxygen, however, resulted in an immediate burst of radical production of magnitude similar to that seen in the wholly aerobic reperfusion experiments. This result clearly demonstrates the involvement of oxygen in radical formation. The signals were identified as either a carbon-centered species or an alkoxy radical, both of which may have been formed by secondary reactions of the initial oxygen radicals with membrane lipids.

4. Effect of radical scavengers

Since ·OH is so highly reactive, there are innumerable compounds that can scavenge ·OH and thus attenuate tissue damage in I/R. However, for practical purposes, we are limited to compounds, which are non-toxic and can therefore be added in sufficient concentration to cardioplegic solutions during reperfusion. Some of the compounds, that have been examined are α-tocopherol, mannitol, allopurinol, oxypurinol, and DMSO.

Increased amounts of α-tocopherol (51% increase) in rat myocardial tissue was found to attenuate lipid peroxidation (as measured by generation of polyunsaturated fatty acids). Contractile dysfunction, excessive accumulation of tissue calcium and release of lactate dehydrogenase after ischemia-reperfusion, were all reduced [1312]. This shows the beneficial effect of α-tocopherol in preventing lipid membrane damage during ischemia-reperfusion.

Mannitol is another frequently used radical scavenger (see Chapter 3). In human patients undergoing myocardial revascularization, mannitol (59.8 mM) protected against damage to mitochondria and resulted in a significant reduction in atrial arrhythmias [1313]. This supports the hypothesis that mannitol reperfusate significantly reduces myocardial damage in patients undergoing open heart surgery.

In Chapter 3, I have discussed the use of DMSO as an effective hydroxyl radical scavenger. DMSO has been used to study the oxygen paradox in perfused rat hearts. Reperfusion damage was assessed by measuring the creatine kinase release. This CK release was greatly reduced by the addition of 10% of DMSO to the reperfusate solution. It was suggested that DMSO alters the response of injured cells to the effects of calcium ions [1314]. Another explanation, of course, is the scavenging of hydroxyl radicals [239]. It appears likely that DMSO exerts its beneficial action in I/R injury via multiple mechanisms.

The CK release has also been used as a measure for reoxygenation damage by Myers et al. [1290]. These authors observed a decrease in CK release in presence of CAT, deferoxamine and allopurinol, but not SOD. These results indicate that ·OH formed in a Fenton reaction ($H_2O_2 + Fe^{2+}$) is involved in reoxygenation damage, in agreement with the DMPO spin trapping studies [1295-1298].

During the ischemic period, ATP is catabolized to hypoxanthine, which, upon reoxygenation is therefore a prime candidate for $O_2^{·-}$ generation. This possible source of $O_2^{·-}$, however, depends on the availability of xanthine oxidase. There is a great difference in xanthine oxidase availability in different species and between organs of the same species. Both the rat and dog heart contain several orders of magnitude higher amounts of xanthine oxidase than rabbit, pig or man [1293]. On the other hand, human intestine and kidney have a higher XO concentration than human heart. Therefore in the case of intes-tinal or kidney ischemia, the addition of XO inhibitors may represent a therapeutic possibility. So while the formation of oxygen radicals via xanthine oxidase may be important in myocardial I/R injury in the dog or rat, it most likely is of only minor importance in the human heart.

Allopurinol, a compound structurally related to hypoxanthine, is an inhibitor of xanthine oxidase. The addition of this compound in myocardial ischemia-reperfusion was found to decrease the EPR signal [1315] and tissue damage [1290,1315,1316]. However, allopurinol and oxypurinol are aromatic compounds related structurally to xanthine. These compounds can not only deactivate xanthine oxidase, but also react with hydroxyl radicals, as was indeed shown by Moorhouse et al. [823]. Therefore, the source of the ROMs responsible for I/R injury may not be the xanthine-xanthine oxidase pathway, but may be produced by activated neutrophils (Chapter 5, phagocytosis). Similar conclusions were reached by Godin and Bhimji [1316] and by Das et al. [1315]. The results of Godin and Bhimji showed the myocardial protection effect of allopurinol in rabbit heart, which did not contain detectable amounts of XO activity. Das et al. studied the effect of allopurinol and oxypurinol on I/R injury in isolated pig heart. Both compounds showed a reduction of EPR signal intensity and prevented I/R damage. However, it was found that neither pig heart nor pig blood contained any XO activity. Allopurinol and oxypurinol inhibited EPR radical signals generated by activated neutrophils [1315].

The inhibition of XO may possibly lead to attenuation of I/R injury, providing that XO is the source of the ROMs produced during reoxygenation. There are multiple sources for ROM formation in I/R injury. If we use a XO inhibitor we only inhibit $O_2^{\cdot-}$ production by this one pathway. Therefore, the scavenging of $O_2^{\cdot-}$ by SOD is a more promising approach to attenuate I/R injury.

5. SOD in ischemia-reperfusion injury

Ever since the discovery of $O_2^{\cdot-}$ and SOD by McCord and Fridovich [83,84] SOD has been used extensively for the identification of $O_2^{\cdot-}$ *in vivo*. In myocardial I/R injury, however, the cardioprotective effect of MnSOD was lost at high doses in the reoxygenated heart [1317,1318]. I have already discussed this paradoxical result in Chapter 8. McCord and coworkers recognized the dual character of the superoxide radical anion, which can act as a chain initiator or a chain terminator in lipid peroxidation [838]. This realization leads to an optimum SOD concentration. Too little SOD or too much SOD will have serious consequences [837].

As I have pointed out before, the study of the abnormal helps us understand the normal. The study of genetically engineered mice is gaining widespread use for studies on the pathogenesis of numerous diseases, such as I/R injury [1319], pulmonary oxygen toxicity [1320], and atherosclerosis [1321]. Transgenic mice overexpressing the copper/zinc superoxide dismutase (CuZnSOD) were protected from the deleterious effects of gut ischemia-reperfusion [1322] and intestinal I/R injury [1323]. Horie et al. [1322] compared the effectiveness of Tg SOD overexpression in attenuating ischemia-reperfusion injury to intravascularly administered CuZnSOD or MnSOD. Exogenously administerd SOD had little or no affect on gut I/R-induced leukostasis or capillary no-reflow in the liver. The authors conclude that cellular localization of SOD activity is an important determinant of the protective actions of this enzyme in experimental models of I/R injury.

Macrophages isolated from the transgenic mice showed diminished capacity to release superoxide radical anion into extracellular fluid and also produced smaller amounts of nitric oxide. This impaired macrophage function could account for the blunted responses of the liver to gut ischemia-reperfusion [1324].

CuZnSOD also protects the brain from I/R-induced damage. CuZnSOD Tg mice have been shown to have smaller brain infarct volumes after focal cerebral ischemia than their wild-type counterparts [1325,1326].

Reperfusion injury can also be protected against by the use of ECSOD (extracellular SOD), which sticks to endothelial cells. A genetically-engineered SOD with properties that mimic those of authentic ECSOD was found to be much more effective (by a factor of 75) in preventing reperfusion injury than MnSOD or CuZnSOD [1327]. The protection was also evident from the decreased LDH (lactate dehydrogenase) release [1327].

6. The role of polymorphonuclear leukocytes in I/R injury

Leukocytes play an important role in I/R injury. Leukocytes accumulate in tissues exposed to I/R, and depletion of circulating neutrophils significantly reduces the microvascular and parenchymal dysfunction associated with I/R [1328]. Leukocytes are implicated in I/R injury to the heart [1328-1331], brain [1332,1333], as well as in numerous other organs [1334].

As I have discussed in Chapter 5, polymorphonuclear leukocytes (PMNs) provide the most abundant source of ROMs. It is therefore not surprising that the accumulation of leukocytes in ischemic tissue leads to the generation of ROMs upon reperfusion, and subsequent tissue damage. The physiological mechanisms of postischemic tissue injury have been reviewed by Granger and Korthuis [1334].

The first step in the inflammatory process is leukocyte-endothelial cell adhesion (LECA). LECA has been identified as playing an important role in many pathologies, such as asthma, pulmonary oxygen toxicity, arthritis, bacterial meningitis, cerebral malaria, and ischemia-reperfusion injury [1335].

The leukocyte-endothelial cell interaction is mediated by various adhesion glycoproteins expressed on the surface of leukocytes. The central role of leukocyte adhesion in the pathogenesis of I/R injury is evident from studies that demonstrate protection against microvascular and/or parenchymal cell dysfunction or necrosis in postischemic tissue by blocking receptors. Monoclonal antibodies directed against the adhesion glycoproteins attenuate I/R - induced myocardial necrosis [1336,1337], as well as I/R induced damage in other organs or tissues [1334]. Results from adhesion molecule-deficient mice provided support for the hypothesis that LECA is important in causing liver dysfunction induced by gut I/R [1319]. The ROMs released by activated leukocytes and endothelial cells have also been implicated in I/R- induced leukocyte accumulation. Exposure of postcapillary venules to exogenous H_2O_2 [1338] or $O_2^{\cdot-}$ [1339,1340] promote leukocyte adhesion. Both SOD and CAT attenuate the leukocyte adhesion observed in mesenteric venules exposed to I/R [1338,1341, 1342].

The mechanism of $O_2^{\cdot-}$ -mediated leukocyte adhesion is unclear, however it has been suggested that NO may contribute to this response [1343]. Inhibition of NO production with analogs of L-arginine (see chapter 5) results in an intense leukocyte adherence response in mesenteric venules [1343-1345], which indicates that NO is an endogenous inhibitor of leukocyte-endothelial adhesion. Based on this hypothesis we would predict that increased formation of $O_2^{\cdot-}$ should lead to increased leukocyte adhesion, since $O_2^{\cdot-}$ deactivates NO by combination to give $^-$OONO [116].

Production of NO by endothelial cells decreases after I/R [1346,1347] and drugs that spontaneously release NO attenuate the increased albumin leakage [1346] in mesenteric venules exposed to I/R and reduce myocardial necrosis and neutrophil accumulation [1348-1350]. The beneficial effects of NO donors have been attributed to their ability to attenuate I/R -induced leukocyte-endothelial adhesion [1346,1348].

However, numerous studies have indicated that NO promotes (rather than protects against) I/R injury. NO synthase inhibitors have been shown to limit neutrophil accumulation and vascular protein leakage in postischemic lung and skeletal muscle [1351], as well as reduce contractile dysfunction and lipid peroxidation associated with myocardial I/R [1352]. The study of Ignarro and coworkers [1352] showed that antioxidants such as mercaptopropionyl glycine and catalase, as well as NO synthase inhibitors afforded similar and complete protection against myocardial reperfusion injury. It was, therefore, suggested that the product of the $O_2^{\cdot-}$- $\cdot$NO combination, the $^-$OONO is the injurious agent [1353].

We thus have a controversy so common to radical chemistry and pathology (NO is good or bad, depending on the circumstances). As I have pointed out in Chapter 7, the combination of $O_2^{\cdot-}$ and NO requires high concentrations of both species (as in activated macrophages), and it also depends on the relative rates of their production. Excess NO or $O_2^{\cdot-}$ act as endogenous modulators of peroxynitrite-induced tissue damage. In addition, as I have discussed, the damaging effect of $^-$OONO depends on its concentration, since any reaction has to compete with the unimolecular isomerization to the inactive NO_3^-. The resolution of the NO controversy in I/R injury will come from a careful analysis of these factors. Several recent reviews have discussed the complex role of NO and ^-OONO in cardiac I/R injury [1354], in brain I/R [1355], and in shock and inflammation [1356].

7. The role of iron in I/R injury

Since the formation of hydroxyl radicals is accelerated by Fe(II) or Cu(I), the control of these metal ions plays an important role in many diseases, including ischemia-reperfusion injury. During cardiac surgery most animal and clinical studies have demonstrated the role of deferoxamine (an iron chelator) in reducing generation of ROMs [1357-1362]. However this effect has not been translated into improved postoperative patient outcome [1363].

As I have discussed (Chapter 3) O_2^- may act as an electron donor (a reducing agent) or as an electron acceptor (an oxidizing agent). Acting as an electron acceptor it initiates lipid peroxidation, causing membrane damage. On the other hand, acting as an electron donor, it can terminate the lipid peroxidation chain [838]. These characteristics, however, do not imply, that whenever O_2^- acts as an electron donor it results in a beneficial action. Superoxide radical anion reduces Fe^{3+} and Cu^{2+} to Fe^{2+} and Cu^+, which are catalysts for $\cdot OH$ formation (Chapter 3).

The importance of iron in oxidative injury has been reviewed by McCord [1364,1365]. There is considerable evidence showing that iron overload is a risk factor in many diseases, such as colorectal and lung cancer [667,668], heart disease [669] and Parkinsons disease (this chapter). Superoxide radical anion causes the release of Fe^{2+} from storage proteins (Chapter 3). As I pointed out in Chapter 6, there is unfortunately some controversy concerning the link between iron stores and heart disease [669, 670] based on epidemiological studies. It has been proposed by Sullivan [1366,1367] that even normal iron sufficiency imposes a risk factor and explains the gender difference in death rates from ischemic heart disease. Healthy young

376

men have serum ferritin levels three or four times those of healthy young women. It has been suggested that menstrual blood loss (resulting in lower serum and tissue ferritin stores) rather than high estrogen levels, may be the protective factor.

Blood donation three times a year will lower the serum ferritin level of a man to that of a young woman [1367]. Medieval physicians prescribed blood letting for almost anything that ailed you. Maybe there is some scientific basis for this procedure, which may turn out not to be as foolish as it first appears. Regular blood letting in men may lower the risk for colorectal and lung cancer, heart disease, and Parkinson's disease.

8. Arachidonic acid metabolism

The metabolism of AA is a complex process, beyond the scope of the present text. A detailed account of AA metabolism can be found in the review by Taylor and Shappel [1294]. The metabolism involves several enzymes, the cyclooxygenase (COX), the 5-lipoxygenase (5-LOX), and the 12-lipoxygenase (12-LOX). These enzymes lead to the formation of the prostaglandins, thromboxanes, and inflammatory leukotrienes. The intermediates in these transformations are epoxides (may attack NH_2 or SH-groups), endoperoxides, and hydroperoxides. I have already discussed the formation and reactions of these products in Chapter 3 (peroxyl radicals) and Chapter 6 (lipid peroxidation). The oxidation of unsaturated fatty acids leads to a number of aldehydes, which are highly reactive (4-hydroxy-2-nonenal from AA), and attract more leukocytes to the affected tissue [590,591] (Chapter 6).

Formation of prostaglandins and lipoxygenase products is regulated not only by the availability of AA as substrate, but also by the presence of peroxides. The cyclooxygenase (COX) activity is stimulated by low concentrations of peroxides (including H_2O_2), but COX activity is inhibited by high concentrations of peroxides. This modulation of COX and LOX enzyme activity by peroxides depends on the prevailing peroxide tone, a balance between the capacity of cells to form or remove peroxides.

Inhibition of the lipoxygenase and cyclooxygenase enzymes by BW755C attenuates leukocyte infiltration into the infarcted myocardium and prevents 12-hydroxyeicosotetraenoic acid formation and significantly reduces infarct size [1331]. This indicates that leukocytes contribute to tissue injury in myocardial ischemia possibly by the release of proinflammatory mediators, such as lipoxygenase products, oxygen radicals, and hydrolytic enzymes. Scavenging of radicals or inhibition of the enzymes involved in AA metabolism may be useful in reducing infarct size.

PMNs contain active lipoxygenase enzymes capable of metabolizing AA to products that are not normally found in the myocardium. Inhibitors of LOX suppress the accumulation of leukocytes into the ischemic myocardium and reduce infarct size [1368]. Agents that block AA metabolism, such as ibuprofen [1369] have been reported to protect myocardium against I/R injury.

9. Adenosine as an endogenous protector in I/R injury

A novel approach to reduce ischemia-reperfusion injury is by a procedure called ischemic preconditioning (IPC). In this procedure a tissue is rendered resistant to the damaging effects of prolonged I/R by prior exposure to brief periods of vascular occlusion [1370]. IPC has been shown to be effective in improving contractile function and reducing necrosis in skeletal muscle and heart [1370-1372]. Studies on the effect of MnSOD on IPC by McCord and coworkers [1373] indicated that IPC is not mediated by $O_2 \cdot^-$ in the isolated rabbit heart. The mechanisms underlying IPC have not been elucidated in most tissues, but there is growing evidence suggesting that preconditioning induces release of adenosine. The endogenous metabolite adenosine has been recognized as a protective agent. Because adenosine is formed during ischemia and protects against the deleterious consequences of reperfusion, it has been called a "retaliatory" metabolite [1374].

Adenosine reduces I/R injury [1371,1375] and adenosine receptor activation reduces leukocyte adhesion to postcapillary venules in tissue subjected to prolonged I/R [1376]. Adenosine modulates neu-trophilic oxidative metabolism and inhibits platelet aggregation and microthrombi formation [1377]. Downey et al. [1371] demonstrated that the protective action of preconditioning is blocked by prior ad-ministration of adenosine-A_1-receptor, but not adenosine-A_2-receptor antagonists. The limitation of infarct size induced by IPC or adenosine-A_1-receptor activation is sustained over a 3-day reperfusion period [1371]. This observation supports the idea that IPC or ade-nosine treatment reduces postischemic tissue injury, rather than de-laying its development.

Adenosine interferes with activated neutrophil function, neutro-phil-endothelial cell adhesion, the production and release of inflam-matory mediators, the expression of adhesion molecules, and it acti-vates antioxidant defenses, thus providing protective effects at mul-tiple levels in the pathogenesis of ischemia-reperfusion [1374].

10. Effect of pH on I/R injury

Ischemia results in accumulation of lactic acid with a drop in pH. Subsequent reperfusion results in hyperoxia in the affected tissue, due to the Bohr effect. High concentrations of CO_2 and H^+, such as occurs in metabolically active tissues, decrease the affinity of hemo-globin for oxygen:

$$Hb + O_2 \underset{\longrightarrow}{\overset{H^+}{\longleftarrow}} HbO_2$$

A low pH shifts the equilibrium to the left, leading upon reperfusion to hyperoxia and subsequent increased formation of superoxide radical anion and increased reperfusion damage.

During myocardial ischemia a decrease in tissue pH occurs and the pH returns to normal after reperfusion. Studies on cultured car-diac myocytes and perfused papillary muscles showed that acido-sis (pH < 7) protected profoundly against cell death during ischemia [1378]. However, the return to normal pH after reperfusion caused myocytes to lose viability. This worsening of injury is a "pH paradox" and was mediated by changes in intracellular pH, since manipulations that caused pH to increase more rapidly after reperfusion accelerated cell killing, whereas manipulations that delayed the increase in pH prevented loss of myocyte viability. Inhibition of the Na^+/H^+ exchange with dimethylamiloride or HOE694 delayed the return to physiologic pH after reperfusion and prevented reperfusion-induced cell killing to cultured myocytes and perfused papillary muscle. These inhibitors of the Na^+/H^+ exchange did not reduce intracellular free Ca^{2+} during reperfusion. On the other hand, reperfusion with dichlorobenzamil, an inhibitor of Na^+/Ca^{2+} exchange decreased free Ca^{2+}, but did not reduce cell killing. These results indicate that the pH paradox is not Ca^{2+}-dependent. It was therefore suggested [1378] that ischemia activates hydrolytic enzymes, such as phospholipases and proteases, whose activity is inhibited at acidic pH. Upon reperfusion the pH rises to normal and thus releases the inhibitory action and thus leads to tissue damage.

H. OXIDATIVE STRESS AND LIFESTYLE

Everything is poison, it just depends on the dose.
Paracelsus

A change in the prooxidant/antioxidant balance can be brought about by genetic abnormalities [974] or by external factors such as bacteria, viruses [1037], toxins [1129], or air pollution [307,1131]. While we can do nothing (at least not yet) about genetic defects, there are some factors that we can control. I have already discussed the effect of smoking on lung diseases (cancer, emphysema, asthma). I shall now discuss how exercise, diet, alcohol consumption and psychological stress affect the prooxidant/antioxidant balance. These lifestyle factors are excellent examples of the importance of balance in everything we do.

380

1. Exercise

*No hay nada que dure mas, que un vago bien cuidado**

Puerto Rican saying

Exercise initiates a complex physiological and biochemical response [1379-1384]. In the present text I would like to discuss how exercise affects the prooxidant/antioxidant balance [1385-1387]. In our exercise conscious society I have to ask the important question: is exercise good or bad for our health? In my opinion no one can give an unequivocal answer. The answer depends on many factors. We are dealing with a complex dynamic system where small changes in initial conditions can have unforeseen consequences. The oxidative stress imposed on a tissue depends on the tissue involved, on the age of the individual, on his or her overall health and training status, on the type of exercise, and on its intensity and duration.

Exercise requires energy, which is obtained through the increased flow of electrons in the mitochondrial respiratory chain leading to increased formation of $O_2\cdot^-$ and H_2O_2 and increased formation of ATP. Exercise exhausts the ATP pool, leading to high levels of ADP, triggering ADP catabolism and the conversion of xanthine dehydrogenase to xanthine oxidase. Xanthine oxidase in turn produces $O_2\cdot^-$. Oxidative stress results in damage to lipids, proteins, and DNA. The biological markers for lipid peroxidation are malonaldehyde (thiobarbituric acid reactive products) or ethane and pentane in the exhaled breath. In addition, a more direct way of measuring lipid peroxidation is by chemiluminescence measurement of lipid hydroperoxides [1388]. DNA damage is assessed by measuring its change in migration in the single cell electrophoresis assay (SCG) [1389] or by formation of 8-OHdG..

The first evidence showing that exercise increases lipid peroxidation in human subjects was presented by Dillard et al. [938], who

---

* There is nothing which lasts longer than a lazy man, who takes care of himself.

observed increased formation of ethane and pentane in exhaled breath (increased lipid peroxidation) in humans undergoing strenuous exercise. The amount of ethane and pentane was decreased in presence of Vitamin E. Subsequently, Davies et al. [1390] showed that rats exercised to exhaustion showed a 2-3-fold increase in the EPR signal in liver, muscle, and tissue homogenates and also an increase in malonaldehyde, indicating that the organ is undergoing oxidative stress.

With a single bout of exercise, lipid peroxidation (malonaldehyde) in rat muscle was proportional to the intensity of the exercise [1391]. With a single bout of exercise it appears that moderate exercise does not induce oxidative stress, but high intensity exercise does [1392]. In a study in men, who exercised by cycling 90 min./ day at 65% V_{2max} (whole maximal oxygen consumption) during 3 consecutive days, no change in the levels of GSH/GSSG or Vitamins C or E was observed, indicating no increased oxidative stress [1393]. This study showed no increased 8-OHdG excretion after a single or repeated bouts of moderate intensity concentric exercise. These results contradict the results of Adelman et al. [867], who suggested that sustained increased metabolic rate during single or successive exercise bouts could increase rates of DNA and RNA oxidation in humans.

Another study showed no signs of oxidative stress (DNA damage in white blood cells) even after long duration, when the intensity was low [1389]. However the same study showed the human subjects undergoing exercise (running on a treadmill) were found to have increased DNA damage (as measured by the single cell gel electrophoresis assay) in the white blood cells. The increased DNA damage was seen 6 hours after the end of the exercise, reaching a maximum after 24 hours, and returned to normal levels after 72 hours , indicating repair [1389].

Oxidative stress appears to be related to exercise intensity. High intensity training induces muscle damage (eccentric more than concentric) [1394] and acute necrotic myopathy [1395]. Marathon running increased excretion of 8-OHdG [1392], but moderate cycling exercise did not show any effect [1393]. Kanter et al. [1396] reported a 77% increase above the resting concentration of TBRS in both plasma and serum from highly trained humans following exhaustive running exercise (80 km).

Other markers for tissue damage are the increased serum levels of creatine kinase (CK) and lactate dehydrogenase (LDH). Exercise leads to oxygen depletion and accumulation of ADP. Skeletal muscle utilizes glucose anaerobically and breaks it down to form lactate. This process is known as glycolysis. Lactate dehydrogenase (LDH) catalyzes the last step in glycolysis, the conversion of pyruvate to lactate:

$$\text{pyruvate} + \text{NADH} + \text{H}^+ \underset{\longleftarrow}{\overset{\text{LDH}}{\longrightarrow}} \text{lactate} + \text{NAD}^+$$

Glycolysis is one mechanism by which many organisms can extract chemical energy from various organic fuels in the absence of oxygen. This process is also known as anaerobic fermentation and is one of the most ancient mechanisms for obtaining energy from nutrient molecules. The increase in these enzymes in the serum supplies the necessary energy when the oxygen supply is exhausted.

ADP is phosphorylated by phosphocreatine :

$$\text{phosphocreatine} + \text{ADP} \underset{\longleftarrow}{\overset{\text{CK}}{\longrightarrow}} \text{creatine} + \text{ATP}$$

This reaction is catalyzed by creatine kinase. At pH 6 in the sarcoplasm the equilibrium is far to the right.

The investigations by Kanter et al. [1396] have shown an increase in these enzymes after exhaustive exercise. The increase in these enzymes in the serum indicates membrane damage, which is caused by lipid peroxidation. A relationship between the increase in CK and LDH and malonaldehyde (a marker for lipid peroxidation) has indeed been established in long distance runners (80 km race) [1396].

Another source for increased ROM-induced damage is the availability of free iron ions. As I have pointed out before (Chapter 3) the initially formed ROMs ($O_2 \cdot^-$ and H_2O_2) are chemically not very reactive and are converted to the more reactive ·OH radical. This transformation requires the presence of catalytic amounts of metal ions (Fe^{2+}, Cu^+). Iron-homeostasis is therefore an important factor in ROM-induced tissue damage. Increased formation of $O_2\cdot^-$ causes release of free Fe-ions from iron-storage proteins (Chapter 3). Anaerobic metabolism leads to local acidosis (lactate accumulation), which leads to muscle soreness. The low pH may cause release of free Fe ions from transferrin [1397].

Jenkins et al. [1398] studied Fe-homeostasis in rats. Both trained and untrained rats showed a considerable increase (factor of 2-3) in loosely bound iron after exhaustive exercise. However, the increase in gastrognemius muscle of trained rats was 33 % lower than in untrained rats. These results show that training enhances a muscle's ability to defend itself against oxidative stress, and may also afford protection against radical-induced damage during exhaustive exercise by reducing the amount of loosely bound iron available for ·OH formation. Nevertheless we have to remember: exercise increases loosely bound iron, thus increasing oxidative stress. Individuals who do not exercise have much lower levels of loosely bound Fe ions.

a. Effect of exercise on the immune system

An important source of ROMs are the phagocytes (Chapter 5). During exercise, tissue damage and local inflammation are observed after running or cycling with leukocytosis and leukocyte activation. [1399-1402]. The leukocyte activation occurs from damage to muscle fibers and connective tissues during muscle contraction. Downhill running (eccentric exercise) is believed to exert greater tension on muscle fibers and adjacent conective tissue than uphill running (concentric exercise) and should therefore lead to more damage and inflammation [1403,1404]. Muscle soreness was only reported after eccentric exercise [1405,1406]. Muscle strain brings about the activation of leukocytes during long duration exercise [1400,1402].

A study by Lewicki et al. [1407] has shown no significant difference between untrained persons and sportsmen with regard to absolute numbers and percentage composition of white blood cells. The results show that long-term intensive training does not affect the quantitative composition of white blood cells. However, the neutrophil bactericidal activity was lower in sportsmen than in the untrained controls at rest. The neutrophils of trained individuals produced less ROMs in response to a stimulus [1408]. These results indicate that trained individuals are more receptive to common infections (viral, bacterial, and parasitic) [1409]. The reduced neutrophil activity in trained athletes may be due to reduced tissue damage upon exercise, since training strengthens muscle fibers and diminishes muscle damage [1410].

During exercise the total number of leukocytes increases proportional to the duration of the exercise [1400]. After a 10 km run the number of leukocytes increased from 6020 ± 640 /μl to 9670 ± 860/ μl. The exercise, however, not only increased the total number of leukocytes, but also increased their activity. After the 10 km jogging exercise the plasma levels of elastase-α-proteinase inhibitor (E-α-PI) increased substantially (ca. 300%). Increased plasma levels of E-α-PI were also observed during bacterial infections [1411]. Thus, the res-ponse of polymorphonuclear leukocytes to physical exercise is similar to their response to infection.

Epidemiological studies have found a correlation between elevated leukocyte counts and the incidence and mortality for myocardial infarction [1412,1413]. Neutrophils play an important role in the pathogenesis of vascular injury [1414]. The increase in PMN elastase concentration upon strenuous exercise may contribute to the prevention of a thrombotic event by breaking down certain coagulation factors [1415] inactivating PAI-1 [1416] and by proteolytic attack on fibrin clots [1417,1418]. On the other hand, elevated PMN elastase levels may, in the long run, accelerate the development of atherosclerotic lesions in the vascular wall. Advanced atherosclerosis has indeed been found at autopsies of men who died from coronary heart disease during strenuous exercise [1419,1420].

b. Antioxidant defenses during exercise

The increased formation of ROMs during exercise is counteracted by antioxidant defenses. These defenses (Chapter 8) consist of SOD, CAT, GSH-Px, and metal ion-complexing proteins (carnosine). The effect of exercise on the level of antioxidant defenses has been studied by many investigators with contradictory results. Whether any change occurs depends on the type of exercise and on its intensity [1421-1428]. Each organ/tissue throughout the body adjusts its antioxidant enzymes to regular exercise in a manner according to its background of metabolism and available antioxidants [1387].

The work of Criswell et al. [1428] showed that induction of SOD and GSH-Px depended on the intensity of the training as well as on the muscle type and the duration of the training studies. The increase in mitochondrial volume associated with exercise training could serve to decrease the stimulus for antioxidant induction over the course of the training protocol.

The induction of antioxidant defenses in skeletal muscle is linked to the increase in radical production ($O_2 \cdot^-$) in the mitochondria associated with exercise [1428-1430]. The mitochondrial version of SOD, the (Mn)SOD, is the primary inducible form of SOD [1428]. It was suggested by Starnes et al. [1430] that the rate of skeletal muscle radical production increases disproportionately with the rate of mitochondrial O_2-consumption such that the ratio of radical production to O_2-consumption increases at higher respiratory rates. It was therefore suggested by Criswell et al. that endurance training involving highly intense exercise bouts (interval training) may provide a greater stimulus for induction of cellular antioxidants compared with low-to-moderate intensity training protocols.

Studies on *Salmonella* and *E. coli* [1431] have shown activation of the oxyR gene by H_2O_2, leading to an increase in CAT and some peroxidases. The signal could be transduced in a short time (5 min.). The exercise-related increase in skeletal muscle and liver antioxidant enzymes activities, could proceed by a similar mechanism.

386

As I have pointed out before (Chapter 8) the antioxidant defenses are not evenly distributed throughout the body. They are produced where they are needed most. Liver and heart have higher antioxidant levels than skeletal muscle. Liver has the highest antioxidant level in the human body. During exercise, the oxygen consumption in heart and liver increases only modestly (4-5 times), whereas O_2-consumption in muscles increases up to 20-fold. Because radical production is proportional to oxygen uptake, the liver and heart may be subjected to only moderate oxidative stress [1424]. Therefore, the modest increase in O_2-consumption in heart and liver during exercise does not induce additional antioxidants, whereas the increase in skeletal muscle is substantial. The effect of exercise on skeletal muscle also depends on the type of muscle studied [1428]. In addition we have to consider the age of the individual undergoing exercise, since protein synthesis decreases with age [1424].

Another defense against ROM-induced tissue damage during exercise is the elimination of Cu and Fe ions in the sweat of athletes [1432]. It has been suggested that excretion reduces the risk of lipid peroxidation. However, I have already pointed out several studies that showed increased levels of lipid peroxidation (MA and ethane and pentane formation) during exhaustive exercise. These results clearly indicate that the defensive mechanisms (induction of antioxidant enzymes and excretion of Cu and Fe ions in the sweat) are not sufficient to counteract increased damage by ROMs.

Vigorous exercise over a lifetime increases the amount of mitochondria [1433] (mitochondrial volume) in muscle and these mitochondria may therefore better withstand the onslaught of ROMs during exercise, but the fact remains that exercise increases oxidative stress. Vigorous exercise increases the amount of ethane and pentane, and MA and 8-OHdG formation. The antioxidant defenses never completely compensate for the increased formation of ROMs during exhaustive exercise. The same conclusion can be reached by considering the effect of Vitamin E on exercise-induced lipid peroxidation. Vitamin E reduced, but did not abolish exercise-induced lipid peroxidation [938,1434]. Adequate Vitamin E is important for membrane integrity during exercise [1422]. The level of vitamin E in plasma has been found to increase during exercise [1435]. It was first suggested that this increase is due to lipolysis, a hypothesis which was later rejected by the same group of investigators [1436]. Due to the importance of Vitamin E it has been suggested that Vitamin E can be replenished by its reaction with ascorbic acid. Although ascorbic acid can restore Vitamin E *in vitro*, its importance *in vivo* is doubtful. For a detailed discussion of this topic see Chapter 8 under Vitamin E.

I have tried to summarize in compact form some of the reactions important during exercise (Fig. 5). The reactions shown have all been discussed in detail in previous chapters. All of these reactions take place under normal (resting) conditions, except during exercise all of these reactions are increased due to the greater demand for oxygen.

c. The pros and cons of exercise

Regular high intensity training may provide the individual with more antioxidant defenses to counteract the increased ROM formation during exercise. However, we should remember that if you do not exercise at all you don't need the increased antioxidants in the first place.

There are both beneficial and deleterious effects of exercise training. It is, like everything else in medicine, a question of balance, which I have discussed on many previous occasions, such as in the case of SOD, ascorbic acid, uric acid, Fe ions, cholesterol and oxygen. We are dealing with a complex dynamic system, where the initial conditions are not precisely defined. Considering biochemical individuality [1437] and the huge number of people exercising, it is not too surprising to find many individuals who show deleterious effects of exercise. People with hypersensitivity of the airways (asthma) have been shown to experience severe asthma attacks during exercise [1438]. We also have to consider environmental factors. Ozone considerably increases pulmonary damage during exercise

388

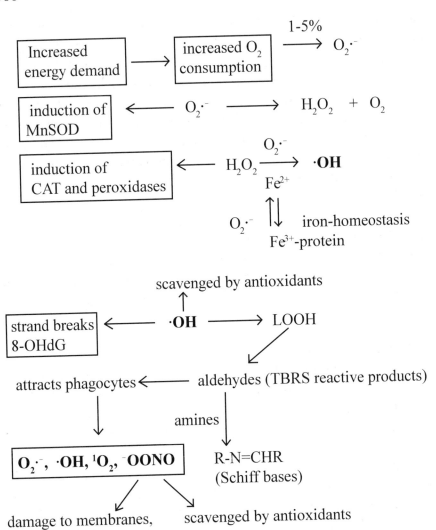

Fig. 5. Some reactions initiated by exercise.

[1439]. Although the absolute risk of cardiovascular complications of exercise is low, the risk of sudden death and cardiac arrest seems to be higher during exercise than at rest. Thompson et al. [1440] estimated the risk of sudden death in jogging to be seven times that in more sedentary activities. Vuori [1441] estimated that the risk of sudden death is 4.5 times higher in prolonged cross-country skiing than during the rest of the day. Exercise consisting of at least 10 miles brisk walking or jogging per week, combined with diet has been shown to have a beneficial effect on the HDL/LDL ratio [1442]. The effect was more pronounced in men than in women.

Everything in excess is bad. Remember Paracelsus: everything is poison, it only depends on the dose. Does regular exercise have any beneficial effect on our longevity? Voluntary exercise did improve survival in rats (i.e., more attained old age), but did not result in an extension of life span [1443]. Food restriction in sedentary rats however considerably increased life span (un vago bien cuidado!). There was no significant difference between voluntary exercising rats and freely eating sedentary rats in the cause of death. This provides evidence that voluntary exercise does not influence the development of disease. Food restriction, on the other hand, resulted in a significant protection against the development of malignancies [1443].

Another important factor in considering the pros and cons of exercise is the age of the individual. Experiments with rats have shown that there exists a "Threshold Age", above which it is disadvantageous for the rat to commence an exercise program. The older rats do not adapt to the increased oxidative stress [1444]. Rats that commence the exercise program prior to 400 days of age had a higher percentage of survival than their controls. In contrast, a lower percentage of survival is found in those rats that commenced exercise beyond 400 days of age. The moral of the story: Old men or women, who never exercised before, should not run a marathon!

2. Alcohol consumption

The consumption of alcoholic beverages represents an excellent example of the wisdom expressed by Paracelsus. High alcohol consumption leads to numerous diseases, like cirrhosis of the liver, and a variety of cancers (oral, pharyngeal, laryngeal, esophageal, and liver) [1445]. However, alcohol taken in moderate amounts (1-3 drinks per day) has been claimed to lower the risk of coronary artery disease (CAD) [1446-1455].

Epidemiological studies have shown that consumption of alcoholic beverages is related to cancer in humans [1445]. However it is less clear if the compound responsible is ethanol. Alcoholic beverages contain a huge number of trace compounds, that may be deleterious or beneficial to our health. Alcoholic beverages contain aldehydes, N-nitrosamines, phenols, flavonoids, tannin, acrolein, and pesticide residues. N-nitrosamine present in some beers are known to be carcinogenic. Flavonoids, on the other hand, are effective antioxidants (Chapter 8). We need a mechanism by which ethanol is metabolized to yield increased amounts of ROMs.

It has been known for some time that in the mitochondria NADH and NADPH catalyze lipid peroxidation in presence of an iron chelate (Fe^{3+}-ATP) [1456,1457]. These results indicate formation of ROMs. In addition to mitochondria, rat liver nuclei and microsomes [1453-1460] have been shown to produce ROMs (as measured by lipid peroxidation) by NADH and NADPH-dependent electron transfer. In all of these subcellular organelles the production of ROMs increased after chronic alcohol treatment.

The mitochondrial respiratory chain produces $O_2\cdot^-$ and H_2O_2 [85,388,389] and the formation of $O_2\cdot^-$ in mitochondria is considerably elevated after acute alcohol ingestion [1461]. The increased formation of ROMs by rat liver mitochondria after chronic alcohol ingestion was demonstrated by Cederbaum and coworkers [1462]. The increase was measured by lipid peroxidation, H_2O_2 and $\cdot$OH formation [1462]. Lipid peroxidation was measured by the standard TBARS-method. Hydrogen peroxide was determined by oxidation of methanol to formaldehyde by catalase-compound I complex [1463] and $\cdot$OH was determined by its reaction with 2-keto-4-thiomethylbutyric acid (KMB) to give ethylene and also by its reaction with DMSO to give formaldehyde (see Chapter 3).

Oxidation of ethanol by alcohol dehydrogenase (ADH) produces NADH in the cytosol [1462]. The outer mitochondrial membrane displays an accumulation of iron ions [1464], which is a required catalyst for lipid peroxidation. We have the following sequence of reactions:

$$CH_3CH_2OH \xrightarrow{\text{ADH}} NADH \xrightarrow{\text{Fe}^{3+}\text{-chelate}} Fe^{2+}\text{-chelate}$$

$$Fe^{2+}\text{-chelate} + O_2 \longrightarrow Fe^{3+}\text{-chelate} + O_2 \cdot^-$$

$$\downarrow \text{disprop.}$$

$$H_2O_2$$

$$O_2\cdot^- + H_2O_2 \xrightarrow{\text{Fe}^{3+}\text{-chelate}} OH^- + \cdot OH + O_2$$

$$Fe^{2+}\text{-chelate} + H_2O_2 \longrightarrow Fe^{3+}\text{-chelate} + OH^- + \cdot OH$$

$$\cdot OH + LH \longrightarrow LOOH \longrightarrow \text{aldehydes}$$

However, alcohol consumed in moderate amounts also has some beneficial effects. Numerous studies have shown that moderate alcohol consumption protects against CAD. Within blood vessels there exists a mechanistic pathway sensitive to low ethanol concentrations, leading to increased expression of alcohol dehydrogenase [1452]. This ADH in turn produces a NADH rich reducing environment that could antagonize lipoprotein oxidation and could thus explain the protective effect of ethanol on atherogenesis.

Another beneficial effect of moderate alcohol consumption is the increase in HDL-cholesterol, which may be of particular significance with regards to atherosclerosis [1452]. The only other method for increasing HDL-cholesterol is through mild exercise or walking combined with diet [1442].

Another mechanism against atherogenesis may involve flavonoids, which are excellent antioxidants. Flavonoids are widely distributed in plants, fruits, vegetables, and beverages, such as tea, coffee, beer and red wine. Red wine is especially rich in flavonoids and *in vitro* inhibition of LDL oxidation by red wine has been demonstrated [1450] and reveratrol has been identified as one of the active ingredients [1451]. The *in vivo* effect of red wine, but not white wine, on the susceptibility of plasma and LDL to lipid peroxidation has been shown to be quite substantial [1455]. Although flavonoids are also present in green tea, the advice to take two glasses of red wine daily instead of two cups of green tea, is more willingly accepted by most people.

3. Diet

The accumulated damage to lipids, proteins, and DNA by ROMs during a lifetime is responsible for aging, cancer, and other degenerative diseases of old age [558,1465]. It appears possible that we can slow the aging process and increase life span by decreasing oxidative stress [1465]. This can be accomplished by lowering ROM formation (metabolic rate) or by increasing the levels of endogenous antioxidants (SOD, CAT, GSH-Px, uric acid, carnosine, melantoin) or by adding dietary antioxidants (Vitamins A, C, E, and selenium). I have discussed the chemistry of these antioxidants in Chapter 8.

Numerous studies have suggested that calorie restriction (CR) markedly extends life span and inhibits carcinogenesis [1466-1470]. The same effect was observed with protein restriction (PR) [1470-1474]. CR and PR have been found to decrease all the biomarkers for oxidative damage, such as lipid peroxidation [1475-1477], lipofuscin accumulation [1475], protein oxidation [1478] and DNA damage [1479-1481]. DNA damage was more extensive in individuals eating few fruits or vegetables [1479,1480].

The question is: is this decrease in oxidative damage due to lower metabolic rate (less ROM generation) or is it due to better antioxidant defenses. Arguments for the latter possibility have been presented by Youngman et al. [1478]. There are many reports of increases in SOD, CAT, GSH-Px by calorie restriction or protein restriction [1475,1476,1482-1484].

Specific metabolic rate in many species has been inversely correlated with life span and oxidative damage to DNA [867]. It has been suggested by Adelman et al. [867] that animals with high metabolic rates produce more ROMs, causing greater damage to macromolecules. However this statement has to be taken with a grain of salt. Results by Viguie et al. [1393] have shown that moderate exercise (which increases metabolic rate) did not increase excretion of 8-OHdG (see under exercise!). While oxygen consumption and metabolic rate may play a role in determining life span and oxidative damage, perhaps the most relevant variable, in terms of extending life span, is the capacity to deal with ROMs.

In an oxygen–respiring organism, cellular homeostasis is continuously challenged by ROMs [1485]. In an old organism, this threat could be more serious because of the increased ROM production and deteriorating defenses during aging [558]. It was suggested by Yu et al. [1485] that food restriction slows the aging process by protecting the self-regulatory mechanisms maintaining cellular homeostasis. Microsomal membranes from control rats were capable of more H_2O_2 production than those of food restricted rats. Laganiere and Yu [1483,1484] showed that liver mitochondrial and microsomal membranes from *ad libitum* fed rats contained more lipid hydroperoxides than food restricted rats. In addition, it has been shown that the levels of reduced glutathione, catalase and GSH-Px were all elevated in food restricted rats. Similar results on catalase activity were reported by Koizumi et al. [1476]. From these increases in antioxidant defenses, it appears reasonable to expect that the internal milieu of the cell is well protected against oxidative damage, in food restricted rats. In addition to better coping with oxidative damage food restricted rats had better DNA repair capabilities [1486].

CR reduces membrane damage [1474,1475,1487,1488]. Tacconi et al. [1488] observed that unlike aged rats on a regular diet, rats on a hypocaloric diet presented brain membrane microviscosity and phospholipid composition similar to those of young rats. This is interesting in relation to the deterioration of memory in aged rats. A lifelong hypocaloric diet acts to prevent age-related memory deficits [1489].

ROM-induced oxidative damage is not only affected by caloric restriction and protein restriction, but also by the fat content of the diet. Djuric et al. [1480] examined DNA damage in nucleated peripheral blood cells in a group of women with a high risk for breast cancer. The levels of 5-hydroxymethyl-uracil was 3-fold higher in the group on a non intervention diet, compared to the low fat diet group. A high fat diet has been associated with increased breast cancer risk [1490,1491]. The oxidative DNA damage may be the mechanistic link, since such damage is associated with tumor promotion [562].

A low protein diet was found to have a beneficial effect on tumor development. It was shown by Youngman and Campbell [1474] that post initiation development of aflatoxin B_1-induced gamma glutamyl-transpeptidase positive (GGT$^+$) hepatic foci were markedly decreased by low protein feeding. Foci development, tumor incidence, tumor size, and the number of tumors per animal were markedly reduced while the time to tumor emergence was increased with low protein feeding.

A study on diet, lifestyle, and mortality in 65 counties in China indicated that age at menarche is significantly prolonged by a low protein diet [1492].

4. Psychological stress

The association between personality and the risk of developing coronary heart disease has been extensively investigated [1493,1494]. It has been shown that certain behaviour patterns (Type A) was related to an increased risk to develop coronary heart disease [1495]. I have already discussed (under atherosclerosis) that mental stress causes a paradoxical vasoconstriction leading to ischemia. A hostile personality is indeed a risk factor (among 245 more) for atherosclerosis [1495].

An interesting hypothesis has been presented by Leeming [1496]. The author claims that the evolutionary changes in our diet, from an ascorbic acid rich and low sodium diet to an ascorbic acid poor and sodium rich diet has contributed to our inability to cope with psychological stress and contributes to many diseases of modern man.

The first link between psychological stress and oxidative stress (as measured by a biological marker), was demonstrated by Adachi et al. [1497]. These authors showed 8-OHdG formation increased in the liver of emotionally stressed rats. The amount of 8-OHdG decreased again 1 hour after cessation of the stressful situation, indicating repair. These results of Adachi et al. may provide a connection between stress and cancer. Although the results show repair after cessation of the stress, the repair system may not be able to cope with continued psychological stress. Psychological stress has indeed been shown to lead to decreased DNA repair in human blood cells [1498]. Another index of oxidative stress (lipid peroxidation) was studied by Kosugi et al. [1499]. Human subjects, who could not sleep all night, showed a considerable increase in thiobarbituric acid-active products (a measure of lipid peroxidation) in the urine.

It is difficult to prove a causal relationship between stress and diseases, that have a long-term development such as cancer. Cancer is caused by numerous factors and its development is modulated by genetic factors, by age, and the state of the immune system. It is therefore not too surprising to find lots of controversy concerning the stress-cancer link [1500,1501]. Several investigators have argued that there is no evidence to support the idea that stress causes cancer [1502,1503]. On the other hand, Riley [1504] has shown that removing mice from the stresses of normal housing conditions may reduce the likelihood of tumor growth. We have to realize that each human is an individual with different initial conditions. What constitutes stress or excess stress varies for each individual [1505].

396

The negative effects of stress on the efficiency of the immune system are well documented [1506]. Stress is known to reduce the activity of natural killer cells [1507,1508]. Cancer may not be caused by stress, but may develop as a consequence of a deficient immune system.

Summary. Some of the guidelines for a longer and healthier life are: a low fat, low protein, low calorie diet with lots of fruits and vegetables. Moderate exercise, moderate alcohol consumption (especially recommended two glasses of red wine per day), no smoking, and avoidance of psychological stress.

Conclusion

... science is really founded on observations rather than upon 'facts', and so is a continually evolving structure.
John D. Barrow in "The World within the World"[1]

In looking through the literature we realize that over the years many "facts" have changed. This is not surprising in such a rapidly expanding area of interdisciplinary research. Although some of the details may change, I hope that the basic concepts, which I have tried to transmit, will endure. We have learned a lot about the formation and the rate constants of these reactive oxygen and nitrogen species, their many competing reactions, and the pros and cons of living in a complex chemical world. I have tried to emphasize the importance of sensitive and reliable analytical techniques. Once these techniques were available, the field experienced explosive growth. Future progress will depend on the development of new non-destructive techniques. A complex chemical system like a cell or the human body can most likely never be completely understood. In order to understand all the details we need to know all the rate constants, concentrations, and specific conditions, which are different for each situation. We are dealing with a complex dynamic system where small changes in initial conditions can have unpredictable consequences.

What do we know for certain? Small species, some of which are oxygen radicals ($O_2^{\cdot-}$/$HO_2\cdot$, $\cdot OH$), some are excited oxygen species ($^1O_2$, excited carbonyls), neutral oxygen derived metabolites (H_2O_2, HOCl), and nitrogen radicals like $\cdot NO$, fulfill very important roles in biological systems. These reactive species are produced by normal metabolism (mitochondrial respiratory chain) as well as by metabolism of xenobiotics, and activated phagocytes. These reactive species are both Dr. Jekyll and Mr. Hyde. Some biological functions are transmission of neuronal messages, control of cardiovascular tone, and elimination of invading bacteria and removal of dead tissue. These reactive species, however, due to their high chemical reactivity also have deleterious consequences, leading to many diseases. These reactive species damage membranes, nuclear and mitochondrial DNA, and many enzymes involved in vital biological functions. The organisms are carrying out a precarious balancing act between prooxidant and antioxidant forces, with the former eventually winning the battle. The formation of superoxide radical and nitric oxide individually has been associated with the development of many pathological conditions. Since superoxide radical is chemically not very "super" the purpose of the superoxide dismutases (SODs) in all aerobic cells has been questioned. Since the discovery of NO it has become apparent that the combination of these two radicals of relatively low reactivity produces the highly destructive peroxynitrite, which can initiate many pathological processes. The field is full of controversies and paradoxes, whose resolutions have advanced our knowledge of the complex world of radical chemistry and pathology.

In summary, I would like to quote I. Fridovich [9]: "This text does not begin to do justice to the state of knowledge of the chemistry, biology and pathology of oxygen metabolites. As in all aspects of science, beauty and perceived complexity increase with increased study. We may confidently expect that this will continue in the future. There will be more".

References

1. Barrow, J. D., *The World within the World*, Oxford University Press, Oxford and New York, 1988, p. 4.

2. Gomberg, M., An instance of trivalent carbon: triphenylmethyl, *J. Am. Chem. Soc.*. 22, 757, 1900.

3. Halliwell, B., Gutteridge, J. M. C., and Cross, C. E., Free radicals, antioxidants, and human disease: where are we now?, *J. Lab. Clin. Med.* 119, 598, 1992.

4. Gutteridge, J. M. C., Free radicals in disease processes: a compilation of cause and consequence, *Free Radical Res. Commun.*, 19, 141, 1993.

5. Culotta, E., and Koshland Jr., D. E., NO News Is Good News, *Science*, 258, 1862, 1992.

6. Biadasz Clerch, L. and Massaro, D. J., in Preface to *Oxygen, Gene Expression, and Cell Function*. L. Biadasz Clerch and D. J. Massaro, eds., Marcel Dekker, New York, 1997.

7. Chance, B., Sies, H. and Boveris, A., Hydroperoxide metabolism in mammalian organs, *Physiol. Rev.*, 59, 527-605, 1979.

8. Hassan, H. M., Cytotoxicity of oxyradicals and the evolution of superoxide dismutases, in *Oxygen, Gene Expression, and Cell Function*, L. Biadasz Clerch and D. J. Massaro, eds., Marcel Dekker, New York, 1997.

9. Fridovich, I., Superoxide radical and superoxide dismutases, *Ann. Rev. Biochem.*, 64, 97-112, 1995.

10. Pauling, L., *Vitamin C and the Common Cold*, W. H. Freeman, San Francisco, 1970.

11. Cameron, E., Pauling, L., and Leibovitz, B., Ascorbic acid and cancer: a review , *Cancer Res.*, 39, 663, 1979.

12. Stich, H. F., Karim, J., Koropatnik, J., and Lo, L., Mutagenic action of ascorbic acid, *Nature*, 260, 722, 1976.

13. Slivka, A., Kang, J., and Cohen, G., Hydroxyl radicals and the toxicity of oral iron, *Biochem. Pharmacol.*, 35, 553, 1986.

14. Halliwell, B., Vitamin C: the key to health or a slow-acting carcinogen?, *Redox Report* 1, 5, 1994.

15. Maskos, Z., Koppenol, W. H., Oxyradicals and multivitamin tablets, *Free Radical Biol. Med.*, 11, 609, 1991.

16. Scott, M. D., Meshnick, S. R., and Eaton, J. W., Superoxide dismutase-rich bacteria. Paradoxical increase in oxidant toxicity, *J. Biol. Chem.*, 262, 3640, 1987.

17. Klebanoff, S. J., Oxygen metabolism and the toxic properties of phagocytes, *Ann. Int. Med.*, 93, 480- 489 (1980).

18. Klebanoff, S. J., Oxygen metabolites from phagocytes, in *Inflammation: Basic Principles and Clinical Correlations*, Second Edition, J. I. Gallin, I. M. Goldstein and R. Snyderman, eds., Raven Press, New York 1992.

19. Watson, J. D., *The double helix*, 1968.

20. Bernstein, C., Sex as a response to oxidative DNA damage, in *DNA and Free Radicals*, B. Halliwell and O. I. Aruoma, eds., Ellis Horwood, New York, 1993.

21. Hoffmann, R., *The Same and Not The Same*, Columbia University Press, New York, 1995.

22. For a detailed discussion of free radical chemistry see *Free Radicals*, Vol. I and II, J. K. Kochi, editor, John Wiley, New York, 1973.

23. Ihde, A. J., The history of free radicals and Moses Gomberg's contributions, *Pure and Appl. Chem.*, 15, 1, 1967.

24. Allen, A. O., *The Radiation Chemistry of Water and Aqueous Solutions*, Van Nostrand-Reinhold, Princeton, NJ, 1969.

25. Draganic, I. G., and Draganic, Z. D., *The Radiation Chemistry of water*, Academic Press, New York, 1979.

26. Russell, G. A., Reactivity, selectivity, and polar effects in hydrogen atom transfer reactions, in *Free Radicals*, Vol. I, Kochi, J. K., Ed., John Wiley, New York, 1973, chap. 7.

27. Selected values taken from: *Handbook of Chemistry and Physics*, 75th edition, CRC Press, Boca Raton, FL. 1995.

28. Koenig, T., The decomposition of peroxides and azoalkanes, in *Free Radicals*, Vol. I, Kochi, J. K., Ed., John Wiley, New York, 1973, chap. 3.

29. Arnold, D. R., Baird, N. C., Bolton, J. R., Brand, J. C. D., Jacobs, P. W. M., de Mayo, P., and Ware, W. R., *Photochemistry, an introduction*, Academic Press, New York, 1974.

30. Adam, W., Saha-Möller, C. R., Schönberger, A., Berger, M., and Cadet, J., Formation of 7,8-dihydro-8-oxoguanine in the 1,2-dioxetane-induced oxidation of calf thymus DNA: evidence for photosensitized DNA damage by thermally generated triplet ketones in the dark, *Photochem. Photobiol.*, 62, 231, 1995.

31. Kochi, J. K., Oxidation-Reduction Reactions of Free Radicals and Metal Complexes, in *Free Radicals*, Vol. I, Kochi, J. K., Ed., John Wiley, New York, 1973, chap. 11.

32. Kochi, J., K., Oxygen Radicals, in *Free Radicals*, Vol. II, Kochi, J. K., Ed., John Wiley, New York, 1973, chap. 23.

33. Pryor, W. A., Free Radical reactions in biological systems, in *Free Radicals in Biology*, Vol.I, Academic Press, New York, 1976.

34. Pryor, W. A., and Hendrickson, W. H., Radical production from the interaction of closed shell molecules. II. The reaction of organic sulfides with *tert.*-butyl peroxybenzoate, *J. Am. Chem. Soc.*, 97, 1580, 1975.

35. Pryor, W. A., and Hendrickson, W. H., Radical production from the interaction of closed shell molecules. III. An isotope effect test for distinguishing S_N2 from electron transfer (ET) reactions, *J. Am. Chem. Soc.*, 97, 1582, 1975.

36. Walling, C., and Heaton, L., Hydrogen bonding and complex formation in solutions of t-butyl-hydroperoxide, *J. Am. Chem. Soc.*, 87, 38, 1965.

37. Lankamp, H., Nauta, W. Th., and McLean, C., A new interpretation of the monomer-dimer equilibrium of triphenylmethyl- and alkylsubstituted-diphenyl methyl-radicals in solution, *Tetrahedron Lett.*, 249, 1968.

38. *Peroxyl Radicals*, Alfassi, Z. B., editor, John Wiley, New York, 1997.

39. Walling, C., Fenton's reagent revisited, *Acc. Chem. Res.*, 8, 125, 1975.

40. Walling, C. and Kato, S., The oxidation of alcohols by Fenton's reagent: the effect of copper ion, *J. Am. Chem. Soc.*, 93, 4275, 1971.

41. Abell, P. L., Addition to multiple bonds, in *Free Radicals*, Vol. II, Kochi, J. K., Ed., John Wiley, New York, 1973, chap 13.

42.   Kharasch, M. S., and Mayo, F. R., The peroxide effect in the addition of reagents to unsaturated compounds. I. The addition of hydrogen bromide to allylbromide, *J. Am. Chem. Soc.*, 55, 2468, 1933.

43.   Dorfman, L. M., Taub, I. A., and Bühler, R. F., Pulse radiolysis studies. I. Transient spectra and reaction rate constants in irradiated aqueous solutions of benzene, *J. Chem. Phys.*, 36, 3051, 1962.

44.   Eberhardt, M. K., Reaction of benzene radical cation with water. Evidence for the reversibility of OH radical addition to benzene, *J. Am. Chem. Soc.*, 103, 3876, 1981.

45.   Walling, C., and Johnson, R. A., Fenton's reagent. V. Hydroxylation and side-chain cleavage of aromatics, *J. Am. Chem. Soc.*, 97, 363, 1975.

46.   Vysotskaya, N. A., and Shevchuk, L. G., Proof of the acid fragmentation of hydroxycyclohexadienyl radicals using oxygen-18, *Zh. Org. Khim.*, 9, 2080, 1973.

47.   Shevchuk, L. G, and Vysotskaya, N. A., A study of the mechanism of the demethylation of anisole and the oxidation of phenoxyacetic acid by peroxides using O^{18}, *Zh. Org. Khim.*, 2, 1229, 1966.

48.   Walling, C., and Camaioni, D. M., Aromatic hydroxylation by peroxydisulfate, *J. Am. Chem. Soc.*, 97, 1603, 1975.

49.   Eberhardt, M. K., The effect of metal ions on the hydroxylation of fluorobenzene and toluene by peroxydisulfate, *J. Org. Chem.*, 42, 832, 1977.

50.   Walling, C., Camaioni, D. M., and Soo Kim, S., Aromatic hydroxylation by peroxydisulfate, *J. Am. Chem. Soc.*, 100, 4817, 1978.

51.   Eberhardt, M. K., Homolytic aromatic hydroxylations via radiolysis of aqueous solutions and via metal ion - oxygen systems, *Rev. Heteroatom Chem.*, Vol. 4, pp.1-26, Shigeru Oae, editor, MYU, Tokyo, 1991.

52.   Jefcoate, C. R. E., Lindsay Smith, J. R., and Norman, R. O. C., Hydroxylation. IV. Oxidation of some benzenoid compounds by Fenton's reagent and the ultraviolet irradiation of hydrogen peroxide, *J. Chem. Soc.*, 1013, 1969.

53.   Christensen, H. C., and Gustafson, R., Radiolysis of aqueous toluene solutions, *Acta Chem. Scand.*, 26, 937, 1972.

54. Christensen, H. C., Sehested, K. and Hart, E. J., Formation of benzyl radicals by pulse radiolysis of toluene in aqueous solutions, *J. Phys. Chem.*, 77, 983, 1973.

55. Eberhardt, M. K. and Martinez, M. I., Radiation-induced homolytic aromatic substitution. V. Effect of metal ions on the hydroxylation of toluene, *J. Phys. Chem.*, 79, 1917, 1975.

56. Buxton, G. V. and Green, J. C., Reactions of some simple - and ß-hydroxyalkyl radicals with Cu^{2+} and Cu^+ ions in aqueous solution, *J. Chem. Soc. Faraday Trans.* I, 74, 697, 1977.

57. Freiberg, M., Mulac, W. A., Schmidt, K. H., and Meyerstein, D., Reactions of aliphatic free radicals with copper cations in aqueous solutions. Part 3. Reactions with cuprous ions: a pulse radiolysis study, *J. Chem. Soc.* Faraday I, 76, 1838, 1980.

58. Eberhardt, M. K., Formation of olefins from alkyl radicals with leaving groups in the ß-position, *J. Org. Chem.*, 49, 3720, 1984.

59. Ryan, D. A., and Espenson, J. H., Reactions involving copper(I) in perchlorate solution. A kinetic study of the reaction of chromium(II) and copper(I) ions, *Inorg. Chem.*, 21, 523, 1982.

60. Pollack, R., *"Signs of Life", The Language and Meanings of DNA,* Houghton Mifflin Company, Boston, New York, p. 8, 1994.

61. Fenton, H. J. H., Oxidation of tartaric acid in presence of iron, *J. Chem. Soc.*, 65, 899, 1894.

62. Haber, F., and Weiss, J. J., The catalytic decomposition of hydrogen peroxide by iron salts, *Proc. Royal Soc. London*, Ser. A, 147, 332, 1934.

63 Lampe, F. W., Field, F. H. and Franklin, J. L., Reactions of gaseous ions. IV. Water, *J. Am. Chem. Soc.*, 79, 6132, 1957.

64. Hart, E. J. and Boag, J. W., Absorption spectrum of the hydrated electron in water and aqueous solutions, *J. Am. Chem. Soc.*, 84, 4090, 1962.

65. Matheson, M. S., and Dorfman, L. M., *Pulse radiolysis*, MIT Press, Cambridge, MA, 1969.

66. Borg, D. C., Application of electron spin resonance in biology, in *Free Radicals in Biology*, W. A. Pryor, ed., Vol.1, pp. 69-147, Academic Press, New York, 1976.

67. Janzen, E. G., A critical review of spin trapping in biological systems, in *Free Radicals in Biology*, W. A. Pryor, ed., Vol. 4, pp. 115-154, Academic Press, New York, 1980.

404

68. Britigan, B. E., Cohen, M. S., and Rosen, G. M., Detection of the production of oxygen-centered free radicals by human neutrophils using spin trapping techniques: a critical perspective, *J. Leukocyte Biol.*, 41, 349-362, 1987.

69. Dixon, W. T., and Norman, R. O. C., Free radicals formed during the oxidation and reduction of peroxides, *Nature* (London), 196, 891, 1962.

70. Khan, A. U. and Kasha, M., Red chemiluminescence of oxygen in aqueous solution, *J. Chem. Phys.*, 39, 2105, 1963.

71. For an account of the history of singlet oxygen research including interesting personal experiences by Kasha, see: Kasha, M., Introductory remarks, The Renascence of Research on Singlet Molecular Oxygen in *Singlet Oxygen*, H. H. Wasserman, R. W. Murray, eds., Organic Chemistry, A Series of Monographs, Vol. 40, Academic Press, 1979.

72. Khan, A. U., Singlet molecular oxygen from superoxide anion and sensitized fluorescence of organic molecules, *Science,* 168, 476, 1970.

73. Russell, G. A., Deuterium-isotope effects in the autoxidation of aralkyl hydrocarbons. Mechanism of the interaction of peroxyl radicals, *J. Am. Chem. Soc.*, 79, 3871, 1957.

74. Howard, J. A. and Ingold, K. U., The self-reaction of sec-butylperoxy radicals. Confirmation of the Russell mechanism, *J. Am. Chem. Soc.*, 90, 1056, 1968.

75. Kopecky, K. R., and Mumford, C., Luminescence in the thermal decomposition of 3,3,4-trimethyl-1,2-dioxetane, *Can. J. Chem.*, 47, 709, 1969.

76. White, E. H., Wieko, J., and Roswell, D. F., Photochemistry without light, *J. Am. Chem. Soc.*, 91, 5194, 1969.

77. White, E. H., Wieko, J., and Wei, C. C., Utilization of chemically generated excited states, *J. Am. Chem. Soc.*, 92, 2169, 1970.

78. Adam, W. and Liu, J. C., An -peroxylactone, *J. Am. Chem. Soc.*, 94, 2894, 1972.

79. White, E. H., Miano, J. D., Watkins, C. J., and Breaux, E. J., Chemically produced excited states, *Angew. Chem. Int. Ed. Engl.*, 13, 229-243, 1974.

80.   Adam, W. and Cilento, G., Four-membered ring peroxidesas excited state equivalents: a new dimension in bioorganic chemistry, *Angew. Chem. Int. Ed. Engl.*, 22, 529, 1983.

81.   White, E. H. and Wei, C. C., A possible role for chemically-produced excited states in biology, *Biochem. Biophys. Res. Commun.*, 39, 1219, 1970.

82.   Cilento, G. and Adam, W., Photochemistry and photobiology without light, *Photochem. Photobiol.*, 48, 361, 1988.

83.   McCord, J. M. and Fridovich, I., The reduction of cytochrome c by milk xanthine oxidase, *J. Biol. Chem.*, 243, 5753, 1968.

84.   McCord, J. M. and Fridovich, I., Superoxide dismutase: An enzyme function for erythrocuprein (hemocuprein), *J. Biol. Chem.*, 244, 6049, 1969.

85.   Boveris, A. and Cadenas, E., Production of superoxide radicals and hydrogen peroxide in mitochondria, in: *Superoxide Dismutase*, vol. 2, L. W. Oberley, ed., CRC Press, Boca Raton, FL, 1982.

86.   Aust, S. D., Roerig, D. L. and Pederson, T. C., Evidence for superoxide generation by NADPH-cytochrome c reductase of rat liver microsomes, *Biochem. Biophys. Res. Commun.*, 47, 1133, 1972.

87.   Furchgott, R. F. and Zawadzki, J. V., The obligaory role of endothelial cells in the relaxation of arterial smooth muscle by acetylcholine, *Nature*, 288, 373, 1980.

88.   Furchgott, R. F., Studies on relaxation of rabbit aorta by sodium nitrite: The basis for the proposal that the acid-activatable factor from bovine retractor penis is inorganic nitrite and the endothelium-derived relaxing factor is nitric oxide radical, in : *Mechanisms of Vasodilation*, P. M. Vanhoutte, ed., Raven Press, New York, pp. 427-435, 1988.

89.   Ignarro, L. J., Buga, G. M., Wood, K. S., Byrns, R. E., and Chauduri, G., Endothelium-derived relaxing factor produced and released by artery and vein is nitric oxide, *Proc. Natl. Acad. Sci. USA*, 84, 9265, 1987.

90.   Palmer, R. M. J., Ferrige, A. G., and Moncada, S., Nitric oxide release accounts for the biological activity of endothelium- derived relaxing factor, *Nature*, 327, 524, 1987.

91. Moncada, S., Palmer, R. M. J., and Higgs, E. A., Nitric oxide: physiology, pathophysiology, and pharmacology, *Pharmacol. Rev.,* 43, 109, 1991.

92. Blough, N. V., and Zafirou, O. C., Reaction of superoxide with nitric oxide to form peroxynitrite in alkaline aqueous solutions, *Inorg. Chem.,* 24, 3502, 1985.

93. Gryglewski, R. J., Palmer, R. M. J., and Moncada, S., O_2^- is involved in the breakdown of endothelium-derived relaxing factor, *Nature,* 320, 454, 1986.

94. McCall, Th. B., Boughton-Smith, N. K., Palmer, R. M. J., Whittle, B. J. R., and Moncada, S., Synthesis of nitric oxide from L-arginine by neutrophils: Release and interaction with superoxide, *Biochem. J.,* 261, 293, 1989.

95. Sturm, R. J., Adams, L. M., Grimes, D., Weichman, B. N., and Rimele, T. J., Inactivation of neutrophil-and endothium-derived relaxing factors (NDRF, EDRF) by superoxide release from rat neutrophils, *FASEB J.,* 3, A 1178, 1989.

96. Kanofsky, J. R., Singlet oxygen production by biological systems, *Chem.-Biol. Interactions,* 70, 1, 1989.

97. Steinbeck, M. J., Khan, A. U. and Karnovsky, M. J., Intracellular singlet oxygen generation by phagocytizing neutrophils in response to particles coated with a chemical trap, *J. Biol. Chem.,* 267, 13425, 1992.

98. Steinbeck, M. J., Khan, A. U. and Karnovsky M. J., Extracellular production of singlet oxygen by stimulated macrophages quantified using 9,10-diphenylanthracene and perylene in a polystyrene film, *J. Biol. Chem.,* 268, 15,649, 1993.

99. Buxton, G. V., Greenstock, C. L., Helman, W. Ph., and Ross, A., Rate constants for reactions of hydroxyl radicals in aqueous solutions, *J. Phys. Chem.* Ref. Data, Vol 17, No 2, 1988.

100. Buxton, G. V., Greenstock, C. L., Helman, W. Ph., and Ross, A., Rate constants for reactions of hydrated electrons in aqueous solutions, *J. Phys. Chem.* Ref. Data, Vol 17, No 2, 1988.

101. *Singlet Oxygen,* H. H. Wasserman, R. W. Murray, eds., Organic Chemistry, A Series of Monographs, Vol. 40, Academic Press, 1979.

102. Harrison, J. E. and Schultz, J., Studies on the chlorinating activity of myeloperoxidase, *J. Biol. Chem.*, 251, 1371, 1976.

103. Steinman, H. M. in, *Superoxide dismutase*, L. W. Oberley, ed. pp. 11-68, CRC Press, Boca Raton, FL, 1982.

104. Beyer, W., Imlay, J. and Fridovich, I., Superoxide dismutases, *Prog. Nucleic Acid Res. Mol. Biol.*, 40, 221, 1991.

105. A discussion by I. Fridovich contra J. A. Fee can be found in: *Oxygen and Oxy-Radicals in Chemistry and Biology*, M. A. J. Rodgers and E. L. Powers, eds., Academic Press, New York, 1981, pp. 197-239.

106. Sawyer, D. T. and Valentine, J. S., How super is superoxide ion?, *Acc. Chem. Res.*, 14, 393, 1981.

107. Fridovich, I., Biological effects of the superoxide radical, *Arch. Biochem. Biophys.*, 247, 1, 1986.

108. Farr, S. B., D'Ari, R., Touati, D., Oxygen-dependent mutagenesis in *Escherichia coli* lacking superoxide dismutase, *Proc. Natl. Acad. Sci. USA*, 83, 8268, 1986.

109. Carlioz, A. and Touati, D., Isolation of superoxide dismutase mutants in *Escherichia coli*: is superoxide dismutase necessary for aerobic life?, *EMBO J.*, 5, 623, 1986.

110. Boveris, A., and Chance, B., The mitochondrial generation of hydrogen peroxide. General properties and effect of hyperbaric oxygen, *Biochem. J.*, 134, 707, 1973.

111. Loschen, G., Azzi, A., Richter, C., and Flohé, L., Superoxide radicals as precursors of mitochondrial hydrogen peroxide, *FEBS Letters*, 42, 68, 1974.

112. Dionisi, O., Galeotti, T., Terranova, T., and Azzi, A., Superoxide radicals and hydrogen peroxide formation in mitochondria from normal and neoplastic tissues, *Biochim. Biophys. Acta*, 403, 292, 1975.

113. Sbarra, A. J., and Karnovsky, M. L., The biochemical basis of phagocytosis. I. Metabolic changes during the ingestion of particles by polymorphonuclear leukocytes, *J. Biol. Chem.*, 234, 1355, 1959.

114. Babior, B. M., Kipnes, R. S., and Curnutte, J. T., Biological defense mechanisms. The production by leukocytes of superoxide, a potential bactericidal agent, *J. Clin. Invest.*, 52, 741, 1973.

408

115. Babior, B. M., Oxidants from phagocytes: agents of defense and destruction, *Blood,* 64, 959, 1984.

116. Huie, R. E., and Padmaja, S., The reaction of NO with superoxide, *Free Radical Res. Commun.,* 18, 195, 1993.

117. Malinski, T., Taha, Z., Nitric oxide release from a single cell measured in situ by a porphyrinic-based microsensor, *Nature,* 358, 676, 1992.

118. Cadenas, E., Giulivi, C., Ursini, F. and Boveris, A., Electronically excited state formation, *Meth. Toxicol.,* 1B, 384, 1994.

119. Boveris, A., Cadenas, E., Reiter, R., Filipkowski, M., Nakase, Y. and Chance, B., Organ chemiluminescence: non-invasive assay for oxidative radical reactions, *Proc. Natl. Acad. Sci. USA,* 77, 347, 1980.

120. Floyd, R. A., Watson, J. J., Wong, P. K., Altmiller, D. H. and Rickard, R. C., Hydroxyl free radical adduct of deoxy-guanosine: a sensitive detection and mechanism of formation, *Free Radical Res. Commun.,* 1, 163, 1986.

121. Floyd, R. A., West, M. S., Enett, K. L., Schneider, J. E., Wong, P. K., Tingey, D. T., and Hogsett, W. E., Conditions influencing yield and analysis of 8-hydroxy-2'-deoxyguanosine in oxidatively damaged DNA, *Anal. Biochem.,* 188, 155, 1990.

122. Floyd, R. A., The role of 8-hydroxyguanine in carcinogenesis, *Carcinogenesis,* 11, 1447, 1990.

123. Kasai, H., Nishimura, S., Formation of 8-hydroxy-deoxy-guanosine in DNA by oxygen radicals and its biological significance, in *Oxidative Stress, Oxidants and Antioxidants,* H. Sies, ed., pp. 99-116, Academic Press, New York 1991.

124. Floyd, R. A., Watson, J. J., Harris, J., West, M., and Wong, P. K., Formation of 8-hydroxydeoxyguanosine, hydroxyl free radical adduct of DNA in granulocytes exposed to the tumor promoter, tetra-deconylphorbol acetate, *Biochem. Biophys. Res. Commun.,* 137, 841, 1986.

125. Aruoma, O. I., Experimental tools in free radical bio-chemistry, in *Free Radicals in Tropical Diseases,* O. I. Aruoma, ed., Harwood Academic Publishers, 1993, chap. 11.

126. Moncada, S. and Higgs, A., The L-arginine-nitric oxide pathway, *N. Engl. J. Med.,* 329, 2002, 1993.

127. Naqui, A., Chance, B. and Cadenas, E., Reactive oxygen intermediates in biochemistry, *Ann. Rev. Biochem.*, 55, 137, 1986.

128. For a detailed review of the huge literature on superoxide radical anion chemistry see Afanas'ev, I. B., *Superoxide Ion: Chemistry and Biological Implications*, Vol I and II, CRC Press, Boca Raton, FL, 1989.

129. Sellers, R. M. and Simic, M. G., Pulse radiolysis study of the reactions of some reduced metal ions with molecular oxygen in aqueous solution, *J. Am. Chem. Soc.*, 98, 6145, 1976.

130. Udenfriend, S., Clark, C. T., Axelrod, J. and Brodie, B. B., Ascorbic acid in aromatic hydroxylation. I. A model system for aromatic hydroxylation, *J. Biol. Chem.*, 208, 731, 1954.

131. Brodie, B. B., Axelrod, J., Shore, P. A. and Udenfriend, S, Ascorbic acid in aromatic hydroxylation. II. Products formed by reaction of substrates with ascorbic acid, ferrous ion, and oxygen, *J. Biol. Chem.*, 208, 741, 1954.

132. Patel, K. B. and Willson, R. L., Semiquinone free radicals and oxygen. Pulse radiolysis study of one electron transfer equilibria, *J. Chem. Soc. Faraday Trans.* 1, 814, 1973.

133. Ilan, Y. A., Czapski, G. and Meisel, D., The one-electron transfer redox potentials of free radicals. I. The oxygen/superoxide system, *Biochim. Biophys. Acta*, 430, 209, 1976.

134. Adams, G. E. and Willson, R. L., Pulse radiolysis studies on the oxidation of organic radicals in aqueous solution, *Trans. Faraday Soc.*, 65, 2981, 1969.

135. Farrington, J. A., Ebert, M., Land, E. J., and Fletcher, K., Bipyridilium quaternary salts and related compounds. V. Pulse radiolysis studies of the reaction of paraquat radical with oxygen. Implications for the mode of action of bipyridil herbicides, *Biochim. Biophys. Acta*, 314, 372, 1973.

136. Marklund, S. and Marklund, G., Involvement of the superoxide anion radical in the autoxidation of pyrogallol and a convenient assay for superoxide dismutase, *Eur. J. Biochem.*, 47, 469, 1974.

137. Heikkila, R. E. and Cohen, G., 6-Hydroxydopamine: evidence for superoxide radical as an oxidative intermediate, *Science*, 181, 456, 1973.

138. Misra, H. P., Generation of superoxide free radical during the autoxidation of thiols, *J. Biol. Chem.*, 249, 2151, 1974.

139. Foote, C. S., Dzakpasu, A. A., and Lin, J. W. P., Chemistry of singlet oxygen. XX. Mechanism of the sensitized photo-oxidation of enamines, *Tetrahedron Lett.*, 1247, 1975.

140. Saito, I., and Matsuura, T., Formation of superoxide ion via one-electron transfer from organic electrondonors to singlet oxygen, in *Oxygen Radicals in Chemistry and Biology*, W. Bors, M. Saran, and D. Tait, eds., Walter deGruyter, Berlin 1984, p. 535.

141. Czapski, G., Levanon, H. and Samuni, A., ESR studies of uncomplexed and complexed $HO_2\cdot$ radical formed in the reaction of H_2O_2 with Ce^{4+}, Fe^{3+} and Ti^{3+} ions, *Isr. J. Chem.*, 7, 375, 1969.

142. Walling, C. and Cleary, M., Oxygen evolution as a critical test in the ferric-ion catalyzed decomposition of hydrogen peroxide, *Int. J. Chem. Kinet.*, 9, 595, 1977.

143. Cabelli, D. E. and Bielski, B. H. J., Kinetics and mechanism for the oxidation of ascorbic acid/ascorbate by $HO_2\cdot/O_2\cdot^-$ radicals. A pulse radiolysis and stopped-flow photolysis study, *J. Phys. Chem.*, 87, 1809, 1983.

144. Rabani, J., Klug-Roth, D. and Lilie, J., Pulse radiolytic investigations of the catalyzed disproportionation of peroxy radicals. Aqueous cupric ions, *J. Phys. Chem.*, 77, 1169, 1973.

145. Rush, J. D. and Bielski, B. H. J., Pulse radiolytic study of the reaction of perhydroxyl/superoxide $O_2\cdot^-$ with iron(II) iron(III) ions. The reactivity of $HO_2\cdot/O_2\cdot^-$ with ferric ions and its implications on the occurrence of the Haber-Weiss reaction, *J. Phys. Chem.*, 89, 5062, 1985.

146. Bielski, B. H. J. and Cabelli, D. E., The role of transition metals in oxy-radicals induced oxidation processes in biological systems, in *Superoxide and Superoxide Dismutase in Chemistry, Biology and Medicine*, G. E. Rotilio, ed., Elsevier, Amsterdam, 1986, p. 3.

147. Ferradini, C., Foos, J., Gilles, L., Haristoy, D. and Pucheault, J., Gamma and pulse radiolysis studies of the reactions between superoxide ions and oxyhemoglobin-methemoglobin system, *Photochem. Photobiol.*, 28, 851, 1978.

148. Sutton, H. C., Roberts, P. B. and Winterbourn, C. C., The rate of reaction of superoxide radical ion with oxyhemoglobin and methemoglobin, *Biochem. J.*, 155, 503, 1976.

149.   Flint, D. H., Tuminello, J. F. and Emptage, M. H., The inactivation of Fe-S cluster containing hydrolyases by superoxide, *J. Biol. Chem.*, 268, 22,369, 1993.

150.   Kou, C. F., Mashino, T. and Fridovich, I., a,b-Dihydroxy isovalerate dehydratase: a superoxide-sensitive enzyme, *J. Biol. Chem.*, 262, 4724, 1987.

151.   Gardner, P. R. and Fridovich, I., Superoxide sensitivity of the *Escherichia coli* 6-phospogluconate dehydratase, *J. Biol. Chem.*, 266, 1478, 1991.

152.   Gardner, P. R., and Fridovich, I., Superoxide sensitivity of the *Escherichia coli* aconitase, *J. Biol. Chem.*, 266, 19,328, 1991.

153.   Woods, S. A., Schwartzbach, S. D. and Guest, J. R., Two biochemically distinct classes of fumarase in *Escherichia coli*, *Biochim. Biophys. Acta*, 954, 14, 1988.

154.   Flint, D. H., Emptage, M. H. and Guest, J. R., Fumarase A from *Escherichia coli*: purification and characterization as an iron-sulfur cluster containing enzyme, *Biochemistry*, 31, 10331, 1992.

155.   Liochev, S. I. and Fridovich, I., Modulation of the fumarases of *Escherichia coli* in response to oxidative stress, *Arch. Biochem. Biophys.*, 301, 379, 1993.

156.   Gardner, P. R. and Fridovich, I., Quinolinate synthatase: The oxygen-sensitive site of de novo NAD^+ biosynthesis, *Arch. Biochem. Biophys.*, 284, 106, 1991.

157.   Liochev.S. I. and Fridovich, I., The role of O_2^- in the production of ·OH: in vitro and in vivo, *Free Radical Biol. Med.*, 16, 29, 1994.

158.   Buettner, G.R., Saran, M. and Bors, W., The kinetics of the reaction of ferritin with superoxide, *Free Radical Res. Commun.*, 2, 369, 1987.

159.   Bielski, B. H. J. and Allen, A. O., Mechanism of the disproportionation of superoxide radicals, *J. Phys. Chem.*, 81, 1048, 1977.

160.   Bielski, B. H. J., Cabelli, D. E., Arudi, R. L., and Ross, A. B., Reactivity of $HO_2·/O_2^-$ radicals in aqueous solution, *J. Phys. Chem.*, Ref. Data 14, 1041, 1985.

161.   Fridovich, I., Superoxide radical: an endogenous toxicant, *Ann. Rev. Pharmacol. Toxicol.*, 23, 239, 1983.

412

162. Pryor, W. A. and Squadrito, G. L., The chemistry of peroxynitrite: a product from the reaction of nitric oxide with superoxide, *Am. J. Physiol.*, 268 (Lung Cell. Mol. Physiol. 12): L699-L722, 1995.

163. Land, E. J. and Swallow, A. J., One-electron reactions in biochemical systems as studied by pulse radiolysis. IV. Oxidation of dihydronicotinamide-adenine dinucleotide, *Biochim. Biophys. Acta*, 234, 34, 1971.

164. Bielski, B. H. J., and Chan, P. C., Enzyme-catalyzed free radical reactions with nicotine-adenine nucleotides. I. Lactate dehydrogenase-catalyzed chain oxidation of bound NADH by superoxide radicals, *Arch. Biochem. Biophys.*, 159, 873, 1973.

165. Bielski, B. H. J. and Chan, P. C., Re-evaluation of the kinetics of lactate dehydrogenase-catalyzed chain oxidation of nicotinamide adenine dinucleotide by superoxide radicals in the presence of ethylenediaminetetraacetate, *J. Biol. Chem.*, 251, 3841, 1976.

166. Bielski, B. H. J. and Chan, P. C., Kinetic study by pulse radiolysis of the lactate dehydrogenase-catalyzed chain oxidation of nicotinamide adenine dinucleotide by $HO_2 \cdot$ and $O_2 \cdot^-$ radicals, *J. Biol. Chem.*, 250, 318, 1974.

167. Fielden, E. M., Roberts, P. B., Bray, R. C., Lowe, D. J., Mautner, G. N., Rotilio, G. and Calabrese, L., Mechanism of action of superoxide dismutase from pulse radiolysis and electron paramagnetic resonance. Evidence that only half the active site functions in catalysis, *Biochem. J.*, 129, 49, 1974.

168. Klug, D., Rabani, J. and Fridovich, I., A direct demonstration of the catalytic action of superoxide dismutase through the use of pulse radiolysis, *J. Biol. Chem.*, 247, 4839, 1972.

169. Rao, P. S. and Hayon, E., Redox potentials of free radicals. IV. Superoxide and hydroperoxy radical: $O_2 \cdot^-$ and $HO_2 \cdot$, *J. Phys.Chem.*, 79, 397, 1975.

170. Miller, R. W., Reactions of superoxide anion, catechols, and cytochrome c, *Can. J. Biochem.*, 48, 935, 1970.

171. Miller, R. W. and Rapp, U., The oxidation of catechols by reduced flavins and dehydrogenases. An electron spin resonance study of the kinetics and initial products of oxidation, *J. Biol. Chem.*, 248, 6084, 1973.

172.   Nishikimi, M. and Machlin, L. J., Oxidation of -tocopherol model compound by superoxide anion, *Arch. Biochem. Biophys.*, 170, 684, 1975.

173.   Thomas, M. J., Mehl, K. S. and Pryor, W. A., The role of superoxide anion in the xanthine-oxidase induced autoxidation of linoleic acid, *Biochem. Biophys. Res. Commun.*, 83, 927, 1978.

174.   Sutherland, M. W. and Gebicki, J. M., A reaction between the superoxide free radical and lipid hydroperoxide in sodium linoleate micelles, *Arch. Biochem. Biophys.* 214, 1, 1982.

175.   Afanas'ev, I. B., Grabovetski, V. V., Kuprianova, N. S. and Gunar, V. I., Kinetics and mechanism of the interaction of O_2^- with ascorbic acid and alpha-tocopherol, in *Superoxide and Superoxide Dismutase in Chemistry, Biology and Medicine*, Rotilio, G., ed., Elsevier, New York, 1986, p.50.

176.   Afanas'ev, I. B., Grabovetski, V. V. and Kuprianova, N. S., Kinetics and mechanism of the reaction of superoxide ion in solution. V. Kinetics and mechanism of the interaction of superoxide ion with vitamin E and ascorbic acid, *J. Chem. Soc.*, Perkin Trans. 2, 281, 1987.

177.   Afanas'ev, I. B., Reactivity of superoxide ion, in *Superoxide ion: Chemistry and Biological Implications*, Vol. I, chap. 4, CRC Press, Boca Raton, FL, 1989.

178.   Afanas'ev, I. B. and Polozova, N. I., Formation of semiquinones in the reaction of O_2^- with hydroquinones, *Zh. Org. Khim.*, 12, 1833, 1976.

179.   Afanas'ev, I. B. and Polozova, N. I., One-electron oxidation of para- and ortho-dihydroxybenzenes by the oxygen radical anion in an aprotic medium, *Zh. Org. Khim.*, 14, 1013, 1978.

180.   Rowley, D. A. and Halliwell, B., Superoxide-dependent formation of hydroxyl radicals in the presence of thiol compounds, *FEBS Letters*, 138, 33, 1982.

181.   Motohashi, N. and Mori, I., Thiol-induced hydroxyl radical formation and scavenger effect of thiocarbamides on hydroxyl radicals, *J. Inorg. Biochem.*, 26, 205, 1986.

182.   van Stevenick, J., van der Zee, J. and Dubbelman, T. M. A. R., Site-specific and bulk-phase generation of hydroxyl radicals in the presence of cupric ions and thiol compounds, *Biochem. J.*, 232, 309, 1985.

183. Florence, T. M., The production of hydroxyl radical from hydrogen peroxide, *J. Inorg. Biochem.*, 22, 221, 1984.

184. Rowley, D. A. and Halliwell, B., Superoxide-dependent formation of hydroxyl radicals from NADH and NADPH in the presence of iron salts, *FEBS Letters*, 142, 39, 1982.

185. Winterbourn, C. C., Comparison of superoxide with other reducing agents in the biological production of hydroxyl radicals, *Biochem. J.*, 182, 625, 1979.

186. Czapski, G., Goldstein, S. and Meyerstein, D., What is unique about superoxide toxicity as compared to other biological reductants: a hypothesis, *Free Radical Res. Commun.*, 4, 231, 1988.

187. Buettner, G. R., Doherty, Th. P. and Bannister, Th. D., Hydrogen peroxide and hydroxyl radical formation by methylene blue in the presence of ascorbic acid, *Radiat. Environ. Biophys.*, 23, 235, 1984.

188. Rowley, D. A. and Halliwell, B., Formation of hydroxyl radicals from hydrogen peroxide and iron salts by superoxide- and ascorbate-dependent mechanisms: relevance to the pathology of rheumatoid disease, *Clin. Sci.*, 64, 649, 1983.

189. von Sonntag, C. and Schuchmann, H.-P., Peroxyl radicals in aqueous solutions, in *Peroxyl Radicals*, Z. Alfassi, ed., John Wiley, New York, 1997, pp.173-234.

190. Ilan, Y., Rabani, J. and Henglein, A., Pulse radiolytic investigations of peroxy radicals produced from 2-propanol and methanol, *J. Phys. Chem.*, 80, 1558, 1976.

191. Bothe, E., Behrens, G. and Schulte-Frohlinde, D., Mechanism of the first order decay of 2-hydroxy-propyl-2-peroxyl radicals and of O_2^- · formation in aqueous solution, *Z. Naturforsch.*, 32b, 886, 1977.

192. Bothe, E., Schuchmann, M. N., Schulte-Frohlinde, D. and von Sonntag, C., HO_2· elimination from -hydroxyalkylperoxyl radicals in aqueous solution, *Photochem. Photobiol.*, 28, 639, 1978.

193. Jin, F., Leitich, J. and v. Sonntag, C., The superoxide radical reacts with tyrosine-derived phenoxyl radicals by addition rather than by electron transfer, *J. Chem. Soc. Perkin Trans.*, 2, 1583, 1993.

194. Pan, X. M., Schuchmann, M. N. and v. Sonntag, C., Oxidation of benzene by the OH radical. A product and pulse radiolysis study in oxygenated aqueous solution, *J. Chem. Soc. Perkin Trans. 2*, 289, 1993.

195.   Merga, G., Schuchmann, H. P., Rao, B. S. M. and v. Sonntag, C.,
        OH radical-induced oxidation of chlorobenzene in aqueous solution
        in the absence and presence of oxygen, *J. Chem. Soc. Perkin
        Trans.* 2, 1097, 1996.

196.   Porter, N. A., Roe, A. N. and McPhail, A. T., Serial cyclization of
        peroxyl free radicals: models for polyolefin oxidation, *J. Am. Chem.
        Soc.*, 102, 7574, 1980.

197.   Roe, A. N., McPhail, A. T. and Porter, N. A., Serial cyclization:
        Studies in the mechanism and stereochemistry of peroxy radical
        cyclyzation, *J. Am. Chem. Soc.*, 105, 1199, 1983.

198.   Porter, N. A., Lehman, L. S., Weber, B. A. and Smith, K. J., Unified
        mechanism for polyunsaturated fatty acid autoxidation. Competition
        of peroxy radical hydrogen atom abstraction, ß-scission and
        cyclization, *J. Am. Chem. Soc.*, 103, 6447, 1981.

199.   Lee, S. H. and Mendenhall, G. D., Relative yields of excited ketones
        from self-reactions of alkoxyl and alkylperoxyl radical pairs, *J. Am.
        Chem. Soc.*, 110, 4318, 1988.

200.   Mendenhall, G. D., Sheng, X. C. and Wilson, T., Yields of excited
        carbonyl species from alkoxyl and from alkylperoxyl radical
        dismutations, *J. Am. Chem. Soc.*, 113, 8976, 1991.

201.   Niu, Q. J. and Mendenhall, G. D., Yields of singlet molecular oxygen
        from peroxyl radical termination, *J. Am. Chem. Soc.*, 114, 165, 1992.

202.   Eberhardt, M. K., Ramirez, G., and Ayala, E., Does the reaction of
        Cu^+ with H_2O_2 give OH radicals? A study of aromatic hydroxylation,
        J. Org. Chem., 54, 5922, 1989.

203.   Tezuka, T. and Narita, N., Hydroxylation of benzene with   -
        azohydroperoxide. A novel route for generation of hydroxyl radical
        and its reaction in anhydrous media, *J. Am. Chem. Soc.*, 101, 7413,
        1979.

204.   Tezuka, T., Narita, N., Ando, W. and Oae, S., Isomer distribution
        ratios of phenols in aromatic hydroxylation with the hydroxyl radical
        generated from   -azohydroperoxide in anhydrous media.
        Comparison with Fenton's reagent, *J. Am. Chem. Soc.*, 103, 3045,
        1981.

205.   Narita, N. and Tezuka, T., On the mechanism of oxidation of
        hydroxycyclohexadienyl radicals with molecular oxygen, *J. Am.
        Chem. Soc.*, 104, 7316, 1982.

416

206. Saito, I., Takayama, M., Matsuura, T., Matsugo, S. and Kawanishi, S., Phthalimide hydroperoxides as efficient photochemical hydroxyl radical generators, a novel DNA-cleaving agent, *J. Am. Chem. Soc.*, 112, 883, 1990.

207. Matsugo, S. and Saito, J., Photochemical cleavage of N-(hydroperoxyalkyl) phthalimides by intramolecular energy transfer, *Tetrahedron Lett.*, 32, 2949, 1991.

208. Boivin, J., Crépon, E. and Zard, S. Z., N-hydroxy-2-pyridinethione: a mild and convenient source of hydroxyl radicals, *Tetrahedron Lett.*, 37, 6869, 1990.

209. Hess, K. M. and Dix, T. A., Evaluation of *N*-hydroxy-2-thiopyridone as a nonmetal dependent source of the hydroxyl radical in aqueous systems, *Anal. Biochem.*, 206, 309, 1992.

210. Reszka, K. J. and Chignell, C. F., Photochemistry of 2-mercaptopyridines. Part 2. An EPR and spin-trapping investigation using 2-methyl-2-nitrosopropane and *aci*-nitromethane as spin traps in aqueous solutions, *Photochem. Photobiol.*, 60, 450, 1994.

211. Reszka, K. J. and Chignell, C. F., Photochemistry of 2-mercaptopyridines. In aqueous and toluene solutions. Part 1. An EPR and spin-trapping investigation using 5,5-dimethyl-1-pyrroline *N*-oxide (DMPO), *Photochem. Photobiol.*, 60, 442, 1994.

212. Epe, B., Ballmaier, D., Adam, W., Grimm, G. N. and Saha-Möller, C. R., Photolysis of N-hydroxypyridinethiones: a new source of hydroxyl radicals for the direct damage of cell-free and cellular DNA, *Nucleic Acids Res.*, 24, 1625, 1996.

213. Adam, W., Ballmaier, D., Epe, B., Grimm, G. N. and Saha-Möller, C. R., *N*-hydroxypyridinethiones as photochemical hydroxyl radical sources for oxidative DNA damage, *Angew. Chem. Int. Ed. Engl.*, 34, 2156, 1995.

214. Adam, W., Cadet, J., Dall'Acqua, F., Epe, B., Ramaiah, D. and Saha-Möller, C. R., Photosensitized formation of 8-hydroxy-2'-deoxyguanosine in salmon testes DNA by furocoumarin hydroperoxides: a novel intercalating "photo-Fenton" reagent for oxidative DNA damage, *Angew. Chem. Int. Ed. Engl.*, 34, 107, 1995.

215. Adam, W., Berger, M., Cadet, J., Dall'Acqua, F., Epe, B., Frank, S., Ramaiah, D., Raoul, S., Saha-Möller, C. R. and Vedaldi, D.,

Photochemistry and photobiology of furocoumarin hydroperoxides derived from imperatorin: novel intercalating photo-Fenton reagents for oxidative DNA modification by hydroxyl radicals, *Photochem. Photobiol.*, 63, 768, 1996.

216. von Sonntag, C., *The Chemical Basis of Radiation Biology*, Taylor and Francis, New York, 1987.

217. Epe, B. and Hegler, J., Oxidative DNA damage: endonuclease fingerprinting, *Methods Enzymol.*, 234, 122, 1994.

218. Winterbourn, C. C., Production of hydroxyl radicals from paraquat radicals and H_2O_2, *FEBS Letters*, 128, 339, 1981.

219. Winterbourn, C. C., Evidence for the production of hydroxyl radicals from the adriamycin semiquinone and H_2O_2, *FEBS Letters*, 136, 89, 1981.

220. Winterbourn, C. C. and Sutton, H. C., Hydroxyl radical production from hydrogen peroxide and enzymatically generated paraquat radicals: catalytic requirements and oxygen dependence, *Arch. Biochem. Biophys.*, 235, 116, 1984.

221. Winterbourn, C. C., Gutteridge, J. M. C. and Halliwell, B., Doxorubicin-dependent lipid peroxidation at low partial pressures of O_2, *J. Free Radical Biol. Med.*, 1, 43, 1985.

222. Beckman, J. S., Beckman, T. W., Chen, J., Marshall, P. A. and Freeman, B. A., Apparent hydroxyl radical production by peroxynitrite: implications for endothelial injury from nitric oxide and superoxide, *Proc. Natl. Acad. Sci. USA*, 87, 1620, 1990.

223. Douki, T. and Cadet, J., Peroxynitrite mediated oxidation of purine bases of nucleosides and isolated DNA, *Free Radical Res.*, 24, 369, 1996.

224. Long, C.A. and Bielski, B. H. J., Rate of reaction of superoxide radical with chloride-containing species, *J. Phys. Chem.*, 54, 655, 1980.

225. Condeias, L. P., Patel, K. B., Stratford, M. R. L. and Wardman, P., Free hydroxyl radicals are formed on reaction between the neutrophil-derived species superoxide anion and hypochlorous acid, *FEBS Letters*, 333, 151, 1993.

226. Janzen, E. G., Substituent effects on electron spin resonance spectra and stability of free radicals, *Acc. Chem. Res.*, 4, 31, 1970.

418

227. Buettner, G. R., Spin trapping: ESR parameters of spin adducts, *Free Radical Biol. Med.*, 3, 259, 1987.

228. Beauchamp, C. O. and Fridovich, I., A mechanism for the production of ethylene from methional: the generation of hydroxyl radical by xanthine oxidase, *J. Biol. Chem.*, 243, 4641, 1970.

229. Eberhardt, M. K., The hydroxyl radical. Formation and some reactions used in its identification, *Trends in Organic Chemistry*, 5, 115-139, 1995.

230. Shiga, T., An electron paramagnetic resonance study of alcohol oxidation by Fenton's reagent, *J. Phys. Chem.*, 69, 3805, 1965.

231. Chandra, M. and Symons, M. C. R., Hydration of spin-trap cations as a source of adducts, *J. Chem. Soc. Chem. Commun.*, p.1301, 1986.

232. Makino, K., Hagiwara, T., Hagi, A., Nishi, M., and Murakami, A., Cautionary note for DMPO spin trapping in the presence of iron ion, *Biochem. Biophys. Res. Commun.*, 172, 1073, 1990.

233. Finkelstein, E., Rosen, G. M., and Rauckman, E. J., Production of hydroxyl radicals by decomposition of superoxide spin-trapped adducts, *Mol. Pharmacol.*, 21, 262, 1982.

234. Hanna, P. M., Kadiiska, M. B. and Mason, R. P., Oxygen-derived free radical and active oxygen complex formation from cobalt(II) chelates in vitro, *Chem. Res. Toxicol.*, 5, 109, 1992.

235. Finkelstein, E., Rosen, G. M., and Rauckman, E. J., Spin trapping of superoxide and hydroxyl radical: practical aspects, *Arch. Biochem. Biophys.*, 200, 1, 1980.

236. Pou, S., Hassett, D. J., Britigan, B. E., Cohen, M. S. and Rosen, G. M., Problems associated with spin trapping oxygen-centered free radicals in biological systems, *Anal. Biochem.*, 177, 1, 1989.

237. Dixon, W. T., Norman, R. O. C., and Buley, A. C., Electron spin resonance studies of oxidation. Part II. Aliphatic acids and substituted acids, *J. Chem. Soc.*, p. 3625, 1964.

238. Gilbert, B. C., Norman, R. O. C., and Sealy, R. C., Electron spin resonance studies. Part XLIII. Reaction of dimethylsulfoxide with the hydroxyl radical, *J. Chem. Soc. Perkin Trans.* II, 303, 1975.

239. Eberhardt, M. K. and Colina, R., The reaction of OH radicals with dimethylsulfoxide. A comparative study of Fenton's reagent and the radiolysis of aqueous dimethylsulfoxide solutions, *J. Org. Chem.*, 53, 1071, 1988.

240. Davies, M. J., Gilbert, B. C. and Norman, R. O. C., Electron spin resonance. Part 67. Oxidation of aliphatic sulphides and sulfoxides by the sulfate radical anion ($SO_4^{-\cdot}$) and of aliphatic radicals by the peroxydisulfate anion ($S_2O_8^{2-}$), *J. Chem. Soc. Perkin Trans. II,* 503, 1984.

241. Klein, S. M., Cohen, G., and Cederbaum, A. I., Production of formaldehyde during metabolism of dimethyl sulfoxide by hydroxyl radical generating systems, *Biochemistry,* 20, 6006, 1981.

242. Klein, S. M., Cohen, G., and Cederbaum, A. I., The interaction of hydroxyl radicals with dimethylsulfoxide produces formaldehyde, *FEBS Letters,* 116, 220, 1980.

243. Repine, J. E., Eaton, J. W., Anders, M. W., Hoidal, J. R., and Fox, R. B., Generation of hydroxyl radical by enzymes, chemicals and human phagocytes in vitro. Detection with the anti-inflammatory agent dimethyl sulfoxide, *J. Clin. Invest.,* 64, 1642, 1979.

244. Li, B., Gutierrez, P. L. and Blough, N. V., Trace determination of hydroxyl radical in biological systems, *Anal. Chem.,* 69, 4295, 1997.

245. Bors, W., Lengfelder, E., Saran, M., Fuchs, C., and Michel, C., Reactions of oxygen radical species with methional: a pulse radiolysis study, *Biochem. Biophys. Res. Commun.,* 70, 81, 1976.

246. Cohen, G. and Cederbaum, A. I., Chemical evidence for production of hydroxyl radicals during microsomal electron transfer, *Science,* 204, 66, 1979.

247. Pryor, W. A. and Tang, R. H., Ethylene formation from methional, *Biochem. Biophys. Res. Commun.,* 81, 498, 1978.

248. Halliwell, B. and Gutteridge, J. M. C., Formation of a thiobarbituric acid-reactive substance from deoxyribose in the presence of iron salts, *FEBS Letters,* 128, 347, 1981.

249. Halliwell, B. and Grootveld, M., Methods for the measurement of hydroxyl radicals in biochemical systems: deoxyribose degradation and aromatic hydroxylation, *Meth. Biochem. Anal.,* 33, 59, 1988, and references cited therein.

250. Halliwell, B., Gutteridge, J. M. C., and Aruoma, O. I., The deoxyribose method: a simple "test-tube" assay for determination of rate constants for reactions of hydroxyl radicals, *Anal. Biochem.,* 165, 215, 1987.

420

251. Winterbourn, C. C. and Sutton, H. C., Iron and xanthine oxidase catalyze formation of an oxidant species distinguishable from OH: comparison with the Haber-Weiss reaction, *Arch. Biochem. Biophys.*, 244, 27, 1986.

252. Halliwell, B. and Grootveld, M., The measurement of free radical reactions in humans. Some thoughts for future experimentation, *FEBS Letters*, 213, 9, 1987.

253. Sun, J. Z., Kaur, H., Halliwell, B., Li. X.-Y., and Bolli, R., Use of aromatic hydroxylation of phenylalanine to measure production of hydroxyl radicals after myocardial ischemia in vivo, *Circ. Res.*, 73, 534, 1993.

254. Dixon, W. T. and Norman, R. O. C., Electron spin resonance studies of oxidation. Part I. Alcohols, *J. Chem. Soc.*, p. 3119, 1963.

255. Dixon, W. T. and Norman, R. O. C., Electron spin resonance studies of oxidation. Part IV. Some benzenoid compounds, *J. Chem. Soc.*, p. 4857, 1964.

256. Sugimoto, H. and Sawyer, D. T., Iron(II)-induced activation of hydrogen peroxide to ferryl ion (FeO^{2+}) and singlet oxygen ($^1O_2$) in acetonitrile: monooxygenations, dehydrogenations, and dioxygenations of organic substrates, *J. Am. Chem. Soc.*, 106, 4283, 1984.

257. Whitburn, K. D., The Interaction of oxymyoglobin with hydrogen peroxide: the formation of ferrylmyoglobin at moderate excesses of hydrogen peroxide, *Arch. Biochem. Biophys.*, 253, 419, 1987.

258. Halliwell, B. and Gutteridge, J. M. C., Role of free radicals and catalytic metal ions in human disease: an overview, *Methods Enzymol.*, 186, 1-85, 1990.

259. Rush, J. D., Maskos, Z. and Koppenol, W. H., Distinction between hydroxyl radical and ferryl species, *Methods in Enzymol.*, 186, 148, 1990.

260. Sawyer, D. T., Sobkowiak, A. and Matsushita, T., Metal [ML_x; M = Fe, Cu, Co, Mn]/hydrperoxide-induced activation of dioxygen for the oxygenation of hydrocarbons: Oxygenated Fenton chemistry, *Acc. Chem. Res.*, 29, 409, 1996.

261. Walling, C., Intermediates in the reactions of Fenton type reagents, *Acc. Chem. Res.*, 31, 155, 1998.

262. MacFaul, P. A., Wayner, D. D. M. and Ingold, K. U., A radical account of oxygenated Fenton Chemistry, *Acc. Chem. Res.*, 31, 159, 1998.

262a. Goldstein, S. and Meyerstein, D., Comments on the mechanism of the Fenton-like reaction, *Acc. Chem. Res.*, 32, 547, 1999.

263. Youngman, R. J. and Elstner, E. F., Oxygen species in paraquat toxicity: the crypto-OH radical, *FEBS Letters*, 129, 265, 1981.

264. Borg, D. C., Schaich, K. M. and Forman, A., Autooxidative cytotoxicity: is there metal-independent formation of hydroxyl radicals? Are there "crypto-hydroxyl" radicals?, in *Oxygen Radicals in Chemistry and Biology*, W. Bors, M. Saran, and D.Tait, eds., Walter deGruyter, Berlin 1984, p.123.

265. Czapski, G., On the use of OH scavengers in biological systems, *Isr. J. Chem.*, 34, 29, 1984.

266. Youngman, R. J., Oxygen activation: is the hydroxyl radical always biologically relevant, *Trends Biochem. Sci.*, 9, 280, 1984.

267. Halliwell, B., Gutteridge, J. M. C. and Blake, D. R., Metal ions and oxygen radical reactions in human inflammatory joint disease,*Phil. Trans. Royal Soc. London* B, 311, 659, 1985.

268. Meneghini, R. and Hoffmann, M. E., The damaging action of hydrogen peroxide on DNA of human fibroblast is mediated by a non-dialyzable compound, *Biochim. Biophys. Acta*, 608, 167, 1980.

269. Stadtman, E. R. and Oliver, C. N., Metal-catalyzed oxidation of proteins, *J. Biol. Chem.*, 266, 2005, 1991.

270. Yamazaki, I. and Piette, L. H., EPR spin-trapping study on the oxidizing species formed in the reaction of the ferrous ion with hydrogen peroxide, *J. Am. Chem. Soc.*, 113, 7588, 1991.

271. Buettner, G. R., In the absence of catalytic metals ascorbate does not autoxidize at pH 7: ascorbate as a test for catalytic metals, *J. Biochem. Biophys. Methods*, 16, 27, 1988.

272. Buettner, G. R., Ascorbate oxidation: UV absorbance of ascorbate and ESR spectroscopy of the ascorbyl radical as assays for iron, *Free Radical Res.Commun.*, 10, 5, 1990.

273. Graf, E., Mahoney, J. R., Bryant, R. G., and Eaton, J. W., Iron-catalyzed hydroxyl radical formation. Stringent requirement for free iron coordination site, *J. Biol. Chem.*, 259, 3620, 1984.

274. Johnson, G. R. A., Nazhat, N. B., and Nazhat, R. A. S., Reaction of the aquocopper(I) ion with hydrogen peroxide: evidence against hydroxyl free radical formation, *J. Chem. Soc. Chem. Commun.*, p. 407, 1985.

275. Meyerstein, D., Trivalent copper. I. A pulse radiolytic study of the chemical properties of the aquo complex, *Inorg. Chem.*, 10, 638, 1971.

276. Masarwa, M., Cohen, H., Meyerstein, D., Hickman, D. L., Bakac, A., and Espenson, J. H., Reactions of low-valent transition-metal complexes with hydrogen peroxide. Are they "Fenton-like" or not? I. The case of Cu^+_{aq} and Cr^{2+}_{aq}, *J. Am. Chem. Soc.*, 110, 4293, 1988.

277. Halliwell, B. and Gutteridge, J. M. C., The importance of free radicals and catalytic metal ions in human diseases, *Molec. Aspects Med.*, 8, 89-193, 1985.

278. Aust, S. D., Morehouse, L. A. and Thomas, C. E., Role of metal ions in oxygen radical reactions, *Free Radical Biol. Med.*, 1, 3, 1985.

279. Rowley, D. A. and Halliwell, B., Formation of hydroxyl radicals from NADH and NADPH in the presence of copper salts, *J. Inorg. Biochem.*, 23, 103, 1985.

280. Rowley, D. A. and Halliwell, B., Superoxide-dependent and ascorbate-dependent formation of hydroxyl radicals in the presence of copper salts: a physiologically significant reaction?, *Arch. Biochem. Biophys.*, 225, 279, 1983.

281. Aruoma, O. I., Halliwell, B., Gajewski, E. and Dizdaroglu, M., Copper-ion-dependent damage to the bases in DNA in the presence of hydrogen peroxide, *Biochem. J.*, 273, 601, 1991.

282. Kadiiska, M. B., Hanna, P. M., Hernandez, L. and Mason, R. P., In vivo evidence of hydroxyl radical formation after acute copper and ascorbic acid intake: electron spin resonance evidence, *Mol. Pharmacol.*, 42, 723, 1992.

283. Hodgson, E. K., and Fridovich, I., The interaction of bovine erythrocyte superoxide dismutase with hydrogen peroxide: inactivation of the enzyme, *Biochemistry*, 14, 5294, 1975.

284. Sato, K., Akaike, T., Kohno, M., Ando, M. and Maeda, H., Hydroxyl radical production by H_2O_2 plus Cu,Zn-superoxide dismutase reflects the activity of free copper released from the oxidatively damaged enzyme, *J. Biol. Chem.*, 267, 25,371, 1992.

285. Rosen, D. R., Siddique, T., Patterson, D. et al., Mutations in Cu/Zn superoxide dismutase gene are associated with familial amyotrophic lateral sclerosis, *Nature*, 362, 59, 1993.

286. Gurney, M. E., Pu, H., Chiu, A. Y. et al., Motor neuron degeneration in mice that express a human Cu/Zn superoxide dismutase mutation, *Science*, 264, 1772, 1994.

287. Robberecht, W., Sapp, P., Viaene, M. K. et al., Cu/Zn superoxide dismutase activity in familial and sporadic amyotrophic lateral sclerosis, *J. Neurochem.*, 62, 384, 1994.

288. Halliwell, B., Free radicals and antioxidants: a personal view, *Nutr. Rev.*, 52, 253, 1994.

289. Spencer, J. P. E., Jenner, A., Aruoma, O. I., Evans, P. J., Kaur, H., Dexter, D. T., Jenner, P., Lees, A. J., Marsden, D. C. and Halliwell, B., Intense oxidative DNA damage promoted by L-DOPA and its metabolites. Implications for neurodegenerative disease, *FEBS Letters*, 353, 246, 1994.

290. Moorhouse, C. P., Halliwell, B., Grootveld, M. and Gutteridge, J. M. C., Cobalt(II) ion as a promoter of hydroxyl radical and possible 'crypto-hydroxyl' radical formation under physiological conditions. Differential effects of hydroxyl radical scavengers, *Biochim. Biophys. Acta*, 843, 261, 1985.

291. Hanna, P. M., Kadiiska, M. B. and Mason, R. P., Oxygen-derived free radical and active oxygen complex formation from cobalt(II) chelates in vitro, *Chem. Res. Toxicol.*, 5, 109, 1992.

292. Kadiiska, M. B., Maples, K. R. and Mason, R. P., A comparison of cobalt(II) and iron(II) hydroxyl and superoxide free radical formation, *Arch. Biochem. Biophys.*, 275, 98, 1989.

293. Eberhardt, M. K., Santos, C. and Soto, M. A., Formation of hydroxyl radicals and Co^{3+} in the reaction of Co^{2+}-EDTA with hydrogen peroxide. Catalytic effect of Fe^{3+}, *Biochim. Biophys. Acta*, 1157, 102, 1993.

294. Cotton, F. A. and Wilkinson, G., *Advanced Inorganic Chemistry*, 5th ed., Wiley, New York, 1988.

295. Bawn, C. F. H. and White, A. G., Reactions of the cobaltic ion. Part I, *J. Chem. Soc.*, 331, 1951.

296. Shi, X. and Dalal, N. S., On the hydroxyl radical formation in the reaction between hydrogen peroxide and biologically generated chromium (V) species, *Arch. Biochem. Biophys.*, 277, 342, 1990.

297. Aiyar, J., Berkovits, H. J., Floyd, R. A. and Wetterhahn, K. E., Reaction of chromium (VI) with hydrogen peroxide in the presence

of glutathione: reactive intermediates and resulting DNA damage, *Chem. Res. Toxicol.*, 3, 595, 1990.

298. Wetterhahn, J. K., Microsomal reduction of the carcinogen chromate produces chromium (IV), *J. Am. Chem. Soc.*, 104, 874, 1982.

299. Weiss, J.J., The catalytic decomposition of hydrogen peroxide on different metals, *Trans. Faraday Soc.*, 31, 1547, 1935.

300. Weitzman, S. A. and Graceffa, P., Asbestos catalyzes hydroxyl and superoxide radical generation from hydrogen peroxide, *Arch. Biochem. Biophys.*, 228, 373, 1984.

301. Eberhardt, M. K., Roman-Franco, A. A. and Quiles, M. R., Asbestos-induced decomposition of hydrogen peroxide, *Environm. Res.*, 37, 287, 1985.

302. Kasai, H. and Nishimura, S., DNA damage induced by asbestos in the presence of hydrogen peroxide, *Gann*, 75, 841, 1984.

303. Kennedy, T. P., Dodson, R., Rao, N. V., Ky, H., Hopkins, C., Baser, M., Tolley, E. and Hoidal, J. R., Dusts causing pneumoconiosis generate ·OH and produce hemolysis by acting as Fenton catalysts, *Arch. Biochem. Biophys.*, 269, 359, 1989.

304. Shull, S., Manohar, M., Marsh, J. P., Janssen, Y. M. W. and Mossman, B. T., Role of iron and reactive oxygen species in asbestos-induced lung injury, in *Free Radical Mechanisms of Tissue Injury*, Mary Treinen Moslen and Charles V. Smith, eds., CRC Press, Boca Raton, FL, 1992.

305. Marsh, J. P., Janssen, Y. M. W. and Mossman, B. T., Ornithine decarboxylase and tumor promotion: a role for oxidants, in *DNA and Free Radicals*, Barry Halliwell and Okezie, I. Aruoma, eds., Ellis Horwood Limited, Chichester, England 1993.

306. Leanderson, P. and Tagesson, C., Mineral fibers, cigarette smoke, and oxidative DNA damage, in *DNA and Free Radicals*, Barry Halliwell and Okezie, I. Aruoma, eds., Ellis Horwood Limited, Chichester, England 1993.

307. Hippeli, S. and Elstner, E. F., Oxygen radicals and air pollution, in *Oxidative Stress*, Helmut Sies, ed., Academic Press, New York, 1991.

308. Brody, A. R., Asbestos exposure as a model of inflammation inducing interstitial pulmonary fibrosis, in *Inflammation: Basic*

Principles and Clinical Correlates, Second Edition, J. I. Gallin, I. M. Goldstein and R. Snyderman, eds., Raven Press, Ltd., New York, 1992.

309.   Lesko, S. A., Lorentzen, R. J. and Ts'o, P. O. P., Role of superoxide in DNA strand scission, *Biochemistry,* 19, 3027, 1980.

310.   Brawn, K. and Fridovich, I., DNA strand scission by enzymatically generated oxygen radicals, *Arch. Biochem. Biophys.,* 206, 414, 1981.

311.   Mello Filho, A. C., Hoffmann, M. E. and Meneghini, R., Cell killing and DNA damage by hydrogen peroxide are mediated by intracellular iron, *Biochem. J.,* 218, 273, 1984.

312.   Thomas, C. E. and Aust, S. D., Reductive release of iron from ferritin by cation free radicals of paraquat and other bipyridyls, *J. Biol. Chem.,* 261, 13,064, 1986.

313.   Monteiro, H. P., Ville, G. F. and Winterbourn, C. C., Release of iron ferritin by semiquinone, anthracyclin, bipyridyl and nitroaromatic radicals, *Free Radical Biol. Med.,* 6, 587, 1989.

314.   Biemond, P., Van Eijk, H. G., Swaak, A. J. G. and Koster, J. F., Iron mobilization from ferritin by superoxide derived from stimulated polymorphonuclear leukocytes. Possible mechanism in inflammation diseases, *J. Clin. Invest.,* 73, 1576, 1984.

315.   Reif, D. W. and Simmons, R. D., Nitric oxide mediates iron release from ferritin, *Arch. Biochem. Biophys.,* 283, 537, 1990.

316.   Boyer, R. F., Clark, H. M. and LaRoche, A. P., Reduction and release of ferritin iron by plant phenolics, *J. Inorg. Biochem.,* 32, 171, 1988.

317.   Szent-Györgyi, A., *The Living State and Cancer,* Marcel Dekker, New York, 1978.

318.   Kasha, M. and Brabham, D. E., Singlet oxygen electronic structure and photosensitization, in *Singlet Oxygen,* H. H. Wasserman and R. W. Murray, eds., Academic Press, New York, 1979, pp.1-32.

319.   Raab, O., Über die Wirkung fluorescirender Stoffe auf Infusorien, *Z. Biol.,* 39, 524, 1900.

320.   Spikes, J. D. and MacKnight, M. L., Photodynamic effects on molecules of biological importance: amino acids, peptides and proteins, *Res. Prog. Org. Biol. Med. Chem.,* 3, 124, 1972.

426

321. Spikes, J. D. and Livingston, R., The molecular biology of photodynamic action: Sensitized photoautoxidation in biological systems, *Adv. Radiat. Biol.*, 3, 29, 1969.

322. Wilson, T. and Hastings, J. W., Chemical and biological aspects of singlet excited molecular oxygen, *Photophysiol.*, 5, 49, 1970.

323. Gallo, U. and Santamaria, L., *Research Progress in Organic Biological and Medicinal Chemistry*, Vol. III, Elsevier, New York, 1972.

324. Foote, C. S., Photosensitized oxidation and singlet oxygen: Consequences in biological systems, in *Free Radicals in Biology*, Vol. II, W. A. Pryor, ed., Academic Press, New York, 1975.

325. Grossweiner, L. I., Photochemical inactivation of enzymes, *Curr. Top. Radiat. Res. Quarterly*, 11, 141, 1976.

326. Murray, R. W., Chemical sources of singlet oxygen, in *Singlet Oxygen*, H. H. Wasserman and R. W. Murray, eds., Academic Press, New York, 1979, pp. 59-112.

327. Kanofsky, J. R., Singlet oxygen production from the reactions of alkylperoxy radicals. Evidence from 1268 nm chemiluminescence, *J. Org. Chem.*, 51, 3386, 1986.

328. Krinsky, N. I., Biological roles of singlet oxygen, in *Singlet Oxygen*, H. H. Wasserman and R. W. Murray, eds., Academic Press, New York, 1979, pp. 597-636.

329. Foote, C. S., Shook, F. C. and Abakerli, R. B., Characterization of singlet oxygen, in: *Oxygen Radicals in Biological Systems*, L. Packer, ed., Academic Press, Orlando, FLA. 1984, pp. 36-47.

330. Khan, A. U., The discovery of the chemical evolution of singlet oxygen. Some current chemical, photochemical, and biological applications, *Int. J. Quantum Chem.* XXXIX, 251, 1991.

331. Bartlett, P. D. and Landis, M. E., The 1,2-dioxetanes, in *Singlet Oxygen*, H. H. Wasserman and R. W. Murray, eds., Academic Press, New York, 1979, pp. 244-283.

332. Gollnick, K. and Kuhn, H. J., Ene-reactions with singlet oxygen, in *Singlet Oxygen*, H. H. Wasserman and R. W. Murray, eds., Academic Press, New York, 1979, pp. 287-419.

333. Trozzolo, A. M., ed., A. M. Trozzolo, and R. W. Murray, Conference Cochairmen : International Conference on Singlet

Molecular Oxygen and its Role in Environmental Sciences, *Ann. N. Y. Acad. Sci.,* 171, pp. 1-302, 1970.

334. Kearns, D. R., Physical and chemical properties of singlet molecular oxygen, *Chem. Rev.,* 71, 395, 1971.

335. *Singlet Oxygen,* H. H. Wasserman and R. W. Murray, eds., Academic Press, New York, 1979.

336. Schenck, G. O., Photosensitized reactions with molecular oxygen, *Naturwissenschaften,* 35, 28, 1948.

337. Schenck, G. O. and Ziegler, K., Photosensitized autoxidation of steroids, *Naturwissenschaften,* 32, 157, 1944.

338. Rosen, H. and Klebanoff, S. J., Formation of singlet oxygen by the myeloperoxidase-mediated antimicrobial system, *J. Biol. Chem.,* 252, 4803, 1977.

339. King, M. M., Lai, E. K. and McCay, P. B., Singlet oxygen production associated with enzyme-catalyzed lipid peroxidation in liver microsomes, *J. Biol. Chem.,* 250, 6496, 1975.

340. Takayama, K., Noguchi, T., Nakano, M. and Migita, T., Reactivities of diphenylfuran (a singlet oxygen trap) with singlet oxygen and hydroxyl radical in aqueous systems, *Biochem. Biophys. Res. Commun.,* 75, 1052, 1977.

341. Eberhardt, M. K., Formation of *cis*-dibenzoylethylene from 2,5-diphenylfuran by Fenton's reagent and by peroxydisulfate. Effect of oxygen, *J. Org. Chem.,* 58, 497, 1993.

342. Kulig, M. J. and Smith, L. L., Sterol metabolism. XXV. Cholesterol oxidation by singlet molecular oxygen, *J. Org. Chem.,* 38, 3639, 1973.

343. Foote, C. S., Abakerli, R. B., Clough, R. L. and Lehrer, R. I., On the question of singlet oxygen production in polymorphonuclear leukocytes, in: *Bioluminescence and Chemiluminescence, Basic Chemistry and Analytical Applications,* M. A. DeLuca and W. D. McElroy, eds., Academic Press, New York, 1981, pp. 81-88.

344. Foote, C. S., Shook, F. C. and Abakerli, R. B., Chemistry of superoxide ion. 4 .Singlet oxygen is not a major product of dismutation, *J. Am. Chem. Soc.,* 102, 2503, 1980.

345. Foote, C. S., Quenching of singlet oxygen, in : *Singlet Oxygen,* H. H. Wasserman and R. W. Murray, eds., Academic Press, New York, 1979, pp. 139-171.

428

346. Guiraud, H. J. and Foote, C. S., Chemistry of superoxide ion. III. Quenching of singlet oxygen, *J. Am. Chem. Soc.*, 98, 1984, 1976.

347. Foote, C. S. and Denny, R. W., Chemistry of singlet oxygen. VII. Quenching by ß-carotene, *J. Am. Chem. Soc.*, 90, 6233, 1968.

348. Schenck, G. O., Gollnick, K. and Neumüller, O. A., Darstellung von Steroid-hydroperoxyden mittels phototoxischer Photosensi-bilatoren, *Justus Liebigs Ann. Chem.*, 603, 46, 1957.

349. Parker, J. G. and Stanbro, W. D., Optical determination of the rates of formation and decay of O_2 ($^1 _g$) in H_2O, D_2O and other solvents, *J. Photochem.*, 25, 545, 1984.

350. Buchko, G. W., Cadet, J., Berger, M. and Ravanat, J.-L., Photooxidation of d(TpG) by phthalocyanines and riboflavin. Isolation and characterization of dinucleoside monophosphates containing the 4R* and 4S* diasteromers of 4,8-dihydro-4-hydroxy-8-oxo-2'-deoxyguanosine, *Nucleic Acid Res.*, 20, 4847, 1992.

351. Ravanat, J.-L. and Cadet, J., Reaction of singlet oxygen with 2'-deoxyguanosine and DNA. Isolation and characterization of the main oxidation products, *Chem. Res. Toxicol.*, 8, 379, 1995.

352. Cadet, J. and Téoule, R., Comparative study of oxidation of nucleic acid components by hydroxyl radicals, singlet oxygen and superoxide anion radicals, *Photochem. Photobiol.*, 28, 661, 1978.

353. Cadet, J., Decarroz, C., Wang, S. Y. and Midden, W. R., Mechanisms and products of photosensitized degradation of nucleic acids and related model compounds, *Isr. J. Chem.*, 23, 420, 1983.

354. Cadet, J., Berger, M., Decarroz, C., Wagner, J. R., van Lier, J. E., Ginot, Y. M. and Vigny, P., Photosensitized reactions of nucleic acids, *Biochimie*, 68, 813, 1986.

355. Sheu, C. and Foote, C. S., Endoperoxide formation in a guanosine derivative, *J. Am. Chem. Soc.*, 115, 10,446, 1993.

356. Sheu, C. and Foote, C. S., Photosensitized oxygenation of a 7,8-dihydro-8-oxoguanosine derivative. Formation of dioxetane and hydroperoxide intermediates, *J. Am. Chem. Soc.*, 117, 474, 1995.

357. Aida, M. and Nishimura, S., An *ab initio* molecular orbital study on the characteristics of 8-hydroxyguanine, *Mutation Res.*, 192, 83, 1987.

358. Oda, Y., Uesugi, S., Ikehara, M., Nishimura, S., Kawase, Y., Ishikawa, H., Inoue, H. and Ohtsuka, E., NMR studies of a DNA

containing 8-hydroxydeoxyguanosine, *Nucleic Acid Res.*, 19, 1407, 1991.

359. Khan, A. U. and Kasha, M., Chemiluminescence arising from simultaneous transitions in pairs of singlet oxygen molecules, *J. Am. Chem. Soc.*, 92, 3293, 1970.

360. Kasha, M. and Khan, A. U., The physics, chemistry, and biology of singlet molecular oxygen, *Ann. N. Y. Acad. Sci.*, 171, 5, 1970.

361. Foote, C. S. and Wexler, S., Olefin oxidations with excited singlet molecular oxygen, *J. Am. Chem. Soc.*, 86, 3879, 1964.

362. Foote, C. S., Wexler, S. and Ando, W., Singlet oxygen. III. Product selectivity, *Tetrahedron Lett.*, 4111, 1965.

363. Kanofsky, J. R., Singlet oxygen production by lactoperoxidase, *J. Biol. Chem.*, 258, 5991, 1983.

364. Khan, A. U., Enzyme generation of singlet ($^1\Delta_g$) molecular oxygen observed directly by 1.0-1.8 m chemiluminescence spectroscopy, *J. Am. Chem. Soc.*, 105, 7195, 1983.

365. Khan, A. U., Myeloperoxidase singlet molecular oxygen generation detected by direct infrared electronic emission, *Biochem. Biophys. Res. Commun.*, 122, 668, 1984.

366. Kanofsky, J. R. and Axelrod, B., Singlet oxygen production by soybean lipoxygenase isozymes, *J. Biol. Chem.*, 261, 1099, 1986.

367. Kanofsky, J. R., Singlet oxygen production from the per-oxidase-catalyzed oxidation of indole-3-acetic acid, *J. Biol. Chem.*, 263, 14,171, 1988.

368. Kanofsky, J. R., Hoogland, H., Wever, R. and Weiss, S. J., Singlet oxygen production by human eosinophils, *J. Biol. Chem.*, 263, 9692, 1988.

369. Hastings, J. W., Bioluminescence, *Ann. Rev. Biochem.*, 37, 597, 1968.

370. Mayo, F. R., The oxidation of unsaturated compounds. V. The effect of oxygen pressure on the oxidation of styrene, *J. Am. Chem. Soc.*, 80, 2465, 1958.

371. Schaap, A. P. and Zaklika, K. A., 1,2-cycloaddition reactions of singlet oxygen, in *Singlet Oxygen,* H. H. Wasserman and R. W. Murray, eds., Academic Press, New York, 1979, pp. 173-242.

372. Schenck, G. O., German Patent 933,925, 1943.

373.   Schönberg, A., *Preparative Organic Photochemistry,* Springer Verlag, Berlin and New York, 1958.

374.   Kearns, D. R., Fenical, W. and Radlick, P., Experimental and quantum chemical investigation of singlet oxygen reactions,*Ann. N. Y. Acad. Sci.,* 171, 34, 1970.

375.   McCapra, F., Application of the theory of electrocyclic reactions of bioluminescence, *Chem. Commun.,* p. 155, 1968.

376.   Foote, C. S. and Lin, J. W.-P., Chemistry of singlet oxygen. VI. Photooxygenation of enamines: evidence for an intermediate (1), *Tetrahedron Lett.,* 3267, 1968.

377.   Rio, G. and Berthelot, J., Photooxidation sensibilisée d'éthyléniques dépourvus d'hydrogène en alpha de la double liaison: cas d'aryléthylènes, *Bull. Soc. Chim. Fr.,* p. 3609, 1969.

378.   McCapra, F. and Hann, R. A., The chemiluminescent reaction of singlet oxygen with 10,10-dimethyl-9,9-biacridylidene, *Chem. Commun.,* p. 442, 1969.

379.   Kopecky, K. R. and Mumford, C., *Abstract 51st Ann. Conf. Chem. Inst. Can.,* 1968, p. 41.

380.   Bartlett, P. D. and Schaap, A. P., Stereospecific formation of 1,2-dioxetanes from cis-and trans-diethoxyethylenes by singlet oxygen, *J. Am. Chem. Soc.,* 92, 3223, 1970.

381.   Bartlett, P. D., Mendenhall, G. D. and Schaap, A. P., Competitive modes of reaction of singlet oxygen, *Ann. N. Y. Acad. Sci.,* 171, 79, 1970.

382.   Adam, W., Andler, S., Nau, W. M. and Saha-Möller, C. R., Oxidative DNA damage by radicals generated in the thermolysis of hydroxymethyl-substituted 1,2-dioxetanes through the cleavage of chemiexcited ketones, *J. Am. Chem. Soc.,*120, 3549, 1998.

383.   Turro, N. J., Lechtken, P., Schuster, G., Orell, J., Steinmetzer, H.-C.and Adam, W., Indirect chemiluminescence by 1,2-dioxetanes. Evaluation of triplet-singlet excitation efficiencies. A long range singlet-singlet energy transfer and an efficient triplet-singlet energy transfer, *J. Am. Chem. Soc.,* 96, 1627, 1974.

384.   Turro, N. J. and Lechtken, P., Molecular photochemistry. LII. Thermal decomposition of tetramethyl-1,2-dioxetane. Selective and efficient chemelectronic generation of triplet acetone, *J. Am. Chem. Soc.,* 94, 2886, 1972.

385. Turro, N. J., Lechtken, P., Shore, N. E., Schuster, G., Steinmetzer, H.-C. and Yekta, A., Tetramethyl-1,2-dioxetane. Experiments in chemiexcitation, chemiluminescence, photochemistry, chemical dynamics, and spectroscopy, *Acc.Chem. Res.,* 7, 97, 1974.

386. Briviba, K., Saha-Möller, C. R., Adam, W. and Sies, H., Formation of singlet oxygen in the thermal decomposition of 3-hydroxymethyl-3,4,4-trimethyl-1,2-dioxetane, a chemical source of triplet-excited ketones, *Biochem. Mol. Biol. Int.,* 38, 647, 1996.

387. Frey, P. A., Vitamins, Coenzymes, and Metal Cofactors, in *Biochemistry,* G. L. Zubay, ed., pp. 278-302, third edition, W.C. Brown Publishers, Dubuque, Iowa, 1993.

388. Forman, H. J. and Boveris, A., Superoxide radical and hydrogen peroxide in mitochondria, in *Free Radicals in Biology,* Vol. V., W. A. Pryor, ed., Academic Press, New York, pp. 65-90, 1982.

389. Parson, W. W., Electron transport and oxidative phosphorylation, in *Biochemistry,* G. L. Zubay, ed., pp. 379-413, third edition, W.C. Brown Publishers, Dubuque, Iowa, 1993.

390. Forman, H. J. and Thomas, M. J., Oxidant production and bactericidal activity of phagocytes, *Ann. Rev. Physiol.,* 48, 669, 1986.

391. Gerschman, K., Gilbert, D. L., Nye, S. W., Dwyer, P. and Fenn, W. O., Oxygen poisoning and X-irradiation: a mechanism in common, *Science,* 119, 623, 1954.

392. Imlay, J. A. and Fridovich, I., Assay of metabolic superoxide production in *Escherichia coli, J. Biol. Chem.,* 266, 6957, 1991.

393. Sies, H., Biochemistry of oxidative stress, *Angew. Chem. Int. Ed. Engl.,* 25, 1058, 1986.

394. Cathcart, R., Schwiers, E., Saul, R. L. and Ames, B. N., Thymine glycol and thymidine glycol in human and rat urine: a possible assay for oxidative DNA damage, *Proc. Natl. Acad. Sci. USA,* 81, 5633, 1984.

395. Ames, B. N., Endogenous DNA damage as related to cancer and aging, *Mutation Research,* 214, 41, 1989.

396. Shigenaga, M. K., Gimeno, C. J. and Ames, B. N., Urinary 8-hydroxy-2'-deoxyguanosine as a biological marker of in vivo oxidative DNA damage, *Proc. Natl. Acad. Sci. USA,* 86, 9697, 1989.

432

397.   Fraga, C. G., Shigenaga, M. K., Park, J. W., Degan, P. and Ames, B. N., Oxidative damage to DNA during aging: 8-hydroxy-2'-deoxyguanosine in rat organ DNA and urine, *Proc. Natl. Acad. Sci. USA*, 87, 4533, 1990.

398.   Ames, B. N. and Shigenaga, M. K., Oxidants are a major contributor to cancer and aging, in *DNA and Free Radicals*, B. Halliwell, O. I. Aruoma, eds, Ellis Horwood, Chichester, England, 1993, pp. 1-15.

399.   Storz, G., Tartaglia, L. A. and Ames, B. N., Transcriptional regulator of oxidative stress-inducible genes: direct activation of oxidation, *Science*, 248, 189, 1990.

400.   Demple, B., Regulation of bacterial oxidative stress genes, *Ann. Rev. Genet.*, 25, 315, 1991.

401.   Sevanian, A., Lipid damage and repair, in *Oxidative Damage and Repair: Chemical, Biological and Medical Aspects*, K. J. A. Davies, ed., Pergamon Press, New York, 1991, pp. 543-549.

402.   Doetsch, P. W., Repair of oxidative DNA damage in mammalian cells, in *Oxidative Damage and Repair: Chemical, Biological and Medical Aspects*, K. J. A. Davies, ed., Pergamon Press, New York, 1991, pp. 192-196..

403.   Turrens, J. F. and Boveris, A., Generation of superoxide anion by the NADPH dehydrogenase of bovine heart mitochondria, *Biochem. J.*, 191, 421, 1980.

404.   Sies, H., Zur Biochemie der Thiolgruppe:Bedeutung des Glutathions, *Naturwissenschaften*, 76, 57, 1989.

405.   Jones, D. P., Eklow, L., Thor, H. and Orrenius, S., Metabolism of hydrogen peroxide in isolated hepatocytes: relative contributions of catalase and glutathione peroxidase in decomposition of endogenously generated H_2O_2, *Arch. Biochem. Biophys.*, 210, 505, 1981.

406.   Buettner, G. R. and Oberley, L. W., Considerations in the spin-trapping of superoxide and hydroxyl radical in aqueous systems using 5,5'-dimethyl-1-pyrroline-1-oxide, *Biochem. Biophys. Res. Commun.*, 83, 69, 1978.

407.   Ueno, I., Kohno, M., Yoshihira, K. and Hirono, I., Quantitative determination of the superoxide radicals in the xanthine oxidase reaction by measurement of the electron spin resonance signal of the superoxide radical spin adduct of 5,5'-dimethyl-1-pyrroline-1-oxide, *J. Pharmaco-Din.*, 7, 563, 1984.

408. Olson, J. S., Ballou, D. P., Palmer, G., and Massey, V., The reaction of xanthine oxidase with molecular oxygen, *J. Biol. Chem.*, 249, 4350, 1974.

409. Porras, A. G., Olson, J. S., and Palmer, G., The reaction of reduced xanthine oxidase with oxygen. Kinetics of peroxide and superoxide formation, *J. Biol. Chem.*, 256, 9096, 1981.

410. Hille, R. and Massey, V., Studies on the oxidative half-reaction of xanthine oxidase, *J. Biol. Chem.*, 256, 9090, 1981.

411. Anderson, R. F., Hille, R., and Massey, V., The radical chemistry of milk xanthine oxidase as studied by radiation chemistry techniques, *J. Biol. Chem.*, 261, 15,870, 1986.

412. Nelson, D. H. and Ruhmann-Wennhold, A., Corticosteroids increase superoxide anion production by rat liver microsomes, *J. Clin. Invest.*, 56, 1062, 1975.

413. Auclair, C., de Prost, D. and Hakim, J., Superoxide anion production by liver microsomes from phenobarbital treated rats, *Biochem. Pharmacol.*, 27, 355, 1978.

414. Massey, V., Strickland, S., Mayhew, S. G., Howell, L. G., Engel, P. C., Matthews, R. G., Schuman, M. and Sullivan, P. A., The production of superoxide anion radicals in the reaction of reduced flavins and flavoproteins with molecular oxygen, *Biochem. Biophys. Res. Commun.*, 36, 891, 1969.

415. Ballou, D., Palmer, G. and Massey, V., Direct demonstration of superoxide anion production during the oxidation of reduced flavin and of its catalytic decomposition by erythrocuprein, *Biochem. Biophys. Res. Commun.*, 36, 898, 1969.

416. Morehouse, L. A., Thomas, C. E. and Aust, S. D., Superoxide generation by NADPH-cytochrome P-450 reductase: the effect of iron chelators and the role of superoxide in microsomal lipid peroxidation, *Arch. Biochem. Biophys.*, 232, 366, 1984.

417. Grover, T. A. and Piette, L. H., Influence of flavin addition and removal on the formation of superoxide by NADPH-cytochrome P-450 reductase: a spin-trap study, *Arch. Biochem. Biophys.*, 212, 105, 1981.

418. Bösterling, B. and Trudell, J. R., Spin trap evidence for production of superoxide radical anions by purified NADPH-cytochrome P-450 reductase, *Biochem. Biophys. Res. Commun.*, 98, 569, 1981.

434

419.   Sligar, S. G., Lipscomb, J. D., Debrunner, P. G. and Gunsalus, I. C., Superoxide production by the autoxidation of cytochrome P450, *Biochem. Biophys. Res. Commun.*, 61, 290, 1974.

420.   Kahl, R., Weiman, A., Weinke, S. and Hildebrandt, A. G., Detection of oxygen activation and determination of the activity of antioxidants toward reactive oxygen species by use of the chemiluminigenic probes luminol and lucigenin, *Arch. Toxicol.*, 60, 158, 1987.

421.   Karnovsky, M. L., Metabolic basis of phagocytic activity, *Physiol. Rev.* 42, 143, 1962.

422.   Klebanoff, S. J., Antimicrobial mechanisms in neutrophilic polymorphonuclear leukocytes, *Semin. Hematol.*, 12, 117, 1975.

423.   Klebanoff, S. J., Oxygen metabolism and the toxic properties of phagocytes, *Ann. Intern. Med.* 93, 480, 1980.

424.   Badwey, J. A. and Karnovsky, M. L., Active oxygen species and the functions of phagocytic leukocytes, *Ann. Rev. Biochem.* 49, 695, 1980.

425.   Root, R. K. and Cohen, M. S., The microbicidal mechanism of human neutrophils and eosinophils, *Rev. of Infect. Diseases*, 3, 565, 1981.

426.   Babior, B. M. and Crowley, C. A., Chronic granulomatous disease and other disorders of oxidative killing by phagocytes in *The Metabolic Basis of Inherited Disease,* fifth edition, J. R. Stanbury, J. B. Wyngaarden, D.S. Fredrickson, J. L. Goldstein, M. S. Brown, eds., McGraw Hill, New York, 1983, pp.1956-1984 .

427.   Babior, B. M., The respiratory burst of phagocytes, *J. Clin. Invest.*, 73, 599, 1984.

428.   Babior, B. M., Oxidants from phagocytes: agents of defense and destruction, *Blood*, 64, 959-966, 1984.

429.   Forman, H. J. and Thomas, M. J., Oxidant production and bactericidal activity of phagocytes, *Ann. Rev. Physiol.*, 48, 669, 1986.

430.   Karnovsky, M. L. and Badwey, J. A., Respiratory burst during phagocytosis: an overview, *Methods Enzymol.*, 132, 353, 1986.

431.   Britigan, B. E., Cohen, M. S. and Rosen, G. M., Detection of the production of oxygen-centered free radicals by human neutrophils using spin trapping techniques: a critical perspective, *J. Leukocyte Biol.*, 41, 349, 1987.

432.   Weiss, S. J., Tissue destruction by neutrophils, *N. Engl. J. Med.* 320, 365, 1989.

433.   Clark, R. A., The human neutrophil respiratory burst oxidase, *J. Infect. Dis.*, 161, 1140, 1990.

434.   Forehand, J. R., Nauseef, W. M., Curnutte, J. T. and Johnston, Jr., R. B., Inherited disorders of phagocyte killing, in *The Metabolic and Molecular Bases of Inherited Disease*, 7th edition, Vol III, C.R. Scriver, D.L. Beaudet, W.S. Sly, D. Valle, eds., McGraw Hill, New York, 1995, pp. 3995-4026.

435.   Klebanoff, S. J., Oxygen metabolites from phagocytes, in *Inflammation: Basic Principles and Clinical Correlations*, second edition, J. I. Gallin, I. M. Goldstein and R. Snyderman, eds., Raven Press, New York, 1992.

436.   Miller, R. A. and Britigan, B. E., The formation and biological significance of phagocyte-derived oxidants, *J. Invest. Med.*, 43, 39, 1995.

437.   Rosen, G. M., Pou, S., Ramos, C. L., Cohen, M. S. and Britigan, B. E., Free radicals and phagocytic cells, *FASEB, J.*, 9, 200, 1995.

438.   Sbarra, A. J., and Karnovsky, M. L., The biochemical basis of phagocytosis. I. Metabolic changes during the ingestion of particles by polymorphonuclear leukocytes, *J. Biol. Chem.*, 234, 1355, 1959.

439.   Iyer, G. Y. N., Islam, M. F., and Quastel J. H., Biochemical aspects of phagocytosis, *Nature*, 192, 535, 1961.

440.   Root, R. K. and Metcalf, J. A., H_2O_2 release from human granulocytes during phagocytosis, *J. Clin. Invest.*, 60, 1266, 1777.

441.   Babior, B. M., Kipnes, R. S. and Curnutte, J. T., Biological defense mechanisms: The production by leukocytes of superoxide, a potential bactericidal agent, *J. Clin. Invest.* 52, 741, 1973.

442.   Patriarca, P., Cramer, R., Moncalvo, S., Rossi, F. and Romeo, D., Enzymatic basis of metabolic stimulation in leukocytes during phagocytosis : the role of activated NADPH oxidase, *Arch. Biochem. Biophys.* 145, 255, 1971.

443.   Gabig, T. G. and Babior, B. M., The O_2^- -forming oxidase responsible for the respiratory burst in human neutrophils. Properties of the solubilized enzyme, *J. Biol. Chem.*, 254, 9070, 1979.

444.   Tauber, A. I. and Babior, B. M., Neutrophil oxygen reduction: The enzymes and the products, *Adv. Free Radical Biol. Med.*, 1, 625, 1985.

445. Rabani, J., Klug, D. and Fridovich, I., Decay of the HO_2· and $O_2^{-·}$ radicals catalyzed by superoxide dismutase, *Isr. J. Chem.*, 10, 1095, 1972.

446. Green, T. R. and Wu, D. E., The $NADPH:O_2$ oxidoreductase of human neutrophils, *J. Biol. Chem.*, 261, 6010, 1986.

447. Follin, P. and Dahlgren, C., Altered $O_2^{-·}/H_2O_2$ production ratio by in vitro and in vivo primed human neutrophils, *Biochem. Biophys. Res. Commun.*, 167, 970, 1990.

448. Khan, A. U., Direct spectral evidence of the generation of singlet molecular oxygen ($^1\Delta_g$) in the reaction of potassium superoxide with water, *J. Am. Chem. Soc.*, 103, 6516, 1981.

449. Repine, J. E., Fox, R. B. and Berger, E. M., Hydrogen peroxide kills *Staphylococcus aureus* by reacting with staphylococcal iron to form hydroxyl radical, *J. Biol. Chem.*, 256, 7094, 1981.

450. Ambruso, D. R. and Johnson, R. B. Jr., Lactoferrin enhances hydroxyl radical production by human neutrophils, neutrophil particulate fractions, and an enzymatic generating system, *J. Clin. Invest.*, 67, 353, 1981.

451. Winterbourn, C. C., Lactoferrin-catalyzed hydroxyl radical production: additional requirement for a chelating agent, *Biochem. J.*, 210, 15, 1983.

452. Aruoma, O. I. and Halliwell, B., Superoxide-dependent and ascorbate-dependent formation of hydroxyl radicals from hydrogen peroxide in the presence of iron: are lactoferrin and transferrin promoters of hydroxyl radical generation?, *Biochem. J.*, 241, 273, 1987.

453. Buettner, G. R. and Oberley, L. W., Considerations in the spin trapping of superoxide and hydroxyl radical in aqueous systems using 5,5'-dimethyl-1-pyrroline-1-oxide, *Biochem. Biophys. Res. Commun.*, 83, 69, 1978.

454. Samuni, A., Black, C. D. V., Krishna, C. M., Malech, H. L., Bernstein, E. F. and Russo, A., Hydroxyl radical production by stimulated neutrophils reappraised, *J. Biol. Chem.*, 263, 13,797, 1988.

455. Ramos, C. L., Pou, S., Britigan, B. E., Cohen, M. S., and Rosen, G. M., Spin trapping evidence for myeloperoxidase-dependent hydroxyl radical formation by human neutrophils and monocytes, *J. Biol. Chem.*, 267, 8307, 1992.

456. Allen, R. C., Stjernholm, R. L. and Steele, R. H. , Evidence for the generation of an electronic excitation state(s) in human polymorphonuclear leukocytes and its participation in bactericidal activity, *Biochem. Biophys. Res. Commun.*, 47, 679, 1972.

457. Allen, R. C., Halide dependence of the myeloperoxidase-mediated antimicrobial system of the polymorphonuclear leukocyte in the phenomenon of electronic excitation, *Biochem. Biophys. Res. Commun.*, 63, 675, 1975.

458. Krinsky, N. I. , Singlet excited oxygen as a mediator of the antibacterial action of leukocytes, *Science,* 186, 363, 1974.

459. Kanofsky, J. R., Wright, J., Miles-Richardson, G. E. and Tauber, A. I., Biochemical requirements for singlet oxygen production by purified human myeloperoxidase, *J. Clin. Invest.*, 74, 1489, 1984.

460. Kettle, A. J. and Winterbourn, C. C., Influence of superoxide on myeloperoxidase kinetics measured with a hydrogen peroxide electrode, *Biochem. J.,* 263, 823, 1989.

461. Khan, A. U., Activated oxygen: singlet molecular oxygen and superoxide anion, *Photochem. Photobiol.*, 28, 615, 1978.

462. Mathews-Roth, M. M., Pathak, M. A., Fitzpatrick, T. B., Harber, L. C. and Kass, E. H., Beta-carotene as a photoprotective agent in erythropoietic protoporphyria, *N. Engl. J. Med.*, 282, 1231, 1970.

463. Baart De La Faille, H., Suurmond, D., Went, L. N., Stevenick, J. and Van Schothorst, A. A., ß-carotene as a treatment for photohypersensitivity due to erythropoietic protoporphyria, *Dermatologica,* 145, 389, 1972.

464. Strauss, R. R., Paul, B. B., Jacobs, A. A. and Sbarra, A. J., Role of the phagocyte in host-parasite interactions. XXVII. Myeloperoxidase-H_2O_2-Cl-mediated aldehyde formation and its relatioship to antimicrobial activity, *Infect. Immun.*, 3, 595, 1971.

465. Thomas, E. L., Myeloperoxidase-hydrogen peroxide-chloride antimicrobial system: effect of exogenous amines on antibacterial action against *Escherichia coli, Infect. Immun.*, 25, 110, 1979.

466. Grisham, M. B., Jefferson, M. M. and Thomas, E. L., Role of monochloramine in the oxidation of erythrocyte hemoglobin by stimulated neutrophils, *J. Biol. Chem.*, 259, 6757, 1984.

467. Grisham, M. B., Jefferson, M. M., Melton, D. F. and Thomas, E. L., Chlorination of endogenous amines by isolated neutrophils.

438

Ammonia-dependent bactericidal, cytotoxic and cytolytic activities of the chloramines, *J. Biol. Chem.*, 259, 10,404, 1984.

468. Zgliczynski, J. M., Stelmaszynska, T., Ostrowski, W., Naskalski, J. and Sznajid, J., Myeloperoxidase of human leukemic leukocytes. Oxidation of amino acids in the presence of hydrogen peroxide, *Eur. J. Biochem.*, 4, 540, 1968.

469. Zgliczynski, J. M., Stelmaszynska, T., Domanski, J. and Ostrowski, W., Chloramines as intermediates of oxidative reaction of amino acids by myeloperoxidase, *Biochim. Biophys. Acta*, 235, 419, 1971.

470. Selvararaj, R. J., Paul, B. B., Strauss, R. R., Jacobs, A. A. and Sbarra, A. J., Oxidative peptide cleavage and decarboxylation by the MPO- H_2O_2-Cl⁻ antimicrobial system, *Infect. Immun.*, 9, 255, 1974.

471. Adeniyi-Jones, S. K. and Karnovsky, M. L., Oxidative decarboxylation of free and peptide-linked amino acids in phagocytizing guinea pig granulocytes, *J. Clin. Invest.*, 68, 365, 1981.

472. Thomas, E. L. and Aune, T. M., Oxidation of *Escherichia coli* sulfhydryl components by the peroxidase-hydrogen peroxide-iodide antimicrobial system, *Antimicrob. Agents Chemother.*, 13, 1006, 1978.

473. Turkall, R. M. and Tsan, M. F., Oxidation of glutathione by the myeloperoxidase system, *J. Reticuloendothel. Soc.*, 31, 353, 1982.

474. Learn, D. B., Fried, V. A. and Thomas, E. L., Taurine and hypotaurine content of human leukocytes, *J. Leukocyte Biol.*, 48, 174, 1990.

475. Weiss, S. J., Klein, R., Slivka, A., and Wei, M., Chlorination of taurine by human neutrophils. Evidence for hypochlorous acid generation, *J. Clin. Invest.*, 70, 598, 1982.

476. Rosen, H. and Klebanoff, S. J., Chemiluminescence and superoxide production by myeloperoxidase-deficient leukocytes, *J. Clin. Invest.*, 58, 50, 1976.

477. Rosen, H. and Klebanoff, S. J., Bactericidal activity of a superoxide anion-generating system. A model for the polymorphonuclear leukocyte, *J. Exp. Med.*, 149, 27, 1979.

478. Stossel, T. P., Mason, R. J. and Smith, A. L., Lipid peroxidation by human phagocytes, *J. Clin. Invest.*, 54, 638, 1974.

479. Cilento, G., Photochemistry in the dark, *Photobiol. Rev.*, 5, 199, 1980.

480. Cilento, G., Electronic excitation in dark biological processes, in *Chemical and Biological Generation of Excited States*, W. Adam and G. Cilento, eds., Academic Press, New York, 1982, pp. 278-307.

481. Cilento, G., Generation of electronically excited triplet species in biochemical systems, *Pure Appl. Chem.* 56, 1179, 1984.

482. Adam, W., Ahrweiler, M., Saha-Möller, C. R., Sauter, M., Schönberger, A., Epe, B., Müller, E., Schiffmann, D., Stopper, H. and Wild, D., Genotoxicity studies of benzofuran dioxetanes and epoxides with isolated DNA, bacteria and mammalian cells, *Toxicol. Lett.*, 67, 41, 1993.

483. Emmert, S., Epe, B., Saha-Möller, C. R., Adam, W. and Rünger, T. M., Assessment of genotoxicity and mutagenicity of 1,2-dioxetanes in human cells using a plasmid shuttle vector, *Photochem. Photobiol.*, 61, 136, 1995.

484. Bechara, E.J.H., Faria Olivera, O.M., Duran, N., Casadei de Baptista, R., and Cilento, G. Peroxidase catalyzed generation of triplet acetone. *Photochem. Photobiol.*, 30, 101, 1979.

485. Baader, W. J., Bohne, C., Cilento, G. and Dunford, H. B., Peroxidase-catalyzed formation of triplet acetone and chemiluminescence from isobutyraldehyde and molecular oxygen, *J. Biol. Chem.*, 260, 10,217, 1985.

486. Haun, M., Duran, N., Augusto, O. and Cilento, G., Model studies of the alpha-peroxidase system: formation of an electronically excited product, *Arch. Biochem. Biophys.*, 200, 245, 1980.

487. Adam, W., Baader, W. J., and Cilento, G., Enols of aldehydes in the peroxidase-promoted generation of excited triplet species, *Biochim. Biophys. Acta*, 881, 330, 1986.

488. Adam, W., Kurz, A. and Saha-Möller, C. R., DNA and 2'-deoxyguanosine damage in the horseradish-peroxidase-catalyzed autoxidation of aldehydes: The search for the oxidizing species, *Free Radical Biol. Med.*, 26, 566, 1999.

489. Nascimento, A. L. T.O., da Fonseca, L. M., Brunetti, I. L., and Cilento, G. Intracellular generation of electronically excited states. Polymorphonuclear leukocytes challenges with a precursor of triplet acetone. *Biochim. Biophys. Acta,*. 881, 337, 1986.

440

490. Nascimento, A. L. T. O. and Cilento, G. Generation of electronically excited states *in situ.* Polymorphonuclear leukocytes treated with phenylacetaldehyde. *Photochem. Photobiol.*, 46, 137, 1987.

491. Adam, W., Boland, W., Hartmann-Schreier, J., Humpf, H.-U., Lazarus, M., Saha-Möller, C. R., Saffert, A. and Schreier, P., -hydroxylation of carboxylic acids with molecular oxygen catalyzed by the α-oxidase of peas (*pisum sativum*): a novel biocatalytic synthesis of enantiomerically pure (R)-2-hydroxy acids, *J. Am. Chem. Soc.*, 120, 11,044, 1998.

492. Medeiros, M. H. G., and Bechara, E. J. H., Chemiluminescent aerobic oxidation of protein adducts with glycolaldehyde catalyzed by horseradish peroxidase, *Arch. Biochem. Biophys.*, 248, 435, 1986.

493. Hamman, J. P., Gorby, D. R. and Seliger, H. H., A new type of biological chemiluminescence: The microsomal chemiluminescence of benzo[a]pyrene arises from the diol epoxide product of the 7,8-dihydrodiol, *Biochem. Biophys. Res. Commun.*, 75, 793, 1977.

494. Thompson, A., Biggley, W. H., Posner, G. H., Lever, J. R., and Seliger, H. H., Microsomal chemiluminescence of benzo[a]pyrene-7,8-dihydrodiol and its synthetic analogues trans- and cis- 1-methoxyvinylpyrene, *Biochim. Biophys. Acta.*, 882, 210, 1986.

495. Hamman, J. P., Seliger, H. H. and Posner, G. H., Specificity of chemiluminescence in the metabolism of benzo[a]pyrene to its carcinogenic diol epoxide, *Proc. Natl. Acad. Sci. USA,* 78, 940, 1981.

496. Seliger, H. H., Thompson, A., Hamman, J. P. and Posner, G. H., Chemiluminescence of benzo[a]pyrene-7,8-diol, *Photochem. Photobiol.*, 36, 359, 1982.

497. Kopecky, K. R., *Chemical and Biological Generation of Excited States*, Academic Press, New York, 1982, pp. 85-114.

498. Boveris, A., Cadenas, E. and Chance, B., Ultraweak chemiluminescence: a sensitive assay for oxidative radical reactions, *Federation Proc.*, 40, 195, 1981.

499. Cadenas, E., Biochemistry of oxygen toxicity, *Ann. Rev. Biochem.*, 58, 79, 1989.

500. Prat., A. G. and Turrens, J. F., Ascorbate-and hemoglobin-dependent brain chemiluminescence, *Free Radical Biol. Med.*, 8, 319, 1990.

501. Di Mascio, P., Catalani, L. H. and Bechara, E. J. H., Are dioxetanes chemiluminescent intermediates in lipoperoxidation?, *Free Radical Biol. Med.*, 12, 471, 1992.

502. Furchgott, R. F., The role of endothelium in the responses of vascular smooth muscle to drugs, *Ann. Rev. Pharmacol. Toxicol.*, 24, 175, 1984.

503. Furchgott, R. F., Zawadzki, J. V. and Cherry, P. D., Role of endothelium in the vasodilator response to acetylcholine, in *Vasodilation*, P. Vanhoutte, I. Leusen, eds., Raven Press, New York, 1981, pp. 49-66.

504. Schröder, H., Noack, E. and Müller, R., Evidence for a correlation between nitric oxide formation by cleavage of organic nitrates and activation of guanylate cyclase, *J. Mol. Cell. Cardiol.*, 17, 931, 1985.

505. Noack, E. and Murphy, M., Vasodilation and oxygen radical scavenging by nitric oxide/EDRF and organic nitrovasodilators, in *Oxidative Stress*, H. Sies, ed., Academic Press, New York, 1991, pp. 445-489.

506   Knowles, R. G. and Moncada, S., Nitric oxide synthases in mammals, *Biochemistry J.*, 298, 249, 1994.

507. Bredt, D. S. and Snyder, S. H., Nitric oxide: A physiologic messenger molecule, *Ann..Rev. Biochem.*, 63, 175, 1994.

508. Griffith, O. W. and Stuehr, D. J., Nitric oxide synthases: Properties and Catalytic Mechanism, *Ann..Rev. Physiol.*, 57, 707, 1995.

509. Bredt, D. S. and Snyder, S. H., Nitric oxide, a novel neuronal messenger, *Neuron*, 8, 3, 1992.

510. Moncada, S., Radomski, M. W. and Palmer, R. M. J., Identification as nitric oxide and role in the control of vascular tone and platelet function, *Biochem. Pharmacol.*, 37, 2495, 1988.

511. Furchgott, R. F. and Vanhoutte, P. M., Endothelium-derived relaxing and contracting factors, *FASEB J.*, 3, 2007, 1989.

512. Ignarro, L. J., Endothelium-derived nitric oxide: actions and properties, *FASEB J.*, 3, 31, 1989.

513. Förstermann, U., Schmidt, H. H. H. W., Pollock, J.S., Sheng, H., Mitchell, J. A., Warner, T. D., Nakane, M., and Murad, F., Isoforms of nitric oxide synthase. Characterization and purification form different cell types, *Biochem. Pharmacol.*, 42, 1849, 1991.

442

514.   Hibbs, Jr. J. B., Vavrin, Z., and Taintor, R. R., L-arginine is required for expression of the activated macrophage effector mechanism causing selective metabolic inhibition in target cells, *J. Immunol.*, 138, 550, 1987.

515.   Marletta, M. A., Yoon, P. S., Iyengar, R., Leaf, C. D. and Wishnok, J. S., Macrophage oxidation of L-arginine to nitrite and nitrate: nitric oxide is an intermediate, *Biochemistry*, 27, 8706, 1988.

516.   Schmidt, H. H. H. W., Seifert, R. and Böhme, E., Formation and release of nitric oxide from human neutrophils and HL-60 cells induced by chemotactic peptide activating factor and leukotriene B$_4$, *FEBS Letters*, 244, 357, 1989.

517.   Wright, D. C., Mulch, A., Busse, R. and Osswald, H., Generation of nitric oxide in human neutrophils, *Biochem. Biophys. Res. Commun.*, 160, 813, 1989.

518.   Palacios, M., Knowles, R. G., Palmer, R. M. J. and Moncada, S., Nitric oxide from L-arginine stimulates the soluble guanylate cyclase in adrenal glands, *Biochem. Biophys. Res. Commun.*, 165, 802, 1989.

519.   Garthwaite, J., Charles, S. L. and Chess-Williams, R., Endothelium-derived relaxing factor release on activation of NMDA receptors suggests role as intercellular messenger in the brain, *Nature*, 336, 385, 1988.

520.   Knowles, R. G., Palacios, M. Palmer, R. M. J. and Moncada, S., Formation of nitric oxide from L-arginine in the central nervous system: a transduction mechanism for stimulation of the soluble guanylate cyclase, *Proc. Natl. Acad. Sci. USA*, 86, 5159, 1989.

521.   Knowles, R. G., Palacios, M. Palmer, R. M. J. and Moncada, S., Kinetic characteristics of nitric oxide synthase from rat brain, *Biochem. J.*, 269, 207, 1990.

522.   Schmidt, H. H. H. W., Wilke, P., Evers, B. and Böhme, E., Enzymatic formation of nitrogen oxides from L-arginine in bovine brain cytosol, *Biochem. Biophys. Res. Commun.*, 165, 284, 1989.

523.   Schmidt, H. H. H. W., Nau, H., Wittfoht, W., Gerlach, J., Prescher, E., Klein, M. M., Niroomand, F. and Böhme, E., Arginine is a physiological precursor of endothelium-derived nitric oxide, *Eur. J. Pharmacol.*, 154, 213, 1988.

524.   Palmer, R. M. J., Rees, D. D., Ashton, D. S. and Moncada, S., L-arginine is the physiological precursor for the formation of nitric oxide in endothelium-dependent relaxation, *Biochem. Biophys. Res. Commun.*, 153, 1251, 1988.

525.   Vallance, P., Collier, J. and Moncada, S., Nitric oxide synthesized from L-arginine mediates endothelium dependent dilatation in human veins in vivo, *Cardiovasc. Res.*, 23, 1053, 1989.

526.   Moncada, S., Palmer, R. M. J. and Higgs, E. A., Biosynthesis of nitric oxide from L-arginine. A pathway for the regulation of cell function and communication, *Biochem. Pharmacol.*, 38, 1709, 1989.

527.   Kwon, N. S., Nathan, C. F., Gilker, C., Griffith, O.W., Matthews, D. E., and Stuehr, D.J., L-Citrulline production from L-arginine by macrophage nitric oxide synthase, *J. Biol. Chem.*, 265, 13,442, 1990.

528.   Ignarro, L. J., Barry, B. K., Gruetter, D. Y., Edwards, J. C., Ohlstein, E. H., Gruetter, C. A. and Baricos, W. H., Guanylate cyclase activation by nitroprusside and nitrosoguanidine is related to formation of S-nitrosothiol intermediates, *Biochem. Biophys. Res. Commun.*, 94, 93, 1980.

529.   Misra, H. P., Inhibition of superoxide dismutase by nitroprusside and electron spin resonance observations on the formation of a superoxide-mediated nitroprusside-nitroxyl free radical, *J. Biol. Chem.*, 259, 12,678, 1984.

530.   Iyengar, R., Stuehr, D. J., and Marletta, M. A., Macrophage synthesis of nitrite, nitrate, and N-nitrosamines: Precursors and role of the respiratory burst, *Biochemistry*, 84, 6369, 1987.

531.   Hibbs, Jr. J. B., Taintor, R. R., Vavrin, Z., and Rachlin, E. M., Nitric oxide: A cytotoxic activated macrophage effector molecule, *Biochem. Biophys. Res. Commun.*, 157, 87, 1988.

532.   Tayeh, M. A. and Marletta, M.A., Macrophage oxidation of L-arginine to nitric oxide, nitrite and nitrate, *J. Biol. Chem.*, 264, 19,654, 1989.

533.   Stuehr, D. J., Kwon, N. S. and Nathan, C. F., FAD and GSH participate in macrophage synthesis of nitric oxide, *Biochem. Biophys. Res. Commun.*, 168, 558, 1990.

444

534. Garthwaite, J., Nitric oxide synthesis linked to activation of excitatory neurotransmitter receptors in the brain, in *Nitric Oxide from L-Arginine: A Bioregulatory System*, S. Moncada and E. A. Higgs, eds. Elsevier, Amsterdam, 1990, pp. 115-137.

535. Bredt, D. S. and Snyder, S. H., Nitric oxide mediates glutamate-linked enhancement of cGMP levels in the cerebellum, *Proc. Natl. Acad. Sci. USA*, 86, 9030, 1989.

536. Miki, N., Kawabe, Y. and Kuriyama, K., Activation of cerebral guanylate cyclase by nitric oxide, *Biochem. Biophys. Res. Commun.*, 75, 851, 1977.

537. Bredt, D. S. and Snyder, S. H., Isolation of nitric oxide synthetase, a calmodulin-requiring enzyme, *Proc. Natl. Acad. Sci. USA*, 87, 682, 1990.

538. Bredt, D. S., Hwang, P.M. and Snyder, S. H., Localization of nitric oxide synthase indicating a neural role for nitric oxide, *Nature*, 347, 768, 1990.

539. Förstermann, U., Gorsky, L. D., Pollock, J. S., Schmitt, H. H. H. W., Heller, M. and Murad, F., Regional distribution of EDRF/NO synthesizing enzymes in rat brain, *Biochem. Biophys. Res. Commun.*, 168, 727, 1990.

540. Ross, C. A., Bredt, D. and Snyder, S. H., Messenger molecules in the cerebellum, *Trends Neurosci.*, 13, 216, 1990.

541. Ashley, R. H., Brammer, M. J. and Marchbanks, R., Measurement of intrasynaptosomal free calcium by using the fluorescent indicator quin-1, *Biochem. J.*, 219, 149, 1984.

542. Olson, D. R., Kon, C. and Breckenridge, B. M., Calcium ion effects on guanylate cyclase of brain, *Life Sci.*, 18, 935, 1976.

543. Moncada, S., Palmer, R. M. J. and Higgs, E. A., The discovery of nitric oxide as the endogenous nitrovasodilator, *Hypertension*, 12, 365, 1988.

544. Ignarro, L. J., Biosynthesis and metabolism of endothelium-derived nitric oxide, *Ann. Rev. Pharmacol. Toxicol.*, 30, 535, 1990.

545. Nathan, C., Nitric oxide as a secretory product of mammalian cells, *FASEB J.*, 6, 3051, 1992.

546. Sanders, K. M. and Ward, S. M., Nitric oxide as a mediator of nonadrenergic noncholinergic neurotransmission, *Am. J. Physiol.*, 262 (*Gastrointest. Liver Physiol.* 25): G379, 1992.

547. Moncada, S. and Higgs, A., The L-arginine-nitric oxide pathway, *N. Engl. J. Med.*, 329, 2002, 1993.

548. Anbar, M., Nitric oxide: a synchronizing chemical messenger, *Experientia*, 51, 545, 1995.

549. Gross, S. S. and Wolin, M. S., Nitric oxide: Pathophysiological Mechanisms, *Ann. Rev. Physiol.*, 57, 737, 1995.

550. Kuo, P. C. and Schroeder, R. A., The emerging multifaceted roles of nitric oxide, *Ann. of Surgery,*. 221, 220, 1995.

551. Loscalzo, J., and Welch, G., Nitric oxide and its role in the cardiovascular system, *Prog. Cardiovasc.Dis..*, 38, 87, 1995.

552. Umans, J. G. and Levi, R., Nitric oxide in the regulation of blood flow and arterial pressure, *Ann. Rev. Physiol.*, 57, 771, 1995.

553. Review of various aspects of NO as a modulator, mediator and effector molecule and its role in various diseases, in *Ann. Med.*, 27, 319-427, 1995.

554. Herzberg, G., Photography of the infrared solar spectrum to wavelength 12,900 A, *Nature(London)*, 133, 759, 1934.

555. Herzberg, G., and Herzberg, L., Fine structure of the infrared of the red spectrum of atmospheric oxygen bands, *Astrophys. J.*, 108, 167, 1948.

556. Herzberg, L., and Herzberg, G., Fine structure of the infrared atmospheric oxygen bands, *Astrophys. J.*, 105, 353, 1947.

557. Kasha, M., Introductory remarks: the renascence of research on singlet molecular oxygen, in *Singlet Oxygen*, H. H. Wasserman, R. W. Murray, eds., Academic Press, 1979.

558. Harman, D., The aging process, *Proc. Natl. Acad. Sci. USA*, 78, 7124, 1981.

559. Yakes, F. M. and Van Houten, B., Mitochondrial DNA damage is more extensive and persists longer than nuclear DNA damage in human cells following oxidative stress, *Proc. Natl. Acad. Sci. USA*, 94, 514, 1997.

560. Slater, T. F., Free radical mechanisms in tissue injury, *Biochem. J.*, 222, 1, 1984.

561. Kappus, H., Lipid peroxidation: mechanisms, analysis, enzymology and biological relevance, in *Oxidative Stress*, H. Sies, ed., Academic Press, 1985, pp. 273-310.

562. Cerutti, P. A., Prooxidant states and tumor promotion, *Science*, 227, 375, 1985.

563. Kagan, V. E., *Lipid Peroxidation in Biomembranes*, CRC Press, Boca Raton, FL, 1988.

564. Comporti, M., Three models of free radical-induced cell injury, *Chem. Biol. Interact.*, 72, 1, 1989.

565. Smith, C. V., Free radical mechanisms of tissue injury, in *Free Radical Mechanisms of Tissue Injury*, M. Treinen Moslen and C. V. Smith, eds., CRC Press, Boca Raton, FL, 1993, pp.1-22.

566. Tappel, A. L., Lipid peroxidation damage to cell components, *Federation Proc.*, 32, 1870, 1973.

567. Mead, J. F., Free radical mechanisms of lipid damage and consequences for cellular membranes, in *Free Radicals in Biology*, Vol. I., W. A. Pryor, ed., Academic Press, New York, 1976, pp. 51-68.

568. Tappel, A. L., Measurement of and protection from *in vivo* lipid peroxidation, in *Free Radicals in Biology*, Vol. IV, W. A. Pryor, ed., Academic Press, New York, 1980, pp. 1-48.

569. Frankel, E. N., Chemistry of free radical and singlet oxidation of lipids, *Prog. Lipid Res.*, 23, 197, 1985.

570. Frankel, E. N., Lipid oxidation, *Prog. Lipid Res.*, 19, 1, 1980.

571. Porter, N., Mechanisms for the autoxidation of polyunsaturated lipids, *Acc. Chem. Res.*, 19, 262, 1986.

572. Gardner, H. W., Oxygen radical chemistry of polyunsaturated fatty acids, *Free Radical Biol. Med.*, 7, 65, 1989.

573. Cheeseman, K. H., Lipid peroxidation and cancer, in *DNA and Free Radicals*, B. Halliwell and O. I. Aruoma, eds., Ellis Horwood Limited, Chichester, England, 1993, pp. 109-144.

574. Trush, M. T. and Kensler, W. K., An overview of the relationship between oxidative stress and chemical carcinogenesis, *Free Radical Biol. Med.*, 10, 201, 1991.

575. Morrero, R. and Marnett, L. J., The role of organic peroxyl radicals in *Carcinogenesis*, in *DNA and Free Radicals*, B. Halliwell and O. I. Aruoma, eds., Ellis Horwood Limited, Chichester, England, 1993, pp. 145-161.

576. Hughes, H., Smith, C. V. and Mitchell, J. R., Quantitation of lipid peroxidation products by gas chromatography-mass spectrometry, *Anal. Biochem.*, 152, 107, 1986.

577. Smith, C. V. and Anderson, R. E., Methods for determination of lipid peroxidation in biological samples, *Free Radical Biol. Med.*, 3, 341, 1987.

578. Esterbauer, H. and Zollner, H., Methods for determination of aldehydic lipid peroxidation products, *Free Radical Biol. Med.*, 7, 197, 1989.

579. Esterbauer, H., Zollner, H. and Schaur, R. J., Aldehydes formed by lipid peroxidation: mechanisms of formation, occurrence and determination, in *Membrane Lipid Peroxidation*, Vol. I, C. Vigo-Pelfrey, ed., CRC Press, FL, 1990, pp. 239-283.

580. Ursini, F., Maiorino, M. and Sevanian, A., Membrane hydroperoxides, in *Oxidative Stress, Oxidants and Antioxidants*, H. Sies, ed., Academic Press, New York, 1991, pp. 319-336.

581. Bast, A., Oxidative stress and calcium homeostasis, in *DNA and Free Radicals*, B. Halliwell and O. I. Aruoma, eds., Ellis Horwood Limited, Chichester, England, 1993, pp. 95-108.

582. Frankel, E. N. and Neff, W. E., Formation of malonaldehyde from lipid oxidation products, *Biochem. Biophys. Acta*, 754, 264, 1983.

583. Esterbauer, H., Aldehydic products of lipid peroxidation, in *Free Radicals, Lipid Peroxidation and Cancer*, D. C. H. McBrien and T. F. Slater, eds., Academic Press, London, 1982, pp. 101-128.

584. Esterbauer, H., Lipid peroxidation products: formation, chemical properties and biological activities, in *Free Radicals in Liver Injury*, G. Poli, K. H. Cheeseman, M. U. Dianzani, and T. F. Slater, eds., IRL Press, Oxford, 1985, pp. 29-47.

585. Zollner, H., Schaur, R. J. and Esterbauer, H., Biological activities of 4-hydroxyalkenals, in *Oxidative Stress: Oxidants and Anti-oxidants*, H. Sies, ed., Academic Press, London, 1991, pp. 337-369.

586. Esterbauer, H., Eckl, P. and Ortner, A., Possible mutagens derived from lipids and lipid precursors, *Mutation Res.*, 238, 223, 1990.

587. Esterbauer, H., Schaur, R. J. and Zollner, H., Chemistry and biochemistry of 4-hydroxynonenal, malondialdehyde and related aldehydes, *Free Radical Biol. Med.*, 11, 81, 1991.

588. Marnett, L. J., Hurd, H. K., Hollstein, M. C., Levin, D. E., Esterbauer, H. and Ames, B. N., Naturally occurring carbonyl compounds are mutagens in the Salmonella tester strain TA 104, *Mutat. Res.*, 148, 25, 1985.

448

589.   Winter, C. K., Segall, H. J. and Haddon, W. F., Formation of cyclic adducts of deoxyguanosine with the aldehydes trans-4-hydroxy-2-hexenal and trans-4-hydroxy-2-nonenal in vitro, *Cancer Res.*, 46, 5682, 1986.

590.   Curcio, M., Toricelli, M. V., Giroud, J. P., Esterbauer, H. and Dianzani, M. U., Neutrophil chemotactic responses to aldehydes, *Res. Commun. Chem. Pathol. Pharmacol.*, 36, 463, 1985.

591.   Curcio, M., DiMauro, C., Esterbauer, H. and Dianzanni, M. U., Chemotactic activity of aldehydes, structural requirements. Role in inflammatory process, *Biomed. Pharmacother.*, 41, 304, 1987.

592.   Mitchel, Y. D. and Petersen, D. R., The oxidation of alpha-beta unsaturated aldehydic products of lipid peroxidation by rat liver aldehyde dehydrogenase, *Toxicol. Appl. Pharmacol.*, 87, 403, 1987.

593.   Mitchel, Y. D. and Petersen, D. R., Oxidation of aldehydic products of lipid peroxidation by rat liver microsomal aldehyde dehydrogenase, *Arch. Biochem. Biophys.*, 269, 11, 1989.

594.   Taylor, A. A. and Shappell, S. B., Reactive oxygen species, neutrophil and endothelial adherence molecules, and lipid-derived inflammatory mediators in myocardial ischemia-reflow injury, in *Free Radical Mechanisms of Tissue Injury,* M. Treinen Moslen and C. V. Smith, eds., CRC Press, Boca Raton, FL, 1993, pp.65-141.

595.   Recknagel, R. O., Glende Jr., E. R., Dolak, J. A. and Waller, R. L., Mechanisms of carbon tetrachloride toxicity, *Pharmacol. Ther.*, 43, 139, 1989.

596.   Poli, G., Cottalasso, D., Pronzato, M. A., Chiarpotto, E., Biasi, F., Corongiu, F. P., Marinari, U. M., Nanni, G. and Dianzani, M. U., Lipid peroxidation and covalent binding in the early functional impairment of liver Golgi apparatus by carbon tetrachloride, *Cell Biochemistry and Function*, 8, 1, 1990.

597.   Poli, G., Albano, E. and Dianzani, M. U., The role of lipid peroxidation in liver damage, *The Chemistry and Physics of Lipids*, 45, 117-142, 1987.

598.   Smith, C. V., Evidence for the participation of lipid peroxidation and iron in diquat-induced hepatic necrosis *in vivo*, *Mol. Pharmacol.*, 32, 417, 1987.

599. Téoule, R. and Cadet, J., Radiation-induced degradation of the base component in DNA and related substances, final products, in *Effects of Ionizing Radiation on DNA*, J. Hüttermann, W. Köhnlein, R. Téoule, and A. J. Bertinchamps, eds., Springer, New York, 1978, pp. 171-203.

600. Téoule, R., Radiation-induced DNA damage and its repair, *Int. J. Radiat. Biol.*, 51, 573, 1987.

601. Steenken, S., Purine bases, nucleosides and nucleotides: aqueous solution redox chemistry and transformation reactions of their radical cations and e⁻ and OH adducts, *Chem. Rev.*, 89, 503, 1989.

602. Dizdaroglu, M., Chemistry of free radical damage to DNA and Nucleoproteins, in *DNA and Free Radicals*, B. Halliwell, O. I. Arouma, edts., Ellis Horwood, New York, 1993, pp. 19-39.

603. Breen, A. P. and Murphy, J. A., Reactions of oxyl radicals with DNA, *Free Radical Biol. Med.*, 18, 1033-1077, 1995.

604. Eberhardt, M. K., Radiation-induced homolytic aromatic hydroxylation. 6. The effect of metal ions on the hydroxylation of benzonitrile, anisole and fluorobenzene, *J. Phys. Chem.*, 81, 1051, 1977.

605. Pryor, W.A., Why is the hydroxyl radical the only radical that commonly adds to DNA? Hypothesis: It has a rare combination of high electrophilicity, high thermochemical reactivity, and a mode of production that can occur near DNA, *Free Radical Biol. Med.*, 4, 219, 1988.

606. Schulte-Frohlinde, D. and v. Sonntag, C., Sugar lesions in cell radiobiology, in *Ionizing Radiation Damage to DNA: Molecular Aspects*, S. S. Wallace and R. B. Painter, eds., Wiley-Liss, New York, 1990, pp. 31-42.

607. Dizdaroglu, M., Schulte-Frohlinde, D. and v. Sonntag, C., Strand breaks and sugar release by -irradiation of DNA in aqueous solution. The effect of oxygen, *Z. Naturforsch., Teil C: BioSciences*, 30, 826, 1975.

608. Dizdaroglu, M., Schulte-Frohlinde, D. and v. Sonntag, C., Strand breaks and sugar release by -irradiation of DNA in aqueous solution, *J. Am. Chem. Soc.*, 97, 2277, 1975.

609. Beesk, F., Dizdaroglu, M., Schulte-Frohlinde, D. and v. Sonntag, C., Radiation-induced DNA strand breaks in deoxygenated aqueous

solutions. The formation of altered sugars as end groups, *Int. J. Radiat. Biol.,* 36, 565, 1979.

610. Dizdaroglu, M., Free radical-induced formation of an 8,5'-cyclo-2'-deoxyguanosine moiety in deoxyribonucleic acid, *Biochem. J.,* 238, 243, 1986.

611. Lafleur, M. V. M., Nieuwint, A. W., Aubry, J. M., Kortbeek, H., Aewert, F. and Joenje, H., DNA damage by chemically generated singlet oxygen, *Free Radical Res. Commun.,* 2, 343, 1987.

612. Floyd, R. A., West, M. S., Eneff, K. L. and Schneider, J. E., Methylene blue plus light mediates 8-hydroxyguanine formation in DNA, *Arch. Biochem. Biophys.,* 273, 106, 1989.

613. Sies, H. and Menck, C. F. M., Singlet oxygen induced DNA damage, *Mutation Res.,* 275, 367, 1992.

614. Epe, B., Mützel, P. and Adam, W., DNA damage by oxygen radicals and excited state species: a comparative study using enzymatic probes in vitro, *Chem. Biol. Interactions,* 67, 149, 1988.

615. Adam, W., Beinhauer, A., Mosandl, T., Saha-Möller. C.R., Vargas, F., Epe, B., Müller., E., Schiffmann, D. and Wild, D., Photobiological studies with dioxetanes in isolated DNA, bacteria, and mammalian cells, *Environ. Health Persp.,* 88, 89, 1990.

616. Oleinick, N. L., Chiu, S., Ramakrishnan, N. and Xuo, L., The formation, identification and significance of DNA-protein cross-links in mammalian cells, *British J. Cancer,* 55 (Suppl. VIII), 135, 1987.

617. Olinski, R., Nackerdien, Z. and Dizdaroglu, M., DNA-protein cross-linking between thymine and tyrosine in chromatin of γ-irradiated or H_2O_2-treated cultured human cells, *Arch. Biochem. Biophys.,* 297, 139, 1992.

618. Moan, J., On the diffusion length of singlet oxygen in cells and tissue, *J. Photochem. Photobiol. B, Biol.* 6, 343, 1990.

619. Epe, B., DNA damage induced by photosensitization, in *DNA and Free Radicals,* B. Halliwell and O. I. Aruoma, eds., Ellis Horwood Limited, Chichester, England, 1993, pp. 41-65.

620. Piette, J., Biological consequences associated with DNA oxidation mediated by singlet oxygen, *J. Photochem. Photobiol. B, Biol.,* 11, 241, 1991.

621. Devasagayam, T. P. A., Steenken, S., Obendorf, M. S. W., Schulz, W. A. and Sies, H., Formation of 8-hydroxy-deoxyguanosine and generation of strand breaks at guanine residues in DNA by singlet oxygen, *Biochemistry*, 25, 6283, 1991.

622. DiMascio, P. and Sies, H., Detection and quantification of singlet oxygen generated by thermolysis of a water-soluble endoperoxide, *J. Am. Chem. Soc.*, 111, 2909, 1989.

623. Boiteux, S., Gajewski, E., Laval, J. and Dizdaroglu, M., Substrate specificity of the *Escherichia coli* Fpg protein (formamidopyrimidine-DNA glycosylase): Excision of purine lesions in DNA produced by ionizing radiation or photosensitization, *Biochemistry*, 31, 106, 1992.

624. Di Mascio, P., Kaiser, S. and Sies, H., Lycopene as the most efficient biological carotenoid singlet oxygen quencher, *Arch. Biochem. Biophys.*, 274, 532, 1989.

625. Conn, P. F., Schalch, W. and Truscott, T. G., The singlet oxygen and carotenoid interaction, *J. Photochem. Photobiol., B. Biol.*, 11, 41, 1991.

626. Kaiser, S., DiMascio, P., Murphy, M. E. and Sies, H., Physical and chemical scavenging of singlet molecular oxygen by tocopherols, *Arch. Biochem. Biophys.*, 277, 101, 1990.

627. DiMascio, P., Devasagayam, T. P. A., Kaiser, S. and Sies, H., carotenoids, tocopherols and thiols as biological singlet molecular oxygen quenchers, *Biochem. Soc. Trans.*, 18, 1054, 1990.

628. Shigenaga, M. K. and Ames, B. N., Assays for 8-hydroxy-2'-deoxyguanosine: a biomarker of in vivo oxidative DNA damage, *Free Radical Biol. Med.*, 10, 211, 1991.

629. Shigenaga, M. K., Park, J. W., Cundy, K. C., Gimeno, C. J. and Ames, B. N., In vivo oxidative DNA damage: measurement of 8-hydroxy-2'-deoxyguanosine in DNA and urine by high-performance liquid chromatography with electrochemical detection, *Meth. Enzymol.* 186, 521, 1990.

630. Park, E. M., Shigenaga, M.K., Degan, P., Korn, T. S., Kitzler, J. W., Wehr, C. M., Kolachana, P. and Ames, B. N., The assay of excised oxidized DNA lesions: Isolation of 8-oxoguanine and its nucleoside derivatives from biological fluids with monoclonal antibody columns, *Proc. Natl. Acad. Sci. USA*, 89, 3375, 1992.

631. Wefers, H., Schulte-Frohlinde, D. and Sies, H., Loss of transforming activity of plasmid DNA (pBR322) in *E. coli* caused by singlet molecular oxygen, *FEBS Letters*, 211, 49, 1987.

632. Di Mascio, P., Wefers, H., Do-Thi, H. P., Lafleur, M. V. M. and Sies, H., Singlet molecular oxygen causes loss of biological activity in plasmid and bacteriophage DNA and induces single-strand breaks, *Biochim. Biophys. Acta*, 1007, 151, 1989.

633. Blazek, E. R., Peak, J. G. and Peak, M. J., Singlet oxygen induces frank strand breaks as well as alkali- and piperidine-labile sites in supercoiled plasmid DNA, *Photochem. Photobiol.*, 49, 607, 1989.

634. Devasagayam, T. P. A., DiMascio, P., Kaiser, S. and Sies, H., Singlet oxygen induced single-strand breaks in plasmid pBR322 DNA: the enhancing effect of thiols, *Biochim. Biophys. Acta*, 1088, 409, 1991.

635. Schneider, J. E., Price, S., Maidt, M. L., Gutteridge, J. M. C. and Floyd, R. A., Methylene blue plus light mediates 8-hydroxy-2-deoxyguanosine formation in DNA preferentially over strand breakage, *Nucleic Acid Res.*, 18, 631, 1990.

636. Cadet, J. and Berger, M., Radiation-induced decomposition of the purine bases with DNA and related model compounds, *Int. J. Radiat. Biol.*, 47, 127, 1985.

637. Dizdaroglu, M., Formation of 8-hydroxyguanine moiety in deoxyribonucleic acid on gamma-irradiation in aqueous solution, *Biochemistry*, 24, 4476, 1985.

638. Dizdaroglu, M., Chemical determination of free radical-induced damage to DNA, *Free Radical Biol. Med.*, 10, 225, 1991.

639. Piette, J., Calberg-Bacq, C. M. and van de Vorst, A., Alteration of guanine residues during proflavin mediated photosensitization of DNA, *Photochem. Photobiol.*, 33, 325, 1981.

640. Cadet, J. and Vigny, P., The photochemistry of nucleic acids, in *Bioorganic Photochemistry, Photochemistry and the Nucleic Acids*, H. Morrison, ed., Vol I, John Wiley, New York, 1990, pp. 1-272.

641. Piette, J., Merville-Louis, M. P. and Decuyper, J., Damage induced in nucleic acids by photosensitization, *Photochem. Photobiol.*, 44, 793, 1986.

642. Lamola, A. A. and Yamane, T., Sensitized photodimerization of thymine in DNA, *Proc. Natl. Acad. Sci. USA,* 58, 443, 1967.

643. Lamola, A. A., Production of pyrimidine dimers in DNA in the dark, *Biochem. Biophys. Res. Commun.,* 43, 893, 1971.

644. Smith, K. C., Spontaneous mutagenesis: Experimental, genetic and other factors, *Mutation Res.,* 277, 139, 1992.

645. Demple, B. and Levin, J. D., Repair systems for radical-damaged DNA, in *Oxidative Stress, Oxidants and Antioxidants,* H. Sies, ed., Academic Press, New York, 1991, pp. 119-154.

646. Ramotar, D. and Demple, B., Enzymes that repair oxidative damage to DNA, in *DNA and Free Radicals,* B. Halliwell and O. I. Aruoma, eds., Ellis Horwood, Chichester, England, 1993, pp. 165-191.

647. Cheng, K. C., Cahill, D. S., Kasai, H., Nishimura, S. and Loeb, L. A., 8-hydroxyguanine, an abundant form of oxidative DNA damage, causes G-T and A-C substitutions, *J. Biol. Chem.,* 267, 166, 1992.

648. Wood, M. L., Dizdaroglu, M., Gajewski, E. and Essigmann, J. M., Mechanistic studies of ionizing radiation and oxidative muta-genesis: Genetic effects of a single 8-hydroxyguanine (7-hydro-8-oxoguanine) residue inserted at a unique site in a viral genome, *Biochemistry,* 29, 7024, 1990.

649. Kasai, H. and Nishimura, S., Hydroxylation of the C-8 position of deoxyguanosine by reducing agents in the presence of oxygen, *Nucleic Acids Symp. Ser.* No. 12, 165, 1983.

650. Kasai, H., and Nishimura, S., Hydroxylation of deoxy-guanosine at the C-8 position by ascorbic acid and other reducing agents, *Nucleic Acids Res.* 12, 2137, 1984.

651. Kasai, H. and Nishimura, S., Hydroxylation of deoxy-guanosine at the C-8 position by polyphenols and aminophenols in the presence of hydrogen peroxide and ferric ion, *Gann,* 75, 565, 1984.

652. Kasai, H., Tanooka, H. and Nishimura, S., Formation of 8-hydroxyguanine residues in DNA by X-irradiation, *Gann,* 75, 1037, 1984.

653. Dizdaroglu, M., Rao, G., Halliwell, B. and Gajewski, E., Damage to the DNA bases in mammalian chromatin by hydrogen peroxide in the presence of ferric and cupric ions, *Arch. Biochem. Biophys.,* 285, 317, 1991.

454

654. Adam, W., Andler, S., Ballmaier, D., Emmert, S., Epe, B., Grimm, G., Mielke, K., Möller, M., Rünger, T. M., Saha-Möller, C. R. and Schönberger, A., Oxidative DNA damage by dioxetanes, photosensotizing ketones, and photo-Fenton reagents, in *Recent Results in Cancer Research,* H. K. Müller-Hermelink, H. G. Neumann, W. Dekant, (eds.), Springer Verlag, Berlin, Heidelberg, Vol. 143, 1997, pp. 21-34.

655. Meneghini, R. and Martins, E. L., Hydrogen peroxide and DNA damage, in *DNA and Free Radicals,* B. Halliwell and O. I. Aruoma, eds., Ellis Horwood Limited, 1993, pp. 83-94.

656. Dizdaroglu, M., Nackerdien, Z., Chao, B. C., Gajewski, E. and Rao, G., Chemical nature of *in vivo* DNA base damage in hydrogen peroxide-treated mammalian cells, *Arch. Biochem. Biophys.,* 285, 388, 1991.

657. Imlay, J. A., Chin, S. M. and Linn, S., Toxic DNA damage by hydrogen peroxide through the Fenton reaction *in vivo* and *in vitro, Science,* 240, 640, 1988.

658. Kasai, H., Crain, P. F., Kuchino, Y., Nishimura, S., Ootsuyama, A. and Tanooka, H., Formation of 8-hydroxyguanine moiety in cellular DNA by agents producing oxygen radicals and evidence for its repair, *Carcinogenesis,* 7, 1849, 1986.

659. Orrenius, S., McConkey, D. J. and Nicotera, P., Mechanisms of oxidant-induced cell damage, in *Oxy-radicals in Molecular Biology and Pathology,* P. Cerutti, I. Fridovich, and J. McCord, eds., Liss, New York, 1988, pp. 327-339.

660. Fraga, C. G. and Tappel, A. L., Damage to DNA concurrent with lipid peroxidation in rat liver slices, *Biochemical J.,* 252, 893, 1988.

661. Kasai, H. and Nishimura, S., Formation of 8-hydroxydeoxy-guanosine in DNA by autoxidized unsaturated fatty acids, in *Medical, Biochemical and Chemical Aspects of Free Radicals,* O. Hayaishi, E.Niki, M. Kondo and T. Yoshikawa, eds., Elsevier Science Publishers, Oxford, UK, 1988, p. 1021.

662. Park, J. W. and Floyd, R. A., Lipid peroxidation products mediate the formation of 8-hydroxydeoxyguanosine in DNA, *Free Radical Biol. Med.,* 12, 245, 1992.

663. Mello Filho, A. C. and Meneghini, R., Iron is the intracellular metal involved in the production of DNA damage by oxygen radicals, *Mutation Res.,* 251, 109, 1991.

664. Nackerdien, Z., Olinski, R. and Dizdaroglu, M., DNA base damage in chromatin of -irradiated cultured human cells, *Free Radical Res. Commun.,* 16, 259, 1992.

665. Halliwell, B., Oxidative DNA damage: meaning and measurement, in *DNA and Free Radicals,* B. Halliwell and O. I. Aruoma, eds., Ellis Horwood Limited, Chichester, England, 1993, pp. 67-82.

666. Richter, C., Park, J. W. and Ames, B. N., Normal oxidative damage to mitochondrial and nuclear DNA is extensive, *Proc. Natl. Acad. Sci. USA,* 85, 6465, 1988.

667. Stevens, R. G., Jones, D. Y., Micozzi, M. S. and Taylor, P. R., Body iron stores and risk of cancer, *N. Engl. J. Med.,* 319, 1047, 1988.

668. Knekt, P., Reunanen, A., Takkunen, H., Aromaa, A., Heliövaara, M. and Hakuline, T., Body iron stores and the risk of cancer, *Int. J. Cancer,* 56, 379, 1994.

669. Salonen, J. T., Nyyssonen, K., Korpela, H., Tuomilehto, J., Seppanen, R. and Salonen, R., High stored iron levels are associated with excess risk of myocardial infarction in eastern Finnish men, *Circulation,* 86, 803, 1992.

670. Sempos, C. T., Looker, A. C., Gillum, R. F. and Makuc, D. M., Body iron stores and the risk of coronary heart disease, *N. Engl. J. Med.,* 330, 1119, 1994.

671. Kaur, H. and Halliwell, B., Evidence for nitric oxide-mediated oxidative damage in chronic inflammation. Nitrotyrosine in serum and synovial fluid from rheumatoid patients, *FEBS Letters,* 350, 9, 1994.

672. Eberhardt, M. K., Radiation-induced homolytic aromatic substitution. III. Hydroxylation and nitration of benzene, *J. Phys. Chem.,* 79, 1067, 1975.

673. Ames, B. N., Cathcart, R., Schwiers, E., Hochstein, P., Uric acid provides an antioxidant defense in humans against oxidant and radical-caused aging and cancer; a hypothesis, *Proc. Natl. Acad. Sci. USA* 78, 6858, 1981.

674. Grootveld, M. and Halliwell, B., Measurement of allantoin and uric acid in human body fluids. A potential index of free radical reactions in vivo?, *Biochem. J.* 243, 803, 1987.

675. Witz, G., Biological interactions of ,ß-unsaturated aldehydes, *Free Radical Biol. Med.,* 7, 333, 1989.

676.   Riely, C. A., Cohen, C. A. and Lieberman, M., Ethane evolution: a new index of lipid peroxidation, *Science*, 183, 208, 1974.

677.   Muller, A. and Sies, H., Assay of ethane and pentane from isolated organs and cells, *Methods in Enzymol.*, 105, 311, 1984.

678.   Gilbert, H. F., Molecular and cellular aspects of thiol/disulfide exchange, *Adv. Enzymol.*, 63, 69, 1989.

679.   DiMonte, D., Bellomo, G., Thor, H., Nicotera, P. and Orrenius, S., Menadione-induced cytotoxicity is associated with protein thiol oxidation and alteration in intracellular Ca^{2+} homostasis, *Arch. Biochem. Biophys.*, 235, 343, 1984.

680.   Smith, C. V., Hughes, H., Lauterburg, B. H. and Mitchell, J. R., Oxidant stress and hepatic necrosis in rats treated with diquat, *J. Pharmacol. Exp. Ther.*, 235, 172, 1985.

681.   Tannenbaum, S. R., Wishnok, J. S. and Leaf, C. D., Inhibition of nitrosamine formation by ascorbic acid, *Am. J. Clin. Nutr.*, 53, 247S, 1991.

682.   Millar, J., The nitric oxide/ascorbate cycle: how neurons may control their own oxygen supply, *Med. Hypothesis*, 45, 21, 1995.

683.   Pryor, W. A., Church, D. F., Govindan, C. K. and Crank, G., Oxidation of thiols by nitric oxide and nitrogen dioxide: synthetic utility and toxicological implications, *J. Org. Chem.*, 47, 156, 1982.

684.   Keilin, D. and Nicholls, P., Reactions of catalase with hydrogen peroxide and hydrogen donors, *Biochim. Biophys. Acta*, 29, 302, 1958.

685.   DeMaster, E. G., Raij, L., Archer, S. L. and Weir, E. K., Hydroxylamine is a vasorelaxant and a possible intermediate in the oxidative conversion of L-arginine to nitric oxide, *Biochem. Biophys. Res. Commun.*, 163, 527, 1989.

686.   Feelisch, M., The biochemical pathways of nitric oxide formation from nitrovasodilators: Appropriate choice of exogenous NO donors and aspects of preparation and handling of aqueous NO solutions, *J. Cardiovasc. Pharmacol.*, 17 (suppl. 3), S25, 1991.

687.   Feelisch, M. and Noack, E., Correlation between nitric oxide formation during degradation of organic nitrates and activation of guanylate cyclase, *Eur. J. Pharmacol.*, 139, 19, 1987.

688.   Kelm, M., Feelisch, M., Spahr, R., Piper, H-M., Noack, E., and Schrader, J., Quantitative and kinetic characterization of nitric oxide

and EDRF released from cultured endothelial cells, *Biochem. Biophys. Res. Commun.*, 154, 236, 1988.

689. Downes, M. J., Edwards, M. W., Elsey, T. S. and Walters, C. L., Determination of a non-volatile nitrosamine by using denitrosation and a chemiluminescence analyser, *Analyst*, 101, 742, 1976.

690. Gryglewski, R. J., Moncada, S. and Palmer, R. M. J., Bioassay of prostacyclin and endothelium-derived relaxing factor (EDRF) from porcine aortic endothelial cells, *Br. J. Pharmacol.*, 87, 685, 1986.

691. Fukuto, J. M., Chemistry of nitric oxide: biologically relevant aspects, *Adv. in Pharmacol.*, 34, 1, 1995.

692. Ford, P. C., Wink, D. A. and Stanbury, D. M., Autoxidation kinetics of aqueous nitric oxide, *FEBS Letters*, 326, 1, 1993.

693. Wink, D. A., Derbyshire, J. F., Nims, R. W., Saavedra, J. E. and Ford, P. C., Reactions of the bioregulatory agent nitric oxide in oxygenated aqueous media: Determination of the kinetics for oxidation and nitrosation by intermediates generated in the NO/O_2 reaction, *Chem. Res. Toxicol.*, 6, 23, 1993.

694. Wink, D. A., Nims, R. W., Darbyshire, J. F., Christodoulou, D., Hanbauer, I., Cox, G. W., Laval, F., Laval, J., Cook, J. A., Krishna, M. C., DeGraff, W. G. and Mitchell, J. B., Reaction kinetics for nitrosation of cysteine and glutathione in aerobic nitric oxide solutions at neutral pH. Insights into the fate and physiological effects of intermediates generated in the NO/O_2 reaction, *Chem. Res. Toxicol.*, 7, 519, 1994.

695. Lewis, R. S., and Deen, W. M., Kinetics of the reaction of nitric oxide with oxygen in aqueous solutions, *Chem. Res. Toxicol.*, 7, 568, 1994.

696. Grisham, M. B., *Reactive Metabolites of Oxygen and Nitrogen in Biology and Medicine*, R. G. Landes Company, Austin, Texas, 1992.

697. Green, L.C., Ruiz de Luzuriaga, K., Wagner, D. A., Rand, W., Istfan, N., Young, V. R. and Tannenbaum, S. R., Nitrate biosynthesis in man, *Proc. Natl. Acad. Sci., USA*, 78, 7764, 1981.

698. Prasad, M. R., Engelman, R. M., Jones, R. M. and Das, D. K., Effects of oxyradicals on oxymyoglobin, *Biochemistry J.*, 263, 731, 1989.

458

699.   Solomons, G. T. W., *Organic Chemistry*, John Wiley, New York, 1976, pp. 831-841.
700.   Wink, D. A., Kasprzak, K. S., Maragos, C. M., Elespurn, R. K., Misra, M., Dunams, T. M., Cebula, T. A., Koch, W. H., Andrews, A. W., Allen, J. S., and Keefer, L. K., DNA deaminating ability and genotoxicity of nitric oxide and its progenitors, *Science* 254, 1001, 1991.
701.   Arroyo, C. M., and Kohno, M., Difficulties encountered in the detection of nitric oxide (NO) by spin trapping techniques. A cautionary note., *Free Radical Res. Commun.* 14, 145, 1991.
702.   Saran, M., Michel, C. and Bors, W., Reaction of NO with $O_2^{-\cdot}$. Implications for the action of endothelial derived relaxation factor (EDRF), *Free Radical Res. Commun.*, 10, 221, 1990.
703.   McDonald, C. C., Phillips, W. D. and Mower, H. F., An electron spin resonance study of some complexes of iron, nitric oxide, and anionic ligands, *J. Am. Chem. Soc.*, 87, 3319, 1965.
704.   Ignarro, L. J., Heme-dependent activation of soluble guanylate cyclase by nitric oxide: Regulation of enzyme activity by porphyrins and metalloporphyrins, *Semin. Hematol.*, 26, 63, 1989.
705.   Ignarro, L. J., Heme-dependent activation of cytosolic guanylate cyclase by nitric oxide: A widespread signal transduction mechanism, *Biochem. Soc. Trans.*, 20(2), 465, 1992.
706.   Taylor, T. G., and Sharma, V. S., Why NO? *Biochemistry*, 31, 2847, 1992.
707.   Doyle, M. P. and Hoekstra, J. W., Oxidation of nitrogen oxides by bound dioxygen in hemoproteins, *J. Inorg. Biochem.*, 14, 351, 1981.
708.   Jameson, G. B. and Ibers, J. A., Biological and synthetic oxygen carriers, in *Bioinorganic Chemistry*, I. Bertini, H. B. Gray, S. J. Lippard and J. S. Valentine, eds., University Science Books, Mill Valley, CA, 1994.
709.   Wade, R. S. and Castro, C. E., Redox reactivity of iron (III) porphyrins and heme proteins with nitric oxide. Nitrosyl transfer to carbon, oxygen, nitrogen and sulfur, *Chem. Res. Toxicol.*, 3, 289, 1990.
710.   Taylor, T. G., Duprat, A. F., and Sharma, V. S., Nitric oxide-triggered heme-mediated hydrolysis: A possible model for biological reactions of NO, *J. Am. Chem. Soc.*, 115, 810, 1993.

711.   Sies, H., Brigelius, R. and Akerboom, T. P. M., Intrahepatic glutathione status, in *Functions of Glutathione: Biochemical, Physiological, Toxicological, and Clinical Aspects*, A. L. Larsen, S. Orrenius, A. Holmgren and B. Mannervik, eds., Raven Press, New York, 1983, pp. 51-64.

712.   Stamler, J. S., Simon, D. I., Osborne, J. A., Mullins, M. E., Jaraki, O., Michel, T., Singel, D. J. and Loscalzo, J., S-nitrosylation of proteins with nitric oxide: Synthesis and characterization of biologically active compounds, *Proc. Natl. Acad. Sci. USA*, 89, 444, 1992.

713.   Saez, G., Thornalley, P. J., Hill, H. A. O., Hems, R. and Bannister, J., The production of free radicals during the autoxidation of cysteine and their effect on isolated rat hepatocytes, *Biochim. Biophys. Acta*, 719, 24, 1982.

714.   Jia, U. and Furchgott, R. F., Inhibition by sulfhydryl compounds of vascular relaxation induced by nitric oxide and endothelium-derived relaxing factor, *J. Pharmacol. Exp. Ther.*, 267, 271, 1993.

715.   Miles, A. M., Gibson, M. F., Kirshna, M., Cook, J. C., Pacelli, R., Wink, D. A. and Grisham, M. B., Effects of superoxide on nitric oxide-dependent N-nitrosation reactions, *Free Radic. Res.*, 23, 379, 1995.

716.   Shelton, J. R. and Kopczewski, R. F., Nitric oxide induced free-radical reactions, *J. Org. Chem.*, 32, 2908, 1967.

717.   Gray, D., Lissi, E. and Heicklen, J., The reaction of hydrogen peroxide with nitrogen dioxide and nitric oxide, *J. Phys. Chem.*, 76, 1919, 1972.

718.   Mikuni, T., Tatsuta, M., and Kamachi, M., Production of hydroxyl free radical by reaction of hydrogen peroxide with N-methyl-N'-nitro-N-nitrosoguanidine, *Cancer Res.*, 45, 6442, 1985.

719.   Mikuni, T., and Tatsuta, M., Production of hydroxyl free radical by exposure of N-methyl-N'-nitro-N-nitroso-guanidine to visible light in the absence of hydrogen peroxide, *Radiat. Res.*, 138, 320, 1994.

720.   Wang, J. M., Lin-Shiau, S. Y. and Lin, J. K., Relaxation of rat thoracic aorta by N-nitroso compounds and nitroprusside and their modifications of nucleic acid bases through release of nitric oxide, *Biochem. Pharmacol.*, 45, 819, 1993.

721.   Nguyen, T., Brunson, D., Crespi, C. L., Penman, B. W., Wishnok, J. S. and Tannenbaum, S. R., DNA damage and mutation in human

460

cells exposed to nitric oxide *in vitro, Proc. Natl. Acad. Sci. USA,* 89, 3030, 1992.

722. Arroyo, P. L., Hatch-Pigott, V., Mower, H. F. and Cooney, R. V., Mutagenicity of nitric oxide and its inhibition by antioxidants, *Mutat. Res.,* 281, 193, 1992.

723. Isomura, K., Chikahira, M., Teranishi, K. and Hamada, K., Induction of mutations and chromosome abberations in lung cells following *in vivo* exposure of rats to nitrogen oxides, *Mutat. Res.,* 136, 119, 1984.

724. Tannenbaum, S. R., Wishnok, J. S., de Rojas-Walker, T., Tamir, S. and Ji, H., Overview of nitric oxide and nitric oxide/superoxide chemical interaction with DNA, *Proc. Am. Assoc. of Cancer Res.,* 34, 583, 1993.

725. Rees, Y. and Williams, G. H., Reactions of organic free radicals with nitrogen oxides, in *Advances in Free Radical Chemistry,* Vol. III, Academic Press, New York, 1969, pp. 199-230.

726 Pryor, W. A. and Lightsey, J. W., Mechanisms of nitrogen dioxide reactions : initiation of lipid peroxidation and the production of nitrous acid, *Science,* 214, 435, 1981.

727. Pryor, W. A., Castle, L., and Church, D. F., Nitrosation of organic hydroperoxides by nitrogen dioxide/dinitrogen tetraoxide, *J. Am. Chem. Soc.,* 107, 211, 1985.

728. Rubanyi, G. M., and Vanhoutte, P. M., Superoxide anions and hyperoxia inactivate endothelium-derived relaxing factor, *Am. J. Physiol.,* 250, H822, 1986.

729. Moncada, S., Palmer, R. M. J. and Gryglewski, R. J., Mechanism of action of some inhibitors of endothelium-derived relaxing factor, *Proc. Natl. Acad. Sci. USA,* 83, 9164, 1986.

730. Kelm, M. and Schrader, J., Control of coronary vascular tone by nitric oxide, *Circ. Res.,* 66, 1561, 1990.

731. Garlick, P. B., Davies, M. J., Hearse, D. J. and Slater, T. F., Direct detection of free radicals in the reperfused rat heart using electron spin resonance spectroscopy, *Circ. Res.,* 61, 757, 1987.

732. Warner, T. D., de Nucci, G. and Vane, J. R., Comparison of the survival of endothelium-derived relaxing factor and nitric oxide within the isolated perfused mesenteric arterial bed of the rat, *Br. J. Pharmacol.,* 97, 777, 1989.

733. Rubanyi, G. M., and Vanhoutte, P. M., Oxygen-derived free radicals, endothelium and responsiveness of vascular smooth muscle, *Am. J. Physiol.*, 250, H815, 1986.

734. Murphy, M. E., and Sies, H., Reversible conversion of nitroxyl anion to nitric oxide by superoxide dismutase, *Proc. Natl. Acad. Sci. USA*, 88, 10,860, 1991.

735. Cocks, T. M., Angus, J. A., Campbell, J. H. and Campbell, G. R., Release and properties of endothelium-derived relaxing factor (EDRF) from endothelial cells in culture, *J. Cell. Physiol.*, 123, 310, 1985.

736. Murray, J. J., Fridovich, I., Makhoul, R. G. and Hagen, P. O., Stabilization and partial characterization of endothelium-derived relaxing factor from cultured bovine aortic endothelial cells, *Biochem. Biophys. Res. Commun.*, 141, 689, 1986.

737. Stamler, J. S., Singel, D. J. and Loscalzo, J., Biochemistry of nitric oxide and its redox-activated forms, *Science*, 258, 1898, 1992.

738. Fukuto, J. M., Hobbs, A. J. and Ignarro, L. J., Conversion of nitroxyl (HNO) to nitric oxide in biological systems: The role of physiological oxidants and relevance to the biological activity of HNO, *Biochem. Biophys. Res. Commun.*, 196, 707, 1993.

739. Hobbs, A. J., Fukuto, J. M. and Ignarro, L. J., Formation of free nitric oxide from L-arginine by nitric oxide synthase: Direct enhancement of generation by superoxide dismutase, *Proc. Natl. Acad. Sci. USA*, 91, 10992, 1994.

740. Pryor, W. A., and Squadrito, G. L., The chemistry of peroxynitrite: a product from the reaction of nitric oxide with superoxide, *Am. J. Physiol.*, 268 (*Lung Cell Mol. Physiol.* 12): L699, 1995.

741. Katsuki, S., Arnold, W., Mittal, C. and Murad, F., Stimulation of guanylate cyclase by sodium nitroprusside, nitroglycerin and nitric oxide in various tissue preparations and comparison to the effects of sodium azide and hydroxylamine, *J. Cyclic Nucleotide Res.*, 3, 23, 1977.

742. Arnold, W. P., Mittal, C. K., Katsuki, S. and Murad, F., Nitric oxide activates guanylate cyclase and increases guanosine-3',5-cyclic monophospate levels in various tissue preparations, *Proc. Natl. Acad. Sci. USA*, 74, 3203, 1977.

462

743. Gruetter, C. A., Barry, B. K., McNamara, D. B., Gruetter, D.Y., Kadowitz, P. K. and Ignarro, L. J., Relaxation of bovine coronary artery and activation of coronary arterial guanylate cyclase by nitric oxide, nitroprusside and a carcinogenic nitrosamine, *J. Cyclic Nucleotide Res.*, 5, 211, 1979.

744. Shikano, K. and Berkowitz, B. A., Endothelium-derived relaxing factor is a selective relaxant of vascular smooth muscle, *J. Pharmacol. Exp. Ther.*, 243, 55, 1987.

745. Shikano, K., Ohlstein, E. H. and Berkowitz, B. A., Differential selectivity of endothelium-derived relaxing factor and nitric oxide in smooth muscle, *Brit. J. Pharmacol.*, 92, 483, 1987.

746. Buga, G. M., Gold, M. E., Wood, K. S., Chauduri, G. and Ignarro, L. J., Endothelium-derived nitric oxide relaxes nonvascular smooth muscle, *Eur. J. Pharmacol.*, 161, 61, 1989.

747. Long, C. J., Shikano, K. and Berkowitz, B. A., Anion exchange resins discriminate between nitric oxide and EDRF, *Eur. J. Pharmacol.*, 142, 317, 1987.

748. Ignarro, L. J. and Gruetter, C. A., Requirement of thiols for activation of coronary arterial guanylate cyclase by glyceryl trinitrate and sodium nitrite, *Biochim. Biophys. Acta,* 631, 221, 1980.

749. Goretski, J. and Hollocher, T. C., Trapping of nitric oxide produced during denitrification by extracellular hemoglobin, *J. Biol. Chem.*, 263, 2316, 1988.

750. Kowaluk, E. A. and Fung, H. L., Spontaneous liberation of nitric oxide cannot account for *in vitro* vascular relaxation by S-nitrosothiols, *J. Pharmacol. Exp. Ther.*, 255, 1256, 1990.

751. Myers, P. R., Minor, Jr., R. L., Guerra, Jr., R., Bates, J. N. and Harrison, D. G., Vasorelaxant properties of the endothelium-derived relaxing factor more closely resemble S-nitroso-cysteine than nitric oxide, *Nature*, 345, 161, 1990.

752. Rubanyi, G. M., Johns, A., Wilcox, D., Bates, F. N. and Harrison, D., Evidence that a S-nitrosothiol, but not nitric oxide, may be identical with endothelium-derived relaxing factor, *J. Cardiovasc. Pharmacol.*, 17 (suppl. 3), S41, 1991.

753. Mathews, W. R. and Kerr, S. W., Biological activity of S-nitrosothiols: The role of nitric oxide, *J. Pharmacol. Exp. Ther.*, 267, 1529, 1993.

754. Bates, J. N., Harrison, D. G., Myers, P. R. and Minor, R. L., EDRF: nitrosylated compound or authentic nitric oxide, *Bas. Res. Cardiol.*, 86 (Suppl. 2), 17, 1991.

755. Stamler, J. S., Jaraki, O., Osborne, J., Simon, D. I., Keaney, J., Vita, J. A., Singel, D., Valeri, C. R. and Loscalzo, J., Nitric oxide circulates in mammalian plasma primarily as an S-nitroso adduct of serum albumin, *Proc. Natl. Acad. Sci. USA*, 89, 7674, 1992.

756. Keaney, Jr., J. F., Simon, D. I., Stamler, J. S., Jaraki, O., Scharfstein, J., Vita, J. A. and Loscalzo, J., NO forms an adduct with serum albumin that has endothelium-derived relaxing factor-like properties, *J. Clin. Invest.*, 91, 1582, 1993.

757. Scharfstein, J. S., Keaney, Jr., J. F., Slivka, A., Welch, G. N., Vita, J. A., Stamler, J. S. and Loscalzo, J., In vivo transfer of nitric oxide between a plasma protein-bound reservoir and low molecular weight thiols, *J. Clin. Invest.*, 94, 1432, 1994.

758. Gaston, B., Reilly, J., Drazen, J., Fackler, J., Ramdev, P., Mullins, M., Sugarbaker, D., Jaraki, O., Singel, D.J., Loscalzo, J. and Stamler, J. S., Endogenous nitrogen oxides and bronchodilator S-nitrosothiols in human airways, *Proc. Natl. Acad. Sci. USA*, 90, 10,957, 1993.

759. Jansen, A., Drazen, J., Osborne, J. A., Brown, R., Loscalzo, J. and Stamler, J. S., The relaxant properties in guinea pig airways of S-nitrosothiols, *J. Pharmacol. Exp. Ther.*, 261, 154, 1992.

760. Wei, E. P., Kontos, H. A., Christman, C. W., DeWitt, D. S. and Povlishock, J. T., Superoxide generation and reversal of acetyl-choline-induced cerebral arteriolar dilation after acute hypertension, *Circ. Res.*, 57, 781, 1985.

761. Pieper, G. M. and Gross, G. J., Oxygen free radicals abolish endothelium-dependent relaxation in diabetic rat aorta, *Am. J. Physiol.*, 255, H825, 1988.

762. Van Benthuysen, K. M., McMurty, I. F. and Horowitz, L. D., Reperfusion after acute coronary occlusion in dogs impairs endothelium-dependent relaxation to acetylcholine and augments contractile reactivity in vitro, *J. Clin. Invest.*, 79, 265, 1987.

763. Freiman, P. C., Mitchell, G. G., Heistad, D. D., Armstrong, M. L., and Harrison, D. G., Atherosclerosis impairs endothelium-dependent vascular relaxation to acetylcholine and thrombin in primates, *Circ. Res.*, 58, 783, 1986.

464

764. Beckman, J. S., Beckman, T. W., Chen, J., Marshall, P. A. and Freeman, B. A., Apparent hydroxyl radical production by peroxynitrite: Implications for endothelial injury from nitric oxide and superoxide, *Proc. Natl. Acad. Sci. USA*, 87, 1620, 1990.

765. Van der Vliet, A., O'Neill, C. A., Halliwell, B., Cross, C. E. and Kaur, H., Aromatic hydroxylation and nitration of phenylalanine and tyrosine by peroxynitrite. Evidence for hydroxyl radical production from peroxynitrite, *FEBS Lett.*, 339, 89, 1994.

766. Maskos, Z., Rush, J. D., and Koppenol, W. H., The hydroxylation of phenylalanine and tyrosine: A comparison with salicylate and tryptophan, *Arch. Biochem. Biophys.*, 296, 521, 1992.

767. Pruetz, W. A., Moenig, H., Butler, J. and Land, E. J., Reactions of nitrogen dioxide in aqueous model systems: Oxidation of tyrosine units in peptides and proteins, *Arch. Biochem. Biophys.*, 243, 125, 1985.

768. Bunbury, D. L., Conditions for the photochemical nitration of benzene, *Can. J. Chem.*, 43, 1714, 1965.

769. Ischiropoulos, H., Zhu, L., Chen, J., Tsai, M., Martin, J. C., Smith, C. D. and Beckman, J. S., Peroxynitrite-mediated tyrosine nitration catalyzed by superoxide dismutase, *Arch. Biochem. Biophys.*, 298, 431, 1992.

770. Uppu, R. M., Squadrito, G. L. and Pryor, W.A., Acceleration of peroxynitrite oxidations by carbon dioxide, *Arch. Biochem. Biophys.*, 327, 335, 1996.

771. Augusto, O., Gatti, R. M., and Radi, R., Spin-trapping studies of peroxynitrite decomposition and of 3-morpholino-sydnonimine N-ethylcarbamide autooxidation: Direct evidence for metal-independent formation of free radical intermediates, *Arch. Biochem. Biophys.*, 310, 118, 1994.

772. Shi, X., Lenhart, A., and Mao, Y., ESR spin trapping investigation on peroxynitrite decomposition: No evidence for hydroxyl radical production, *Biochem. Biophys. Res. Commun.*, 203, 1515, 1994.

773. Koppenol, W. H., Moreno, J. J., Pryor, W. A., Ischiropoulos, H. and Beckman, J. S., Peroxynitrite, a cloaked oxidant formed by nitric oxide and superoxide, *Chem. Res. Toxicol.*, 5, 834, 1992.

774. Merenyi, G. and Lind, J., Thermodynamics of peroxynitrite and its CO_2 adduct, *Chem. Res. Toxicol.*, 10, 1216, 1997.

775. Koppenol, W. H. and Kissner, R., Can O=NOOH undergo homolysis? *Chem. Res. Toxicol.*, 11, 87, 1998.

776. Pryor, W. A. and Hendrickson, Jr. W. H., Reaction of nucleophiles with electron acceptors by S_N2 or electron transfer (ET) Mechanisms: tert-butyl peroxybenzoate / dimethyl sulfide and benzoyl peroxide / N,N - dimethylaniline systems. *J. Am. Chem. Soc.*, 105, 7114, 1983.

777. Pryor, W. A. and Hendrickson, Jr. W. H., The mechanism of radical production from the reaction of N,N-dimethylaniline with benzoyl-peroxide, *Tetrahedron Lett.*, 24, 1459, 1983.

778. Van Der Vliet, A., Smith, D., O'Neill, C.A., Kaur, H., Darley-Usmar, V., Cross, C. E., and Halliwell, B., Interactions of peroxynitrite with human plasma and its constituents: oxidative damage and antioxidant depletion, *Biochem. J.*, 303, 295, 1994.

779. Radi, R., Beckman, J. S., Bush, K. M., and Freeman, B. A. , Peroxynitrite oxidation of sulfhydryls. The cytotoxic potential of superoxide and nitric oxide, *J. Biol. Chem.*, 266, 4244, 1991.

780. Bartlett, D., Church, D. F., Bounds, P. L. and Koppenol, W. H., The kinetics of the oxidation of L-ascorbic acid by peroxynitrite, *Free Radical Biol. Med.*, 18, 85, 1995.

781. Hogg, N., Joseph, J. and Kalyanaraman, B., The oxidation of - tocopherol and trolox by peroxynitrite, *Arch. Biochem. Biophys.*, 314, 153, 1994.

782. Pryor, W. A., Jin, X., Squadrito, G. L., One- and two-electron oxidations of methionine by peroxynitrite, *Proc. Natl. Acad. Sci. USA*, 91, 11173, 1994.

783. Packer, J. E., Slater, T. F. and Willson, R. L., Direct observation of a free radical interaction between vitamin E and vitamin C, *Nature*, 278, 737, 1979.

784. Liebler, D., The role of metabolism in the antioxidant function of vitamin E, *Crit. Rev. Toxicol.* 23, 147, 1993.

785. Radi, R., Beckman, J. S., Bush, K. M., and Freeman, B. A., Peroxynitrite-induced membrane lipid peroxidation: The cytotoxic potential of superoxide and nitric oxide, *Arch. Biochem. Biophys.*, 288, 481, 1991.

786. Squadrito, G. L. and Pryor, W. A., The formation of peroxynitrite in vivo from nitric oxide and superoxide, *Chem. Biol. Int.*, 96, 203, 1995.

466

787. Ischiropoulos, H., Zhu, L., Beckman, J. S., Peroxynitrite formation from macrophage-derived nitric oxide, *Arch. Biochem. Biophys.*, 298, 446, 1992.

788. Malinski, T., and Ziad, T., Nitric oxide release from a single cell measured in situ by a porphyrinic-based microsensor, *Nature*, 358, 676, 1992.

789. Malinski, T., Bailey, F., Zhang, Z. G., and Chopp, M., Nitric oxide measured by a porphyrinic microsensor in rat brain after transient middle cerebral artery occlusion, *J. Cereb. Blood Flow Metab.* 12, 355, 1993.

790. Miles, A. M., Bohle, D. S., Glassbrenner, P. A., Hansert, B., Wink, D. A. and Grisham, M. B., Modulation of superoxide-dependent oxidation and hydroxylation reactions by nitric oxide, *J. Biol. Chem.*, 271, 40, 1996.

791. Beckman, J. S., Chen, J., Ischiropoulos, H. and Crow, J. P., Oxidative chemistry of peroxynitrite, *Methods Enzymol.*, 233, 229, 1994.

792. Kanner, J., Harel, S. and Granit, R., Nitric oxide as an antioxidant, *Arch. Biochem. Biophys.*, 289, 130, 1991.

793. Wink, D. A., Cook, J. A., Pacelli, R., Liebman, J., Krishna, M. C., and Mitchell, J. B., Nitric oxide protects against cellular damage by reactive oxygen species, *Toxicol. Lett.* 82/83, 221, 1995.

794. Maragos, C. M., Morley, D., Wink, D. A., Dunams, T. M., Saavedra, J. E., Hoffman, A., Bove, A. A., Isaac, L., Hrabie, J. A. and Keefer, L. K., Complexes of ·NO with nucleophiles as agents for the controlled biological release of nitric oxide. Vasorelaxant effects, *J. Med. Chem.*, 34, 3242, 1991.

795. Wink, D. A., Hanbauer, I., Krishna, M. C., DeGraff, W., Gamson, J. and Mitchell, J. B., Nitric oxide protects against cellular damage and cytotoxicity from reactive oxygen species, *Proc. Natl. Acad. Sci. USA*, 90, 9813, 1993.

796. Wink, D. A., Cook, J. A., Krishna, M. C., Hanbauer, I., DeGraff, W., Gamson, J. and Mitchell, J. B., Nitric oxide protects against alkyl peroxide-mediated cytotoxicity: further insights into the role nitric oxide plays in oxidative stress, *Arch. Biochem. Biophys.*, 319, 402, 1995.

797. Rubbo, H., Radi, R., Trujillo, M., Telleri, R., Kalyanaraman, B., Barnes, S., Kirk, M. and Freeman, B. A., Nitric oxide regulation of superoxide and peroxynitrite-dependent lipid peroxidation, *J. Biol. Chem.*, 269, 26066, 1994.

798. Padmaja, S. and Huie, R. E., The reaction of nitric oxide with organic peroxyl radicals, *Biochem. Biophys. Res. Commun.*, 195, 539, 1993.

799. Chin, J. H., Azhar, S. and Hoffman, B. B., Inactivation of endothelial derived relaxing factor by oxidized lipoproteins, *J. Clin. Invest.*, 89, 10, 1992.

800. Mügge, A., Elwell, J. H., Peterson, T. E., Hofmeyer, T. G., Heistad, D. D. and Harrison, D. G., Chronic treatment with polyethylene-glycolated superoxide dismutase partially restores endothelium-dependent vascular relaxations in cholesterol-fed rabbits, *Circ. Res.*, 69, 1293, 1991.

801. Ohara, Y., Peterson, T. E. and Harrison, D. G., Hypercholesterolemia increases endothelial superoxide anion production, *J. Clin. Invest.*, 91, 2546, 1993.

802. Hogg, N., Kalyanaraman, B., Joseph, J., Struck, A. and Parthasarathy, S., Inhibition of low-density lipoprotein oxidation by nitric oxide. Potential role in atherogenesis, *FEBS Letters*, 334, 170, 1993.

803. Clancy, R. M., Leszczynska-Piziak, J. and Abramson, S. B., Nitric oxide, an endothelial cell relaxation factor, inhibits neutrophil superoxide anion production via a direct action on the NADPH oxidase, *J. Clin. Invest.*, 90, 1116, 1992.

804. Stadtman, E. R., Protein oxidation and aging, *Science*, 257, 1220, 1992.

805. Stubbe, J. A., Protein radicals involvement in biological catalysis?, *Ann. Rev. Biochem.*, 58, 257, 1989.

806. Stubbe, J. A., Radicals in biological catalysis, *Biochemistry*, 27, 3893, 1988.

807 Thelander, L. and Reichard, P., Reduction of ribonucleotides, *Ann. Rev. Biochem.*, 48, 133, 1979.

808. Reichard, P. and Ehrenberg, A., Ribonucleotide reductase-a radical enzyme, *Science*, 221, 514, 1983.

468

809.   Graeslund, A., Sahlin, M. and Sjoeberg, B. M., The tyrosyl free radical in ribonucleotide reductase, *Environm. Health Persp.*, 64, 139, 1985.

810.   Feng Li, Y., Heelis, P. F. and Sancar, A., Active site of DNA photolyase: tryptophan-306 is the intrinsic hydrogen atom donor essential for flavin radical photoreduction and DNA repair in vitro, *Biochemistry*, 30, 6322, 1991.

811.   Sivaraja, M., Goodin, D. B., Smith, M. and Hoffman, B. M., Identification by ENDOR of Trp[191] as the free radical site in cytochrome c peroxidase compound ES, *Science*, 245, 738, 1989.

812.   Eiserich, J. P., Butler, J., van der Vliet, A., Cross, C. E. and Halliwell, B., Nitric oxide rapidly scavenges tyrosine and tryptophan radicals, *Biochem. J.*, 310, 745, 1995.

813.   Nordlund, P., Sjoeberg, B. M. and Eklund, H., Three-dimensional structure of the free radical protein of ribonucleotide reductase, *Nature*, 345, 593, 1990.

814.   Lepoivre, M., Fieschi, F., Coves, J., Thelander, L. and Fontecave, M., Inactivation of ribonucleotide reductase by nitric oxide, *Biochem. Biophys. Res. Commun.*, 179, 442, 1991.

815.   Lepoivre, M., Flaman, J. M., Bobe, P., Lemaire, G. and Henry, Y., Quenching of the tyrosyl free radical of ribonucleotide reductase by nitric oxide, *J. Biol. Chem.*, 269, 21,891, 1994.

816.   Lepoivre, M., Chenais, B., Yapo, A., Lemaire, G., Thelander, L. and Tenu, J. P., Alterations of ribonucleotide reductase activity following induction of the nitrite-generating pathway in adenocarcinoma cells, *J. Biol. Chem.*, 265, 14,143, 1990.

817.   Kwon, N. S., Stuehr, D. J. and Nathan, C. F., Inhibition of tumor cell ribonucleotide reductase by macrophage-derived nitric oxide, *J. Exp. Med.*, 174, 761, 1991.

818.   Nathan, C. F. and Hibbs, Jr., J. B., Role of nitric oxide synthesis in macrophage antimicrobial activity, *Curr. Opinion Immunol.*, 3, 65, 1991.

819.   Balentine, J., *Pathology of Oxygen Toxicity*, Academic Press, New York, 1982.

820.   Turrens, J. F., Freeman, B. A., Levitt, J. G. and Crapo, J. D., The effect of hyperoxia on superoxide production by lung submitochondrial particles, *Arch. Biochem. Biophys.*, 217, 411, 1982.

821. Jamieson, D., Chance, B., Cadenas, E. and Boveris, A., The relation of free radical production to hyperoxia, *Ann. Rev. Physiol.*, 48, 703, 1986.

822. Halliwell, B. and Gutteridge, J. M. C., The antioxidants of human extracellular fluids, *Arch. Biochem. Biophys.* 280, 1, 1990.

823. Moorhouse, P. C., Grootveld, M., Halliwell, B., Quinlan, J. G. and Gutteridge, J. M. C., Allopurinol and oxypurinol are hydroxyl radical scavengers, *FEBS Letters*, 213, 23, 1987.

824. Kohen, R., Yamamoto, Y., Cundy, K. C. and Ames, B. N., Antioxidant activity of carnosine, homocarnosine, and anserine present in muscle and brain, *Proc. Natl. Acad. Sci. USA*, 85, 3175, 1988.

825. Ames, B. N., Cathcart, R., Schwiers, E., Hochstein, P., Uric acid provides an antioxidant defense in humans against oxidant and radical-caused aging and cancer; a hypothesis, *Proc. Natl. Acad. Sci. USA* 78, 6858, 1981.

826. Grootveld, M. and Halliwell, B., Measurement of allantoin and uric acid in human body fluids. A potential index of free radical reactions in vivo?, *Biochem. J.* 243, 803, 1987.

827. Goldblatt, H. and Cameron, G., Induced malignancy in cells from rat myocardium subjected to intermittent anaerobiosis during long propagation in vitro, *J. Exp. Med.*, 97, 525, 1953.

828. Levine, M., New concepts in the biology and biochemistry of ascorbic acid, *N. Engl. J. Med.*, 314, 892, 1986.

829. Fraga, C. G., Motchnik, P. A., Shigenaga, M. K., Helbock, H. J., Jacob, R. A. and Ames, B. N., Ascorbic acid protects against endogenous oxidative DNA damage in human sperm, *Proc. Natl. Acad. Sci. USA*, 88, 11,003, 1991.

830. Rothstein, J. O., Bristol, L. A., Hosler, B., Brown, Jr., R. H., Kuncl, R. W., Chronic inhibition of superoxide dismutase produces apoptotic death of spinal neurons, *Proc. Natl. Acad. Sci. USA* 91, 4155, 1994.

831a. Joester, K. E., Jung, G., Weber, U. and Weser, U., Superoxide dismutase activity of Cu^{2+}-amino acid chelates, *FEBS Lett.*, 25, 25, 1972.

831b. Brigelius, R., Hartmann, H. J., Bors, W., Saran, M., Lengfelder, E. and Weser, U., Superoxide dismutase activity of $Cu(Tyr)_2$ and Cu,

Co-erythrocuprein, *Hoppe-Seyler's Z. Physiol. Chem.,* 356, 739, 1975.

831c. Voelter, W., Sokolowski, G., Weber, U. and Weser, U., The initial binding of Cu(II) to some amino acids and dipeptides: A ^{13}C nuclear-magnetic-resonance study, *Eur. J. Biochem.,* 58, 159, 1975.

832. Boveris, A. and Cadenas, E., Cellular sources and steady-state levels of reactive oxygen species, in *Oxygen, Gene Expression, and Cell Function.* L. Biadasz Clerch and D. J. Massaro, eds., Marcel Dekker, New York, 1997.

833. Kono, Y. and Fridovich, I., Superoxide radical inhibits catalase, *J. Biol. Chem.,* 257, 5751, 1982.

834. Metodiewa, D. and Dunford, H. B., The reactions of horseradish peroxidase, lactoperoxidase and myeloperoxidase with enzymatically generated superoxide, *Arch. Biochem. Biophys.* 272, 245, 1989.

835. Scott, M. D., Eaton, J. W., Kuypers, F. A., Chiu, D. T. Y., and Lubin, B. H., Enhancement of erythrocyte superoxide dismutase activity: effects on cellular oxidant defense, *Blood,* 74, 2542, 1989.

836. Liochev, S. I. and Fridovich, I., Effects of overproduction of superoxide dismutase on the toxicity of paraquat towards *Escherichia coli, J. Biol. Chem.,* 266, 8747, 1991.

837. McCord, J. M., Superoxide radical: controversies, contradictions, and paradoxes, *Proc. Soc. Exp. Biol. Med.,* 209, 112, 1995.

838. Nelson, S. K., Bose, S. K. and McCord, J. M., The toxicity of high-dose superoxide dismutase suggests that superoxide can both initiate and terminate lipid peroxidation in the reperfused heart, *Free Radical Biol. Med.,* 16, 195, 1994.

839. Hesse, H., *The Glass Bead Game,* translated by Richard and Clara Winston, Henry Holt, New York, 1990, p. 379.

840. Afanas'ev, I. B., *Superoxide Ion: Chemistry and Biological Implications,* Vol. I, CRC Press, Boca Raton, FL, 1989, pp. 173-176.

841. Afanas'ev, I. B., *Superoxide Ion: Chemistry and Biological Implications,* Vol. I, CRC Press, Boca Raton, FL, 1989, pp. 225-236.

842. Yamazaki, I. and Piette, L. H., Mechanism of free radical formation and disappearance during the ascorbic acid oxidase and peroxidase reactions, *Biochim. Biophys. Acta,* 50, 62, 1961.

843. Iyanagi, T., Yamazaki, I. and Anan, K. F., One-electron oxidation-reduction properties of ascorbic acid, *Biochim. Biophys. Acta,* 806, 255, 1985.

844. Buettner, G., Ascorbate autoxidation in the presence of iron and copper chelates, *Free Radical Res. Commun.,* 1, 349, 1985.

845. Buettner, G. R., Use of ascorbate as test for catalytic metals in simple buffers, *Meth. Enzymol.* 186, 125, 1990.

846. Mushran, S. P. and Agrawal, M. C., Mechanistic studies on the oxidation of ascorbic acid, *J. Sci. Ind. Res.,* 36, 274, 1977.

847. Laurence, G. S. and Ellis, K. J., The detection of a complex intermediate in the oxidation of ascorbic acid by ferric ion, *J. Chem. Soc., Dalton Trans.,* p. 1667, 1972.

848. Keypour, H., Silver, J., Wilson, M. T. and Hamed, M. Y., Studies on the reactions of ferric iron with ascorbic acid : a study of solution chemistry using Mössbauer spectroscopy and stopped-flow techniques, *Inorg. Chim. Acta,* 125, 97, 1986.

849. Nishikimi, M., Oxidation of ascorbic acid with superoxide anion generated by the xanthine-xanthine oxidase system, *Biochem. Biophys. Res. Commun.* 63, 463, 1975.

850. Laroff, G. P., Fessenden, R. W. and Schuler, R. H., The electron spin resonance spectra of radical intermediates in the oxidation of ascorbic acid and related substances, *J. Am. Chem. Soc.,* 94, 9062, 1972.

851. Sapper, H., Kang, S. D., Paul, H. H. and Lohmann, W., The reversibility of the vitamin C redox system: electrochemical reasons and biological aspects, *Z. Naturforsch.,* 37c, 942, 1982.

852. Nienhuis, A. W., Vitamin C and iron, *N. Engl. J. Med.* 304, 170, 1981.

853. Halliwell, B. and Gutteridge, J. M. C., Lipid peroxidation, oxygen radicals, cell damage and antioxidant therapy, *Lancet,* 1, 1396, 1984.

854. Halliwell, B. and Gutteridge, J. M. C., Oxygen radicals and the nervous system, *Trends Neurosci.,* 8, 22, 1985.

855. Campbell, G. D., Steinberg, M. H. and Bower, J. D., Ascorbic acid-induced hemolysis in G-6-PD deficiency, *Ann. Int. Med.,* 82, 810, 1975.

856. Baader, S. L., Bruchelt, G., Carmine, T. C., Lode, H. N., Rieth, A. G. and Niethammer, D., Ascorbic acid mediated iron release from

cellular ferritin and its relation to the formation of DNA strand breaks in neuroblastoma cells, *J. Cancer Res. Clin. Oncol.*, 120, 415, 1994.

857.   Frei, B., England, L. and Ames, B. N., Ascorbate is an outstanding antioxidant in human plasma, *Proc. Natl. Acad. Sci. USA*, 86, 6377, 1989.

858.   Hornig, D., Distribution of ascorbic acid, metabolites and analogues in man and animals, *Ann. N.Y. Acad. Sci.*, 258, 103-118, 1975.

859.   Lambertsen, C. J., Transport of oxygen, carbon dioxide and inert gases by the blood, in *Medical Physiology*, V. B. Mountcastle, ed., Vol. 2, Mosby, St. Louis, 1980, p. 1722.

860.   Maples, K. R. and Mason, R. P., Free radical metabolite of uric acid, *J. Biol. Chem.*, 263, 1709, 1988.

861.   Hooper, D. C., Spitsin, S., Kean, R. B., Champion, J. M., Dickson, G. M., Chaudury, I. and Koprowski, H., Uric acid, a natural scavenger of peroxynitrite, in experimental allergic encephalomyelitis and multiple sclerosis, *Proc. Natl. Acad. Sci. USA*, 95, 675, 1998.

862.   Whiteman, M. and Halliwell, B., Protection against peroxynitrite-dependent tyrosine nitration and alpha 1-antiproteinase inactivation by ascorbic acid. A comparison with other biological antioxidants, *Free Radical Res.*, 25, 275, 1996.

863.   Kaur, H. and Halliwell, B., Action of biologically-relevant oxidizing species upon uric acid. Identification of uric acid oxidation products, *Chem. Biol. Interact.*, 73, 235, 1990.

864.   Davies, K. J. A., Sevanian, A., Muakkassah-Kelly, S. F. and Hochstein, P., Uric acid-iron ion complexes, *Biochem. J.*, 235, 747, 1986.

865.   Sevanian, A., Davies, K. J. A. and Hochstein, P., Conservation of vitamin C by uric acid in blood, *J. Free Radical Biol. Med.*, 1, 117, 1985.

866.   Howell, R. R. and Wyngaarden, J. B., On the mechanism of peroxidation of uric acids by hemoproteins, *J. Biol. Chem.*, 235, 3544, 1960.

867.   Adelman, R., Saul, R. L. and Ames, B. N., Oxidative damage to DNA: Relation to species metabolic rate and life span, *Proc. Natl. Acad. Sci. USA*, 85, 2706, 1988.

868. Ames, B. N. and Gold, L. S., Chemical Carcinogenesis: Too many rodent carcinogens, *Proc. Natl. Acad. Sci. USA,* 87, 7772, 1990.

869. Parkhouse, W. S., McKenzie, D. C., Hochachka, P. W. and Ovalle, W. K., Buffering capacity of deproteinized human vastus lateralis muscle, *J. Appl. Physiol.,* 58, 14, 1985.

870. Procter, P. , Similiar functions of uric acid and ascorbate in man?, *Nature,* 228, 868, 1970.

871. Tan, D. X., Chen, L. D., Poeggeler, B., Manchester, L. C. and Reiter, R. J., Melatonin: a potent endogenous hydroxyl radical scavenger, *Endocrinol. J.,* 1, 57, 1993.

872. Reiter, R. J., Tan, D. X., Poeggeler, B., Menendez-Pelaez, A., Chen, L. D. and Saarela, S., Melatonin as a free radical scavenger: Implications for aging and age-related diseases, *Ann. N. Y. Acad. Sci.,* 719, 1, 1994. .

873. Reiter, R. J., Carneiro, R. C. and Oh, C.-S., Melatonin in relation to cellular antioxidative defense mechanisms, *Horm. Metab. Res.,* 29, 363,1997.

874. Marshall, K.A., Reiter, R. J., Poeggeler, B., Aruoma, O. I. and Halliwell, B., Evaluation of the antioxidative activity of melatonin in vitro, *Free Radical Biol. Med.,* 21, 307, 1996.

875. Chan, T. Y. and Tang, P. L., Characterization of the antioxidant effects of melatonin and related indoleamines in vitro, *J. Pineal Res.,* 20, 187, 1996.

876. Cagnoli, C. M., Atabay, C., Kharlamova, E. and Manev, H., Melatonin protects neurons from singlet oxygen-induced apoptosis, *J. Pineal Res.,* 18, 222, 1995.

877. Miller, J.W., Selhub, J. and Joseph, J. A., Oxidative damage caused by free radicals produced during catecholamine autoxidation: Protective effects of O-methylation and melatonin, *Free Radical Biol. Med.,* 21, 241, 1996.

878. Pappolla, M. A., Sos, M., Beck, R. A., Hickson-Bick, D. L. M., Reiter, R. J., Efthimiopoulos, S. and Robakis, N. K., Melatonin prevents death of neuroblastoma cells exposed to Alzheimer amyloid protein, *J. Neurosci.,* 17, 1683, 1997.

879. Melchiorri, D., Reiter, R. J., Attia, A. M., Hara, M., Burgos, A. and Sewerynek, E., Potent protective effects of melatonin on in vivo paraquat-induced oxidative damage in rats, *Life Sci.,* 56, 83, 1995.

474

880. Melchiorri, D., Reiter, R. J., Sewerynek, E., Hara, M., Chen, L. D. and Nistico, G., Paraquat toxicity and oxidative damage: Reduction by melatonin, *Biochem. Pharmacol.,* 51, 1095, 1996.

881. Tan, D. X., Poeggeler, B., Reiter, R. J., Chen, L. D., Chen, S., Manchester, L. C.and Barlow-Walden, L. R., The pineal hormone melatonin inhibits DNA-adduct formation induced by the chemical carcinogen safrole in vivo, *Cancer Lett.,* 70, 65, 1993.

882. Tan, D. X., Reiter, R. J., Chen, B., Poeggeler, B., Manchester, L. C. and Barlow-Walden, L. R., Both physiological and pharmacological levels of melatonin reduce DNA adduct formation induced by the carcinogen safrole, *Carcinogenesis,* 15, 215, 1994.

883. Reiter, R. J., Melatonin: the chemical expression of darkness, *Mol. Cell. Endocrinol.,* 79, C153, 1991.

884. Ingold, K. U., Webb, A. C., Witter, D., Burton, G. W., Metcalfe, T. A. and Muller, D. P. R., Vitamin E remains the major lipid-soluble chain-breaking antioxidant in human plasma even in individuals suffering severe vitamin E deficiency, *Arch. Biochem. Biophys.,* 259, 224, 1987.

885. Burton, G. W. and Ingold, K. U., Autoxidation of biological molecules. 1. The antioxidant activity of vitamin E and related chain-breaking phenolic antioxidants in vitro,*J. Am. Chem. Soc.,* 103, 6472, 1981.

886. Burton, G. and Ingold, K. U., Vitamin E: Application of the principles of physical organic chemistry to the exploration of its structure and function, *Acc. Chem. Res.,* 19, 194, 1986.

887. Castle, L. and Perkins, M. J., Inhibition kinetics of chain-breaking phenolic antioxidants in SDS micelles. Evidence that intermicellar diffusion rates may be rate-limiting for hydrophobic inhibitors such as α-tocopherol, *J. Am. Chem. Soc.,* 108, 6381, 1986.

888. Pryor, W. A., Strickland, T. and Church, D. F., Comparison of the efficiencies of several natural and synthetic antioxidants in aqueous sodium dodecyl sulfate micelle solutions,*J. Am. Chem. Soc.,* 110, 2224, 1988.

889. Barclay, L. R. C., Baskin, K. A., Locke, S. J. and Vindquist, M. R., The antioxidant activities of phenolic antioxidants in free radical peroxidation of phospholipid membranes, *Can. J. Chem.,* 68, 2258, 1990.

890.   Bowry, V. W. and Ingold, K. U., The unexpected role of vitamin E (α-tocopherol) in the peroxidation of human low-density lipoprotein, *Acc. Chem. Res.*, 32, 27, 1999, and references cited therein.

891.   Nishikimi, M. and Machlin, L. J., Oxidation of α-tocopherol model compound by superoxide anion, *Arch. Biochem. Biophys.*, 170, 684, 1975.

892.   Afanas'ev, I. B., *Superoxide Ion: Chemistry and Biological Implications,* Vol. I, chap. 10, CRC Press, Boca Raton, FL, 1989, pp. 234-242.

893.   Niki, E., Saito, M., Kawakami, A. and Kamiya, Y., Inhibition of oxidation of methyl linoleate in solution by vitamin E and vitamin C, *J Biol. Chem.*, 259, 4177, 1984.

894.   Doba, T., Burton, G. W. and Ingold, K. U., Antioxidant and co-antioxidant activity of vitamin C. The effect of vitamin C, either alone or in the presence of vitamin E or a water-soluble vitamin E analogue, upon the peroxidation of aqueous multilamellar phospholipid liposomes, *Biochim. Biophys. Acta*, 835, 298, 1985.

895.   Pryor, W. A., Kaufman, M. J. and Church, D. F., Autoxidation of micelle-solubilized linoleic acid. Relative inhibitory efficiencies of ascorbate and ascorbyl palmitate, *J. Org. Chem.*, 50, 281, 1985.

896.   Wayner, D. D. M., Burton, G. W. and Ingold, K. U., The antioxidant efficiency of vitamin C is concentration-dependent, *Biochim. Biophys. Acta,* 884, 119, 1986.

897.   Bendisch, A., Machlin, L. J., Scandurra, O., Burton, G. W. and Wayner, D. D. M., *Adv. Free Radical Biol. Med.*, 2, 419, 1986.

898.   Burton, G. W., Joyce, A. and Ingold, K. U., Is vitamin E the only lipid-soluble, chain-breaking antioxidant in human blood plasma and erythrocyte membranes?, *Arch. Biochem. Biophys.*, 221, 281, 1983.

899.   Burton, G. W., Wronska, U., Stone, L., Foster, D. O. and Ingold, K. U., Biokinetics of dietary RRR-α-tocopherol in the male guinea pig at three dietary levels of vitamin C and two levels of vitamin E. Evidence that vitamin C does not "spare" vitamin E in vivo, *Lipids,* 25, 199, 1990.

900.   Harats, D., Ben-Naim, M., Dabach, Y., Hollander, G., Havivi, E., Stein, O. and Stein, Y., Effect of vitamin C and E supplementation on susceptibility of plasma lipoproteins to peroxidation induced by acute smoking, *Atherosclerosis,* 85, 47, 1990.

476

901. Maiorino, M., Zamburlini, A., Roveri, A. and Ursini, F., Prooxidant role of vitamin E in copper-induced lipid peroxidation, *FEBS Lett.*, 330, 174, 1993.

902. Yoshida, Y., Tsuchiya, J. and Niki, E., Interaction of α-tocopherol with copper and its effect on lipid peroxidation, *Biochim. Biophys. Acta*, 1200, 85, 1994.

903. Lynch, S. M. and Frei, B., Reduction of copper, but not iron, by human low density lipoprotein (LDL), *J. Biol. Chem.*, 270, 5158, 1995.

904. Kontush, A., Meyer, S., Finckh, B., Kohlschütter, A. and Beisiegel, U., α-Tocopherol as a reductant for Cu(II) in human lipoproteins, *J. Biol. Chem.*, 271, 11,106, 1996.

905. Proudfoot, J. M., Croft, K. D., Puddey, 1. B. and Beilin, L. J., The role of copper reduction by α-tocopherol in low-density lipoprotein oxidation, *Free Radical Biol. Med*, 23, 720, 1997.

906. Neuzil, J., Thomas, S. R. and Stocker, R., Requirement for promotion, or inhibition by α -tocopherol of radical-induced initiation of plasma lipoprotein lipid peroxidation, *Free Radical Biol. Med*, 22, 57, 1997, and references cited therein.

907. Husain, S. R., Cillard, J. and Cillard, P., Hydroxyl radical scavenging activity of flavonoids, *Phytochemistry*, 26, 2489, 1987.

908. Robak, J. and Gryglewski, R. J., Flavonoids are scavengers of superoxide anion, *Biochem. Pharmacol.*, 37, 837, 1988.

909. Torel, J., Cillard, J. and Cillard, P., Antioxidant activity of flavonoids and reactivity with peroxyl radical, *Phytochemistry*, 25, 383, 1986.

910. Sorata, Y., Takahama, U. and Kimura, M., Protective effect of quercetin and rutin on photosensitized lysis of human erythrocytes in the presence of hematoporphyrin, *Biochim. Biophys. Acta*, 799, 313, 1984.

911. Takahama, U., Suppression of lipid photoperoxidation by quercetin and its glycosides in spinach chloroplasts, *Photochem. Photobiol.*, 38, 363, 1983.

912. Das, M. and Ray, P. K., Lipid antioxidant properties of quercetin in vitro, *Biochem. Int.*, 17, 203, 1988.

913. Laughton, M. J., Halliwell, B., Evans, P. J. and Hoult, J. R., Antioxidant and prooxidant actions of the plant phenolics quercetin, gossypol and myricetin, *Biochem. Pharmacol.*, 38, 2859, 1989.

914. Bindoli, A., Cavallini, L. and Siliprandi, N., Inhibitory action of silymarin on lipid peroxide formation in rat liver mitochondria and microsomes, *Biochem. Pharmacol.*, 26, 2405, 1977.

915. Janero, D. R. and Burghardt, B., Protection of rat myocardial phospholipid against peroxidative injury through superoxide-(xanthine oxidase)-dependent iron-promoted Fenton chemistry by the male contraceptive gossypol, *Biochem. Pharmacol.*, 37, 3335, 1988.

916. Srivastava, A. K. and Padmanaban, G., Gossypol mediated DNA degradation, *Biochem. Biophys. Res. Commun.* 146, 1515, 1987.

917. Bhat, R. and Hadi, S. M., DNA breakage by tannic acid and Cu(II): sequence specificity of the reaction and involvement of active oxygen species, *Mutation Res.*, 313, 39, 1994.

918. Rahman, A., Sahabuddin, S. M., Hadi, S. M., Parish, J. H. and Ainley, K., Strand Scissions of DNA induced by quercetin and Cu(II): role of Cu(I) and oxygen free radicals in the reaction, *Carcinogenesis*, 10, 1833, 1989.

919. Ames, B. N., Dietary carcinogens and anti-carcinogens (oxygen radicals and degenerative diseases), *Science*, 221, 1256, 1983.

920. Stich, H. F., The beneficial and hazardous effects of some simple phenolic compounds, *Mutation Res.*, 259, 307, 1991.

921. Di Mascio, P., Kaiser, S. and Sies, H., Lycopene as the most efficient biological carotenoid singlet oxygen quencher, *Arch. Biochem. Biophys.*, 274, 532, 1989.

922. Di Mascio, P., Murphy, M. E. and Sies, H., Antioxidant defense systems: the role of carotenoids, tocopherols, and thiols, *Am. J. Clin. Nutrition*, 53, 194S, 1991.

923. Krinsky, N. I. and Deneke, S. M., Interaction of oxygen and oxy-radicals with carotenoids, *J. Natl. Cancer Inst.*, 69, 205, 1982.

924. Terao, J., Antioxidant activity of ß-carotene-related carotenoids in solution, *Lipids,* 24, 659, 1989.

925. Krinsky, N. I., Antioxidant functions of ß-carotene, *Free Radical. Biol. Med.*, 7, 617, 1989.

926. Packer, J. E., Mahood, J. S., Mora-Arellano, V. O., Slater, T. F., Willson, R. L. and Wolfenden, B. S., Free radicals and singlet oxygen scavengers: Reaction of a peroxy-radical with beta-carotene,

diphenylfuran and 1,4-diazobicyclo-(2,2,2)-octane, *Biochem. Biophys Res. Commun.*, 98, 901, 1981.

927. Burton, G. W. and Ingold, K. U., ß-carotene: an unusual type of lipid antioxidant, *Science*, 224, 569, 1984.

928. Ozhogina, O. A., Kasaikina, O. T., ß-carotene as an interceptor of free radicals, *Free Radical Biol. Med.* 19, 575, 1995.

929. Halliwell, B. and Gutteridge, J. M.C., Oxygen toxicity, oxygen radicals, transition metals and disease, *Biochem. J.*, 219, 1, 1984.

930. Gutteridge, J. M. C., Antioxidant properties of ceruloplasmin towards iron- and copper-dependent oxygen radical formation, *FEBS Letters*, 157, 37, 1983.

931. Gutteridge, J. M. C., Patterson, S. K., Segal, A. W. and Halliwell, B., Inhibition of lipid peroxidation by the iron-binding protein lactoferrin, *Biochem. J.*, 199, 259, 1981.

932. Lovstad, R. A., Copper catalyzed oxidation of ascorbate (vitamin C). Inhibitory effect of catalase, superoxide dismutase, serum proteins (ceruloplasmin, albumin, apotransferrin) and amino acids, *Int. J. Biochem.* 19, 309, 1987.

933. Aruoma, O. I. and Halliwell, B., Superoxide-dependent and ascorbate-dependent formation of hydroxyl radicals from hydrogen peroxide in the presence of iron: are lactoferrin and transferrin promoters of hydroxyl radical generation?, *Biochem .J.*, 241, 273, 1987.

934. Gutteridge, J. M. C., Hill, C. and Blake, D. R., Copper stimulated phospholipid membrane peroxidation: antioxidant activity of serum and synovial fluid from patients with rheumatoid arthritis, *Clin. Chim. Acta*, 139, 85, 1984.

935. Gutteridge, J. M. C., Richmond, R. and Halliwell, B., Inhibition of the iron-catalyzed formation of hydroxyl radicals from superoxide and of lipid peroxidation by desferrioxamine, *Biochem. J.*, 184, 469, 1979.

936. Wasserman, H. W. and Lipshutz, B. H., Reactions of singlet oxygen with heterocyclic systems, in *Singlet Oxygen*, H. H. Wasserman and R. W. Murray, eds., Academic Press, New York, 1979, pp. 429-509.

937. Brown, C. E., Interactions among carnosine, anserine, ophidin and copper in biochemical adaptation, *J. Theor. Biol.*, 88, 245, 1981.

938. Dillard, C. J., Litov, R. E., Savin, W. M., Dumelin, E. E. and Tappel, A. L., Effects of exercise, vitamin E, and ozone on pulmonary function and lipid peroxidation, *J. Appl. Physiol. Respir. Environ. Exercise Physiol.*, 45, 927, 1978.

939. Byers, T. and Perry, G., Dietary carotenes, vitamin C, and vitamin E as protective antioxidants in human cancers, *Ann. Rev. Nutr.*, 12, 139, 1992.

940. Clark, L. C., Cantor, K. P. and Allaway, W. H., Selenium in forage crops and cancer mortality in US counties, *Arch Environ Health*, 46, 37, 1991.

941. Schrauzer, G. N., White, D. A., Schneider C. J., Cancer mortality correlation studies. III. Statistical associations with dietary selenium intakes, *Bioinorg. Chem.*, 7, 23, 1977.

942. Schrauzer, G. N., White, D. A., Schneider C. J., Cancer mortality correlation studies. IV. Associations with dietary intakes and blood levels of certain trace elements, notably selenium-antagonists. *Bioinorg Chem.*, 7, 35, 1977.

943. Shamberger, R. H., Tytko, S. A. and Willis, C. E., Antioxidants and cancer. VI. Selenium and age-adjusted human cancer mortality. *Arch. Environ. Health*, 31, 231, 1976.

944. El-Bayoumy, K., Upadhyaya, P., Chae, Y. H., Sohn, O. S., Rao, C. V., Fiala, E. and Reddy, B. S., Chemoprevention of cancer by organoselenium compounds, *J. Cell Biochem.*, 22 (suppl.), 92, 1995.

945. Hocman, G., Chemoprevention of cancer: selenium, *Int. J. Biochem.*, 20, 123, 1988.

946. Yu, S. Y., Mao, B. L., Xiao, P., Yu, W. P., Wang, Y. L., Huang, C. Z., Chen, W. Q. and Xuan, X. Z., Interventional trial with selenium for the prevention of lung cancer among tin miners in Yunnan, China: a pilot study, *Biol. Trace Elem. Res.*, 24, 105, 1990.

947. Yu, S. Y., Chu, Y. J., Gong, X. L. and Hou, C., Regional variation of cancer mortality incidence and its relation to selenium level in China, *Biol. Trace Elem. Res.*, 7, 21, 1985.

948. Chou, Y. L., Liu, Q. Y., Hou, C. and Yu, S. Y., Blood selenium concentration in residents of areas in China having a high incidence of lung cancer, *Biol. Trace Elem. Res.*, 6, 133, 1984.

480

949.   Heffner, J. E. and Repine, J. E., State of the art. Pulmonary strategies of antioxidant defense, *Ann. Rev. Resp. Dis.*, 140, 531, 1989.

950.   Cantin, A. M., North, S. L., Hubbard, R. C. and Crystal, R. G., Normal epithelial lining fluid contains high levels of glutathione, *J. Appl. Physiol.*, 63, 152, 1987.

951.   Marklund, S. L., Westman, N. G., Lundgren, E. and Roos, G., Copper and zinc-containing superoxide dismutase, manganese-containing superoxide dismutase, catalase and glutathione peroxidase in normal and neoplastic cell lines and normal human tissues, *Cancer Res.*, 42, 1955, 1982.

952.   Fridovich, I. and Freeman, B., Antioxidant defenses in the lung, *Ann. Rev. Physiol.*, 48, 693, 1986.

953.   Avissar, N., Finkelstein, J. N., Horowitz, S., Willey, J. C., Coy, E., Frampton, M. W., Watkins, R.H., Khullar, P., Xu, Y-L., Cohen, H. J., Extracellular glutathione peroxidase in human lung epithelial lining fluid and in lung cells, *Am. J. Physiol.*, 270, L173, 1996.

954.   Zachara, B. A., Marchaluk-Wisniewska, E., Maciag, A., Peplinski, J., Skokowski, J., and Lambrecht, W., Decreased selenium concentration and glutathione peroxidase activity in blood and increase of these parameters in malignant tissue of lung cancer patients, *Lung*, 175, 321, 1997.

955.   Clark, L. C., Combs, G. F. Jr., Turnbull, B. W. et al., Effects of selenium supplementation for cancer prevention in patients with carcinoma of the skin, *JAMA*, 276, 1957, 1996.

956a.   Giovannucci, E, Epidemiologic characteristics of prostate cancer, *Cancer*, 75, 1766, 1995.

956b.   Giovannucci, E., Rimm, E. B., Wolk, A., Ascherio, A., Stampfer, M. J., Colditz, G. A. and Willett, W. C., Calcium and fructose intake in relation to risk of prostate cancer, *Cancer Res.*, 58, 442, 1998.

957.   Reichel, H., Koeffler, P. and Norman, A. W., The role of the vitamin D endocrine system in health and disease, *N. Engl. J. Med.*, 320, 980, 1989.

958.   Stampfer, M. J., Hennekens, C.H., Manson, J.E., Colditz, G.A., Rosner, B.and Willett, W. C., Vitamin E consumption and the risk of coronary disease in women, *N. Engl. J. Med.*, 328, 1444, 1993.

959.   Rimm, E.B., Stampfer, M.J., Ascherio, A., Giovannucci, E., Colditz, G.A. and Willett, W.C., Vitamin E consumption and the risk of coronary heart disease in men, *N. Engl. J. Med.*, 328, 1450, 1993.

960.   Hertog, M. G. L., Feskens, E. J. M., Hollman, P. C. H., Katan, M. B. and Kromhout, D., Dietary antioxidant flavonoids and the risk of coronary heart disease. The Zutphen elderly study, *Lancet*, 342, 1007, 1993.

961.   Byers, T., Vitamin E supplements and coronary heart disease, *Nutr. Rev.* 51, 333, 1993.

962.   Steinberg, D., Antioxidant vitamins and coronary heart disease (editorial), *N. Engl. J. Med.*, 328, 1487, 1993.

963.   Peto, R., Doll, R., Buckley, J. D. and Sporn, M. B., Can dietary beta-carotene materially reduce human cancer rates? *Nature*, 290, 201, 1981.

964.   Connett, J. E., Kuller, L. H., Kjelsberg, M. O., Polk, B. F., Collins, G., Rider, A. and Hulley, S. B., Relationship between carotenoids and cancer. The multiple risk factor intervention trial (MRFI) study, *Cancer*, 64, 126, 1989.

965.   Mathews-Roth, M. M., Carotenoids and cancer prevention-experimental and epidemiological studies, *Pure Appl. Chem.*, 57, 717, 1985.

966.   Krinsky, N. I., Effects of carotenoids in cellular and animal systems, *Am. J. Clin. Nutr.*, 53, 238S, 1991.

967.   Ziegler, R. G., Vegetables, fruits, and carotenoids and the risk of cancer, *Am. J. Clin. Nutr.*, 53, 251S, 1991.

968.   Mathews-Roth, M. M. and Krinsky, N. I., Carotenoids affect development of UV-B induced skin cancer, *Photochem. Photobiol.*, 46, 507, 1987.

969.   Mathews-Roth, M. M. and Krinsky, N. I., Carotenoids dose levels and protection against UV-B induced skin tumor, *Photochem. Photobiol.*, 42, 35, 1985.

970.   Heinonen, O. P. and Albanes, D., The effect of vitamin E and beta carotene on the incidence of lung cancer and other cancers in male smokers, *N. Engl. J. Med.*, 330, 1029, 1994.

971.   Yong, L. C., Brown, C. C., Schatzkin, A., Dresser, C. M., Slesinski, M. J., Cox, C. S. and Taylor, P. R., Intake of vitamins E, C. and A and risk of lung cancer. The NHANES I epidemiologic follow-up

482

study. First National Health and Nutrition Examination Survey, *Am. J. Epidemiol.*, 146, 231, 1997.

972.　Gonzalez, M. J., Riordan, N. H., Matos, M. I. and Arguelles, M., Antioxidants and cancer: a brief discussion on controversies, contradictions and paradoxes, *J. Orthomol. Med.*, 12, 145, 1997.

973　Living in a chemical world. Occupational and environmental significance of industrial carcinogens, *Ann. N. Y. Acad. Sci.*, 534, 1988.

974.　Smith, K. C., Spontaneous mutagenesis: Experimental, genetic and other factors, *Mutation Res.*, 277, 139-162, 1992.

975.　Glickman, B. W., Burns, P. A. and Fix, D. F., Mechanisms of spontaneous mutagenesis: clues from altered mutational specificity in DNA repair-defective strains, in: *Antimutagenesis and Anticarcinogenesis Mechanisms*, D. M. Shankel, P. E. Hartman, T. Kada and A. Hollaender, eds., Plenum Press, New York, 1986, pp. 259-281.

976.　Winterbourn, C. C. and Stern, A., Human red cells scavenge extracellular hydrogen peroxide and inhibit formation of hypochlorous acid and hydroxyl radical, *J. Clin. Invest.*, 80, 1486, 1987.

977.　Keaney, J. F. jr. and Vita, J. A., Atherosclerosis, oxidative stress, and antioxidant protection in endothelium-derived relaxing factor action, *Prog. Cardiovasc. Dis.*, 38, 129, 1995.

978.　Halliwell, B. and Gutteridge, J. M. C., Lipid peroxidation, oxygen radicals, cell damage, and antioxidant therapy, *Lancet*, 1, 1396, 1984.

979.　Gutteridge, J. M. C. and Stocks, J., Peroxidation of cell lipids, *Med. Lab. Sci.*, 33, 281, 1976.

980.　Barber, A. A., Mechanisms of lipid peroxide formation in rat tissue homogenates, *Radiat. Res., Suppl.*, 3, 33, 1963.

981.　Barber, A. A. and Bernheim, F., Lipid peroxidation: its measurement, occurrence and significance in animal tissues, *Adv. Gerontol. Res.*, 2, 355, 1967.

982.　Sohal, R. S. and Wolfe, L. S., Lipofuscin: characteristics and significance, *Prog. Brain Res.*, 70, 171, 1986.

983.　Tappel, A. L., Lipid peroxidation and fluorescent molecular damage to membranes, in *Pathobiology of Cell Membranes, Vol 1*, B. F.

Trump and A. V. Arsila, eds., Academic Press, New York, 1975, pp. 145-170.

984. Tappel, A. L., Measurement and protection from in vivo lipid peroxidation, in *Free Radicals in Biology, Vol. 4,* W. A. Pryor, ed., Academic Press, New York, 1980, pp. 1-47.

985. Sohal, R. S., Peters, P. D. and Hall, T. A., Structure, origin, composition and age-dependence of mineralized dense bodies (concretions) in the midgut of the adult housefly, *Musca domestica, Tissue Cell,* 9, 87, 1977.

986. Armstrong, D., Free radical involvement in the formation of lipopigments, in *Free Radicals in Molecular Biology, Aging and Disease,* D. Armstrong, R. S. Sohal, R. G. Cutler and T. F. Slater, eds., Raven Press, New York, 1984, pp. 129-141.

987. Barden, H., Further histochemical studies characterizing the lipofuscin component of human neuromelanin, *J. Neuropathol. Exp. Neurol.,* 37, 437, 1978.

988. Barden, H., Acid-fast staining of oxidized neuromelanin and lipofuscin in the human brain, *J. Neuropathol. Exp. Neurol.,* 38, 453, 1979.

989. Thaw, H. H., Brunk, U. T. and Collins, P. V., Influence of oxygen tension, pro-oxidants and antioxidants on the formation of lipid peroxidation products (lipofuscin) in individual cultivated human glial cells, *Mech. Aging Develop.,* 24, 211, 1984.

990. Sohal, R. S. and Allen, R. G., Relationship between metabolic rate, free radicals, differentiation and aging: a unified theory, in *The Molecular Basis of Aging,* A. D. Woodhead, A. D. Blackett and A. Hollaender, eds., Plenum Press, New York, 1985, pp. 75-104.

991. Davis, R. A. and Fraenkel, G. F., The oxygen consumption of flies during flight, *J. Exp. Biol.,* 17, 402, 1940.

992. Sohal, R. S., Muller, A., Koletzko, B. and Sies, H., Effect of age and ambient temperature on n-pentane production in adult housefly, *Musca domestica, Mech. Aging Develop.,* 29, 317, 1985.

993. Kny, W., Über die Verteilung des Lipofuschsins in der Skelettmuskulatur in ihrer Beziehung zur Funktion, *Virch. Arch. Pathol. Anat. Physiol.,* 229, 468, 1937.

994. Friede, R. L., The relation of the formation of lipofuscin to the distribution of oxidative enzymes in the human brain, *Acta Neuropathol.,* 2, 113, 1962.

484

995.   Lyman, C. P., O'Brian, R. C., Green, G. C. and Papafrangos, E.D., Hibernation and longevity in the Turkish hamster *Mesocricetus brandti, Science,* 221, 668, 1981.

996.   Papafrangos, E. D. and Lyman, C. P., Lipofuscin accumulation and hibernation in Turkish hamster, *Mesocritus brandti, J. Gerontol.,* 37, 417, 1982.

997.   Sohal, R. S., Aging changes in insect flight muscle, *Gerontology,* 22, 317, 1976.

998.   Sohal, R. S. and Weindruch, R., Oxidative stress, caloric restriction and aging, *Science,* 273, 59, 1996.

999.   Dix, D. and Cohen, P., On the role of aging in cancer incidence, *J. Theor. Biol.,* 83, 163, 1980.

1000.   Ames, B. N., Shigenaga, M. K. and Hagen, T. M., Oxidants, antioxidants, and the degenerative diseases of aging, *Proc. Natl. Acad. Sci., USA,* 90, 7915, 1993.

1001.   Oberley, L. W. and Oberley, T. D., Free radicals, cancer, and aging, in *Free Radicals, Aging, and Degenerative Diseases,* J. E. Johnson, Jr., R. Walford, D. Harman and J. Miquel, eds. Alan, R. Liss, New York, 1986, pp. 325-371.

1002.   Schulz-Allen, M.-F., *Aging and Human Longevity,* Birkhäuser, Boston, 1997.

1003.   Trush, M. A. and Kensler, T. W., An overview of the relationship between oxidative stress and chemical carcinogenesis, *Free Radical Biol. Med.,* 10, 201, 1991.

1004.   Ames, B. N., Gold, L. S. and Willett, W. C., The causes and prevention of cancer, *Proc. Natl. Acad. Sci., USA,* 92, 5258, 1995.

1005.   Ts'o, P. O. P., Caspary, W. J. and Lorentzen, R. J., The involvement of free radicals in chemical carcinogenesis, in *Free Radicals in Biology,* W. A. Pryor, ed., Vol. III, Academic Press, San Diego, 1977, p. 251.

1006.   Floyd, R. A., Watson, J. J., Harris, J., West, M. S. and Wong, P. K., Formation of 8-hydroxydeoxyguanosine, hydroxyl free radical adduct of DNA in granulocytes exposed to the tumor promoter tetradeconyl phorbolacetate, *Biochem. Biophys. Res. Commun.,* 137, 841, 1986.

1007.   Kasai, H., Nishimura, S., Kurokawa, Y. and Hayashi, Y., Oral administration of the renal carcinogen, potassium bromate, specifically produces 8-hydroxydeoxyguanosine in rat target organ DNA, *Carcinogenesis,* 8, 1959, 1987.

1008. Fiala, E. S., Conaway, C. C. and Mathis, J. E., Oxidative DNA and RNA damage in the liver of sprague-dawley rats treated with the hepatocarcinogen 2-nitropropane, *Cancer Res.,* 49, 5518, 1989.

1009. Kasai, H., Crain, P. F., Kuchino, Y., Nishimura, S., Ootsuyama, A. and Tanooka, H., Formation of 8-hydroxyguanine moiety in cellular DNA by agents producing oxygen radicals and evidence for its repair, *Carcinogenesis,* 7, 1849, 1987.

1010. Kohda, K., Tada, M., Kasai, H., Nishimura, S. and Kawazoe, Y., Formation of 8-hydroxyguanine residues in cellular DNA exposed to the carcinogen 4-nitro-quinoline-1-oxide, *Biochem. Biophys. Res. Commun.,* 139, 626, 1986.

1011. Fucci, L., Oliver, C. N., Coon, M. J. and Stadtman, E. R., Inactivation of key metabolic enzymes by mixed-function oxidation reactions: Possible implication in protein turnover and aging, *Proc. Natl. Acad. Sci.,* 80, 1521, 1983.

1012. Weitzman, S. A. and Gordon, L. I., Inflammation and cancer: role of phagocyte-generated oxidants in carcinogenesis, *Blood,* 76, 655, 1990.

1013. Collins, R. H., Feldman, M., and Fordtran, J. S.,Colon cancer, displasia and surveillance in patients with ulcerative colitis: a critical review, *N. Engl. J. Med.,* 316, 1654, 1987.

1014. Ferguson, A. R., Associated bilharziosis and primary malignant disease of the urinary bladder, with observations on a series of forty cases, *J. Pathol. Bacteriol.,* 16, 76, 1911.

1015. Correa, P., A human model of gastric carcinogenesis, *Cancer Res.,* 48, 3554, 1988.

1016. Konstaninides, A., Smulow, J. B. and Sonnenschein, C., Tumorigenesis at a predetermined oral site after one intraperitoneal injection of N-nitroso-N-methylurea, *Science,* 216, 1235, 1982.

1017. Waalkes, M. P., Rehm, S., Kasprzak, K. S. and Issaq, H. J., Inflammatory, proliferative, and neoplastic lesions at the site of metallic ear tags in Wistar [Cr:WI BR] rats, *Cancer Res.,* 47, 2445, 1987.

1018. Sieweke, M. H., Stoker, A. W. and Bissell, M. J., Evaluation of the cocarcinogenic effect of wounding in Rous sarcoma virus tumorigenesis, *Cancer Res.,* 49, 6419, 1989.

1019. Argyris, T. S. and Slaga, T. J., Promotion of carcinomas by repeated abrasion in initiated skin of mice, *Cancer Res.*, 41, 5793, 1981.

1020. Davis, C. P., Cohen, M. S., Gruber, M. B., Anderson, M. D. and Warren, M. M., Urothelial hyperplasia and neoplasia: a response to chronic urinary tract infections in rats, *J. Urol.*, 132, 1025, 1984.

1021. Pozharisski, K. M., The significance of non-specific injury in colon carcinogenesis in rats, *Cancer Res.*, 35, 3824, 1975.

1022. Barthold, S. W., The role of non-specific injury in colon carcinogenesis, in *"Experimental Colon Carcinogenesis"*, H. Autrup, G. M. Williams, eds., CRC Press, Boca Raton, FL, 1983, p. 185.

1023. Weitzman, S. A. and Stossel, T. P., Mutation caused by human phagocytes, *Science,* 212, 546, 1981.

1024. Weitzman, S. A. and Stossel, T. P., Effects of oxygen radical scavengers and antioxidants on phagocyte-induced mutagenesis, *J. Immunol.,* 128, 2770, 1982.

1025. Barak, M., Ulitzur, S. and Merzbach, D., Phagocytosis induced mutagenesis in bacteria, *Mutat. Res.,* 121, 7, 1983.

1026. Roman-Franco, A. A., Non-enzymatic extramicrosomal bio-activation of chemical carcinogens by phagocytes: A proposed new pathway, *J. Theor. Biol.,* 97, 543, 1982.

1027. Dix, T. A. and Marnett, L. J., Metabolism of polycyclic aromatic hydrocarbon derivatives to ultimate carcinogens during lipid peroxidation, *Science,* 221, 77, 1983.

1028. Burdon, R. H., Gill, V. and Rice-Evans, C., Oxidative stress and tumor cell proliferation, *Free Radical Res. Commun.,* 11, 65, 1990.

1029. Henderson, B. E., Ross, R. and Bernstein, L., Estrogens as a source of human cancer: The Richard and Hinda Rosenthal Foundation Award Lecture, *Cancer Res.,* 48, 246, 1988.

1030. Preston-Martin, S., Pike, M. C., Ross, R. K., Jones, P. A. and Henderson, B. E., Increased cell division as a cause of human cancer, *Cancer Res.,* 50, 7415, 1990.

1031. Dunsford, H. A., Sell, S. and Chisari, F. V., Hepatocarcinogenesis due to chronic liver cell injury in hepatitis B virus transgenic mice, *Cancer Res.,* 50, 3400, 1990.

1032. Blaser, M. J., *Heliobacter pylori* and the pathogenesis of gastroduodenal inflammation, *J. Infect. Dis.* 161, 1626, 1991.

1033. Parsonnet, J., Friedman, G. D., Vandersteen, D. P., Chang, Y., Vogelman, J. H., Orentreich, N. and Sibley, R. K., *Heliobacter pylori* infection and the risk of gastric carcinoma, *N. Engl. J. Med.*, 325, 1127, 1991.

1034. Diehl, A. K., Gallstone size and the risk of gallbladder cancer, *JAMA*, 250, 2323, 1983.

1035. Landino, L. M., Crews, B. C., Timmons, M. D., Morrow, J. D. and Marnett, L. J., Peroxynitrite, the coupling product of nitric oxide and superoxide, activates prostaglandin biosynthesis, *Proc. Natl. Acad. Sci., USA*, 93, 15069, 1996.

1036. Clancy, R. M. and Abramson, S. B., Nitric oxide: a novel mediator of inflammation, *Proc. Soc. Exp. Biol. Med.*, 210, 93, 1995.

1037. Akaike, T., Suga, M. and Maeda, H., Free radicals in viral pathogenesis: molecular mechanisms involving superoxide and NO, *Proc. Soc. Exp. Biol. Med.*, 217, 64, 1998.

1038. Umezawa, K., Akaike, T., Fujii, S., Suga, M., Setoguchi, K., Ozawa, A. and Maeda, H., Induction of nitric oxide synthesis and xanthine oxidase and their role in the antimicrobial mechanism against *Samonella typhimurium* in mice, *Infect. Immunol.*, 65, 2932, 1997.

1039. Umezawa, K., Ohnishi, N., Tanaka, K., Kamiya, S., Koga, Y., Nakazawa, H. and Ozawa, A., Granulation in livers of mice infected with *Samonella typhimurium* is caused by superoxide released from host phagocytes, *Infect. Immunol.*, 63, 4402, 1995.

1040. Wright, P. F., Respiratory diseases, in *Viral Pathogenesis*, N. Nathanson, R. Ahmed, F. Gonzalez-Scarano, D. E. Griffin, K. V. Holmes, F. A. Murphy and H. L. Robinson, eds., Lippincott-Raven Publishers, 1997, pp. 703-711.

1041. Liu, R. H. and Hotchkiss, J. H., Potential genotoxicity of chronically elevated nitric oxide: a review, *Mutat. Res.*, 339, 73, 1995.

1042. Vuillaume, M., Reduced oxygen species, mutation, induction, and cancer initiation, *Mutat. Res.*, 186, 43, 1987.

1043. Ohshima, H. and Bartsch, H., Chronic infections and inflammatory processes as cancer risk factors: possible role of nitric oxide in carcinogenesis, *Mutat. Res.*, 305, 253, 1994.

1044. Inoue, S. and Kawanishi, S., Oxidative DNA damage induced by simultaneous generation of nitric oxide and superoxide, *FEBS Letters*, 371, 86, 1995.

488

1045.   Webb, P. M. and Forman, D., Heliobacter pylori as a risk factor for cancer, *Baillieres Clin. Gastroenterol.*, 9, 563, 1995.

1046.   Forman, D., Heliobacter pylori and gastric cancer, *Scand. J. Gastroenterol. Suppl.*, 215, 48, 1996.

1047.   Crabtree, J. E., Immune and inflammatory responses to Heliobacter pylori infection, *Scand. J. Gastroenterol. Suppl.* 215, 3, 1996.

1048.   Götz, J. M., van Kan, C. I., Verspaget, H. W., Biemond, I., Lamers, C. B. and Veenendaal, R. A., Gastric mucosal superoxide dismutase in heliobacter pylori infection, *Gut*, 38, 502, 1996.

1049.   Ravindranath, V., Shivakumar, B. R. and Anandatheerthavaranda, H. K., Low glutathione levels in brain regions of aged rats, *Neurisci. Lett.*, 101, 187, 1989.

1050.   Phebus, L. A., Perry, K. W., Clemens, J. A. and Fuller, R. W., Brain anoxia releases striatal dopamine in rats, *Life Sci.*, 38, 2447, 1986.

1051.   Halliwell, B., Oxidants and the central nervous system: some fundamental questions, *Acta Neurol. Scand.*, 126, 23, 1989.

1052.   Halliwell, B., Reactive oxygen species and the central nervous system, *J. Neurochem.*, 59, 1609, 1992.

1053.   Evans, P. H., Free radicals in brain metabolism and pathology, *Br. Med. Bull.*, 49, 577, 1993.

1054.   Jenner, P., Oxidative damage in neurodegenerative disease, *Lancet*, 344, 796, 1994.

1055.   Gutteridge, J. M. C., Hydroxyl radicals, iron, oxidative stress, and neurodegeneration, *Ann. N. Y. Acad. Sci.*, 738, 201, 1994.

1056.   Knight, J. A., Reactive oxygen species and the neurodegenerative disorders, *Ann. Clin. Lab. Sci.*, 27, 11, 1997.

1057.   Simonian, N. A. and Coyle, J. T., Oxidative stress in neurodegenerative diseases, *Ann. Rev. Pharmacol. Toxicol.*, 36, 83, 1996.

1058.   Giulian, D., Ameboid microglia are effectors of inflammation in the central nervous system, *J. Neurosci. Res.*, 18, 155, 1987.

1059.   Dickson, D. W., Mattiace, L. A., Kure, K., Hutchins, K., Lyman, W. D. and Brosnan, C. F., Microglia in human disease, with an emphasis on aquired immune deficiency syndrome, *Lab. Invest.*, 64, 135, 1991.

1060.   Colton, C. A. and Gilbert, D. L., Production of superoxide anions by a CNS macrophage, the microglia, *FEBS Letters*, 223, 284, 1987.

1061.   Colton, C. A., Snell, J., Chernyshev, O. and Gilbert, D. L., Induction of superoxide anion and nitric oxide production in cultured microglia, *N. Y. Acad. Sci.*, 738, 54, 1994.

1062.   Boje, K. M. and Arora, K., Microglial-produced nitric oxide and reactive nitrogen oxides mediate neuronal cell death, *Brain Res.*, 587, 250, 1992.

1063.   Chao, C. C., Hu, S., Molitor, T. W., Shaskan, E. G. and Peterson, P. K., Activated microglia mediate neuronal cell injury via a nitric oxide mechanism, *J. Immunol.*, 149, 2736, 1992.

1064.   Merrill, J. E., Ignarro, L. J., Sherman, M. P., Melinek, J. and Lane, T. E., Microglial cell cytotoxicity of oligodendrocytes is mediated through nitric oxide, *J. Immunol.*, 151, 2132, 1993.

1065.   McGeer, P. L., Itagaki, S., Boyes, B. E. and McGeer, E. G., Reactive microglia are positive for HLA-DR in the substantia nigra of Parkinson's and Alzheimer's disease brains, *Neurol.*, 38, 1285, 1988.

1066.   Djerassi, C., *NO*, University of Georgia Press, Athens, GA, 1998.

1067.   McNamara, J. O. and Fridovich, I., Did radicals strike Lou Gehrig?, *Nature*, 362, 20, 1993.

1068.   Rosen, D. R., Bowling, A. C., Patterson, D. et al., A frequent ala4 to val superoxide dismutase-1 mutation is associated with a rapidly progressive familial amyotrophic lateral sclerosis, *Hum. Mol. Genet.*, 3, 981, 1994.

1069.   Deng, H. X., Hentati, A., Tainer, J. A. et al., Amyotrophic lateral sclerosis and structural defects in Cu,Zn superoxide dismutase, *Science*, 261, 1047, 1993.

1070.   Katzman, R., Alzheimer's disease, *N. Engl. J. Med.*, 314, 964, 1986.

1071.   Dowson, J. H., Neuronal lipofuscin accumulation in ageing and Alzheimer's dementia: a pathogenic mechanism?, *Brit. J. Psychiatry*, 140, 142, 1982.

1072.   Jeandel, C., Nicolas, M. B., Dubois, F., Nabet-Belleville, F., Penin, F. and Cuny, G., Lipid peroxidation and free radical scavengers in Alzheimer's disease, *Gerontology*, 35, 275, 1989.

1073.   West, M. J., Coleman, P. D., Flood, D. G. and Troncoso, J. C., Differences in the pattern of hippocampal neuronal loss in normal ageing and Alzheimer's disease, *Lancet*, 344, 769, 1994.

490

1074.   St. George-Hyslop, P. H., Tanzi, R. E., Polinsky, R. J. et al., The genetic defect causing familial Alzheimer's disease maps on chromosome 21, *Science,* 235, 885, 1987.

1075.   Somerville, M. J., Percy, M. P., Bergeron, C., Yoong, L. K. K., Grima, E. A. and McLachlan, D. R. C., Localization and quantitation of 68 kDa neurofilament and superoxide dismutase-1 mRNA in Alzheimer's brain, *Mol. Brain Res.,* 9, 1, 1991.

1076.   Yankner, B., Dawes, L., Fisher, S., Villa-Komaroff, L., Oster-Granite, M. et al., Neurotoxicity of a fragment of the amyloid precursor associated with Alzheimer's disease, *Science,* 243, 417, 1989.

1077.   Butterfield, D. A., Hensley, K., Harris, M., Mattson, M. and Carney, J., ß-Amyloid peptide free radical fragments initiate synaptosomal lipoperoxidation in a sequence-specific fashion: implications to Alzheimer's disease, *Biochem. Biophys. Res. Commun.,* 200, 710, 1994.

1078.   Behl, C., Davis, J., Cole, G. M. and Schubert, D., Vitamin E protects nerve cells from amyloid ß protein toxicity, *Biochem. Biophys. Res. Commun.,* 186, 944, 1992.

1079.   Behl, C., Davis, J., Lesley, R. and Schubert, D., Hydrogen peroxide mediates amyloid ß protein toxicity, *Cell,* 77, 817, 1994.

1080.   Meda, L., Cassatella, M., Szendrei, G., Otvos, L., Baron, P. et al. Activation of microglial cells by ß-amyloid protein and interferon-, *Nature,* 374, 647, 1995.

1081.   Hensley, K., Carney, J. M., Mattson, M. P., Aksenova, M., Harris, M., Wu, J. F., Floyd, R. A. and Butterfield, D. A., A model for ß-amyloid aggregation and neurotoxicity based on free radical generation by the peptide: relevance to Alzheimer's disease, *Proc. Natl. Acad. Sci. USA,* 91, 3270, 1994.

1082.   Smith, C. D., Carney, J. M., Starke-Reed, P. E., Oliver, C. N., Stadtman, E. R., Floyd, R. A. and Markesbury, W. R., Excess brain protein oxidation and enzyme dysfunction in normal aging and Alzheimer's disease, *Proc. Natl. Acad. Sci. USA,* 88, 10,540, 1991.

1083.   Subbarao, K. V., Richardson, J. S. and Ang, L. C., Autopsy samples of Alzheimer's cortex show increased peroxidation *in vitro, J. Neurochem.,* 55, 342, 1990.

1084. Palmer, A. M. and Buerns, M. A., Selective increase in lipid peroxidation in the inferior temporal cortex in Alzheimer's disease, *Brain Res.*, 645, 338, 1994.

1085. Mecocci, P., MacGarvey, U. and Beal, M. F., Oxidative damage to mitochondrial DNA is increased in Alzheimer's disease, *Ann. Neurol.*, 36, 747, 1994.

1086. Good, P. F., Werner, P., Hsu, A., Olanow, C. W. and Perl, D. P., Evidence for neuronal oxidative damage in Alzheimer's disease, *Am. J. Pathol.*, 149, 21, 1996.

1087. Ciccone, C. D., Free-radical toxicity and antioxidant medications in Parkinson's disease, *Physical Ther.*, 78, 313, 1998.

1088. Rodgers, J., Luber-Narod, J., Styren, S. D. and Civin, W. H., Expression of immune system-associated antigens by cells of the central nervous system: relationship to the pathology of Alzheimer's disease, *Neurobiol. Aging*, 9, 339, 1988.

1089. Kuiper, M. A., Visser, J. J., Bergmans, P. L. M., Scheltens, P. and Wolters, E. C., Decreased cerebrospinal fluid nitrate levels in Parkinson's disease, Alzheimer's disease and multiple system atrophy patients, *J. Neurol. Sci.*, 121, 46, 1994.

1090. Barford, P. A., Blair, J. A., Eggar, C., Hamon, C., Morar, C. and Whitburn, S. B., Tetrahydrobiopterin metabolism in the temporal lobe of patients dying with senile dementia of Alzheimer type, *J. Neurol. Neurosurg. Psychiatr.*, 47, 736, 1984.

1091. Nagatsu, T., Yamaguchi, T., Kato, T., Sugimoto, T., Matsuura, S., Akino, M., Nagatsu, I., Iitsuka, R. and Narabayashi, H., Biopterin in human brain and urine from controls and parkinsonian patients: application of a new radioimmunoassay, *Clin. Chim. Acta*, 109, 305, 1981.

1092. Heinzel, B. M. J., Klatt, P., Böhme, E. and Mayer, B., Ca^{2+}/calmodulin -dependent formation of hydrogen peroxide by brain nitric oxide synthase, *Biochem. J.*, 281, 627, 1992.

1093. Sinet, P. M., Metabolism of oxygen derivatives in Down's syndrome, *Ann. N. Y. Acad. Sci.*, 396, 83, 1982.

1094. Antila, E. and Westermarck, T., On the etiopathogenesis and therapy of Down's syndrome, *Int. J. Dev. Biol.*, 33, 183, 1989.

1095. Bö, L., Dawson, T. M., Wesselingh, S., Mörk, S., Choi, S., Kong, P. A., Hanley, D. and Trapp, B. D., Induction of nitric oxide synthase in demyelinating regions of multiple sclerosis brains, *Ann. Neurol.*, 36, 778, 1994.

1096. Lin, R. F., Lin, T. S., Tilton, R. G. and Cross, A. H., Nitric oxide localized to spinal cords of mice with experimental allergic encephalomyelitis: an electron paramagnetic resonance study, *J. Exp. Med.*, 178, 643, 1993.

1097. Sherman, M. P., Griscavage, J. M. and Ignarro, L. J., Nitric oxide-mediated neuronal injury in multiple sclerosis, *Med. Hypothesis*, 39, 143, 1992.

1098. Bagasra, O., Michaels, F. H., Zheng, Y. M., Bobroski, L. E., Spitsin, S. V., Fu, Z. F., Tawandros, R. and Koprowski, H., Activation of the inducible form of nitric oxide synthase in the brains of patients with multiple sclerosis, *Proc. Natl. Acad. Sci. USA*, 92, 12,041, 1995.

1099. Giovannoni, G., Heales, S. J. R., Silver, N. C., O'Riordan, J., Miller, R. F., Land, J. M., Clark, J. B. and Thompson, E. J., Raised serum nitrate and nitrite levels in patients with multiple sclerosis, *J. Neurol. Sci.*, 145, 77, 1997.

1100. Xiao, B. G., Zhang, G. X., Ma, C. G. and Link, H., The cerebrospinal fluid from patients with multiple sclerosis promotes neuronal and oligodendrocyte damage by delayed production of nitric oxide in vitro, *J. Neurol. Sci.*, 142, 114, 1996.

1101. Yamashita, T., Ando, Y., Obayashi, K., Uchino, M. and Ando, M., Changes in nitrite and nitrate (NO_2^-/NO_3^-) levels in cerebrospinal fluid of patients with multiple sclerosis, *J. Neurol. Sci.*, 153, 32, 1997.

1102. Ebers, G. C., Sadovnik, A. D., Risch, N. J. and the Canadian Collaborative Study Group. Familial aggregation in multiple sclerosis is genetic, *Nature*, 377, 150, 1995.

1103. Sadovnik, A. D., Ebers, G. C., Dyment, D. A. and Risch, N. J., Evidence for a genetic basis of multiple sclerosis, *Lancet*, 347, 1728, 1996.

1104. Olson, R. C., A proposed role for nerve growth factor in the etiology of multiple sclerosis, *Med. Hypothesis*, 51, 493, 1999.

1105. Cross, A. H., Misko, T. P., Lin, R. F., Hickey, W. F., Trotter, J. L. and Tilton, R. G., iNOS antagonists ameliorate clinical symptoms of allergic encephaomyelitis in animals, *J. Clin. Invest.*, 93, 2984, 1994.

1106. Toshniwal, P. K. and Zarling, E. J., Evidence for increased lipid peroxidation in multiple sclerosis, *Neurochem. Res.*, 17, 205, 1992.

1107. Adams, C. W. M., Perivascular iron deposition and other vascular damage in multiple sclerosis, *J. Neurol. Neurosurg. Psychiatr.*, 51, 260, 1988.

1108. Clemens, J. A. and Phebus, L. A., Dopamine depletion protects striatal neurons from ischemia-induced cell death, *Life Sci.*, 42, 707, 1988.

1109. Jenner, P., Oxidative stress as a cause of Parkinson's disease, *Acta Neurol. Scand.*, 84, Suppl.136, 6, 1991.

1110. Martilla, R. J., Lorentz, H. and Rinne, U. K., Oxygen toxicity protecting enzymes in Parkinson's disease, *J. Neurol. Sci.*, 86, 321, 1988.

1111. Saggu, H., Cooksey, J., Dexter, D. et al., A selective increase in particulate superoxide dismutase activity in parkinsonian substantia nigra, *J. Neurochem.*, 53, 692, 1989.

1112. Ceballos, I., Lafon, M., Javoy-Agid, F., Hirsch, E., Nicole, A., Sinet, P. M. and Agid, Y., Superoxide dismutase and Parkinson's disease, *Lancet*, 335, 1035, 1990.

1113. Hirsch, E., Graybiel, A. M. and Agid, Y., Melanized dopaminergic neurons are differentially susceptible to degeneration in Parkinson's disease, *Nature*, 334, 345, 1988.

1114. Sofic, E., Lange, K. W., Jellinger, K. and Riederer, P., Reduced and oxidized glutathione in the substantia nigra of patients with Parkinson's disease, *Neurosci. Lett.*, 142, 128, 1992.

1115. Fahn, S. and Cohen, G., The oxidant stress hypothesis in Parkinson's disease: evidence supporting it, *Ann. Neurol.*, 32, 804, 1992.

1116. Ambani, I. M., VanWoert, M. H. and Murphy, S., Brain peroxidase and catalase in Parkinson's disease, *Arch. Neurol.*, 32, 114, 1975.

1117. Perry, T. L., Godin, D. V. and Hansen, S., Parkinson's disease: a disorder due to nigral glutathione deficiency?, *Neurosci. Lett.*, 33, 305, 1982.

1118. Lange, K. W., Youdim, M. B. H. and Riederer, P., Neurotoxicity and neuroprotection in Parkinson's disease, *J. Neural Transm.*, 38 (Suppl.), 27, 1992.

1119. Sian, J., Dexter, D. T., Lees, A. J. et al, Alterations in glutathione levels in Parkinson's disease and other neurodegerative disorders affecting the basal ganglia, *Ann. Neurol.*, 36, 348, 1994.

1120. Graham, D. C., On the origin and significance of neuromelanin, *Arch. Pathol. Lab. Med.,* 103, 359, 1979.

1121. Adams, J. D. and Odunze, I. N., Oxygen free radicals and Parkinson's disease, *Free Radical Biol. Med.,* 10, 161, 1991.

1122. Dexter, D. T., Wells, F. R., Agid, F., Agid, Y., Lees, A. J., Jenner, P. and Marsden, C. D., Increased nigral iron content in postmortem parkinsonian brain, *Lancet,* 2 (8569), 1219, 1987.

1123. Dexter, D. T., Wells, F. R., Lees, A. J., Agid, F., Agid, Y., Jenner, P. and Marsden, C. D., Increased nigral iron content and alterations in other metal ions occurring in brain in Parkinson's disease, *J. Neurochem.,* 52, 1830, 1989.

1124. Sofic, E., Riederer, P., Heinsen, H., Beckman, H., Reynolds, G. P., Hebenstreit, G. and Youdim, M. B. H., Increased iron(III) and total iron content in post mortem substantia nigra of parkinsonian brain,*J. Neural Transm.,* 74, 199, 1988.

1125. Dexter, D. T., Carter, C. J., Wells, F. R., Javoy-Agid, F., Agid, Y., Lees, A., Jenner, P. and Marsden, C. D., Basal lipid peroxidation in substantia nigra is increased in Parkinson's disease,*J. Neurochem.,* 52, 381, 1989.

1126. Molina, J. A., Jiménez-Jiménez, F. J., Navarro, J. A. et al., Plasma levels of nitrates in patients with Parkinson's disease, *J. Neurol. Sci.,* 127, 87, 1994.

1127. Youdim, M. B. H., Lavie, L. and Riederer, P., Oxygen free radicals and neurodegeneration in Parkinson's disease: a role for nitric oxide, *Ann. N. Y. Acad. Sci.,* 738, 64, 1994.

1128. Ciccone, C. D., Free-radical toxicity and antioxidant medications in Parkinson's disease, *Physical Ther.,* 78, 313, 1998.

1129. Smith, L. L., The response of the lung to foreign compounds that produce free radicals, *Ann. Rev. Physiol.,* 48, 681, 1986.

1130. Doelman, C. J. A. and Bast, A., Oxygen radicals in lung pathology, *Free Radical Biol. Med.,* 9, 381, 1990.

1131. Menzel, D. B., The toxicity of air pollution in experimental animals and humans: the role of oxidative stress, *Toxicol. Lett.,* 72, 269, 1994.

1132. Pryor, W. A., Dooley, M. M. and Church, D. F., Human -1-proteinase inhibitor is inactivated by exposure to side-stream smoke, *Toxicol. Lett.,* 28, 65, 1985.

1133.　Church, D. F. and Pryor, W. A., Free radical chemistry of cigarette smoke and its toxicological implications, *Environ. Health Persp.,* 64, 111, 1985.

1134.　Moldéus, P., Cotgreave, I. A. and Berggren, M., Lung protection by a thiol-containing antioxidant: N-acetylcysteine, *Respiration,* 50, 31, 1986.

1135.　Cantin, A. M., North, S. L., Hubbard, R. C. and Crystal, R. G., Normal alveolar epithelial lining fluid contains high levels of glutathione, *J. Appl. Physiol.,* 63, 152, 1987.

1136.　Cantin, A.M., North, S. L. and Crystal, R. G., Glutathione is present in normal alveolar epithelial lining fluid in sufficient concentration to provide antioxidant protection to lung parenchymal cells, *Am. Rev. Resp. Dis.,* 131, A372, 1985.

1137.　Wright, C. E., Tallan, H. H. and Lin, Y. Y., Taurine: biological update, *Ann. Rev. Biochem.,* 55, 427, 1986.

1138.　Toth, K. M., Clifford, D. P., Berger, E. M., White, C. W. and Repine, J. E., Intact human erythrocytes prevent hydrogen peroxide-mediated damage to isolated perfused rat lungs and cultured bovine pulmonary endothelial cells, *J. Clin. Invest.,* 74, 292, 1984.

1139.　Van Asbeck, B. S., Hoidal, J., Vercelotti, G. M., Schwartz, B. A., Moldow, C. F. and Jacob, H. S., Protection against lethal hyperoxia by tracheal insufflation of erythrocytes: role of red cell glutathione, *Science,* 227, 756, 1985.

1140.　Wali, R. K., Jaffe, S., Kumar, D., Sorgente, N. and Kalra, V. K., Increased adherence of oxidant-treated human and bovine erythrocytes to cultured endothelial cells, *J. Cell Physiol.,* 133, 25, 1987.

1141.　Toth, K. M., Berger, E. M., Beehler, C. J. and Repine, J. E., Erythrocytes from cigarette smokers contain more glutathione and catalase and protect endothelial cells from hydrogen peroxide better than do erythrocytes from nonsmokers, *Am. Rev. Resp. Dis.,* 134, 281, 1986.

1142.　Mann, J., Hoidal, J. R., Rao, N. V., McCusker, K. T. and Kennedy, T. P., Erythrocytes protect the ischemic lung from oxidant injury during reperfusion (abstract), *Am. Rev. Respir. Dis.,* 137, 369, 1988.

1143.   Mengel, C. E., Kann, H. E., Smith, W. W. and Horton, B. D., Effects of in vivo hyperoxia on erythrocytes. I. Haemolysis in mice exposed to hyperbaric oxygenation, *Proc. Soc. Exp. Biol. Med.*, 116, 259, 1964.

1144.   Del Principe, D., Menichelli, A., De Matteis, W., Di Corpo, M. L., Di Giulio, S. and Finazzi-Agro, A., Hydrogen peroxide has a role in the aggregation of human platelets, *FEBS Letters*, 185, 142, 1985.

1145.   Ramos Martinez, J. I., Launay, J. M. and Dreux, C., A sensitive fluorimetric microassay for the determination of glutathione peroxidase activity. Application to human blood platelets, *Anal. Biochem.*, 98, 154, 1979.

1146.   Pietra, G. G., Ruttner, J. R., Wust, W. and Gling, W., The lung after trauma and shock. Fine structure of alveolar-capillary barrier in 23 autopsies, *J. Trauma*, 21, 454, 1981.

1147.   Fantone, J. C., Kunkel, R. G. and Kinnes, D. A., Potentiation of - naphthylthiourea-induced lung injury by prostaglandin E and platelet depletion, *Lab Invest.*, 50, 703, 1984.

1148.   McGarrity, S. T., Stephenson, A. H., Hyers, T. M. and Webster, R.O., Inhibition of neutrophil superoxide anion generation by platelet products: role of adenine nucleotides, *J. Leukocyte Biol.*, 44, 411, 1988.

1149.   Ward, P. A., Cunningham, T. W., McCulloch, K. K., Phan, S. H., Powell, J. and Johnson, K. J., Platelet enhancement of $O_2^{\cdot-}$ responses in stimulated human neutrophils, *Lab Invest.*, 58, 37, 1988.

1150.   Berger, E. M., Beehler, C. H., Harada, R. N. and Repine, J. E., Phagocytic cells as scavengers of hydrogen peroxide (abstract), *Clin. Res.*, 35, 170A, 1987.

1151.   Freeman, B. A. and Crapo, J. D., Hyperoxia increases oxygen radical production in rat lungs and lung mitochondria, *J. Biol. Chem.*, 256, 10,986, 1981.

1152.   Turrens, J. F., Freeman, B. A. and Crapo, J. D., Hyperoxia increases H_2O_2 release by lung mitochondria and microsomes, *Arch. Biochem. Biophys.*, 217, 411, 1982.

1153.   Freeman, B. A., Topolsky, M. K. and Crapo, J. D., Hyperoxia increases radical production in rat lung homogenates, *Arch. Biochem. Biophys.*, 216, 477, 1982.

1154. Turrens, J. F., Freeman, B. A., Levitt, J. G. and Crapo, J. D., The effect of hyperoxia on superoxide production by lung submitochondrial particles, *Arch. Biochem. Biophys.*, 217, 401, 1982.

1155. Cross, C. E. and Hasegawa, G., Enhanced lung toxicity of O_2 in selenium-deficient rats, *Res. Commun. Chem. Pathol. Pharmacol.*, 16, 695, 1977.

1156. Deneke, S. M., Lynch, B. A. and Sanberg, B. L., Transient depletion of lung glutathione by diethylmaleate enhances oxygen toxicity, *J. Appl. Physiol.*, 85, 571, 1985.

1157. Crapo, J. D. and Tierney, D. F., Superoxide dismutase and pulmonary oxygen toxicity, *Am. J. Physiol.*, 226, 1401, 1974.

1158. Kimball, R. E., Reddy, K., Pierce, T. H., Schwartz, L. W., Mustafa, M. G. and Cross, C. E., Oxygen toxicity: augmentation of antioxidant defense mechanisms in rat lung, *Am. J. Physiol.*, 230, 1425, 1976.

1159. Crapo, J. D., Barry, B. E., Foscue, H. A. and Shelburne, J., Structural and biochemical changes in rat lungs occurring during exposures to lethal and adaptive doses of oxygen, *Am. Rev. Respir. Dis.*, 122, 123, 1980.

1160. Yusa, T., Crapo, J. D. and Freeman, B. A., Hyperoxia enhances lung and liver nuclear superoxide generation, *Biochim. Biophys. Acta*, 798, 167, 1984.

1161. Rodell, T. C., Cheronis, J. C., Ohnemus, C. L., Piermattei, D. J. and Repine, J. E., Xanthine oxidase mediates elastase-induced injury to isolated lungs and endothelium, *J. Appl. Physiol.*, 63, 2159, 1987.

1162. Frank, L. and Massaro, D., Oxygen toxicity, *Am. J. Med.*, 69, 117, 1980.

1163. Rinaldo, J. E., English, D., Levine, J., Stiller, R. and Henson, J., Increased intrapulmonary retention of radiolabeled neutrophils in early oxygen toxicity, *Am. Rev. Respir. Dis.*, 137, 345, 1988.

1164. Shasby, D. M., Fox, R. B., Harada, R. N. and Repine, J. E., Reduction of the edema of acute hyperoxic lung injury by granulocyte depletion, *J. Appl. Physiol.*, 52, 1237, 1982.

1165. Parrish, D. A., Mitchele, B. C., Henson, P. M. and Larsen, G. L., Pulmonary response of fifth component of complement-sufficient and -deficient mice in hyperoxia, *J. Clin. Invest.*, 74, 956, 1984.

1166. Haugaard, N., Oxygen poisoning. XI. The relation between inactivation of enzymes by oxygen and essential sulfhydryl groups, *J. Biol. Chem.*, 164, 265, 1946.

1167. Jocelyn, P. C., *Biochemistry of the SH Group*, Academic Press, London, 1972.

1168. Phillips, P. G. and Tsan, M. F., Hyperoxia causes increased albumin permeability of cultured endothelial monolayers, *J. Appl. Physiol.*, 64, 1196, 1988.

1169. Willett, W. C., Green, A., Stampfer, M. J., Speizer, F. E., Colditz, G. A., Rosner, B., Monson, R. R., Stason, W. and Hennekens, C.H., Relative and absolute excess risk of coronary heart disease among women who smoke cigarettes, *N. Engl. J. Med.*, 317, 1303, 1987.

1170. Janoff, A., Carp, H., Laurent, P. and Raju, L., The role of oxidative processes in emphysema, *Am. Rev. Resp. Dis.*, 127, S31, 1983.

1171. Hatch, G. E., Asthma, inhaled oxidants, and dietary antioxidants, *Am. J. Clin. Nutr.*, 61 (Suppl.), 625S, 1995.

1172. Loeb, L. A., Ernster, V. L., Warner, K. E., Abbotts, J. and Laszlo, J., Smoking and lung cancer: an overview, *Cancer Res.*, 44, 5940, 1984.

1173. Rylander, R., Environmental tobacco smoke and lung cancer, *Europ. J. Resp. Dis.*, 65, 127, 1984.

1174. Hoshino, E., Shariff, R., van Gossum, A., Allard, J. P., Pichard, C., Kurain, R. and Jeejeebhoy, K. N., Vitamin E suppresses increased lipid peroxidation in cigarette smokers, *J. Parenteral Enteral. Nutr.*, 14, 300, 1990.

1175. Brown, K. M., Morrice, P. C. and Duthie, G. G., Vitamin E supplementation suppresses indexes of lipid peroxidation and platelet counts in blood of smokers and non-smokers, but plasma lipoprotein concentrations remain unchanged, *Am. J. Clin. Nutr.*, 60, 383, 1994.

1176. Loft, S., Vistisen, K., Ewertz, M., Tjønneland, A., Overvad, K. and Poulsen, H. E., Oxidative DNA damage estimated by 8-hydroxydeoxyguanosine excretion in humans: influence of smoking, gender and body mass index, *Carcinogenesis*, 13, 2241, 1992.

1177. Loft, S., Astrup, A., Buemann, B. and Poulsen, H. E., Oxidative DNA damage correlates with oxygen consumption in humans, *FASEB J.*, 8, 534, 1994.

1178. Nakayama, T., Kaneko, M., Kodama, M. and Nagata, C., Cigarette smoke induces DNA single strand breaks in human cells, *Nature,* 314, 462, 1985.

1179. Pacht, E. R., Kaseki, H., Mohammed, J. R., Cornwell, D. G. and Davis, W. B., Deficiency of vitamin E in the alveolar fluid of cigarette smokers. Influence on alveolar macrophage cytotoxicity, *J. Clin. Invest.,*77, 789, 1986.

1180. Bui, M. H., Sauty, A., Collet, F. and Leuenberger, P., Dietary vitamin C intake and concentrations in the body fluids and cells of male smokers and nonsmokers, *J. Nutr.,* 122, 312, 1992.

1181. Gabor, S. and Anca, Z., Effects of asbestos on lipid peroxidation in the red cells, *Brit. J. Ind. Med.,* 32, 39, 1975.

1182. Vallyathan, V., Shi, X., Dala, N. S., Irr, W. and Castranova, V., Generation of free radicals from freshly fractured silica dust. Potential role in acute silica-induced lung injury, *Am. Rev. Resp. Disease,* 138, 1213, 1988.

1183. Carp, H., Miller, F., Hoidal, J. R. and Janoff, A., Potential mechanism of emphysema: alpha-1-proteinase inhibitor recovered from lungs of cigarette smokers contain oxidized methionine and has decreased elastase inhibiting activity, *Proc. Natl. Acad. Sci. USA,* 79, 2041, 1982.

1184. Schwartz, J. and Weiss, S. T., Dietary factors and their relationship to respiratory symptoms: NHANES II, *Am. J. Epidemiol.,* 132, 67, 1990.

1185. Schwartz, J. and Weiss, S. T., Relationship between dietary vitamin C intake and pulmonary function in the First National Health and Nutrition Examination Survey (NHANES I), *Am. J. Clin. Nutr.,* 59, 110, 1994.

1186. Olusi, S. O., Ojutiku, O. O., Jessop, W., J. and Iboko, M. I., Plasma and white blood cell ascorbic acid concentrations in patients with brochial asthma, *Clin. Chem. Acta,* 92, 161, 1979.

1187. Barnes, P. J., Reactive oxygen species and airway inflammation, *Free Radical Biol. Med.,* 9, 235, 1990.

1188. Frigas, E. and Gleich, G. J., The eosinophil and the pathology of asthma, *J. Allergy Clin. Immunol.,* 77, 527, 1986.

1189. Shult, P. A., Graziano, F. M. and Busse, W. W., Enhanced eosinophil luminol-dependent chemiluminescence in allergic rhinitis,*J. Allergy Clin. Immunol.,* 77, 702, 1986.

500

1190. Cluzel, M., Damon, M., Chanez, P. et al., Enhanced alveolar cell luminol-dependent chemiluminescence in asthma, *J. Allergy Clin. Immunol.*, 80, 195, 1987.

1191. Kelly, C. A., Ward, C., Stenton, S. C., Bird, G., Hendrick, D. J. and Walters, E. H., Numbers and activity of cells obtained at bronchoalveolar lavage in asthma, and their relationship to airway responsiveness, *Thorax*, 43, 684, 1988.

1192. Gaston, B., Drazen, J., Chee, C. B. E., Wohl, M. E. B. and Stamler, J. S., Expired nitric oxide (NO) concentrations are elevated in patients with reactive airways disease, *Endothelium*, 1, 87, 1993.

1193. Goldstein, J. L. and Brown, M. S., The low density lipoprotein pathway and its relation to atherosclerosis, *Ann. Rev. Biochem.*, 46, 897, 1977.

1194. Ross, R. and Glomset, J. A., The pathogenesis of atherosclerosis (part I), *N. Engl. J. Med.*, 95, 369, 1976.

1195. Brown, M. S. and Goldstein, J. L., Lipoprotein metabolism in the macrophage: implications for cholesterol deposition in atherosclerosis, *Ann. Rev. Biochem.*, 52, 223, 1983.

1196. Parthasarathy, S., Fong, L. G. and Steinberg, D., Oxidative modification of low density lipoprotein and atherosclerosis: concepts and consequences, in : *Lipid Peroxidation in Biological Systems*, A. Sevanian, I. L. Champaign, eds, *Am. Oil Chemists Soc.*, 1987, p. 225.

1197. Gebicki, J. M., Jürgens, G. and Esterbauer, H., Oxidation of low density lipoprotein in vitro, in *Oxidative Stress*, H. Sies, ed., Academic Press, London, 1991, pp. 371-397.

1198. Esterbauer, H., Gebicki, J., Puhl, H. and Jürgens, G., The role of lipid peroxidation and antioxidants in oxidative modification of LDL, *Free Radical Biol. Med.*, 13, 341, 1992.

1199. Parthasarathy, S., Steinberg, D. and Witztum, J. L., The role of oxidized low-density lipoproteins in the pathogenesis of atherosclerosis, *Ann. Rev. Med.*, 43, 219, 1992.

1200. Ross, R., The pathogenesis of atherosclerosis: a perspective for the 1990s, *Nature*, 362, 801, 1993.

1201. Halliwell, B., Oxidation of low-density lipoproteins: a question of initiation, propagation, and the effect of antioxidants, *Am. J. Clin. Nutr.*, 61 (suppl), 670S, 1995.

1202. Berliner, J. A. and Heinecke, J. W., The role of oxidized lipoproteins in atherogenesis, *Free Radical Biol. Med,* 20, 707, 1996.

1203. Brown, M. S. and Goldstein, J. L., Receptor-mediated control of cholesterol metabolism, *Science,* 191, 150, 1976.

1204. Geer, J. C., McGill, H. C. and Strong, J. P., The fine structure of human atherosclerotic lesions, *Am. J. Pathol.,* 38, 263, 1961.

1205. Schaffner, T., Taylor, K., Bartucci, E. J., Fischer-Dzoga, K., Beeson, J. H., Glagov, S. and Wissler, R. W., Arterial foam cells with distinctive immunomorphologic and histochemical features of macrophages, *Am. J. Pathol.,* 100, 57, 1980.

1206. Mitchinson, M. J., Ball, R. Y., Carpenter, K.L.H. and Parums, D. V., Macrophages and ceroid in atherosclerosis, in *Hyperlipidemia and Atherosclerosis,* K. E. Suckling, P. H. E. Groot, eds., Academic Press, London, 1988, pp. 117-134.

1207. Goldstein, J. L., Ho, Y. K., Basu, S. K. and Brown, M. S., Binding site on macrophages that mediates uptake and degradation of acetylated low density lipoprotein, producing massive cholesterol deposition, *Proc. Natl. Acad. Sci. USA,* 76, 333, 1979.

1208. Steinberg, D., Parthasarathy, S., Carew, T. E., Khoo, J. C. and Witztum, J. L., Beyond cholesterol: modifications of low density lipoprotein that increase its atherogenicity, *N. Engl. J. Med.,* 320, 915, 1989.

1209. Brown, M. S. and Goldstein, J. L., Familial hypercholesterolemia: defective binding of lipoproteins to cultured fibroblasts associated with impaired regulation of 3-hydroxy-3-methylglutaryl coenzyme A reductase activity, *Proc. Natl. Acad. Sci. USA,* 71, 788, 1974.

1210. Steinbrecher, U. P., Zhang, H. and Lougheed, M., Role of oxidatively modified LDL in atherosclerosis, *Free Radical Biol. Med.,* 9, 155, 1990.

1211. Henriksen, T., Mahoney, E, M. and Steinberg, D., Enhanced macrophage degradation of low density lipoprotein previously incubated with cultured endothelial cells: recognition by receptors for acetylated low density lipoproteins, *Proc. Natl. Acad. Sci. USA,* 78, 6499, 1981.

1212. Steinbrecher, U. P., Parthasarthy, S., Leake, D. S., Witztum, J. L. and Steinberg, D., Modification of low density lipo-protein by endothelial cells involves lipid peroxidation and degradation of low

502

density lipoprotein phospholipids, *Proc. Natl. Acad. Sci. USA,* 81, 3883, 1984.

1213. Morel, D. W., DiCorleto, P. E. and Chisolm, G. M., Endothelial and smooth muscle cells alter low density lipoprotein in vitro by free radical oxidation, *Arteriosclerosis,* 4, 357, 1984.

1214. van Hinsberg, V. W. M., Scheffer, M., Havekes, L. and Kemper, H. J. M., Role of endothelial cells and their products in the modification of low density lipoproteins, *Biochim. Biophys. Acta,* 878, 49, 1986.

1215. Heinecke, J. W., Baker, L., Rosen, H. and Chait, A., Superoxide-mediated modification of low density lipoprotein by arterioal smooth muscle cells, *J. Clin. Invest.,* 77, 757, 1986.

1216. Hiramatsu, K., Rosen, H., Heinecke, J. W., Wolfbaur, G. and Chait, A., Superoxide initiates oxidation of low density lipoprotein by human monocytes, *Arteriosclerosis,* 7, 55, 1986.

1217. Cathcart, M. K., Morel, D. W. and Chisolm, G. M., Monocytes and neutrophils oxidize low density lipoprotein making it cytotoxic, *J. Leukocyte Biol.,* 38, 341, 1985.

1218. Parthasarathy, S., Printz, D. J., Boyd, D., Joy, L. and Steinberg, D., Macrophage oxidation of low density lipoprotein generates a modified form recognized by the scavenger receptor, *Arteriosclerosis,* 6, 505, 1986.

1219. Leake, D. S. and Rankin, S. M., The oxidative modification of low density lipoproteins by macrophages, *Biochem. J.,* 270, 741, 1990.

1220. Kalyanaraman, B., Antholine, W. E. and Parthasarathy, S., Oxidation of low density lipoprotein by Cu^{2+} and lipoxygenase: an electron spin resonance study, *Biochim. Biophys. Acta,* 1035, 286, 1990.

1221. Kuzuya, M., Yamada, K., Hayashi, T., Funaki, C., Naito, M., Asai, K. and Kuzuya, F., Oxidation of low density lipoprotein by copper and iron in phosphate buffer, *Biochim. Biophys. Acta,* 1084, 198, 1991.

1222. Kuzuya, M., Yamada, K., Hayashi, T., Funaki, C., Naito, M., Asai, K. and Kuzuya, F., Role of lipoprotein-copper-peroxidation of low-density lipoprotein, *Biochim. Biophys. Acta,* 1123, 334, 1992.

1223. Fogelman, A, M., Shechter, I., Saeger, J., Hokom, M., Child, J. S. and Edwards, P. A., Malondialdehyde alteration of low density lipoprotein leads to cholesteryl ester accumulation in human monocyte-macrophages, *Proc. Natl. Acad. Sci. USA,* 77, 2214, 1980.

1224.   Henriksen, T., Mahoney, E, M. and Steinberg, D., Enhanced macrophage degradation of biologically modified low density lipoprotein, *Arteriosclerosis,* 3, 149, 1983.

1225.   Henriksen, T., Mahoney, E, M. and Steinberg, D., Interactions of plasma lipoproteins with endothelial cells, *Ann. N. Y. Acad. Sci.,* 401, 102, 1982.

1226.   Steinbrecher, U. P., Role of superoxide in endothelial-cell modification of low density lipoprotein, *Biochim. Biophys. Acta,* 959, 20, 1988.

1227.   Heinecke, J. W., Rosen, H. and Chait, A., Iron and copper promote modification of low density lipoprotein in human arterial smooth muscle cells in culture, *J. Clin. Invest.,* 74, 1890, 1984.

1228.   Hiramatsu, K., Rosen, H., Heinecke, J. W., Wolfbaur, G. and Chait, A., Superoxide initiates oxidation of low density lipoprotein by human monocytes, *Arteriosclerosis,* 7, 55, 1987.

1229.   Rosen, G. M. and Freeman, B. A., Detection of superoxide generated by endothelial cells, *Proc. Natl. Acad. Sci. USA,* 81, 7269, 1984.

1230.   Parthasarathy, S., Wieland, E. and Steinberg, D. A., A role for endothelial cell lipoxygenase in the oxidative modification of low density lipoprotein, *Proc. Natl. Acad. Sci. USA,* 86, 1046, 1989.

1231.   O'Leary, V. J., Darley-Usmar, V. M., Russell, L. J. and Stone, D., Prooxidant effects of lipoxygenase-derived peroxides on the copper-initiated oxidation of low density lipoprotein, *Biochem. J.,* 282, 631, 1992.

1232.   Chan, P. C., Peller, O. G. and Kesner, L., Copper(II) catalyzed lipid peroxidation in liposomes and erythrocyte membranes, *Lipids,* 17, 331, 1982.

1233.   Thomas, C. E. and Jackson, R. L., Lipid hydroperoxide involvement in copper-dependent and independent oxidation of low density lipoproteins, *J. Pharmacol. Exp. Ther.,* 256, 1182, 1990.

1234.   Salonen, J. T., Salonen, R., Seppanen, K., Kantola, M., Suntioninen, S. and Korpela, H., Interactions of serum copper, selenium, and low density lipoprotein cholesterol in atherogenesis, *Brit. Med. J.,* 302, 756, 1991.

504

1235.   Bedwell, S., Dean, R. T. and Jessup, W., The action of defined oxygen -centered free radicals on human low density lipoprotein, *Biochem. J.*, 262, 707, 1989.

1236.   Gebicki, J. M. and Bielski, B. H. J., Comparison of the capacities of the perhydroxyl and the superoxide radicals to initiate chain oxidation of linoleic acid, *J. Am. Chem. Soc.*, 103, 7020, 1981.

1237.   Kuzuya, M., Naito, M., Funaki, C., Hayashi, T., Asai, K. and Kuzuya, F., Lipid peroxide and transition metals are required for the toxicity of oxidized low density lipoprotein to cultured endothelial cells, *Biochim. Biophys. Acta*, 1096, 155, 1991.

1238.   Savenkova, M. I., Mueller, D. M. and Heinecke, J. W., Tyrosyl radical generated by myeloperoxidase is a physiological catalyst for the initiation of lipid peroxidation in low density lipoprotein,*J. Biol. Chem.*, 269, 20394, 1994.

1239.   Daugherty, A., Dunn, J. L., Rateri, D. L. and Heinecke, J. W., Myeloperoxidase , a catalyst for lipoprotein oxidation, is expressed in human atherosclerotic lesions,*J. Clin. Invest.*, 94, 437, 1994.

1240.   Ylä-Herttuala, S., Rosenfeld, M. E., Parthasarathy, S., Glass, C. K., Sigal, E., Witztum, J. L. and Steinberg, D, Colocalization of 15-lipoxygenase mRNA and protein with epitopes of oxidized low density lipoprotein in macrophage-rich areas of atherosclerotic lesions, *Proc. Natl. Acad. Sci. USA*, 87, 6959, 1990.

1241.   Kühn, H., Belkner, J., Zaiss, S., Fahrenklemper, T. and Wohlfeil, S., Involvement of 15-lipoxygenase in early stages of atherogenesis, *J. Exp. Med*, 179, 1903, 1994.

1242.   Ylä-Herttuala, S., Luoma, J., Vita, H., Hiltunen, T., Sisto, T. and Nikkari, T., Transfer of 15-lipoxygenase gene into rabbit iliac arteries results in the appearance of oxidation-specific lipid-protein adducts characteristic of oxidized low density lipoprotein, *J. Clin. Invest.*, 95, 92, 1995.

1243.   Darley, V. S., Mar, V. M., Hogg, N., O'Leary, V. J. and Moncada, S., The simultaneous generation of superoxide and nitric oxide can initiate lipid peroxidation in human LDL, *Free Radical Res. Commun.*, 17, 19, 1992.

1244.   Radi, R., Beckman, J. S , Bush, K. M. and Freeman, B. A., Peroxynitrite-induced membrane lipid peroxidation: The cytotoxic potential of superoxide and nitric oxide,*Arch. Biochem. Biophys.*, 288, 481, 1991.

1245. Graham, A. N., Hogg, N., Kalyanaraman, B., O'Leary, V. J., Darley-Usmar, V. and Moncada, S., Peroxynitrite modification of LDL leads to recognition by the macrophage scavenger receptor, *FEBS*, 330, 181, 1993.

1246. Beckman, J. S., Ye, Y. Z., Anderson, P. G., Chen, J., Acavitti, M. A., Tarpey, M. M. and White, C. R., Extensive nitration of protein tyrosines in human atherosclerosis detected by immunohisto-chemistry, *Biol. Chem. Hoppe-Seyler*, 375, 81, 1994.

1247. Cushing, S. D., Berliner, J. A., Valente, A. J., Territo, M. C., Navab, M., Parhami, F., Gerrity, R., Schwartz, C. J. and Fogelman, A. M., Minimally modified low density lipoprotein induces monocyte chemotactic protein-1 in human endothelial cells and smooth muscle cells, *Proc. Natl. Acad. Sci. USA*, 87, 5134, 1990.

1248. Berliner, J. A., Territo, M. C., Sevanian, A., Ramin, S., Kim, J. A., Bamshad, B., Esterson, M. and Fogelman, A. M., Minimally modified low density lipoprotein stimulates monocyte endothelial interactions, *J. Clin. Invest.*, 85, 1260, 1990.

1249. Frostegård, J., Haegerstrand, A., Gidlund, M. and Nilsson, J., Biologically modified LDL increases the adhesive properties of endothelial cells, *Atherosclerosis*, 90, 119, 1991.

1250. Gaziano, J. M., Hatta, A., Flynn, M., Johnson, E. J., Krinsky, N. I., Ridker, P. M., Hennekens, C. H. and Frei, B., Supplementation with ß-carotene in vivo and in vitro does not inhibit low density lipoprotein oxidation, *Atherosclerosis*, 112, 187, 1995.

1251. Princen, H. M. G., van Poppel, G., Vogelezang, C., Buytenhek, R. and Kok, F. J., Supplementation with vitamin E but not ß-carotene in vivo protects low density lipoprotein from lipid peroxidation in vitro, *Atheroscler.Thromb.*, 12, 554, 1992.

1252. Reaven, P. D., Khouw, A., Beltz, W. F., Parthasarathy, S. and Witztum, J. L., Effect of dietary antioxidant combinations in humans. Protection by vitamin E but not by ß-carotene, *Atheroscler. Thromb.*, 13, 590, 1993.

1253. Jialal, I., Vega, G. L. and Grundy, S. M., Physiologic levels of ascorbate inhibit oxidative modification of low density lipoprotein, *Atherosclerosis*, 82, 185, 1990.

1254. Esterbauer, H., Striegl, G., Puhl, H., Oberreither, S., Rotheneder, M., El-Saadani, M. and Jürgens, G., The role of vitamin E and

506

carotenoids in preventing oxidation of low density lipoprotein, *Ann. N. Y. Acad. Sci.,* 570, 254, 1989.

1255. Prasad, K. and Kalra, J., Oxygen free radicals and hypercholesterolemic atherosclerosis: effect of vitamin E, *Am. Heart J.,* 125, 958, 1993.

1256. Wojcicki, J., Rozewicka, B., Barcew-Wisziewska, B. et al., Effect of selenium and vitamin E on the development of experimental atherosclerosis in rabbits, *Atherosclerosis,* 87, 9, 1991.

1257. Keaney, Jr., J. F., Gaziano, J. M., Xu, A., Frei, B., Curran-Celentano, J., Shwaery, G. T., Loscalzo, J. and Vita, J. A., Low-dose -tocopherol improves and high dose -tocopherol worsens endothelial vasodilator function in cholesterol-fed rabbits, *J. Clin. Invest.,* 93, 844, 1994.

1258. Williams, R. J., Motteram, J. M., Sharp, C. H. and Gallagher, P. J., Dietary vitamin E and the attenuation of early lesion development in modified Watanabe rabbits, *Atherosclerosis,* 94, 153, 1992.

1259. Retsky, K. L., Freeman, M. W. and Frei, B., Ascorbic acid oxidation products protect human low-density lipoprotein against atherogenic modification, *J. Biol. Chem.,* 268, 1304, 1993.

1260. Ginter, E., Babala, J. and Cervex, J., The effect of chronic hypovitaminosis C on the metabolism of cholesterol and atherogenesis in guinea pigs, *J. Atheroscler. Res.,* 10, 341, 1968.

1261. Kimura, H., Yamada, Y., Morita, Y., Ikeda, H. and Matsuo, T., Dietary ascorbic acid depresses plasma and low density lipoprotein lipid peroxidation in genetically scorbutic rabbits, *J. Nutr,* 122, 1904, 1992.

1262. Carew, T. E., Schwenke, D. C. and Steinberg, D., Antiatherogenic effect of probucol unrelated to its hypercholesterolemic effect: evidence that antioxidants in vivo can selectively inhibit low density lipoprotein degradation in macrophage-rich fatty streaks and slow progression of atherosclerosis in the Watanabe hertable hyperlipidemic rabbit, *Proc. Natl. Acad. Sci. USA,* 84, 7725, 1987.

1263. Gey, K. F. and Puska, P., Plasma vitamins E and A inversely correlated to mortality from ischemic heart disease in cross-cultural epidemiology, *Ann. N.Y. Acad. Sci.,* 570, 268, 1989.

1264. Luc, G. and Fruchart, J. C., Oxidation of lipoproteins and atherosclerosis, *Am. J. Clin. Nutr.* 53, *Suppl.,* 206S, 1991.

1265. Parthasarathy, S., Young, S. G., Witztum, J. L., Pitmann, R. C. and Steinberg, D., Probucol inhibits oxidative modification of low density lipoprotein, *J. Clin. Invest.,* 77, 641, 1986.

1266. Jessup, W., Rankin, S. M., De Whalley, C. V., Hoult, J. R.., Scott, J. and Leake, D. S., -Tocopherol consumption during low density lipoprotein oxidation, *Biochem. J.,* 265, 399, 1990.

1267. Esterbauer, H., Puhl, H., Dieber-Rotheneder, M., Waeg, G. and Rabl. H., Effect of antioxidants in oxidative modification of LDL, *Ann. Med.,* 23, 573, 1991.

1268. Regnström, J., Nilsson, J.,Tornvall, P., Landon, C. and Hamsten, A., Susceptibility to low density lipoprotein oxidation and coronary atherosclerosis in man, *Lancet,* 339, 1183, 1992.

1269. Frankel, E. N., Kanner, J., German, J. B., Parks, E. and Kinsella, J. E., Inhibition of oxidation of human low density lipoproteins by phenolic substances in red wine, *Lancet,* 341, 454, 1993.

1270. Parthasarathy, S., Barnett, J. and Fong, L. G., High-density lipoprotein inhibits the oxidative modification of low density lipoprotein, *Biochim. Biophys. Acta,* 1044, 275, 1990.

1271. Miller, G. J., High density lipoproteins and atherosclerosis, *Ann. Rev. Med.,* 31, 97, 1980.

1272. Hopkins, P. N. and Williams, R. R., A survey of 246 suggested coronary risk factors, *Atherosclerosis,* 40, 1, 1981.

1273. Krinsky, N. I., Actions of carotenoids in biological systems, *Ann. Rev. Nutr.,* 13, 561, 1993.

1274. Bowry, V. W., Ingold, K. U. and Stocker, R., Vitamin E in human low- density lipoprotein. When and how this antioxidant becomes a pro-oxidant, *Biochem. J,* 288, 341, 1992.

1275. Bowry, V. W. and Stocker, R., Tocopherol-mediated peroxidation. The prooxidant effect of vitamin E on the radical-initiated oxidation of human low density lipoprotein, *J. Am. Chem. Soc.,* 115, 6029, 1993.

1276. Massey, J. B., Kinetics of transfer of -tocopherol between model and native plasma lipoproteins, *Biochim. Biophys. Acta,* 793, 387, 1984.

1277. Esterbauer, H., Striegl, G., Puhl, H. and Rotheneder, M., Continuous monitoring of in vitro oxidation of human low density lipoprotein, *Free Radical Res. Commun.,* 6, 67, 1989.

508

1278. Suarna, C., Dean, R. T., May, J. and Stocker, R., Human atherosclerotic plaque contains both oxidized lipids and relatively large amounts of -tocopherol and ascorbate, *Arterioscler. Thromb.* 15, 1616, 1995.

1279. Freiman, P. C., Mitchell, G. G., Heistad, D. D., Armstrong, M. L. and Harrison, D. G., Atherosclerosis impairs endothelium-dependent vascular relaxation to acetylcholine and thrombin in primates,*Circ. Res.,* 58, 783, 1986.

1280. Bossaller, C., Habib, G. B., Yamamoto, H., Williams, C., Wells, S. and Henry, P. D., Impaired muscarinic endothelium-dependent relaxation and cyclic guanosine-5'-monophosphate formation in atherosclerotic human coronary artery and rabbit aorta, *J. Clin. Invest.,* 79, 170, 1987.

1281. Ludmer, P. L., Selwyn, A. P., Shook, T. L., Wayne, R. R., Mudge, G. H., Alexander, W. and Ganz, P., Paradoxical vasoconstriction induced by acetylcholine in atherosclerotic coronary arteries, *N. Engl. J. Med.,* 315, 1046, 1986.

1282. Ardlie, N. G., Selley, M. L. and Simons, L. A., Platelet activation by oxidatively modified low density lipoprotein, *Atherosclerosis,* 76, 117, 1989.

1283. Minor, R. L., Myers, P. R., Guerra, R., Bates, J. N. and Harrison, D. G., Diet induced atherosclerosis increases the release of nitrogen oxides from rabbit aorta, *J. Clin. Invest.,* 86, 2109, 1990.

1284. Henriksen, T., Evenson, S. A. and Carlander, B., Injury to human endothelial cells in culture induced by low density lipoproteins, *Scand. J. Clin. Lab. Invest.,* 39, 361, 1979.

1285. Jain, S. K., The accumulation of malondialdehyde, a product of fatty acid peroxidation, can disturb aminophospholipid organization in the membrane bilayer of human erythrocytes, *J. Biol. Chem.,* 259, 3391, 1984.

1286. Cooke, J. P., Singer, A. H., Tsao, P., Zera, P., Rowan, R. A. and Billingham, M. E., Antiatherogenic effects of L-arginine in the hypercholesterolemic rabbit,*J. Clin. Invest.,* 90, 1168, 1992.

1287. Yeung, A. C., Vekshtein, V. I., Krantz, D. S., Vita, J. A., Ryan, Jr., T. J., Ganz, P. and Selwyn, A. P., The effect of atherosclerosis on the vasomotor response of coronary arteries to mental stress, *N. Engl. J. Med.,* 325, 1551, 1991.

1288. Deanfield, J. E., Shea, M., Kensett, M., Horlock, P., Wilson, R. A., de Landsheere, C. M. and Selwyn, A. P., Silent myocardial ischemia due to mental stress, *Lancet*, 2 (8410), 1001, 1984.

1289. Burton, K. P., Superoxide dismutase enhances recovery following myocardial ischemia, *Am. J. Physiol.*, 248, H637, 1985.

1290. Myers, C. L., Weiss, S. J., Kirsh, M. M. and Shlafer, M., Involvement of hydrogen peroxide and hydroxyl radical in the 'oxygen paradox': Reduction of creatine kinase release by catalase, allopurinol or deferoxamine, but not by superoxide dismutase, *J. Mol. Cell Cardiol.*, 17, 675, 1985.

1291. McCord, J. M., Oxygen-derived free radicals in post-ischemic tissue injury, *N. Engl. J. Med.*, 312, 159, 1985.

1292. Nakazawa, H., Ban, K., Ichimori, K., Minezaki, K., Okino, H., Masuda, T., Aoki, N. and Hori, S., The link between free radicals and myocardial injury, *Jpn. Circ. J.*, 52, 646, 1988.

1293. Omar, B., McCord, J. M. and Downey, J., Ischaemia-reperfusion, in *Oxidative Stress*, H. Sies, ed., Academic Press, 1991, pp. 493-527.

1294. Taylor, A. A. and Shappell, S. B., Reactive oxygen species, neutrophil and endothelial adherence molecules, and lipid-derived inflammatory mediators in myocardial ischemia-reflow injury, in *Free Radical Mechanisms of Tissue Injury*, M. Treinen Moslen, C. V. Smith, eds., CRC Press, Boca Raton, FL, 1992, pp. 65-142.

1295. Kramer, J. H., Arroyo, C. M., Dickens, B. F. and Weglicki, W. B., Spin-trapping evidence that graded myocardial ischemia alters post-ischemic superoxide production, *Free Radical Biol. Med.*, 3, 153, 1987.

1296. Culcasi, M., Pietri, S. and Cozzone, P. J., Use of 3,3,5,5-tetramethyl-1-pyrroline-1-oxide spin trap for the continuous flow ESR monitoring of hydroxyl radical generation in the ischemic and reperfused myocardium, *Biochem. Biophys. Res. Commun.*, 164, 1274, 1989.

1297. Zweier, J. L., Measurement of superoxide-derived free radicals in the reperfused heart, *J. Biol. Chem.*, 263, 1353, 1988.

1298. Zweier, J. L., Kuppusamy, P., Williams, R., Rayburn, B. K., Smith, D., Weisfeldt, M. L. and Flaherty, J. T., Measurement and characterization of postischemic free radical generation in the isolated perfused heart, *J. Biol. Chem.*, 264, 18,890, 1989.

510

1299. Garlick, P. B., Davies, M. J., Hearse, D. J. and Slater, T. F., Direct detection of free radicals in the reperfused rat heart using electron spin resonance spectroscopy, *Circ. Res.*, 61, 757, 1987.

1300. Bolli, R., Patel, B. S., Jeroudi, M. O., Lai, E. K. and McCay, P. B., Demonstration of free radical generation in "stunned" myocardium of intact dogs with the use of the spin trap alpha-phenyl-N-tert.-butyl nitrone, *J. Clin. Invest.*, 82, 476, 1988.

1301. Shen, A. C. and Jennings, R. B., Kinetics of calcium accumulation in acute myocardial ischemic injury, *Am. J. Pathol.*, 67, 441, 1972.

1302. Fleckenstein, A., Janke, J., Doring, H. J. and Leder, O., Myocardial fiber necrosis due to intracellular Ca overload: a new principle in cardiac pathophysiology, *Recent Adv. Stud. Cardiac Struct. Metab.*, 4, 563, 1974.

1303. Poole-Wilson, P. A., Harding, D. P., Bourdillon, P. D. V. and Tones, M. A., Calcium out of control, *J. Mol. Cell Cardiol.*, 16, 175, 1984.

1304. Choi, D. W., Calcium-mediated neurotoxicity: relationship to specific channel types and role in ischemic damage, *Trends Neurosci.*, 11, 465, 1988.

1305. Siesjo, B. K., Historical Overview. Calcium, ischemia, and death of brain cells, *Ann. N. Y. Acad. Sci.*, 522, 638, 1988.

1306. Choi, D. W., Calcium: still center-stage in hypoxic-ischemic neuronal death, *Trends Neurosci.*, 18, 58, 1995.

1307. Kristian, T. and Siesjo, B. K., Calcium in ischemic cell death, *Stroke*, 29, 705, 1998.

1308. Jennings, R. B., Hawkins, H. K., Lowe, J. E., Hill, M. C., Klotman, S. and Reimer, K. A., Relation between high energy phosphate and lethal injury in myocardial ischemia in the dog, *Am. J. Pathol.*, 92, 187, 1978.

1309. Buja, L. M., Hagler, H. K. and Willerson, J. T., Altered calcium homeostasis in the pathogenesis of myocardial ischemic and hypoxic injury, *Cell Calcium*, 9, 205, 1988.

1310. Zweier, J. L., Flaherty, J. T. and Weisfeldt, M. L., Direct measurement of free radical generation following reperfusion of ischemic myocardium, *Proc. Natl. Acad. Sci. USA*, 84, 1404, 1987.

1311. Nakazawa, H., Ischimori, K., Shinozaki, Y., Okino, H. and Hori, S., Is superoxide demonstration by electron-spin resonance spectroscopy really superoxide? *Am. J. Physiol.*, 255, H213, 1988.

1312. Massey, K. D. and Burton, K. P., α-Tocopherol attenuates myocardial membrane-related alterations resulting from ischemia and reperfusion, *Am. J. Physiol.,* 256, H1192, 1989.

1313. Ferreira, R., Burgos, M., Llesuy, S., Molteni, L., Milei, J., Gonzalez Flecha, H. and Boveris, A., Reduction of reperfusion injury with mannitol cardioplegia, *Ann. Thorac. Surg.,* 48, 77, 1989.

1314. Ganote, C. E., Simms, M. and Safavi, S., Effects of dimethyl-sulfoxide (DMSO) on the oxygen paradox in perfused rat hearts, *Am. J. Pathol.,* 109, 270, 1982.

1315. Das, D. K., Engelman, R. M., Clement, R., Otani, H., Prasad, M. R. and Rao, P. S., Role of xanthine oxidase inhibitor as free radical scavenger: a novel mechanism of action of allopurinol and oxypurinol in myocardial salvage, *Biochem. Biophys. Res. Commun.,* 148, 314, 1987.

1316. Godin, D. V. and Bhimji, S., Effects of allopurinol on myocardial ischemic injury induced by coronary artery ligation and reperfusion, *Biochem. Pharmacol.,* 36, 2101, 1987.

1317. Omar, B. A. and McCord, J. M., The cardioprotective effect of Mn-superoxide dismutase is lost at high dose in the postischemic isolated rabbit heart, *Free Radical Biol. Med.,* 9, 473, 1990.

1318. Omar, B. A., Gad, N. M., Jordan, M. C., Striplin, S. P., Russell, W. J., Downey, J. M. and McCord, J. M., Cardioprotection by Cu,Zn-superoxide dismutase is lost at high dose in the reoxygenated heart, *Free Radical Biol. Med.,* 9, 465, 1990.

1319. Horie, Y., Wolf, R., Anderson, D. C. and Granger, D. N., Hepatic leukostasis and hypoxic stress in adhesion molecule deficient mice after gut ischemia-reperfusion, *J. Clin. Invest.,* 99, 781, 1997.

1320. Shanley, P. F., White, C. W., Avraham, K. B., Groner, Y. and Burke, T. J., Use of transgenic animals to study disease models: hyperoxic lung injury and ischemic acute renal failure in "high SOD" mice, *Ren. Fail.,* 14, 391, 1992.

1321. Henninger, D. D., Gerritsen, M. E. and Granger, D. N., Low-density lipoprotein receptor knockout mice exhibit exaggerated microvascular responses to inflammatory stimuli, *Circ. Res.,* 81, 274, 1997.

1322. Horie, Y., Wolf, R., Flores, S. C., McCord, J. M., Epstein, C. J. and Granger, D. N., Transgenic mice with increased copper/zinc-superoxide dismutase activity are resistant to hepatic leukostasis and

512

capillary no-reflow after gut ischemia/reperfusion, *Circ. Res.*, 83, 691, 1998.

1323.   Deshmukh, D. R., Mirochnitchenko, O., Ghole, V. S., Agnese, D., Shah, P. C., Reddell, M., Brolin, R. E. and Inouye, M., Intestinal ischemia and reperfusion injury in transgenic mice overexpressing copper-zinc superoxide dismutase, *Am. J. Physiol.*, 273, C1130, 1997.

1324.   Mirochnitchenko, O. and Inouye, M., Effect of overexpression of human Cu,Zn superoxide dismutase in trangenic mice on macrophage functions, *J. Immunol.*, 156, 1578, 1996.

1325.   Yang, G., Chan, P. H., Chen, J., Carlson, E., Chen, S. F., Weinstein, P., Epstein, C. J. and Kamii, H., Human copper-zinc superoxide dismutase transgenic mice are highly resistant to reperfusion injury after focal cerebral ischemia, *Stroke,* 25, 165, 1994.

1326.   Kinouchi, H., Epstein, C. J., Mizui, T., Carlson, E., Chen, S. F. and Chan, P. H., Attenuation of focal cerebral ischemic injury in transgenic mice overexpressing CuZn superoxide dismutase, *Proc. Natl. Acad. Sci. USA,* 88, 11158, 1991.

1327.   McCord, J. M., unpublished results, presented at a workshop on Free Radicals and Disease Processes at the Ponce School of Medicine, Ponce, Puerto Rico, Feb. 1-3, 1999.

1328.   Romson, J. L., Hook, B. G., Kunkel, S. L., Abrams, G. D. and Schork, M. A., Reduction of the extent of ischemic myocardial injury by neutrophil depletion in the dog, *Circ.,* 67, 1016, 1983.

1329.   Kofsky, E. R., Julia, P. L., Buckberg, G. D., Quillen, J. E. and Acar, C., Studies of controlled reperfusion after ischemia. XXII. Reperfusate composition: effects of leukocyte depletion of blood and blood cardioplegic reperfusates after acute coronary occlusion, *J. Thor. Cardiovasc. Surg.,* 101, 350, 1991.

1330.   Lucchesi, B. R., Mickelson, J. K., Homeister, J. W. and Jackson, C. V., Interaction of formed elements of blood with the coronary vasculature in vivo, *Fed. Proc.,* 46, 63, 1987.

1331.   Mullane, K. M., Read, N. and Salmon, J. A., Role of leukocytes in acute myocardial infarction in anaesthesized dogs: relationship to myocardial salvage by anti-inflammatory drugs, *J. Pharmacol. Exp. Ther.,* 228, 510, 1984.

1332. Kochanek, P. M. and Hallenbeck, J. M., Polymorphonuclear leukocytes and monocytes/macrophages in the pathogenesis of cerebral ischemia and stroke, *Stroke,* 23, 1367, 1992.
1333. Shiga, Y., Onodera, H., Kogure, K., Yamasaki, Y. and Yashima, Y., Neutrophil as a mediator of ischemic edema formation in the brain, *Neurosci. Lett.,* 125, 110, 1991.
1334. Granger, D. N. and Korthuis, R. J., Physiological mechanisms of postischemic tissue injury, *Ann. Rev. Physiol.,* 57, 311, 1995.
1335. Korthuis, R. J., Anderson, D. C. and Granger, D. N., Role of neutrophil-endothelial cell adhesion in inflammatory disorders, *J. Crit.Care,* 9, 47, 1994.
1336. Ma, X. L., Tsao, P. S. and Lefer, A. M., Antibody to CD-18 exerts endothelial and cardiac protective effects in myocardial ischemia and reperfusion, *J. Clin. Invest.,* 88, 1237, 1991.
1337. Gomoll, A. W., Lekich, R. F. and Grove, R. I., Efficacy of a monoclonal antibody (MoAb 603) in reducing myocardial injury from ischemia/reperfusion in the ferret, *J. Cardiovasc. Pharmacol.,* 17, 873, 1991.
1338. Suzuki, M., Inauen, W., Kvietys, P. R., Grisham, M. B., Meininger, C. et al. , Superoxide mediates reperfusion-induced leukocyte-endothelial cell interactions, *Am. J. Physiol.,* 257, H1740, 1989.
1339. Del Maestro, R. F., Planker, M. and Arfors, K. E., Evidence for the participation of superoxide radical anion in altering the adhesive interaction between granulocytes and endothelium in vivo, *Int, J. Microcirc. Clin. Exp.,* 1, 105, 1982.
1340. Gaboury, J., Anderson, D. C. and Kubes, P., Molecular mechanisms involved in superoxide-induced leukocyte-endothelial cell interactions in vivo, *Am. J. Physiol.,* 266 (2 Pt. 2), H637, 1994.
1341. Granger, D. N., Benoit, J. N., Suzuki, M. and Grisham, M. B., Leukocyte adherence to venular endothelium during ischemia-reperfusion, *Am. J. Physiol.,* 257, G683, 1989.
1342. Suzuki, M., Grisham, M. B. and Granger, D. N., Leukocyte-endothelial cell interactions: role of xanthine oxidase-derived oxidants, *J. Leukocyte Biol.,* 50, 488, 1991.
1343. Kubes, P., Susuki, M. and Granger, D. N., Nitric oxide: an endogenous modulator of leukocyte adhesion, *Proc. Natl. Acad. Sci. USA,* 88, 4651, 1991.

514

1344. Arndt, H., Smith, C. W. and Granger, D. N., Leukocyte-endothelial cell adhesion in spontaneously hypersensitive and normal rats, *Hypertension*, 21, 667, 1993.

1345. Kurose, I., Kubes, P., Wolf, R., Anderson, D. C., Paulson, J. C. et al., Inhibition of nitric oxide production: mechanisms of vascular albumin leakage, *Circ. Res.*, 73, 164, 1993.

1346. Kurose, I., Wolf, R., Grisham, M. B. and Granger, D. N., Modulation of ischemia/reperfusion-induced microvascular dysfunction by nitric oxide, *Circ. Res.*, 74, 376, 1994.

1347. Ma, X. L., Weyrich, A. S., Lefer, D. J. and Lefer, A. M., Diminished basal nitric oxide release after myocardial ischemia and reperfusion promotes neutrophil adherence to coronary endothelium, *Circ. Res.*, 72, 403, 1993.

1348. Cooke, J. P. and Tsao, P. S., Cytoprotective effects of nitric oxide, *Circulation*, 88, 2451, 1993.

1349. Lefer, A. M., Siegfried, M. R. and Ma, X. L., Protection of ischemia-reperfusion injury by sydnonimine NO donors via inhibition of neutrophil-endothelium interaction, *J. Cardiovasc. Pharmacol.*, 22 (Suppl 7), S27, 1993.

1350. Lefer, D. J., Nakanishi, K., Johnston, W. E. and Vinten-Johansen, J., Antineutrophil and myocardial protecting actions of a novel nitric oxide donor after acute myocardial ischemia and reperfusion in dogs, *Circ.*, 88, 2337, 1993.

1351. Seekamp, A., Mulligan, M. S., Till, G. O. and Ward, P. A., Requirements for neutrophil products and L-arginine in ischemia-reperfusion injury, *Am. J. Pathol.*, 142, 1217, 1993.

1352. Matheis, G., Sherman, M. P., Buckberg, G. D., Haybron, D. M., Young, H. H. and Ignarro, L. J., Role of L-arginine-nitric oxide pathway in myocardial reoxygenation injury, *Am. J. Physiol.*, 262, H616, 1992.

1353. Beckman, J. S. and Crow, J. P., Pathological implications of nitric oxide formation, *Biochem. Soc. Trans.*, 21, 330, 1993.

1354. Nonami, Y., The role of nitric oxide in cardiac ischemia-reperfusion injury, *Jpn. Circ. J.*, 61, 119, 1997.

1355. Szabo, C., Physiological and pathophysiological roles of nitric oxide in the central nervous system, *Brain Res. Bull.*, 41, 131, 1996.

1356. Szabo, C., The pathophysiological role of peroxynitrite in shock, inflammation, and ischemia-reperfusion injury, *Shock*, 6, 79, 1996.

1357. Menasche, P., Grousset, C., Gauduel, Y., Mouas, C. and Piwnica, A., Prevention of hydroxyl radical formation: a critical concept for improving cardioplegia. Protective effects of deferoxamine, *Circulation*, 76, V180, 1987.

1358. Menasche, P., Pasquier, C., Belluci, S., Lorente, P., Jaillon, P. and Piwnica, A., Deferoxamine reduces neutrophil-mediated free radical production during cardiopulmonary bypass in man, *J. Thorac. Cardiovasc. Surg.*, 96, 582, 1988.

1359. Bolli, R., Patel, B. S., Zhu, W. X., O'Neill, P. G., Hartley, C. J., Charlat, M. L. and Roberts, R., The iron chelator desferrioxamine attenuates postischemic ventricular dysfunction, *Am. J. Physiol.*, 253, 1372, 1987.

1360. Ferreira, R., Burgos, M., Milei, J., Llesuy, S., Molteni, L., Hourquebi, H. and Boveris, A., Effect of supplemented cardioplegic solution with deferoxamine on reperfused human myocardium, *J. Thorac. Cardiovasc. Surg.*, 100, 708, 1990.

1361. Williams, R. E., Zweier, J. L. and Flaherty, J. T., Treatment with deferoxamine during ischemia improves functional and metabolic recovery and reduces reperfusion-induced oxygen radical generation in rabbit hearts, *Circulation*, 83, 1006, 1991.

1362. Drossos, G., Lazou, A., Panagopoulos, P. and Westaby, S., Deferoxamine cardioplegia reduces superoxide radical production in human myocardium, *Ann. Thorac. Surg.*, 59, 169, 1995.

1363. Bel, A., Martinod, E. and Menasche, P., Cardioprotective effect of desferrioxamine, *Acta Haematol.*, 95, 63, 1996.

1364. McCord, J. M., Is iron sufficiency a risk factor in ischemic heart disease?, *Circulation*, 83, 1112, 1991.

1365. McCord, J. M., Iron, free radicals, and oxidative injury, *Semin. Hematol.*, 35, 5, 1998.

1366. Sullivan, J. L., Iron and the sex difference in heart disease risk, *Lancet*, 1 (8233), 1293, 1981.

1367. Sullivan, J. L., The iron paradigm of ischemic heart disease, *Am. Heart J.*, 117, 1177, 1989.

1368. Mullane, K. M., Salmon, J. A. and Kraemer, R., Leukocyte-derived metabolites of arachidonic acid in ischemia-induced myocardial injury, *Fed. Proc.*, 46, 2422, 1987.

516

1369. Flynn, P. J., Becker, W. K., Vercellotti, G. M.,Weisdorf, D. J., Craddock, P. R., Hammerschmidt, D. E., Lillehei, R. C. and Jacob, H. S., Ibuprofen inhibits granulocyte responses to inflammatory mediators. A proposed mechanism for reduction of experimental myocardial infarct size, *Inflammation*, 8, 33, 1984.

1370. Murry, C. E., Jennings, R. B. and Reimer, K. A., Preconditioning with ischemia : a delay of lethal cell injury in ischemic myocardium, *Circ.*, 74, 1124, 1986.

1371. Downey, J. M., Liu, G. S. and Thornton, J. D., Adenosine and the anti-infarct effects of preconditioning, *Cardiol. Res.*, 27, 3, 1993.

1372. Van Winkle, D. M., Chien, G. L., Wolff, R. A., Soifer, B. E., Kuzume, K. and Davis, R. F., Cardioprotection provided by adenosine receptor activation is abolished by blockage of the K_{ATP} channel, *Am. J. Physiol.*, 266, H829, 1994.

1373. Omar, B. A., Hanson, A. K., Bose, S. K. and McCord, J. M., Ischemic preconditioning is not mediated by free radicals in the isolated rabbit heart, *Free Radical Biol. Med.*, 11, 517, 1991.

1374. Bouma, M. G., van den Wildenberg, F. A. and Buurman, W. A., The anti-inflammatory potential of adenosine in ischemia-reperfusion injury: established and putative beneficial actions of a retaliatory metabolite, *Shock*, 8, 313, 1997.

1375. Ely, S. W. and Berne, R. M., Protective effects of adenosine in myocardial ischemia, *Circulation*, 85, 893, 1992.

1376. Grisham, M. B., Hernandez, L.A. and Granger, D. N., Adenosine inhibits ischemia-reperfusion-induced leukocyte adherence and extravasation, *Am. J. Physiol.*, 257, H1334, 1989.

1377. Cronstein, B. N., Levin, R. I., Belanoff, J., Weissman, G. and Hirschhorn, R., Adenosine: an endogenous inhibitor of neutrophil-mediated injury to endothelial cells, *J. Clin. Invest.*, 78, 760, 1986.

1378. Lemasters, J. J., Bond, J. M., Chacon, E., Harper, I. S., Kaplan, S. H., Ohata, H., Trollinger, D. R., Herman, B. and Cascio, W. E., The pH paradox in ischemia-reperfusion injury to cardiac myocytes, *EXS*, 76, 99, 1996.

1379. Holloszy, J. O. and Booth, F. W., Biochemical adaptations to endurance exercise in muscle, *Ann. Rev. Physiol.*, 38, 273, 1976.

1380. Vuori, I., Feasibility of long-distance (20-90 km) ski hikes as a mass sport for middle-aged and old people, in *Guide to Fitness after*

Fifty, R. Harris, L. J. Frankel, eds., Plenum Press, New York, 1977, pp. 95-142.

1381. Jenkins, R. R., The role of superoxide dismutase and catalase in muscle fatigue, in *Biochemistry of Exercise,*Human Genetics, Champaign, IL, 1983, pp. 467-471.

1382. Packer, L., Oxygen radicals and antioxidants in endurance exercise, in *Biochemical Aspects of Physical Exercise,* G. Benzi, L. Packer and N. Siliprandi, eds., Elsevier Science Publishers, New York, 1986, pp. 73-92.

1383. Booth, F. W. and Thomason, D. B., Molecular and cellular adaptation of muscle in response to exercise: perspectives of various models, *Physiol. Rev.,* 71, 541, 1991.

1384. *Exercise and Oxygen Toxicity,* C. K. Sen, L. Packer and O. Hänninen, eds., Elsevier, 1994.

1385. Jenkins, R. R., Free radical chemistry. Relationship to exercise, *Sports Med.,* 5, 156, 1988.

1386. Sjödin, B., Westing, Y. H. and Apple, F. S., Biochemical mechanisms for oxygen free radical formation during exercise, *Sports Med.,* 10, 236, 1990.

1387. Møller, P., Wallin, H. and Knudsen, L. E., Oxidative stress associated with exercise, psychological stress and life-style factors, *Chem. Biol. Int.,* 102, 17, 1996.

1388. Frei, B., Yamamoto, Y., Niclas, D. and Ames, B. N., Evaluation of an isoluminal chemiluminescence assay for the detection of hydroperoxides in human blood plasma,*Anal. Biochem.,* 175, 120, 1988.

1389. Hartmann, A., Plappert, U., Raddatz, K., Grünert-Fuchs, M. and Speit, G., Does physical activity induce DNA damage, *Mutagenesis,* 9, 269, 1994.

1390. Davies, K. J. A., Quintanilha, A. T., Brooks, G. A. and Packer, L., Free radicals and tissue damage during exercise, *Biochem. Biophys. Res. Commun.,* 107, 1198, 1982.

1391. Alessio, H. M., Goldfarb, A. H. and Cutler, R. G., MDA content increases in fast- and slow-twitch skeletal muscle with intensity of exercise in a rat, *Am. J. Physiol.,* 255, C874, 1988.

1392. Alessio, H. M., Exercise-induced oxidative stress,*Med. Sci. Sports Exer.,* 25, 218, 1993.

518

1393. Viguie, C. A., Frei, B., Shigenaga, M. K., Ames, B. N., Packer, L. and Brooks, G. A., Antioxidant status and indexes of oxidative stress during consecutive days of exercise, *J. Appl. Physiol.*, 75, 566, 1993.

1394. Clarkson, P. and Ebbeling, C., Investigation of serum creatine kinase variability after muscle-damaging exercise, *Clin. Sci.*, 75, 257, 1988.

1395. Vihko, V., Salminen, A. and Rantamaki, J., Oxidative and lysosomal capacity in skeletal muscle of mice after endurance training of different intensities, *Acta Physiol. Scand.*, 104, 74, 1978.

1396. Kanter, M. M., Lesmes, G. R., Kaminsky, L. A., LaHamsalger, J. and Nequin, N. D., Serum creatine kinase and lactate dehydrogenase changes following an eighty kilometer race, *Eur J. Appl. Physiol.*, 57, 69, 1988.

1397. Wolbarsht, M. L. and Fridovich, I., Hyperoxia during reperfusion is a factor in reperfusion injury, *Free Rad. Biol. Med.*, 6, 61, 1989.

1398. Jenkins, R. R., Krause, K. and Schofield, L. S., Influence of exercise on clearance of oxidant stress products and loosely bound iron, *Med. Sci. Sports Exerc.*, 25, 213, 1993.

1399. Forster, N. K., Martyn, J. B., Rangno, R. E., Hogg, J. C. and Pardy, R. L., Leukocytosis of exercise: role of cardiac output and catecholamines, *J. Appl. Physiol.*, 61, 2218, 1986.

1400. Schaefer, R. A., Kokoi, K., Heidland, A. and Plass, R., Jogger's leukocytes, *N. Engl. J. Med.*, 316, 223, 1987.

1401. Pincemail, J., Camus, G., Roesgen, A., Dreezen, E., Bertrand, Y., Lismonde, M., Deby-Dupont, G. and Deby, C., Exercise induces pentane production and neutrophil activation in humans. Effect of propanolol, *Eur. J. Appl. Physiol.*, 61, 319, 1990.

1402. Hansen, J., Wilsgård, L. and Østerud, B., Biphasic changes in leukocytes induced by strenuous exercise, *Eur. J. Appl. Physiol.*, 62, 157, 1991.

1403. Camus, G., Pincemail, J., Ledent, M., Juchmes-Ferir, A., Lamy, M., Deby-Dupont, G. and Deby, C., Plasma levels of polymorphonuclear elastase and myeloperoxidase after uphill walking and downhill running at similar energy cost, *Int. J. Sports Med.*, 13, 443, 1992.

1404. Smith, L. L., McCammon, M., Smith, S., Chamness, M., Israel, R. G. and O'Brian, K. F., White blood cell response to uphill walking and downhill jogging at similiar metabolic loads, *Eur. J. Appl. Physiol.,* 58, 833, 1989.

1405. Smith, L. L., Acute inflammation: the underlying mechanism in delayed onset muscle soreness?, *Med. Sci. Sports Exer.,* 23, 542, 1991.

1406. Armstrong, R. B., Mechanisms of exercise induced delayed onset of muscular soreness: a brief review, *Med. Sci. Sports Exerc.,* 16, 529, 1984.

1407. Lewicki, R., Tchorzewski, H., Denys, A., Kowalska, M. and Golinska, A., Effect of physical exercise on some parameters of immunity in conditioned sportsmen, *Int. J. Sports Med.,* 8, 309, 1987.

1408. Smith, J. A., Telford, R. D., Mason, I. B. and Weidemann, M. J., Exercise, training and neutrophil microbicidal activity, *Int. J. Sports Med.* 11, 179, 1990.

1409. Jokl, E., The immunological status of athletes, *J. Sports Med.,* 14, 165, 1974.

1410. Ebbeling, C. B. and Clarkson, P. M., Exercise-induced muscle damage and adaptation, *Sports Med.,* 7, 207, 1989.

1411. Jochum, M., Duswald, K. H., Neumann, S., Witte, J. and Fritz, H., Proteinases and their inhibitors in septicemia-basic concepts and clinical implications, *Adv. Exp. Med. Biol.,* 167, 391, 1984.

1412. Friedman, G. D., Klatsky, A. L. and Siegelaub, A. B., The leukocyte count as a predictor of myocardial infarction, *N. Engl. J. Med.,* 290, 1275, 1974.

1413. Prentice, R. L., Szatrowski, T. P., Fujikura, T., Kato, H., Mason, M. W. and Hamilton, H. H., Leukocyte counts and coronary heart disease in a Japanese cohort, *Am. J. Epidemiol.,* 116, 496, 1982.

1414. Wedmore, C. V. and Williams, T. J., Control of vascular permeability by polymorphonuclear leukocytes in inflammation, *Nature,* 289, 646, 1981.

1415. Schmidt, W., Egbring, R. and Havemann, K., Effect of elastase-like and chymotrypsin-like neutral proteases from human granulocytes on isolated clotting factors, *Thromb. Res.,* 6, 315, 1975.

1416. Machovich, R. and Owen, W. G., Elastase-dependent plasmin expression as an alternative pathway for plasminogen activation, *Thromb. Haemost.,* 62, 329, 1989.

1417. Plow, E. F., The major fibrinolytic proteases of human leukocytes, *Biochim. Biophys. Acta,* 630, 47, 1980.

1418. Plow, E. F., Leukocyte elastase release during blood coagulation. A potential mechanism for activation of the alternative fibrinolytic pathway, *J. Clin. Invest.,* 69, 564, 1982.

1419. Thompson, P. D., Funk, E. J., Carleton, R. A. and Sturner, W. Q., Incidence of death during jogging in Rhode Island from 1975 through 1980, *JAMA,* 247, 2535, 1982.

1420. Vuori, I., The cardiovascular risks of physical activity, *Acta Med. Scand.(Suppl.),* 711, 205, 1985.

1421. Alessio, H. M. and Goldfarb, A. H., Lipid peroxidation and scavenger enzymes during exercise: adaptive response to training, *J. Appl. Physiol.,* 64, 1333, 1988.

1422. Quintanilha, A. T., Effects of physical exercise and/or vitamin E tissue oxidation mechanism, *Biochem. Soc. Trans.,* 12, 403, 1984.

1423. Sen, C. K., Marin, E., Kretzschmar, M. and Hänninen, O., Skeletal muscle and liver glutathione homeostasis in response to training, exercise, and immobilization, *J. Appl. Physiol.,* 73, 1265, 1992.

1424. Ji, L. L., Antioxidant enzyme response to exercise and aging, *Med. Sci. Sports Exer.,* 25, 225, 1993.

1425. Tiidus, P. M. and Houston, M. E., Antioxidant and oxidative enzyme adaptations to vitamin E deprivation and training, *Med. Sci. Sports Exer.,* 26, 354, 1994.

1426. Higuchi, M., Cartier, L., Chen, M. and Holloszy, J. O., Superoxide dismutase and catalase in skeletal muscle: adaptive response to exercise, *J. Gerontol.,* 40, 281, 1985.

1427. Laughlin, M. H., Simpson, T., Sexton, W. L., Brown, O. R., Smith, J. K. and Korthuis, R. J., Skeletal muscle oxidative capacity, antioxidant enzymes, and exercise training, *J. Appl. Physiol.,* 68, 2337, 1990.

1428. Criswell, D., Powers, S., Dodd, S., Lawler, J., Edwards, W., Renshler, K. and Grinton, S., High intensity training-induced changes in skeletal muscle antioxidant enzyme activity, *Med. Sci. Sports Exer.,* 25, 1135, 1993.

1429. Ji, L. L., Dillon, D. and Wu, E., Alteration of antioxidant enzymes with aging in rat skeletal muscle and liver, *Am. J. Physiol.,* 258, R918, 1990.

1430. Starnes, J., Cantu, G., Farrar, R. and Kehrer, J., Skeletal muscle lipid peroxidation in exercised and food restricted rats during aging, *J. Appl. Physiol.,* 67, 69, 1989.
1431. Storz, G., Tartaglia, L. A. and Ames, B. N., Transcriptional regulator of oxidative stress-inducible genes: direct activation by oxidation, *Science,* 248, 189, 1990.
1432. Gutteridge, J. M. C., Rowley, D. A., Halliwell, B., Cooper, D. F. and Heeley, D. M., Copper and iron complexes catalytic for oxygen radical reactions in sweat from human athletes, *Clin. Chim. Acta,* 145, 267, 1985.
1433. Davies, K. J. A., Packer, L. and Brooks, G. A., Biochemical adaptation of mitochondria, muscle, and whole-animal respiration to endurance training, *Arch. Biochem. Biophys.,* 209, 539, 1981.
1434. Brady, P. S., Brady, L. J. and Ullrey, D. E., Selenium, vitamin E and the response to swimming stress in the rat, *J. Nutr.,* 109, 1103, 1979.
1435. Pincemail, J., Deby, C., Camus, G., Pirnay, F. , Bouchez, R., Massaux, L. and Goutier, R., Tocopherol mobilization during intense exercise, *Eur. J. Appl. Physiol. Occup. Physiol.,* 57, 189, 1988.
1436. Camus, G., Pincemail, J., Roesgen, A., Dreezen, E., Sluse, F. E. and Deby, C., Tocopherol mobilization during dynamic exercise after beta-adrenergic blockade, *Arch. Int. Physiol. Biochim.,* 98, 121, 1990.
1437. Williams, R. J., *Biochemical Individuality,* John Wiley, New York, 1963.
1438. McFadden, E. R. and Gilbert, I. A., Exercise-induced asthma, *N. Engl. J. Med.,* 330, 1362, 1994.
1439. Mautz, W. J., McClure, T. R., Reischl, P., Phalen, R. F. and Crocker, T. T., Enhancement of ozone-induced lung injury by exercise, *J. Toxicol. Environ. Health,* 16, 841, 1985.
1440. Thompson, P. D., Funk, E. J., Carleton, R. A. and Sturner, W. Q., Incidence of death during jogging in Rhode Island from 1975 through 1980, *JAMA,* 247, 2535, 1980.
1441. Vuori, I., The cardiovascular risks of physical activity, *Acta Med. Scand.,* (suppl.), 711, 205, 1985.
1442. Stefanick, M. L., Mackey, S., Sheehan, M., Ellsworth, N., Haskell, W. L. and Wood, P. D., Effects of diet and exercise in men and

postmenopausal women with low levels of HDL cholesterol and high levels of LDL cholesterol, *N. Engl. J. Med.*, 339, 12, 1998.

1443. Holloszy, J. O., Smith, E. K., Wining, M. and Adams, S., Effect of voluntary exercise on longevity of rats, *J. Appl. Physiol.*, 59, 826, 1985.

1444. Edington, D. W., Cosmas, A. C. and McCafferty, W. B., Exercise and longevity: evidence for a threshold age, *J. Gerontol.*, 27, 341, 1972.

1445. Blot, W. J., Alcohol and cancer, *Cancer Res. (Suppl.)*, 52, 2119s, 1992.

1446. Hennekens, C. H., Rosner, B. and Cole, D. S., Daily alcohol consumption and fatal coronary heart disease, *Am. J. Epidemiol.*, 107, 196, 1978.

1447. Stampfer, M. J., Colditz, G. A., Willett, W. C., Speizer, F. E. and Hennekens, C. H., A prospective study of moderate alcohol consumption and the risk of coronary disease and stroke in women, *N. Engl. J. Med.*, 319, 267, 1988.

1448. Rimm, E. B., Giovanucci, E. L., Willett, W. C., Colditz, G. A., Ascherio, A., Rosner, B. and Stampfer, M. J., Prospective study of alcohol consumption and risk of coronary disease in man, *Lancet*, 338, 464, 1991.

1449. Klatsky, A. L., Armstrong, M. A. and Friedman, G. D., Risk of cardiovascular mortality in alcohol drinkers, ex-drinkers and nondrinkers, *Am. J. Cardiol.*, 66, 1237, 1990.

1450. Frankel, E. N., Waterhouse, A. L. and Kinsella, J. E., Inhibition of oxidation of human low density lipoprotein by phenolic substances in red wine, *Lancet*, 341, 454, 1993.

1451. Frankel, E. N., Waterhouse, A. L. and Kinsella, J. E., Inhibition of human LDL oxidation by resveratrol, *Lancet*, 341, 1103, 1993.

1452. Srivastava, L. M., Vasisht, S., Agarwal, D. P. and Goedde, H. W., Relation between alcohol intake, lipoproteins and coronary heart disease: the interest continues, *Alcohol and Alcoholism*, 29, 11, 1994.

1453. Kiechl, S., Willeit, J., Egger, G., Oberhollenzer, M. and Aichner, F., Alcohol consumption and carotid atherosclerosis: evidence of dose-dependent atherogenic effects. Results from the Bruneck study, *Stroke*, 25, 1593, 1994.

1454. Bello, A. T., Bora, N. S., Lange, L. G. and Bora, P. S., Cardioprotective effects of alcohol mediation by human vascular alcohol dehydrogenase, *Biochem. Biophys. Res. Commun.*, 203, 1858, 1994.

1455. Fuhrman, B., Lavy, A. and Aviram, M., Consumption of red wine with meals reduces the susceptibility of human plasma and low-density lipoprotein to lipid peroxidation, *Am. J. Clin. Nutr.*, 61, 549, 1995.

1456. Pfeifer, P. M. and McCay, P. B., Reduced triphosphopyridine nucleotide oxidase-catalyzed alterations of membrane phospholipids, *J. Biol. Chem.*, 247, 6763, 1972.

1457. Takayanagi, R., Takeshige, K. and Minakami, S., NADH-and NADPH-dependent lipid peroxidation in bovine heart submitochondrial particles, *Biochem. J.*, 192, 853, 1980.

1458. Reitz, R. C., A possible mechanism for the peroxidation of lipids due to chronic ethanol ingestion, *Biochem. Biophys. Acta*, 380, 145, 1975.

1459. Cederbaum, A. I., Oxygen radical generation by microsomes: role of iron and implications for alcohol metabolism and toxicity, *Free Radical Biol. Med.*, 7, 559, 1989.

1460. Dicker, E. and Cederbaum, A. I., Increased NADH-dependent production of reactive oxygen intermediates by microsomes after chronic ethanol consumption: comparisons with NADPH, *Arch. Biochem. Biophys.*, 293, 274, 1992.

1461. Ribiere, C., Sabourault, D., Saffar, C. and Nordmann, R., Mitochondrial generation of superoxide free radicals during acute alcohol intoxication in the rat, *Alcohol and Alcoholism, Suppl.* 1, 241, 1987.

1462. Kukielka, E., Dicker, E. and Cederbaum, A. I., Increased production of reactive oxygen species by rat liver mitochondria after chronic alcohol treatment, *Arch. Biochem. Biophys.*, 309, 377, 1994.

1463. Hildebrandt, A. G., Roots, E., Tjoe, M. and Heinemeyer, G., Hydrogen peroxide in hepatic microsomes, *Meth. Enzymol.*, 52, 342, 1978.

1464. Romslo, I. and Flatmark, T., Energy-dependent accumulation of iron by isolated rat liver mitochondria, *Biochim. Biophys. Acta*, 347, 160, 1974.

524

1465. Harman, D., The aging process: Major risk factor for disease and death, *Proc. Natl. Acad. Sci. USA,* 88, 5360, 1991.

1466. Tannenbaum, A., The genesis and growth of tumors. II. Effects of caloric restriction per se, *Cancer Res.,* 2, 460, 1942.

1467. Weindruch, R., Walford, R. L., Fligiel, S and Guthrie, D., The retardation of aging in mice by dietary restriction: longevity, cancer, immunity and lifetime energy intake, *J. Nutr.,* 116, 641, 1986.

1468. Masoro, E. J., Yu, B. P. and Bertrand, H. A., Action of food restriction in delaying the aging process, *Proc. Natl. Acad. Sci. USA,* 79, 4239, 1982.

1469. Yu, B. P., Masoro, E. J. and Mahan, C. A., Nutritional influences on aging of Fischer 344 rats: I. Physical, metabolic, and longevity characteristics, *J. Gerontol.,* 40, 657, 1985.

1470. Kumar, S. P., Roy, S. J., Tokumo, K. and Reddy, B. S., Effect of different levels of calorie restriction on azoxymethane-induced colon carcinogenesis in male F 344 rats, *Cancer Res.,* 50, 5761, 1990.

1471. Goodrick, C. L., Body weight increment and length of life: the effect of genetic constitution and dietary protein, *J. Gerontol.,* 33, 184, 1978.

1472. Leto, S., Kokkonen, G. and Barrows, C., Dietary protein, life span, and biochemical variables in female mice, *J. Gerontol.,* 31, 144, 1976.

1473. Youngman, L. D. and Campbell, T. C., High protein intake promotes the growth of hepatic preneoplastic foci in Fischer #344 rats: evidence that early remodeled foci retain the potential for future growth, *J. Nutr.,* 121, 1454, 1991.

1474. Youngman, L. D. and Campbell, T. C., Inhibition of aflatoxin B_1-induced gamma-glutamyltranspeptidase positive (GGT^+) hepatic preneoplastic foci and tumors by low protein diets: evidence that altered GGT^+ foci indicate neoplastic potential, *Carcinogenesis,* 13, 1607, 1992.

1475. De, A. K., Chipalkaatti, S. and Aiyar, A. S., Some biological parameters of aging in relation to dietary protein, *Mech. Aging Dev.,* 21, 37, 1983.

1476. Koizumi, A., Weindruch, R. and Walford, R. L., Influences of dietary restriction and age on liver enzyme activities and lipid peroxidation in mice, *J. Nutr.,* 117, 361, 1987.

1477. Kim, J. W. and Yu, B. P., Characterization of age-related malondialdehyde oxidation: the effect of modulation by food restriction, *Mech. Aging Dev.,* 50, 277, 1989.

1478. Youngman, L. D., Park, J. Y. and Ames, B. N., Protein oxidation associated with aging is reduced by dietary restriction of protein or calories, *Proc. Natl. Acad. Sci., USA,* 89, 9112, 1992.

1479. Simic, M. G. and Bergtold, D. S., Dietary modulation of DNA damage in human, *Mutat. Res.* 250, 17, 1991.

1480. Djuric, Z., Heilbrun, L. K., Reading, B. A., Boomer, A., Valeriote, F. A. and Martino, S., Effects of a low-fat diet on levels of oxidative damage to DNA in human peripheral nucleated blood cells, *J. Natl. Cancer Inst.,* 83, 766, 1991.

1481. Chung, M. H., Kasai, H., Nishimura, S. and Yu, B. P., Protection of DNA damage by dietary restriction, *Free Radical Biol. Med.,* 12, 523, 1992.

1482. Lammi-Keefe, C. J., Swan, R. B. and Hegarty, P. V. J., Effect of level of dietary protein and total or partial starvation on catalase and superoxide dismutase activity in cardiac and skeletal muscles in young rats, *J. Nutr.,* 114, 2235, 1984.

1483. Laganiere, S. and Yu, B. P., Effect of chronic food restriction in aging rats. I. Liver subcellular membranes, *Mech. Aging Dev.,* 48, 207, 1989.

1484. Laganiere, S. and Yu, B. P., Effect of chronic food restriction in aging rats II: Liver cytosolic antioxidants and related enzymes, *Mech. Aging Dev.,* 48, 221, 1989.

1485. Yu, B. P., Lee, D. W., Marler, C. G. and Choi, J. H., Mechanism of food restriction: Protection of cellular homeostasis, *Proc. Soc. Exp. Biol. Med.,* 193, 13, 1990.

1486. Weraarchakul, N., Strong, R., Wood, W. G. and Richardson, A., The effect of aging and dietary restriction on DNA repair, *Exp. Cell Res.,* 181, 197, 1989.

1487. Kim, J. W. and Yu, B. P., Characterization of age-related malondialdehyde oxidation: the effect of modulation by food restriction, *Mech. Ageing Dev.,* 50, 277, 1989.

1488. Tacconi, M. T., Lligona, L., Salmona, M., Pitsikas, N. and Algeri, S., Aging and food restriction: effect on lipids of cerebral cortex, *Neurobiol. Aging,* 12, 55, 1991.

526

1489. Pitsikas, N. and Algeri, S., Deterioration of spatial and nonspatial reference and working memory in aged rats: protective effect of life-long calorie restriction, *Neurobiol. Aging,* 13, 369, 1992.

1490. Schatzkin, A., Greenwald, P., Byar, D. P. and Clifford, C. K., The dietary fat-breast cancer hypothesis is alive,*JAMA,* 261, 3284, 1989.

1491. Howe, G. R., Hirohata, T., Hislop, T. G. et al., Dietary factors and risk of breast cancer. Combined analysis of 12 case-control studies, *J. Natl. Cancer Inst.,* 82, 561, 1990.

1492. Chen, J., Campbell, T. C., Li, J. and Peto, R., *Diet, Life Style and Mortality in China: A Study of the Characteristics of 65 Chinese Counties,* Oxford Univ. Press, Oxford, UK, 1990.

1493. Menninger, K. A. and Menninger, W. C., Psychoanalytical observations in cardiac disorders,*Am. Heart J.,* 11, 10, 1936.

1494. Friedman, M. and Rosenman, M. W., Association of specific overt behavior patterns with blood and cardiovascular findings, *JAMA,* 169, 1286, 1959.

1495. Deary, I. J., Fowkes, F. G. R., Donnan, P. T. and Housley, E., Hostile personality and risks of peripheral arterial disease in the general population, *Psychosomatic Med.,* 56, 197, 1994.

1496. Leeming, R. A., From the common cold to cancer: how evolution and the modern lifestyle appear to have contributed to such eventualities, *Med. Hypotheses,* 43, 145, 1994.

1497. Adachi, S., Kawamura, K. and Takemoto, K., Oxidative damage of nuclear DNA in liver of rats exposed to psychological stress, *Cancer Res.,* 53, 4153, 1993.

1498. Kiecolt-Glaser, J. K., Stephens, R. E., Lipetz, P.D., Speicher, C. E. and Glaser, R., Distress and DNA repair in human lymphocytes,*J. Behavioral Med.,* 8, 311, 1985.

1499. Kosugi, H., Enomoto, H., Ishizuka, Y. and Kikugawa, K., Variations in the level of urinary thiobarbituric acid reactant in healthy humans under different physiological conditions, *Biol. Pharm. Bull.,* 17, 1645, 1994.

1500. Selye, H. and Tuchweber, B., Stress in relation to aging and disease, in *Hypothalamus, Pituitary and Aging,* A. Everitt, J. Burgess, eds., C. C. Thomas, Springfield, 1976.

1501. Rosch, P. C., Stress and Cancer, in *Psychosocial Stress and Cancer,* C. L. Cooper, ed., John Wiley, New York, 1984, I, pp. 3-19.

1502.   Fox, B. H., Psychosocial factors in the immune system in cancer, in *Psychoneuroimmunology,* R. Ader, ed., Academic Press, New York, 1981, pp. 103-158.

1503.   Bammer, K., Stress, Spread and Cancer, in *Stress and Cancer,* K. Bammer and B. H. Newberry, eds., C. H. Hogrefe, Toronto, 1982

1504.   Riley, V., Psychoneuroendocrine influences on immuno-competence and neoplasia, *Science,* 217, 1100, 1981.

1505.   McEwen, B.S. and Stellar, E., Stress and the individual: Mechanisms leading to disease, *Arch. Intern. Med.,* 153, 2093, 1993.

1506.   Solomon, G., Psychoneuroimmunology: Interaction between control nervous system and immune system, *J. Neurosci. Res.,* 18, 1, 1987.

1507.   Rodin, J., Health, Control and Aging, in *Aging and the Psychology of Control,* M. Baltes and P. B. Baltes, eds., Lawrence Erlbaum, Hillsdale, N. J, 1986, pp. 83-92.

1508.   Levy, S., Herberman, R., Lippman, M. and d'Angelo, T., Correlation of stress factors with sustained depression of natural killer cell activity and predicted prognosis in patients with breast cancer, *J. Clin. Oncol.,* 5, 348, 1987.

Author Index

A

Abakerli, R. B., 329,343,344
Abbotts, J., 1172
Abell, P. L., 41
Abrams, G. D., 1328
Abramson, S. B., 803,1036
Acar, C., 1329
Acavitti, M. A., 1246
Adachi, S., 1497
Adam, W.,30,78,80,82,382,
   383,386,482, 483,487,488,
   491,614,615,654
Adams, C. W. M., 1107
Adams, G. E. , 134
Adams, J. D., 1121
Adams, S., 1443
Adelman, R., 867
Adeniyi-Jones, S. K., 471
Aewert, F., 611
Afanas'ev, I. B.,128,175,176,
   177,178,179,840, 841,892
Agarwal, D. P.,1452
Agid, F., 1122,1123
Agid, Y., 1113,1122,1123,1125
Agnese, D., 1323
Agrawal, M. C., 846
Ahrweiler, M., 482
Aichner, F., 1453
Aida, M., 347
Ainley, K., 918
Aiyar, A. S., 1475
Aiyar, J., 297
Akaike, T., 284,1038
Akerboom, T. P. M., 711
Akino, M., 1091

Aksenova, M., 1081
Albanes, D., 970
Albano, E., 597
Alessio, H. M., 1391,1392,
   1421
Alexander, W., 1281
Algeri, S., 1488,1489
Alfassi, Z. B., 38
Allard, J. P., 1174
Allaway, W. H., 940
Allen, A. O., 24,159
Allen, J. S., 700
Allen, R. C., 456,457
Allen, R. G., 990
Ambani, I. M., 1116
Ambruso, D. R., 450
Ames, B. N., 385-399,588,
   628,629,630,666,673,824,
   825,829, 857,867,868,919,
   1000,1004,1388,1393,
   1431,1478
Anan, K. F., 843
Anandatheerthavaranda, H. K.,
   1049
Anbar, M., 548
Anca, Z., 1181
Anders, M. W., 243
Anderson, D. C., 1319,1335,
   1340,1345
Anderson, M. D., 1020
Anderson, P. G., 1246
Anderson, R. E., 577
Anderson, R. F., 411
Andler, S., 382,654
Ando, M., 284,1101
Ando, W., 204,362
Ando, Y., 1101

Debrunner, P. G., 419
Deby, C., 1401,1403,1435, 1436
Deby-Dupont, G., 1401,1403
Decarroz, C., 353
Decuyper, J., 641
Deen, W. M., 695
Degan, P., 397,630
DeGraff, W., 795,796
DeGraff, W. G., 694
Del Maestro, R. F., 1339
Del Principe, D., 1144
DeMaster, E. G., 685
Demple, B., 400,645,646
Deneke, S. M., 923,1156
Deng, H. X., 1069
Denny, R. W., 347
Denys, A., 1407
Derbyshire, J. F., 693
Deshmukh, D. R., 1323
Devasagayam, T. P. A., 621, 627,634
DeWitt, D. S. , 760
Dexter, D. T., 1111,1119, 1122, 1123
Di Corpo, M. L., 1144
Di Giulio, S., 1144
Di Mascio, P., 501,622, 624,626,627,632, 634, 921,922
Dianzani, M. U., 590,591,596
Dickens, B. F., 1295
Dicker, E., 1460,1462
Dickson, D. W., 1059
Dickson, G. M., 861
DiCorleto, P. E., 1213
Dieber-Rotheneder, M., 1267
Diehl, A. K., 1034

Dillard, C. J., 938
Dillon, D., 1429
DiMauro, C., 591
DiMonte, D., 679
Dionisi, O., 112
Dix, D., 999
Dix, T. A., 209,1027
Dixon, W. T., 69,237,254
Dizdaroglu, M., 281,602, 607,608,609, 610, 617, 623,637,638,648,653,656, 664
Djerassi, C., 1066
Djuric, Z., 1480
Do-Thi, H. P., 632
Doba, T., 894
Dodd, S., 1428
Doelman, C. J. A., 1130
Doetsch, P. W., 402
Doherty, Th. P., 187
Dolak, J. A., 595
Doll, R., 963
Domanski, J., 469
Donnan, P. T., 1495
Dooley, M. M., 1132
Dorfman, L. M., 43,65
Doring, H. J., 1302
Douki, T., 223
Downes, M. J., 689
Downey, J. M., 1293,1318, 1371
Dowson, J. H., 1071
Doyle, M. P., 707
Draganic, I. G., 25
Draganic, Z. D., 25
Drazen, J., 758,759,1192
Dreezen, E., 1401,1436
Dresser, C. M., 971

540

Huang, C. Z., 946
Hubbard, R. C., 950,1135
Hughes, H., 576,680
Huie, R. E., 116,798
Hulley, S. B., 964
Humpf, H.-U., 491
Hurd, H. K., 588
Husain, S. R., 907
Hutchins, K., 1059
Hwang, P.M., 538
Hyers, T. M., 1148

I

Ibers, J. A., 708
Iboko, M. I., 1186
Ichimori, K., 1292
Ignarro, L. J., 89,512,528,
   544,704,705,738,739,743,
   746,748,1064, 1097,1352
Ihde, A. J., 23
Iitsuka, R., 1091
Ikeda, H., 1261
Ikehara, M., 358
Ilan, Y., 190
Ilan, Y. A., 133
Imlay, J. A., 392,657
Inauen, W., 1338
Ingold, K. U., 74,262,884,
   885,886,890,894,896, 898,
   899,927,1274
Inoue, H., 358
Inoue, S., 1044
Inouye, M., 1324
Irr, W., 1182
Isaac, L., 794
Ischimori, K., 1311
Ischiropoulos, H., 769,773,
   787,791

Ishikawa, H., 358
Ishizuka, Y., 1499
Islam, M. F., 439
Isomura, K., 723
Issaq, H. J., 1017
Istfan, N., 697
Itagaki, S., 1065
Iyanagi, T., 843
Iyengar, R., 515,530
Iyer, G. Y. N., 439

J

Jackson, C. V., 1330
Jackson, R. L., 1233
Jacob, H. S., 1139,1369
Jacob, R. A., 829
Jacobs, A. A., 464,470
Jacobs, P. W. M., 29
Jaffe, S., 1140
Jaillon, P., 1358
Jain, S. K., 1285
Jameson, G. B., 708
Jamieson, D., 821
Janero, D. R., 915
Janke, J., 1302
Janoff, A., 1170,1183
Jansen, A., 759
Janssen, Y. M. W., 305
Janzen, E. G., 67,226
Jaraki, O., 712,755,756,758
Javoy-Agid, F., 1112
Jeandel, C., 1072
Jeejebhoy, K. N., 1174
Jefcoate, C. R. E., 52
Jefferson, M. M., 466,467
Jellinger, K., 1114
Jenkins, R. R., 1385,1398
Jenner, A., 289

Jenner, P., 289,1054,
1109,1122,1123
Jennings, R. B., 1301,1308,
1370
Jeroudi, M. O., 1300
Jessop, W. J., 1186
Jessup, W., 1235,1266
Ji, H., 724
Ji, L. L., 1429
Jia, U., 714
Jialal, I., 1253
Jiménez-Jiménez, F. J., 1126
Jin, F., 193
Jin, X., 782
Jocelyn, P. C., 1167
Jochum, M., 1411
Joenje, H., 611
Joester, K. E., 831a
Johns, A., 752
Johnson, E. J., 1250
Johnson, G. R. A., 274
Johnson, K. J., 1149
Johnson, R. A., 45
Johnson, R. B. jr., 434,450
Johnston, W. E., 1350
Jokl, E., 1409
Jones, D. P., 405
Jones, D. Y., 667
Jones, P. A., 1030
Jones, R. M., 698
Jordan, M. C., 1318
Joseph, J., 781,802
Joseph, J. A., 877
Joy, L., 1218
Juchmes-Ferir, A., 1403
Julia, P. L., 1329
Jung, G., 831a
Jürgens, G., 1197,1198,1254

K

Kadiiska, M. B., 234,282,291
Kadowitz, P. K., 743
Kagan, V. E., 563
Kahl, R., 420
Kaiser, S., 624,626,627,634,
921
Kalra, J., 1255
Kalra, V. K., 1140
Kalyanaraman, B., 781,797,
802,1220,1245
Kamachi, M., 718
Kamii, H., 1325
Kaminsky, L. A., 1396
Kamiya, S., 1039
Kamiya, Y., 893
Kaneko, M., 1178
Kang, J., 13
Kang, S. D., 851
Kann, H. E., 1143
Kanner, J., 792,1269
Kanofsky, J. R., 96,327,363,
366,367,368,459
Kantola, M., 1234
Kaplan, S. H., 1378
Kappus, H., 561
Karim, J., 12
Karnovsky M. J., 97,98
Karnovsky, M. L., 421,430,
438,471
Kasai, H., 123,302,647, 649,
650,651, 652,658,661,
1007,1009,1010, 1481
Kasaikina, O. T., 928
Kaseki, H., 1179
Kasha, M., 70,71,318,359,
360,557

550

Ramaiah, D. , 214,215
Ramakrishnan, N., 616
Ramdev, P., 758
Ramin, S., 1248
Ramirez, G., 202
Ramos Martinez, J. I., 1145
Ramos, C. L., 437,455
Ramotar, D., 646
Rand, W., 697
Rangno, R. E., 1399
Rankin, S. M., 1219,1266
Rantamaki, J., 1395
Rao, C. V., 944
Rao, G., 653,656
Rao, N. V., 1142
Rao, P. S., 169,1315
Raoul, S., 215
Rapp, U., 171
Rateri, D. L., 1239
Rauckman, E. J., 233,235
Ravanat, J.-L., 350,351
Ravindranath, V., 1049
Ray, P. K., 912
Rayburn, B. K., 1298
Read, N., 1331
Reading, B. A., 1480
Reaven, P. D., 1252
Recknagel, R. O., 595
Reddell, M., 1323
Reddy, B. S., 944,1470
Reddy, K., 1158
Rees, D. D., 524
Rees, Y., 725
Regnström, J., 1268
Rehm, S., 1017
Reichard, P., 808
Reichel, H., 957
Reif, D. W., 315

Reilly, J., 758
Reimer, K. A., 1308,1370
Reischl, P., 1439
Reiter, R., 119
Reiter, R. J., 871,872,873,
    874,878,879,880,881,882,
    883
Reitz, R. C., 1458
Renshler, K., 1428
Repine, J. E., 243,449,949,
    1138,1141,1150,1161,1164
Reszka, K. J., 210
Retsky, K. L., 1259
Reunanen, A., 668
Reynolds, G. P., 1124
Ribiere, C., 1461
Rice-Evans, C., 1028
Richardson, A., 1486
Richardson, J. S., 1083
Richmond, R., 935
Richter, C., 666
Rickard, R. C., 120
Rider, A., 964
Ridker, P. M., 1250
Riederer, P.,1114,1118,
    1124,1127
Riely, C. A., 676
Rieth, A. G., 856
Riley, V., 1504
Rimele, T. J., 95
Rimm, E. B., 956b, 959,1448
Rinaldo, J. E., 1163
Rinne, U. K., 1110
Rio, G., 377
Riordan, N. H., 972
Risch, N. J., 1102
Robak, J., 908
Robakis, N. K., 878

Salminen, A., 1395
Salmon, J. A., 1331,1368
Salmona, M., 1488
Salonen, J. T., 669,1234
Salonen, R.669,1234
Samuni, A., 141,454
Sanberg, B. L., 1156
Sancar, A., 810
Sanders, K. M., 546
Santamaria, L., 323
Santos, C., 293
Sapp, P., 287
Sapper, H., 851
Saran, M., 158,702, 831b
Sato, K., 284
Saul, R. L., 394
Sauter, M., 482
Sauty, A., 1180
Savenkova, M.I. , 1238
Savin, W. M., 938
Sawyer, D. T., 106,256,260
Sbarra, A. J., 113,438,464, 470
Scandurra, O., 897
Schaap, A. P., 371,380,381
Schaefer, R. A., 1400
Schaffner, T., 1205
Schaich, K. M., 264
Schalch, W., 625
Scharfstein, J., 756,757
Schatzkin, A., 971,1490
Schaur, R. J., 579,585,587
Scheffer, M., 1214
Scheltens, P., 1089
Schenck, G. O., 336,337, 348,372
Schiffmann, D., 482,615
Schmidt, H. H. H. W., 513,

516,522,523
Schmidt, W., 1415
Schneider, J. E., 121,612, 635
Schofield, L. S., 1398
Schönberg, A., 373
Schönberger, A., 30,482, 654
Schork, M. A., 1328
Schrader, J., 688,730
Schrauzer, G. N., 941,942
Schreier, P., 491
Schubert, D., 1078,1079
Schuchman, H.-P., 189
Schuchmann, M. N., 192, 194
Schuler, R. H., 850
Schulte-Frohlinde, D., 191, 192,606, 607,608, 609, 631
Schultz, J., 102
Schulz, W. A., 621
Schulz-Allen, M.-F, 1002
Schuman, M., 404
Schuster, G., 383,385
Schwartz, B. A., 1139
Schwartz, C. J., 1247
Schwartz, J., 1185
Schwartz, L. W., 1158
Schwartzbach, S. D., 153
Schwenke, D. C., 1262
Schwiers, E., 394,673,825
Scott, J., 1266
Scott, M. D., 16,835
Sealy, R. C., 238
Seekamp, A., 1351
Segal, A. W., 931
Segall, H. J., 589

Sinet, P. M., 1093,1112
Singel, D. J., 712,737,755
Singer, A. H., 1286
Sisto, T., 1242
Sivaraja, M., 811
Sjödin, B., 1386
Sjoeberg, B. M., 809,813
Skokowski, J., 954
Slaga, T. J., 1019
Slater, T. F., 560,731,783,
    926, 1299
Slesinski, M. J., 971
Sligar, S. G., 419
Slivka, A., 13,475,757
Sluse, F. E., 1436
Smith, A. L., 478
Smith, C. D., 769,1082
Smith, C. V., 565,577,598,680
Smith, C. W., 1344
Smith, D., 778,1298
Smith, E. K., 1443
Smith, J. A., 1408
Smith, J. K., 1427
Smith, K. C., 644
Smith, K. J., 198
Smith, L. L., 342,1129,
    1404,1405
Smith, M., 811
Smith, S., 1404
Smith, W. W., 1143
Smulow, J. B., 1016
Snell, J., 1061
Snyder, S. H., 507,509, 535,
    537,538,540
Sofic, E., 1114,1124
Sohal, R. S., 982,985,990,
    992,997,998
Sohn, O. S., 944

Soifer, B. E., 1372
Sokolowski, G., 831c
Solomon, G., 1506
Solomons, G. T. W., 699
Somerville, M. J., 1075
Sonnenschein, C., 1016
Soo Kim, S., 50
Sorata, Y., 910
Sorgente, N., 1140
Sos, M., 878
Soto, M. A., 293
Spahr, R., 688
Speicher, C. E., 1498
Speit, G., 1389
Speizer, F. E., 1169,1447
Spencer, J. P. E., 289
Spikes, J. D., 320,321
Spitsin, S., 861
Spitsin, S. V., 1098
Sporn, M. B., 963
Squadrito, G. L., 162,740,
    770,782,786
Srivastava, A. K., 916
Srivastava, L. M., 1452
St. George-Hyslop, P. H.,
    1074
Stadtman, E. R., 269,804,
    1011,1082
Stamler, J. S., 712,737,755,
    756,757,758,759,1192
Stampfer, M. J., 956b,
    958,959,1169,1447,
    1448
Stanbro, W. D., 349
Stanbury, D. M., 692
Starke-Reed, P. E., 1082
Starnes, J., 1430
Stason, W., 1169

Subject Index

580

584

Selegiline,
    monoamine oxidase inhibitor in PD, 334
Singlet oxygen,
    reactivity and electronic structure, 99-101
Singlet oxygen and DNA damage, 191-193
    formation of 4, 8-dihydro-4-hydroxy-8-oxo-2'-deoxyguanosine, 106, 107, 191
    formation of 8-hydroxydeoxyguanosine, 191-193
Singlet oxygen, formation
    bimolecular reaction of peroxyl radicals, 56-58, 101
    by phagocytes, 109, 139-142
    chemical formation, 101, 108, 109
    dismutation of superoxide radical anion, 101
    photosensitization, 99, 100
    reaction of HOCl with superoxide radical anion, 65, 95, 101, 134, 136
    reaction of HOONO with H_2O_2, 107, 108
    reaction of ClO$^-$ with H_2O_2, 108
Singlet oxygen, identification of
    chemiluminescence, 103
    effect of D_2O on singlet oxygen reactions, 105, 190, 191
    quenching, 104, 191
    reaction with cholesterol, 102, 103
    reaction with 2'-deoxyguanosine, 106, 107
    reaction with 9, 10-diphenylanthracene, 105, 106
    reactions with furan derivatives, 102
Singlet oxygen, reactions
    with alkyl substituted olefins, 102, 110
    with cyclic conjugated dienes, 102
Smoking,
    and Vitamin C, 340
    and Vitamin E, 340
    increased lipid peroxidation, 340
    increased 8-OHdG excretion, 340
S-nitrosoalbumin
    storage vessel for NO, 235
    transnitrosylation, 236

590